donkey = stupid.

They asked on ... ... make a Motorway.

He said we get a donkey and put paint
with a hose on. Then where the
donkey goes we make it a road/motorway.
They ask what if you haven't got a
donkey.

he said. We get an engineer.

Vibrational Mechanics

Nonlinear Dynamic Effects,
General Approach, Applications

Vibrational Mechanics

Nonlinear Dynamic Effects,
General Approach, Applications

Iliya I Blekhman

Russian Academy of Sciences
St. Petersburg

Translated by
Minna Perelman

 **World Scientific**
Singapore • New Jersey • London • Hong Kong

Published by

World Scientific Publishing Co. Pte. Ltd.

P O Box 128, Farrer Road, Singapore 912805

USA office: Suite 1B, 1060 Main Street, River Edge, NJ 07661

UK office: 57 Shelton Street, Covent Garden, London WC2H 9HE

British Library Cataloguing-in-Publication Data
A catalogue record for this book is available from the British Library.

VIBRATIONAL MECHANICS
Nonlinear Dynamic Effects, General Approach, Applications

ISBN 981-02-3890-8

Printed in Singapore by Uto-Print

Dedicated to the memory of Nina

Preface to the Russian Edition

This book has two aims. The first aim is to describe a number of wonderful phenomena caused by the action of vibration on nonlinear mechanical systems. Here is a short list of such phenomena: vibrational displacement — the directed on the average "slow" motion or the change in the state of the system affected by "fast" undirected on the average mechanical actions (vibration); the change in the physical and mechanical properties of a body under vibration with regard to slow actions; the transformation of equilibrium states, in particular — their stabilization and destabilization under the action of vibration; the change in free oscillation frequencies of the system, caused by vibration; vibrational maintenance of rotation and self-synchronization of unbalanced rotors having an unexpected analogue in the peculiarities of the motion of celestial bodies; the seeming change in both magnitude and direction of the gravitational force. That enumeration is far from being complete. Many of those phenomena are widely used in technology and engineering. Still more effective use of such phenomena is expected in the nearest future. The author hopes this book will contribute to it.

The second, and perhaps the main aim of the book is to propose a general mechanical and mathematical approach to the description and investigation of this class of phenomena — the approach which may be called *vibrational mechanics* . Vibrational mechanics is a mechanics for the observer interested only in the "slow" motions of the system. Such motions appear together with the fast motions in a non-linear system under vibration and are, as a rule, of the main practical interest. It has been found that this observer, whom we call the observer **V** to distinguish him from the "ordinary" observer whom we will call the observer **O**, in order to make a correct description of the behaviour of the system must add to all the slow forces, acting upon the system, certain supplementary slow forces, called *vibrational forces* . These forces are calculated according to certain rules. Thus, the vibrational mechanics is in some sense analogous to the mechanics of a relative motion.

All the above-mentioned phenomena can be easily and naturally explained by the appearance of vibrational forces. On the other hand, the disregard of those forces leads to misunderstanding and errors, which happened in the past and are still happening now. We will discuss them in this book.

Vibrorheology is an important part of vibrational mechanics. It mainly studies the change caused by vibration in the rheological characteristics of bodies with regard to the slow actions, and it also studies the corresponding slow motions.

Vibrational mechanics and vibrorheology play an important role in the new section of the applied theory of oscillations which has been formed in recent years — the *theory of vibrational processes and vibrational devices* . This theory studies the regularities of excitation and effects of vibration in different mechanical systems. It also embraces the theory of the machines in which vibration is used for beneficial purposes.

A few more words about the positions of the observers **O** and **V** . - positions which are widely used in this book and in some illustrations are marked by

symbols   and   . The letters **O** and **V** are the initial letters of the words "ordinary" and "vibrational". In the world, perceived by the observer O, the ordinary laws of mechanics are valid in their direct form. Making and solving differential equations of motion, based on those laws, that observer gives correct descriptions of the phenomena, considered in the book. But his descriptions are rather intricate, with unnecessary details, as a result of which the interpretation of the results is sometimes quite embarrassing.

The observer **V** , as was mentioned, deliberately "does not notice" any fast forces or fast motions, though he does not forget that they do exist in reality. Staying within the frames of the "ordinary" mechanics, to make his descriptions correct, he must add to all the slow forces also the vibrational forces. The world of the observer **V** is much simpler than that of the observer **O** . In particular, the multidimensional system may be seen by him as a system of much smaller dimensions, an essentially non-conservative system may seem to him to be conservative, the discontinuous system may be seen by him as continuous or "smooth" and so on. As was mentioned, for the observer **V** vibrational mechanics and vibrorheology are mechanics and rheology respectively . In connection with the mistakes, which are often made when analyzing such phenomena, we can also speak of the position of the observer **W** (the first letter of the word "wrong"). He fails to take into account certain rather important circumstances. This observer either does not notice vibration and the fast forces, acting in the system, or does not take into consideration any possible consequences of their presence. So when considering the slow motions, he does not take into account a possibility of the appearance of vibrational forces. The world of this observer is full of "miracles", riddles and paradoxes. Trying to explain them he sometimes begins to doubt the validity of the basic laws of mechanics — the law of conservation of energy, the law of action and reaction. He believes that under the action of vibration the weight of the body changes, that the velocity of the center of inertia of the system may change only at the expense of internal forces, etc.

However, the position of the observer **W** may sometimes also prove useful. In inventors' activities a temporal disregard of the laws of physics and mechanics, which hinder in attaining their aim, is an effective method of finding new technical solutions. We mean here the so called *fantastic analogy*, proposed by Gordon [176].

Now a few words about the terms *vibrational mechanics* and *vibrorheology*. Their introduction seems to be justified, first, by the peculiarity of the class of phenomena that are to be considered and, secondly, by the availability of a common methodical approach to the investigation of these effects. These terms, used by the author in his presentations and publications met no objections on the part of his colleagues [89, 92, 95, 100]. Equations, describing the slow motions, are called by us *the main equations of vibrational mechanics*. It is surprising and remarkable that for a comparatively wide class of systems these equations can be written in the form, typical of the potential system with dissipation, while the initial system is essentially non-potential. Systems of this type are named by us *potential on the average dynamic systems* . A number of important practical problems, in particular problems on synchronization (resonances) under the rotation of solid bodies, are reduced to studying those systems.

Vibrational mechanics is considered in this book as a special case of a more general concept which can be named *mechanics of systems with hidden motions*. Both in the process of mechanical-mathematical modeling real systems and in striving to simplify the models, one has a natural desire "not to see", "to partly ignore" certain components of motions of the system and even some degrees of freedom which seem to be of minor importance. The question arises: by which mechanics should the observer be guided lest he should come into conflict with reality? The answer is analogous to the previous one: certain additional forces should be introduced into the equations, describing the motions which are to be considered. Just like above, the mechanics of systems with partly ignored motions can be interpreted from the points of view of the three observers **O′**, **V′** and **W′** similar to observers **O** , **V** and **W** . It is remarkable that vibrational mechanics is connected with both the classic works on mechanics and modern investigations. By the former we mean the works by Thomson, Tait, and Routh on mechanics of systems with cyclic coordinates, the works by Reynolds on the theory of turbulence, by Poincare and Lyapunov on celestial mechanics and theory of the stability of motion, by Krylov, Bogolyubov, Mitropolsky and Malkin on methods of averaging and the theory of periodic solutions of differential equations with a small parameter. By the latter we mean the works by Kapitsa and his followers on the behavior of a pendulum with a vibrating axis of suspension, on the theory of vibrational processes and devices and also on the problem of resonances in the Solar system. That common character of ideas and their continuity has been systematically reflected in this book.

The first part of the book is devoted to the theoretical basis of vibrational mechanics, in particular to the description of the general approach, mentioned above, which is mainly based on the so called *method of direct separation of motions* . The main statements have been formulated here as theorems. It should be noted however that the elaboration of the mathematical apparatus of vibrational mechanics cannot as yet be considered completed. Other chapters of the book are devoted to numerous applications of vibrational mechanics.

The book is provided with a detailed table of contents and detailed captions to figures. It seems to us both of them will be helpful to the readers.

The book is meant first of all for the specialists in both theoretical and applied mechanics, in the theory of non-linear oscillations and vibrational technology. It may also prove to be useful for mathematicians involved in the theory of ordinary differential equations as a source of some new problems.

While working at the concept and apparatus of vibrational mechanics I had a happy chance to discuss the main points with V.I.Babitsky, V.V.Beletsky, V.V.Koslov, M.Z. Kolovsky, P.S.Landa, L.G.Loytsyansky, Ja.G.Panovko, A.A.Pervosvansky, V.V.Rumyantsev, O.V.Savinov, G.Ju.Stepanov, and K.V.Frolov. My young colleagues E.B.Kremer, O.Z.Malakhova, A.V.Petchenev. A.Ya. Fidline and N.P.Yaroshevich both on my request and on their own initiative made a number of complicated researches the results of which have been reflected in the book. Besides, O.Z.Malakhova read the manuscript very attentively and helped to eliminate some drawbacks. While reviewing the book, important remarks were made by M.Z.Kolovsky. I express my deep gratitude to those scientists.

I appreciate the unfailing understanding and support of my colleagues and the management of the Research and Design Institute of Mechanical Treatment of Mineral Resources ("Mekhanobr", St-Petersburg) which made it possible for me to devote myself to this most interesting work.

I am greatly thankful to my wife Nina Granat who passed away when the work on the book was practically completed. A gifted researcher herself (the reader will find some results of her work in the book), Nina selflessly released me from any domestic care. To Nina's fond memory this book is dedicated.

November 29, 1991, St.Petersburg

I.Blekhman

Preface to the English Edition

I am thankful to the World Scientific Publishing Co. for the opportunity to present my book to the English readers.

Five years have passed since the publication of this book in Russia [125]. The ideas and results stated there have been essentially developed and supplemented during that time. In this English edition of the book the author tended to at least partially reflect those developments:

- item 3.3 has been added, which is devoted to the characteristics of the advantages and limitations of the approach and methods stated in the book;

- item 5.3 has been added, which gives a short review of the results referring to the stability of composite pendulums with a vibrating axis of suspension, including the follower-loaded ones;

- item 7.8 was supplemented by the general definition of synchronization;

- item 10.6, devoted to the vibrational pumps and coaches has been considerably enlarged;

- a new chapter had been introduced. This chapter (chapter 18) contains some ideas on the problem of controlling the properties of the nonlinear systems by means of vibration. Here the new approach to the creation of the materials with given dynamic properties — the so called dynamic materials — is discussed most thoroughly;

- the bibliography contains additional references to both the latest investigations and to those made before, which unfortunately escaped the attention of the author when the book was being prepared.

In 1996 the author delivered a PhD course of lectures on vibrational mechanics at the Department of Solid Mechanics, headed by Professor Pauli Pederson, at the Technical University of Denmark (DTU).

Then similar lectures were delivered at the Institute of the Problems of Mechanical Engineering of the Russian Academy of Sciences; at the St.Petersburg Technical University (the former Polytechnical Institute) and at the Marine Technical University.

Professors D. A. Indeitsev, M. Z. Kolovsky and S. V. Sorokin, who had invited me, and their disciples as well as their colleagues at the DTU made an attentive and interested audience. The contact with them proved to be both pleasant and useful. Some of them, including Professors J. J. Thomsen, S. V. Sorokin and their pupils J. S. Jensen, D. M. Tchernyak and E. C. Miranda found it interesting to make their own investigations of that problem. References to their investigations and short reviews of them are given in the book.

The author's contacts with Professors A. L. Fradkov, K. A. Lurie, and L. Sperling were both very useful and very pleasant.

I am deeply grateful to all those scientists.

I want to thank Associate Professor Minna Perelman for her careful translation in the process of which some fragments of the text were improved.

I am also thankful to my colleague Larissa Titova for her valuable assistance in preparing the manuscript.

I am grateful to Professor Mikhail Levinshtein for his helpful advice.

July 30, 1999, St.Petersburg

I.Blekhman

Contents

Part I

Fundamentals of Theory of Vibrational Mechanics

Chapter 1

Introduction. Subject–Matter of Vibrational Mechanics

1.1 Oscillatory Processes in Nature and Engineering

Oscillatory processes are characteristic of all organic and inorganic nature, from cell to community of organisms, from atom to galaxies. They play an appreciable role in the psycho-neurological life of a human being, and even in the sphere of social phenomena. That is why the philosophic significance of the science of vibrations is absolutely doubtless. A lot of bright ideas, formulated by great thinkers of the past and by our contemporaries could be given here (see, e.g., [21, 355, 188, 202]). The question why nature often "prefers" oscillations to a monotonous flow of processes deserves discussion. It is worth discussing not only from the physical and biological points of view, but also from the philosophic position. But so far, that has not yet been done. We may just guess that oscillatory processes can be characterized by a certain expedience and sometimes they can even be optimal.

1.2 Damaging Vibration and Useful Vibration. Vibrational Engineering

Perhaps just due to the lack of understanding the optimal features of vibration, the latter has long been considered a harmful factor, causing breakdowns, accidents and professional diseases [a]. It is only since the beginning of this century that a period of a rapid development of vibrational technology has started.

---

[a]By *vibration* we mean here mechanical oscillations whose period is much shorter than that at which the motion of the system is being considered, and whose swing is far smaller than the characteristic size of the system.

Without that technology, a number of most important industries, such as mining, the processing of natural resources, chemical technology, metallurgy, the manufacture of building materials, and the erecting of various constructions would have been absolutely unthinkable.[b]

The diversity of trends in using vibration is reflected in the titles of sections of the book [114]: "Vibration shifts", "Vibration transforms (in Vibrorheology)", "Vibration separates and classifies", "Vibration intensifies the processes and the treatment of workpieces", "Vibration consolidates - vibration destroys", "Vibration unites (self-synchronization of unbalanced rotors)", "Vibration maintains rotation - vibration retards rotation", "Vibration cancels vibration - vibration intensifies vibration (the generalized principle of autobalancing)", "Vibration helps in measuring - vibration hinders measuring", "Vibration cures - vibration causes diseases". The title of a popular book by Goncharevich is also remarkable:"Vibration as a nonstandard way" [209].

The use of vibration made it possible to literally revolutionize many industries, providing a great technical and economical effect. Potentialities, however, have not yet been exhausted. The application of vibrational technology seems to be most promising in the future.

Apart from the book [114], there is an extensive literature in Russian (monographs and reference books) devoted to general and special problems of the use of vibration in technology (see, e.g., [46, 84, 145, 192, 207, 208, 209, 215, 226, 259, 301, 302, 306, 315, 323, 364, 400, 402, 420, 417, 440, 443, 445, 454, 475, 550, 159, 227]). Achievements of vibrational technology are widely reflected in volume 4 of a six-volume reference book "Vibration in technology" [564] issued by the publishing house "Mashinostroenie" in 1978–81.

It should be noted that while in general certain advantages of oscillatory processes have not yet been revealed or understood properly (as compared to those going on monotonously) they can, as a rule, be easily seen in vibrational processes and devices. So, e.g. the efficiency of the use of vibration in enriching the minerals is quite clear. It is often based on the fact that under vibration, forces of the type of dry friction, preventing the separating processes under the action of weak factors (such as the difference in density of the particles), act as if they were transformed into forces of viscous friction, under which weak factors can be displayed.

---

[b]Certain examples of using vibration have been known since ancient times: it was used in screening dry substances, in building and even in medicine.

1.3   The Theory of Vibrational Processes and Devices — a New Section of the Applied Theory of Oscillations

The necessity to develop and improve vibrational technology on the one hand, and the need to interpret and describe mathematically a number of specific phenomena, connected with the effect of vibration on mechanical systems on the other hand, have recently resulted in the appearance of a new and fast — developing branch of the applied theory of oscillations — the *theory of vibrational processes and devices* [364]. A dominant role in creating this theory was played by the scientists of the former USSR. And that is not surprising, since that theory is based on the fundamentals of the theory of non-linear oscillations and stability of motion, in whose development the Russian and Ukrainian physicists, mechanics and mathematicians played the leading part. We mean the fundamental works by Lyapunov, Mandelshtam, Papalexi, Andronov, Vitt, Krylov, Bogolyubov, Mitropolsky and others.

The theory of vibrational processes and devices studies the regularities of excitation of vibration and its action in various mechanical systems; it also includes the theory of machines where vibration is useful.

Despite the fact that physical oscillatory systems are nonlinear, a number of applied problems of the theory of mechanical oscillations can be very well analyzed in a linear setting of the problem, i.e. without considering any nonlinear factors.

As is known, the effect of external vibration on linear systems has been on the whole investigated exhaustively; the main quantitative regularities being grouped together around the phenomenon of resonance. However, even those rather simple regularities have not yet been fully taken advantage of in vibrational engineering (unlike the electrical- or radio engineering) ,to say nothing of nonlinear oscillations. The latter are characterized by extraordinary qualitative diversity. When being investigated, they always manifest new remarkable effects. So, quite recently it has been discovered that on the one hand rather simple nonlinear systems, having just one and a half degree of freedom, may display quite a complicated chaotic behavior [195, 271, 409, 410, 447, 231], while on the other hand complex systems with a very great number of degrees of freedom may display a coordinated (synchronous, coherent) behavior.

An important peculiarity of nonlinear systems lies in the fact that their oscillations needn't necessarily come from "outside". They may spring up and be steadily maintained within the oscillatory system. Here we are dealing with self-excited oscillations in whose investigation the Russian scientific school has also played an outstanding role.

The action of vibration in nonlinear mechanical systems often leads to peculiar and sometimes quite unexpected results. These effects, on the one hand, can be used in technology, the principles of action of quite a number of most efficient machines being based on them, on the other hand, the same effects may be the cause of undesirable and even disastrous situations.

The aim of this book is to consider a large group of phenomena that attend the action of vibration in nonlinear mechanical systems and to propose a unitary efficient approach and a mathematical apparatus for their description and investigation. This approach is named by us vibrational mechanics and the main method of solving the respective problems is called by us a method of direct separation of motions.

1.4   On the Effects Caused by the Action of Vibration in Non-linear Oscillatory Systems

The book mainly discusses the following four groups of effects which appear under vibration in non-linear mechanical systems.

1. **The change in the behaviour of oscillatory systems and mechanisms under the action of vibration.** This group of effects comprises such cases as the disappearance of the former equilibrium positions and former types of motions of the system and the appearance of the new positions of equilibrium and new types of motion; the change in the character of the equilibrium positions (i.e. of their stability or instability); the change in frequencies of free small oscillations near the positions of stable equilibrium; effects of a vibrational link, in particular the self-synchronization of unbalanced rotors (vibro-exciters); the effect of vibrational maintenance and retarding of rotation of the unbalanced rotors, a peculiar behavior of the so called "oscillatory systems with limited excitation" and some others.

2. **Effects of vibrational displacement and drift.** This group comprises effects of vibrational transportation of solid and granular materials, effects of vibro-dipping and vibro-extracting of grooves, piles and shells; separation of particles of the material according to their properties on the vibrating surfaces and in the oscillating vessels with granular material or with liquid; the appearance of slow flows of granular materials or streams of liquid in vibrating vessels (particularly a phenomenon of vibro-bunkering — filling up the bunker with granular materials from below upwards); the drift and location of particles in heterogeneous vibrational fields; mutual microshifts and wear of contiguous parts in the nominally fixed joints; the drift of pointers of the devices and of the axes of gyroscopes under the action of vibration of the foundation they are

installed on.

3. **Vibrorheological effects.** These effects denote the change, caused by vibration, in the rheological properties of bodies with regard to slow actions, or, as it is sometimes said, they denote the seeming change in the rheological properties of bodies under vibration. We mean here the seeming transformation of dry friction into viscous friction (pseudo-liquefaction) which occurs under vibration; the reduction of dry friction coefficients, the seeming change in the viscosity coefficient (the classical example is the transfer from the laminar flow of liquid to the turbulent stream); the effect of vibro-creep and many others.

4. **The appearance of an intensive mechanical interaction between particles and volumes in the multicmponent systems.** This group of effects comprises the loosening of granular materials in the vibrating pan chutes and vessels — the formation of the so-called vibro-boiling layer; the appearance of intensive relative vibartions of solid particles, different in size and density, in the vibrating liquid or in the granular material, etc. It is natural that such effects facilitate the intensification of chemical reactions, being a basis for the use of vibration for the fine grinding of the material and also for the abrasive processing of articles.

It should be noted that this classification is to a certain extent arbitrary. It is easy to notice that some effects can belong to two groups. The effects of the fourth group can sometimes be interpreted by means of linear models. Should we add to the listed effects the fundamental phenomena of resonance and of self-excited oscillations which had already been mentioned by us, we will obtain an almost exhaustive list of effects used in vibrational technology.

A man who first comes across these effects is greatly impressed by them. One cannot very well stay indifferent, seeing that as a result of but a slightly appreciable vibration the upper position of the pendulum becomes stable, a heavy metal ball "flows up" in a layer of a sand, a pile quite easily goes down into the ground under the effect of its own weight, a heavy body or a layer of granular material moves upward along a slope, the rotation of a rotor is maintained steadily with the electric engine being switched off, etc. Very often and, as we will see later, not without a ground, we gain an impression that a gravitational force has changed its direction, and that a well-known statement that it is impossible to accelerate or retard the motion of the center of masses of the system only at the expense of its inner forces has become invalid, that the law of mechanics of action and reaction has lost its validity, that the essentially non-conservative system behaves as if it were conservative, etc.

No wonder that the above effects time and again gave rise to delusions, including the "overthrow" of the laws of mechanics (see 1.7). Though, to do

justice it should be mentioned that those "subverters" sometimes quite accidentally made useful intricate inventions, stimulating interesting investigations
(just like their predecessors — inventors of the "perpetual mobile").

1.5 Vibrational Mechanics and Vibrational Rheology. Observer O and Observer V

Most of the enumerated effects are characterized by the fact that the motion
which appears in the system under vibration can be presented as a sum of two
components — the "fast", "vibrational" component and the "slow" component
which changes very little in one period of vibration, and it is the slow motion
that is of special interest in the overwhelming majority of cases. Let us imagine
that there is an observer who does not notice (or does not want to notice)
either those fast (as a rule small) motions or fast forces. This observer is
either wearing special glasses which do not let him see the fast motions of the
system, or he may be watching the motion in the stroboscopic (i.e. interrupted)
light, the frequency of flashes being equal to that of vibration. This observer
V, unlike the ordinary observer **O** who "sees everything", will notice only the
slow component of motion, and lest he should contradict the laws of mechanics,
he will have to explain all those paradoxical effects by the appearance of certain
additional slow forces or moments acting together with the ordinary slow forces.
We will call them after Kapitsa [257, 258] "vibrational forces". From the point
of view of that "biassed" observer it is those forces that cause the above effects
on which the technical use of vibration is based.

In terms of differential equations it looks like this. Let the motion of the
system be described by the equation [c]

$$m\ddot{x} = F(\dot{x}, x, t) + \Phi(\dot{x}, x, t, \omega t), \qquad (1.5.1)$$

where m is the mass, x is the coordinate, F is the "slow" force, and Φ is
the fast force. Differentiation with respect to the "ordinary", "slow" time t is
designated by a dot. The force Φ unlike F depends not only on t, but also on
the fast time ωt, proportional to a "large parameter" ω that is the frequency
of vibration. In the simplest case the function is a periodic function ωt with a
period 2π. Let the motion be presented as

$$x = X(t) + \psi(t, \omega t), \qquad (1.5.2)$$

---

[c]We consider here a system with one degree of freedom, the generalization of a system
with many degrees of freedom presenting no special difficulty (see chapter 3).

where X is the slow component, and ψ is the fast component (not necessarily small compared to X). Then to the observer **V** who does not see the fast motions ψ and the force Φ it will seem that the slow motion can be described by the equation

$$m\ddot{X} = F(\dot{X}, X, t) + V(\dot{X}, X, t), \qquad (1.5.3)$$

where V is the vibrational force.

As will be shown in chapter 3, equation (1.5.3) as well as the expression for the vibrational force can, under certain conditions, be obtained analytically, the appearance of that force explaining all the paradoxical effects listed above.

Thus, we come to the statement, in many respects similar to the well-known theorem of the mechanics of relative motion. According to that theorem, the observer, connected with the coordinate system moving with acceleration, must add the forces of inertia to all the ordinary forces exerted upon the system. In our case, the observer **V** who does not see either the fast forces or the fast motions, must add vibrational forces to all ordinary forces. While in the mechanics of relative motion the addition of forces of inertia to all the ordinary forces seems to be a kind of a fine for the use of the noninertial (i.e. moving with acceleration) coordinate system, in our case the addition of vibrational forces is a fine for the ignoring of the fast (usually small) motions of the system (see section 6 where it is discussed in detail).

On this account, the mechanics which the observer **V** is guided by (the observer who does not see any fast forces or fast motions) will be called by us *vibrational mechanics*. As was already mentioned, a certain basis for distinguishing vibrational mechanics is the fact that it helps to explain and describe a great number of processes taking place under the action of vibration on non-linear mechanical systems. The book describes such processes. The equations of type (1.5.3) will be called by us the main equation of *vibrational mechanics*.

As is established in chapter 3, equation (1.5.3) has been obtained as a result of averaging in a certain way the initial differential equation (1.5.1), the vibrational force reflecting the effects, "accumulated" on account of the action of vibration upon the non-linear system.

By *vibrorheology* (or more exactly by *macrovibrorheology*) we mean a section of mechanics, and at the same time of vibrational mechanics and of rheology, which studies the change, caused by vibration, in the rheological properties of bodies with regard to slow forces, and also the corresponding slow motions of bodies[d]. In other words, macrovibrorheology can be defined as rheology for the observer **V**.

---

[d]This definition has been given in publications [95, 99, 110, 114], though the term "vibrorheology" had been proposed earlier; Rebinder must have been the first to use it.

When considering vibrorheological effects, we will call the equation of slow motions of type (1.5.3), i.e. the main equation of vibrational mechanics, the *vibrorheological equation*.

The study of specific features of the behavior of many-phase systems under vibration can be referred to macrovibrorheology.

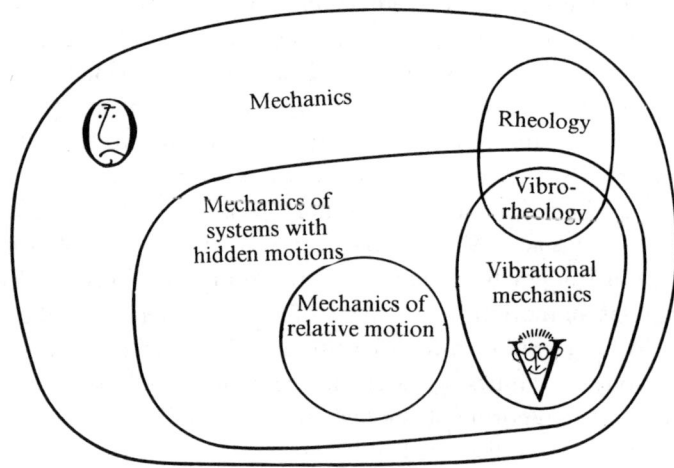

Figure 1.1. The relation between mechanics, mechanics of systems with hidden motion, mechanics of a relative motion, rheology, vibrational mechanics and vibrorheology.

Vibrorheology is related to vibrational mechanics in the same way as rheology is related to mechanics, and vibrorheology is related to rheology — just like vibrational mechanics is related to mechanics. This circumstance is shown schematically in Fig. 1.1. Though vibrorheological effects (as well as those merely rheological) are not necessarily of a purely mechanical nature: they can be essentially connected with thermal phenomena, with chemical transformations etc. Therefore rheology is often considered to be a branch of physics.

What has been said here is shown schematically in Fig. 1.2. In the left part of Fig. 1.2,a a scheme of the initial system is indicated, i.e. the system which is seen by the observer **O**. This observer points out that the system is acted upon by a slow force $F(\dot{x}, x, t)$ and by a fast force $\Phi(\dot{x}, x, t, \omega t)$; and he perceives the

---

It should be noted however that this term is sometimes used with another meaning (see, say, [208, 420]. We must also emphasize that the given definition refers to macrovibrorheology; see section 12.2

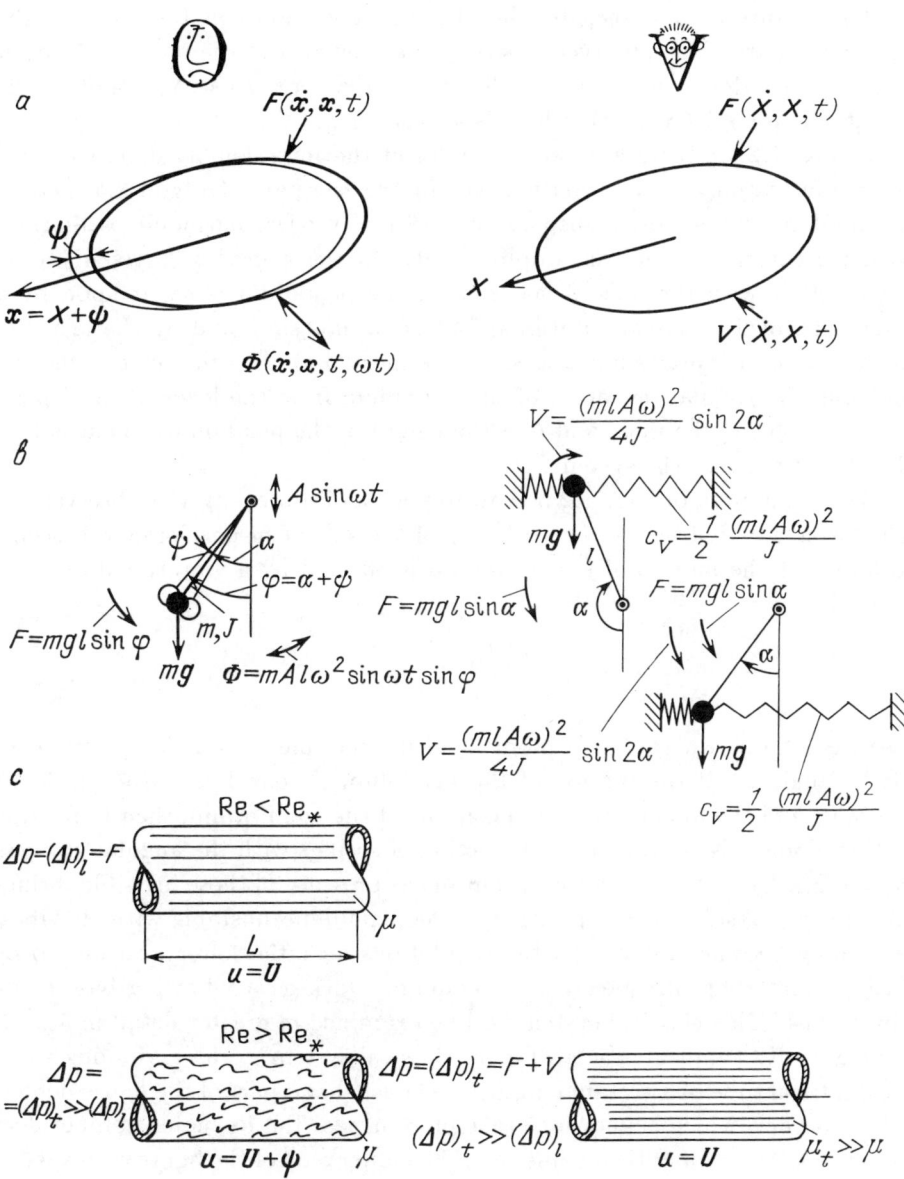

Figure 1.2. The picture seen by the observer **O** (in the left-hand part) and that seen by the observer **V** (in the right-hand part).

motion of the system x as consisting of two components — the slow component X and the fast component ψ.

The picture as it is seen by the observer $\mathbf{V}$ is shown in Fig. 1.2,a in the right-hand part. This observer sees only the slow part of the motion X and it seems to him that it is caused by the initial slow force $F(\dot{X}, X, t)$ and by the vibrational force $V(\dot{X}, X, t)$ (which is also slow).

In Fig. 1.2,b,c there are two examples of the intentionally different character, illustrating the above situation. In the left part of Fig. 1.2,b there is a pendulum whose axis of suspension performs vertical harmonic oscillations with a frequency ω and an amplitude A. In the coordinate system, moving together with the axis of suspension, the pendulum is acted upon by a "slow" moment of the gravitational force $F = mgl \sin\varphi$ and by the fast moment $\Phi = mA\omega^2 \sin\omega t \sin\varphi$ caused by the inertial force in the relative motion (φ being the angular deviation of the pendulum from the lower vertical position, m — being the mass and l — the length of the pendulum). That is how the observer $\mathbf{O}$ sees the system.

The right part of Fig. 1.2,b indicates a picture seen by the observer $\mathbf{V}$. The latter does not notice the vibration of the axis of suspension and it seems to him that the motion of the pendulum is affected by a vibrational moment applied to it:

$$V = \frac{1}{4} \frac{(mlA\omega)^2}{J} \sin 2\alpha$$

(see item 4.1.5) where J is a moment of inertia, and α is a slow component of the angle φ. If the motion of the pendulum is considered near the lower ($\alpha_1 = 0$) and the upper ($\alpha_2 = \pi$) positions of the equilibrium, then the action of this moment is equivalent to the action of springs with the angular stiffness $c_v = 1/2(mlA\omega)^2/J$. It is natural that in the presence of those invisible springs the upper ("upside-down") position of the pendulum, unstable without vibration, may become stable, and the oscillations near the lower position go on with a greater eigenfrequency, the pendulum clock on a vibrating base being always fast. This classical system will be examined in greater detail in 5.1

Figure 1.2,b indicates a "still more classical system": water, running along a cylindrical tube of a certain length L. The left-hand part of the figure depicts a laminar flow which is known to take place in case the Reynolds number does not exceed a certain critical value Re_*. The lower part of the figure corresponds to the turbulent regime when $Re > Re_*$. This regime is characterized by a rather complicated pulsation of the fluid, appearing "autonomously" without any external action.

As a result, to provide the same discharge of water, it is necessary instead of the pressure difference $\Delta p = (\Delta p)_l$ to have a much greater pressure difference $\Delta p = (\Delta p)_t$. The observer **O** quite justly explains it by the fact that part of the pressure is wasted in the turbulent pulsation. As for the observer **V** (see the right-hand part of Fig. 1.2,b, he does not notice the pulsation of water and he will still see the averaged motion as laminar as before. He will explain the necessity of a greater pressure difference Δp by an abrupt growth of the viscosity coefficient of water μ. As Novozhilov very expressively remarked, for that observer the water seemed to have been transformed into treacle.

The book contains many other examples of a similar kind.

It is essential that the equation of vibrational mechanics (1.5.3), unlike the initial equation (1.5.1), does not contain any information which would be excessive for finding a slow component; that is why equation (1.5.3) is simpler than equation (1.5.1). It is also remarkable that the equation of vibrational mechanics may correspond to the conservative system, while the initial system is essentially non-conservative (see chapter 4). Similarly, the non-continuous system (see chapter 14) is answered by the "smooth" system, and the system with many degrees of freedom is answered by a system with a far smaller number of degrees of freedom (see section 6.3).

1.6   Vibrational Mechanics as a Section of the Mechanics of Systems with Hidden Motions

Vibrational mechanics and the mechanics of the relative motion can be regarded within the framework of a wider conception which can be named *conception of partial ignoring of motions*, or *the mechanics of systems with hidden motions* [108, 114]. And we can say that the mechanics of relative motion represents mechanics for the observer who does not see the relative motion of the system, while the vibrational mechanics represents mechanics for the observer who does not see any fast motions or fast forces (see the scheme in Fig. 1.1). The difference however lies in the fact that in the first case the "hidden"("ignored") motion is considered to be known, while in the second case it is still to be determined.

Chapter 3 of this book gives the proof of the main thesis of the conception under discussion, which is in fact almost obvious: the presence of hidden motions is as a rule connected with the necessity to add to the equation of motions, that are being considered, some additional forces which in the general case depend on the constants of integration. Certain elements of this conception can be found in the classical works by Routh, Thomson and Tait, devoted to the dynamics of systems with cyclic coordinates (see 2.2.2).

The situation when certain degrees of freedom are ignored is quite common. It is met almost always. This situation should also be included into the mechanics of systems with partially ignored motions. In this sense the applied mechanics is in fact a mechanics of systems with hidden motions (see Fig. 1.1). It is known that in many cases such a "neglect" of the degrees of freedom makes it possible to simplify the investigation to a considerable extent without making any bad mistakes. In other cases, on the contrary, such ignoring of the degrees of freedom leads to wrong results (see 1.7).

A distinct demarcation between the above-mentioned opposite cases is a very important and interesting mathematical problem, as well as the demarcation between certain intermediate cases. Some investigations in this direction have already been made [41, 42, 307, 413, 414, 415, 526]; the problem however is still far from being exhausted. Just like the ignoring of fast motions in vibrational mechanics, an intended ignoring of either essential or weakly affecting degrees of freedom can prove to be a very effective method. (If necessary, expressions for the additional forces, either approximate, or exact, can also be included into equations of the motions under consideration). In case some motions or degrees of freedom are ignored unconsciously, there may often appear illusions that laws of mechanics are violated. Below we will touch upon the corresponding and very instructive paradoxes and mistakes connected with the misinterpretation of the behavior of non-linear mechanical systems under the action of vibration.

1.7   Mistakes and Paradoxes, connected with the Interpretation of the Phenomena under Consideration. The Observer W

Numerous mistakes, made when studying and interpreting the above effects, are connected with the following two objective circumstances.

First, as it was already mentioned, all the non-linear effects listed in 1.4 seem paradoxical, at least on the face of it.

Secondly, these paradoxes are often conditioned by a "strong" qualitatively appreciable action of some "weak" factors. As a rule, the weak actions result in weak changes in the behavior of the system. Therefore cases when it is not so seem to be strange. And they are really surprising and worth paying special attention, since more than once they stimulated new discoveries and helped to get a better understanding of things. It is quite natural that the outstanding mathematicians, mechanics and physicists developed special mathematical conceptions for the analysis of such situations. We mean in particular certain

cases in the theory of non-linear differential equations with a small parameter
and also *the bifurcation theory*. Many statements of that theory are now sometimes referred to the so called *theory of catastrophes*. Such situations are also
known in the theory of the stability of motion which is closely connected with
those two mentioned above.

Chemistry and biology are not exceptions either: small quantities of substances, introduced into a reactor or into a living organism, may lead to quite
significant consequences. Similar facts are known in sociology too [e].

As was mentioned in the preface, the errors we are discussing now can be
connected with the wrong position of the observer **W** (the first letter of the
word "Wrong".) We will consider this position along with the positions of the
observers **O** and **V**.

The most unforgivable mistakes are to our mind caused by the violation of
a very important principle of methods, known under the name of "Ockham's
razor" [f]. This principle can be formulated as follows: " Do not introduce
any new essence without necessity." In our case the violation of this principle
consists in the fact that instead of a thorough analysis of the phenomena, one
doubts the validity of the main laws of mechanics. Unfortunately, there are
very many cases of such mistakes. It is enough to recall the discussion of the
so called problem of inertioid — the coach, supposedly set in motion only at
the account of its internal forces alone, or the discussion of Dean's notorious
effect which we will speak about later.

Doubts in the validity of laws of mechanics and the desire to correct them.
are based, as a rule, on four facts.

1) The fact that **the degrees of freedom or motions of the system which seem
minor, but are in fact quite important, are not taken into consideration.** As a
result no additional forces, in particular vibrational forces, are being considered
in the equation of motion. That leads to paradoxes when compared with the
experiments where those forces are involved. So for instance, a body with
the internal degree of freedom, lying on a rough vibrating surface, changes its
direction of motion when the frequency of vibration is changed (See Fig. 2.2,*c*).
In case we do not take into consideration this degree of freedom, this effect will
seem inexplicable. The motion of such a system in the direction opposite to that
of the pulses (See Fig. 2.2,*d*) will also seem quite inexplicable. Meanwhile the

---

[e] In one of stories by Ray Bradbury the adequate idea is brightly illustrated in a fantastic
form: in a prehistoric forest a hero crushed a butterfly with his boot, several thousand
years later that resulted... in the victory of a most reactionary candidate at the presidential
elections.

[f] William Ockham (1285-1349) is the outstanding English philosopher and logician of
Middle Ages.

observer **V** will easily explain everything by determining the vibrational force
V and stating that it changes its direction with the change of the frequency.

2) The fact that the presence or the effect of the external forces is not taken
into consideration. Then there is an illusion that the position of the center of
inertia and the value of the moment of momentum may be changed due to the
action of the internal forces alone.

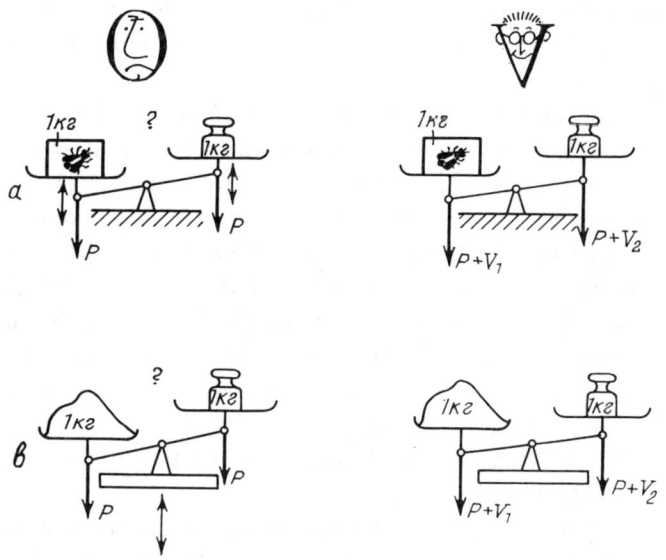

Figure 1.3. Violation of the balance of the sensible scales caused by vibration. To
the left there are pictures seen by the observer **O**, to the right — those, seen by the
observer **V**.

As an example we will consider very sensitive scales shown schematically
in Fig. 1.3,a. A box with a fly in it is counterbalanced by a weight lying on
the other scale. Will the balance be changed if the fly begins flying within
the box? The author was asked that question at the entrance exam to the
University, and he gave the answer the examiners had been expecting: no, the
balance will not be changed, because the inner forces cannot change the angular
momentum. That is true in case we do not take into account the presence of
the outer force, and namely, the air resistance to the motion of the scale or if
we assume that the scales are placed in vacuum. Without those assumptions

the balance may change! It will change due to "aerodynamic asymmetry" of the system: it is natural to assume that the air resistance to the motion of the scales is, first, not the same when moving upward and downward, and secondly, that the difference for each scale is not the same. Then when due to the flying of the fly the scales are oscillating near the equilibrium position, the average force V_1 acting on the left scale (i.e. the vibrational force) will differ from the vibrational force V_2 acting on the right scale. As a result the balance will be broken

Academician Konstantinov demonstrated an experiment in which instead of a fly there was an ordinary electric bell, balanced on the scales. When the bell was switched on, the balance was broken. But it would be practically retained when the bell was placed under a cup from which the air was pumped out. Thus, here too, the main reason of breaking the balance was the presence of aerodynamic forces, arising due to the vibration of the mobile elements of the bell. As was justly mentioned by Stepanov [514], the effect could also be observed when the bell was switched on in the vacuum, due to the vibration of mobile parts of the scales. In other words, for the sake of purity of the experiment, not only the bell, but the whole device should be placed into a vacuum.

Let us consider another example: the same scales where some load on the left scale is exactly balanced by a weight on the right scale (Fig. 1.3,b). Should vertical vibration be imparted to the base on which the scales are installed, the balance may be changed for the same reason as in the previous case.A case is known to the author when a researcher who observed such a system came to the conclusion that vibration changes the weight of a body.

Finally a third example which is quite a curious case. It happened when the idea of cosmic ship-graviflier, proposed by Beletsky and Ghiverts ([56] also see Fig. 10.16 and 10.5.4.) was being discussed. The authors declared that if inside the cosmic ship, revolving, say, as a kind of a satellite of the Earth along the elliptic orbit, two load-dumb-bells were periodically brought together and moved apart, then, in principle, the satellite might be accelerated to such an extent, that it might go beyond the area of the attraction of the Earth. As is said in the book [56], one of the readers when he learned of that idea declared that that was impossible in view of the theorem about the motion of an isolated system, and that the ignorance of the authors was scandalous. But that reader was certainly mistaken: both the ship and the loads of the dumb-bells were acted upon by an external force - the gravitational force. It was the non-linear dependence of that force on this distance from the center of the Earth that caused the above effect. It is not difficult to show that the manipulation with the weights requires a certain expenditure of energy and that it results in the appearance of the vibrational forces similar to traction. From the point of view of the observer **V** it is this force that causes the evolution of the orbit of the ship (see 10.5.4). To justify that reader to a certain extent, it should be noted that in the authors' first publications the circumstances, causing that effect had not been explained clearly enough.

3) The fact that **the driving vibrational force, arising at the vibrational transformation of the non-linear resistance, in particular of the dry friction into the viscous friction, is not taken into account.** In numerous articles and text-books authors write that dry friction is transformed into viscous friction as a result of the vibrational effect. They are right, but not quite. They often forget that along with the vibrational transformation of dry friction, as a rule there also appears another force, called by us *the driving vibrational force* (see 9.22 and chapter 11). This force appears due to the "asymmetry" of the system and due to the vibrational effect. It is this force that causes a number of effects, united by the notion of vibrational displacement and drift. Part III of this book is devoted to the analysis and applications of such effects.

The fact that the driving vibrational force is not taken into consideration results in serious mistakes. The author knows cases when, while designing precision instruments, the engineers tried to use vibration to liquidate the dead zone, caused by dry friction. But doing so they did not think of a possible appearance of a driving vibrational force which caused systematic mistakes in the readings of the devices (see chapter 11). They had either to give up using vibration altogether, or to introduce special correcting devices.

Another example is the so called "potential theory of vibrational separation of particles of the granular material". According to that theory, the granular materials when under vibration, tend toward the state corresponding to the minimum of the potential energy of the totality of its particles in the gravitational or any other potential field. Thus the possibility of the appearance of a driving vibrational force is not taken into consideration. Meanwhile that force (which is sometimes potential too) leads to very important effects going beyond the scope of the potential theory (e.g. "floating up" of a big heavy particle in a vibrating medium of light small particles, in particular — a steel ball in a layer of sand). These effects are discussed in detail in 10.2.2 and 10.2.5.

Along with that, the consideration and control of a driving vibrating force afford a number of important technological and technical opportunities discussed by us in part III of this book.

It should be emphasized that the driving vibrating force may also appear when vibration affects systems with other characteristics of forces, resistant to the motion, including those with smooth characteristics.

4) The fact of **absolutization of the position of the observer V** . What we mean by it is the ascribing of physical nature to those changes in the characteristics of the bodies forming the system which are discovered by the observer **V** who "does not notice" the fast forces or fast motions; these characteristics, however, remain unchanged for the ordinary observer **O** . So sometimes the liquefaction of the granular materials under vibration is spoken about, though in fact it is the pseudo-liquefaction that takes place: for the observer **O** friction remains the Coulomb friction or similar to that.

Absolutization of the position of the observer **V** takes place in the above-mentioned cases of the broken balance of the scales and of the steel ball floating up in the layer of sand when those facts were ascribed to the change in the weight of the body, caused by vibration. The story with Dean's machine refers to the same category of mistakes. That funny story is described in detail in the bright article by Stepanov [514] that was already mentioned by us and in the book by Goncharevich [209].

Chapter 2

On the Mechanics of Systems with Hidden Motions

2.1 General Statements and Main Equations; Theorem 1

The above-mentioned analogy between the vibrational mechanics and mechanics of relative motion is not accidental. It will be shown that they can be considered as special cases of the general conception, named by us the *conception of a partial ignoring of motions* or the *mechanics of systems with hidden motions* [108, 119].

The essence of this conception lies in the following. Let the motion of a dynamic system be described by a system of differential equations which can be presented as

$$\mathbf{m}(\mathbf{x})\ddot{\mathbf{x}} = \mathbf{R}(\dot{\mathbf{x}}, \mathbf{x}, t) \tag{2.1.1}$$

where $\mathbf{x}$ is the n-dimensional vector-column of the generalized efficients, $\mathbf{m}(\mathbf{x})$ is the nondegenerate positive[a] symmetrical $n \times n$ - matrix of the inertial coordinates, $\mathbf{R}(\dot{\mathbf{x}}, \mathbf{x}, t)$ is the vector of forces. We will assume that

$$x_1 = X_1 + \psi_1, \ldots, \ x_k = X_k + \psi_k; \ x_{k+1} = X_{k+1}, \ldots, \ x_{k+l} = X_{k+l};$$
$$x_{k+l+1} = \overline{\psi}_{k+l+1}, \ldots, \ x_n = \overline{\psi}_n \tag{2.1.2}$$

and will name $X_1, \ldots, X_{k+l}$ the *explicit (taken into account) motions* , and $\psi_1, \ldots, \psi_k; \ \overline{\psi}_{k+l+1}, \ldots, \overline{\psi}_n$ — the *hidden (ignored) motions* . The generalized coordinates $x_{k+1}, \ldots, x_{k+l}$ will be called *explicit*, and $x_1, \ldots, x_k$ will be called *partially hidden* , while $x_{k+l+1} = \overline{\psi}_{k+l+1}, \ldots, x_n = \overline{\psi}_n$ will be called the *hidden generalized coordinates*. In accordance with equation (1.1.2) we will pass

---

[a]By *positive* here and below we mean a symmetrical square matrix corresponding to a positive quadratic form of a definite sign.

over to the new generalized coordinates $X_1, \ldots, X_{k+l}; \overline{\psi}_{k+l+1}, \ldots, \overline{\psi}_n$ and the excess variables $\psi_1, \ldots, \psi_k$ will be considered either the prearranged functions of time, or the variables satisfying certain k mutually independent additional relationships

$$A_s(\mathbf{X}, \boldsymbol{\Psi}) = 0 \qquad (s = 1, \ldots, k) \tag{2.1.3}$$

where $\mathbf{X} = (X_1, \ldots, X_{k+l})$ and $\boldsymbol{\Psi} = (\psi_1, \ldots, \psi_k; \overline{\psi}_{k+l+1}, \ldots, \overline{\psi}_n)$ are the $(k+l)$- and $(n-l)$-dimensional vector-columns respectively.

Relationships (1.1.3) can to a certain extent be preassigned arbitrarily. These equalities are quite important since they determine the principle according to which the ignored motions are separated from those taken account of in the first k relationships (1.1.3). Here to make our reasoning more simple we will assume that relationships (1.1.3) are finite, solvable for $\psi_1, \ldots, \psi_k$ but in principle they may also be certain differential or integro-differential equations (see, say, 3.2).

Let

$$\mathbf{m_0}(\mathbf{X_0})\ddot{\mathbf{X}}_0 = \mathbf{R_0}(\dot{\mathbf{X}}_0, \mathbf{X_0}, t) \tag{2.1.4}$$

be a system of equations of the $2(k+l)$-th order, composed under the assumption that the hidden motions are absent ($\mathbf{m_0}$ being a positive symmetrical $(k+l) \times (k+l)$-matrix). We will call this system a *simplified system* . We substitute expressions (1.1.2) into equations (1.1.1). Using $(n-k-l)$ of those equations and k relationships (1.1.3), under the assumptions made, it is always possible to find the derivatives $\dot{\psi}_1, \ldots, \dot{\psi}_k, \ \dot{\psi}_{k+l+1}, \ldots, \dot{\psi}_n$ and eliminate them from the other $(k+l)$ equations. Using relationships (1.1.3), we can also eliminate the variables $\psi_1, \ldots \psi_k$ and $\dot{\psi}_1, \ldots, \dot{\psi}_k$ from those equations. As a result, the indicated $(k+l)$ equations can be written in the form

$$\mathbf{m_0}(\mathbf{X})\ddot{\mathbf{X}} = \mathbf{R_0}(\dot{\mathbf{X}}, \mathbf{X}, t) + \mathbf{V}_1(\dot{X}, X, \dot{\overline{\Psi}}, \overline{\Psi}, t) \tag{2.1.5}$$

where $\overline{\boldsymbol{\Psi}} = (\overline{\psi}_{k+l+1}, \ldots, \overline{\psi}_n)$. Equations (1.1.5) differ from (1.1.4) by the presence of the term $\mathbf{V}_1$, which will be named by us *vector of additional forces*.

Together with other differential equations and relationships (1.1.3), equations (1.1.5) form a system equivalent to the initial system (1.1.1) in view of (1.1.2). If these other differential equations could be fully integrated, taking account of relationships (1.1.3), then the functions $\overline{\boldsymbol{\Psi}} = \overline{\boldsymbol{\Psi}}(\dot{\mathbf{X}}, \mathbf{X}, t, \overline{\mathbf{C}})$ which depend on $2(n-k-l)$ arbitrary constants would be determined.

$\overline{\mathbf{C}} = (\overline{C}_{2(k+l)+1}, \ldots, \overline{C}_{2n})$; these constants are determined by the initial conditions for the functions $\overline{\psi}_{k+l+1}, \ldots, \overline{\psi}_n$. As a result, system (1.1.5) will be written as

$$\mathbf{m_0}(\mathbf{X})\ddot{\mathbf{X}} = \mathbf{R_0}(\dot{\mathbf{X}}, \mathbf{X}, t) + \mathbf{V}_2(\dot{X}, X, t, \overline{\mathbf{C}}). \tag{2.1.6}$$

The systems of type (1.1.6) can be also arrived at by the following reasoning. Let $x_s = x_s(t, C_1, \ldots, C_{2n}), s = 1, \ldots, n$ be the general solution of the initial system (1.1.1), i.e. the functions, containing independently $2n$ constants and satisfying the system (1.1.1). Then from the $2(k+l)$ equalities $x_1 = x_1(t, C_1, \ldots, C_{2n}), \ldots, x_{k+l} = x_{k+l}(t, C_1, \ldots, C_{2n}); \dot{x}_1 = \dot{x}_1(t, C_1, \ldots, C_{2n}), \ldots, x_{k+l} = x_{k+l}(t, C_1, \ldots, C_n)$ it is possible to find $2(k+l)$ constants $C_1, \ldots, C_{2(k+l)}$. Substituting the expressions for these constants via $t, x_1, \ldots, x_{k+l}; \dot{x}_1, \ldots, \dot{x}_{k+l}$ and $C_{2(k+l)+1}, \ldots, C_{2n}$ into the corresponding $(k+l)$ equations (1.1.1) and using relations (1.1.2) and (1.1.3), we will come right to the equations of type (1.1.6).

But these general deductions and transformations have not reduced of course system (1.1.1) of the $2n$–th order to the system of $2(k+l)$–th order: the elimination of $2(n - k - l)$ constants $\overline{\mathbf{C}}$ from (1.1.6) will again bring it to the system of the $2n$–th order as it must. And in general, systems (1.1.5) and (1.1.6) together with the differential equations and relations supplementing them (1.1.3) are not simpler than the initial system (1.1.1) and contain all information about it. Therefore the transfer to system (1.1.5) and (1.1.6) is justified, for instance, if this information is excessive: the explicit motions are of utmost interest while the hidden motions affect the explicit motions but weakly, and their influence may be considered only approximately. Then systems (1.1.5) and (1.1.6) may prove to be much more convenient.

However, on close inspection, it may be useful to ignore the hidden motions, amounting their presence to the action of certain additional forces, which is quite usual and convenient in mechanics; that is what we have for instance in the mechanics of relative motion.

The conception of ignoring certain motions is especially advantageous in the case when the additional forces $\mathbf{V}_2$ can be thought of as independent of the constant $\overline{\mathbf{C}}$. We mean here the situation typical of the systems that are met in particular in vibrational mechanics when studying motions, asymptotically stable with regard to the hidden generalized coordinates $\overline{\psi}_{k+l+1}, \ldots, \overline{\psi}_n$ and the velocities $\dot{\overline{\psi}}_{k+l+1}, \ldots, \dot{\overline{\psi}}_n$ within all the range of change of the variables[b]. In this case we may "almost forget" about the existence of the hidden generalized coordinates in the system: as time passes, the system "forgets" the corresponding initial conditions, the values $\dot{\overline{\Psi}}$ and $\overline{\Psi}$ in equations (1.1.5) become concrete functions of time. Then equations (1.1.5) and (1.1.6) acquire

---

[b]The main ideas of the theory of stability of motion with regard to some part of variables, are stated and developed in the monographs by Rumyantsev and Osiraner [466] and by Vorotnikov [576].

the following form:

$$\mathbf{m_0}(\mathbf{X})\ddot{\mathbf{X}} = \mathbf{R_0}(\dot{\mathbf{X}}, \mathbf{X}, t) + \mathbf{V}(\dot{X}, X, t). \qquad (2.1.7)$$

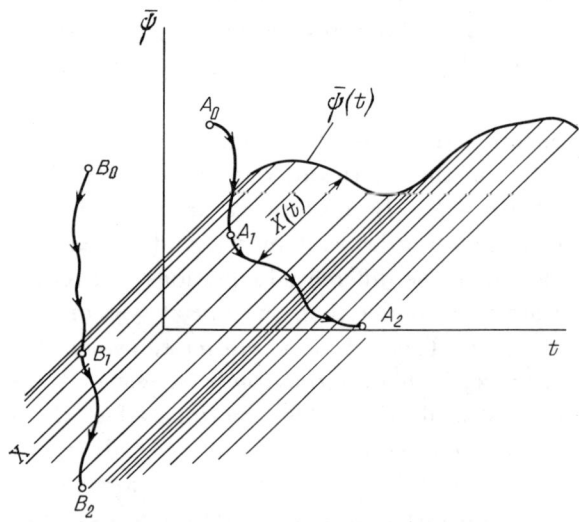

Figure 2.1. The phase space of the system (the scheme).The case, when the phase trajectories "are attracted" to the cylindrical surface $\overline{\psi} = \overline{\psi}(t)$

The hidden motions are represented in these equations only as the expressions for the additional forces $\mathbf{V}$. When there are several above-mentioned asymptotically stable motions, the expressions for $\mathbf{V}$ will be different for every type of that motion. The advantages of the transfer from the initial equations (1.1.1) to the equations for explicit motions are in this case especially great. Then the equations under consideration are reduced in order, which will be illustrated below by many examples.

Figure 2.1 shows schematically the situation adequate to this case. In the phase space $X, \overline{\psi}, t$ we see the trajectories $A_0, A_1, \ldots$ and $B_0, B_1, \ldots$ of the system going from certain points A_0 and B_0. The two trajectories, presented there and all the other trajectories are attracted to the cylindrical surface $\overline{\psi} = \overline{\psi}(t)$. After reaching the points A_1 and B_1 the motion can be thought of as taking place either over the plane or in close vicinity of it. It is natural, of course, that the equations of motion of the representative point over the mentioned surface differ from the equations of motion of the point in the space $X, \overline{\psi}, t$.

Mark that under the formulated conditions and provided the transformation from the initial variables $\dot{\mathbf{x}}, \mathbf{x}$ to the variables $\dot{\mathbf{X}}, \mathbf{X}, \dot{\overline{\mathbf{\Psi}}}, \overline{\mathbf{\Psi}}$ and vice versa, defined by equalities (1.1.2) and (1.1.3), is unambiguous and continuous in the vicinity of the motion under consideration, there is a full agreement between the properties of the stability of motions of the initial system (1.1.1) with regard to the variables $\dot{\mathbf{x}}$ and $\mathbf{x}$ and those of the motions of system (1.1.7) with regard to the variables $\dot{\mathbf{X}}$ and $\mathbf{X}$.

It should be noted that the problem is reduced to the equations of type (1.1.7) also in the cases when we are interested in the additional forces in the motions, corresponding to quite definite initial conditions. It can be exemplified by a problem about the motion of a ball in a viscous incompressible liquid. This problem is considered in 2.2.4.

Equations of type (1.1.5)–(1.1.7) are the main equations of the mechanics of systems with hidden motions.They testify an almost obvious and very essential thesis, named by us the *main theorem of mechanics of systems with hidden motions*.

T h e o r e m I. *The differential equations of explicit motions differ from the equations of the simplified system by the appearance of certain additional forces, dependent in the general case on both the explicit- and the hidden generalized coordinates and velocities or on the explicit generalized coordinates and velocities and the corresponding number of the constants of integration. In the case of the asymptotic stability of motion with regard to the hidden generalized coordinates and velocities at any* $\mathbf{X(t)}$ *of the region under consideration, the dependence of the additional forces on the indicated constants as time passes becomes inessential, and these forces can be considered dependent on the explicit motions alone. In case of the unambiguity and continuity of the transformation from the initial variables* $\dot{\mathbf{x}}, \mathbf{x}$ *to the variables* $\dot{\mathbf{X}}, \mathbf{X}, \dot{\overline{\mathbf{\Psi}}}, \overline{\mathbf{\Psi}}$ *and vice versa in the vicinity of the motions under consideration, there is a full agreement between the properties of the stability of motions of the initial system (1.1.1) with regard to the variables* $\dot{\mathbf{x}}$ *and* $\mathbf{x}$ *and system (1.1.7) with regard to the variables* $\dot{\mathbf{X}}$ *and* $\mathbf{X}$.

It should be emphasized that the appearance of additional forces in the equations of explicit motions defies explanation unless we consider the presence of hidden motions. It is natural that in case of hidden motions the main laws and statements of mechanics for the explicit motions do not hold, or are held very roughly if the additional forces in the corresponding equations are not taken into account. As was already mentioned, this circumstance has more than once been the cause of paradoxes and errors and even caused doubts of the validity of laws of mechanics. On the other hand, an intended ignoring of inessential motions and degrees of freedom makes it possible to simplify greatly investigations. This ignoring is in fact a necessary (and even inevitable) element in the process of simulating the system. But if the essential degrees of freedom are ignored, that may cause not only the quantitative errors, but also wrong conclusions of a qualitative nature, so the stable motions can be mistaken for the unstable and vice versa (see e.g.[426, 428]).

In this connection two very important questions arise:

1) Is it permissible to ignore the hidden motions in particular the hidden degrees of freedom, i.e. instead of equations (1.1.1) or (1.1.5)–(1.1.7) consider the simplified equations (1.1.4)?

2) How can we in practice obtain expressions, at least approximate expressions, for the additional forces $\mathbf{V}$?

The first question is closely connected with the problems of simulation, identification and decomposition of dynamic systems. To consider it efficiently one may use the methods of a small parameter, in particular, methods of the theory of singular perturbations, the Poincare method, methods of averaging, and the method of integral diversity. Unfortunately, comparatively few investigations have been devoted to a systematic study of this question from the general positions of mechanics. The following publications can be referred to: [41, 42, 307, 388, 413, 414, 415, 526]. Important methodological theses are given in [413, 482, 483]. As for the second question, this book is mainly devoted to its solution for the problems of vibrational mechanics. In the general form it is studied in chapters 2 and 3. The next section is mostly devoted to other problems in form of examples and special cases.

2.2   Special Cases, Examples

2.2.1   *Mechanics of the Relative Motion*

The name of this item is somewhat arbitrary. We mean the case when $k = n$ and $\psi_1, \ldots, \psi_k$ are the preassigned functions of time, so that $x_s = X_s + \psi_s(t)$. The initial equation (1.1.1) looks in this case as

$$\mathbf{m}(\mathbf{X} + \boldsymbol{\Psi})(\ddot{\mathbf{X}} + \ddot{\boldsymbol{\Psi}}) = \mathbf{R}(\dot{\mathbf{X}} + \dot{\boldsymbol{\Psi}}, \mathbf{X} + \boldsymbol{\Psi}, t). \qquad (2.2.1)$$

Multiplying this equation from the left by the ungenerated matrix $\mathbf{m}(\mathbf{X})\mathbf{m}^{-1}(\mathbf{X} + \boldsymbol{\Psi})$ we can write it in form of equation (1.1.5) (in this case

$\mathbf{m_0} = \mathbf{m}$ and $\mathbf{R} = \mathbf{R_0}$):

$$\mathbf{m(X)\ddot{X}} = \mathbf{R(\dot{X}, X}, t) + \mathbf{V}_1(\mathbf{\dot{X}, X, \ddot{\Psi}, \dot{\Psi}, \Psi}, t) \qquad (2.2.2)$$

where

$$\mathbf{V}_1(\mathbf{\dot{X}, X, \ddot{\Psi}, \dot{\Psi}, \Psi}, t) = \mathbf{m(X)m^{-1}(X + \Psi)R(\dot{X} + \dot{\Psi}, X + \Psi}, t)$$
$$-\mathbf{m(X)\ddot{\Psi} - R(\dot{X}, X}, t) \qquad (2.2.3)$$

is the additional force, corresponding to the force, of inertia in the relative motion.

2.2.2 A Solid Body with the Internal (Hidden) Degrees of Freedom

Quite a number of the main physical effects, caused by the presence of the hidden degrees of freedom, can be illustrated by an example of the motion of a solid body, possessing the internal (hidden) degrees of freedom. The role of the corresponding system is however more than just illustrative: it is important for celestial mechanics, for the theory of cosmic flights, the theory of vibrational devices, the theory of systems with gyroscopes, the theory of dynamics of vehicles and coaches and even for the theory of certain circus feats. Here we will briefly dwell on the effects of external harmonic exciting forces, i.e. on the cases which are directly concerned with the main subject-matter of this book. A number of such systems will be discussed in detail later.

As a most simple case we will consider a linear system, i.e. an absolutely solid body whose mass is m_1 and which can move freely along the axis x and inside which there is another absolutely solid body whose mass is m_2 (Fig. 2.2,a). This body is connected with the first body by means of an elastic element whose stiffness is c and which can move in the same direction. The external body is acted upon by a harmonic exciting force $F\sin(\omega t)$. This system is the simplest model of a solid body being deformed, it describs its behavior under dynamic effects. The degree of freedom, connected with the mobility of the mass m_2, will be thought of as hidden.

If the frequency of the free oscillations $\lambda = \sqrt{c/m} \gg \omega$ ($m = m_1 m_2/(m_1 + m_2)$ is the associated mass), i.e. if at the preassigned m_1, m_2, and ω the stiffness c is large enough, the system will perform oscillations as a single absolutely solid body whose mass is $m_1 + m_2$ according to the equation

$$(m_1 + m_2)\ddot{X}_1 = F\sin\omega t. \qquad (2.2.4)$$

Here $x_1 = X_1$ is the coordinate of the external body (the explicit motion), coinciding in this case with the coordinate of the internal body (the hidden motion). Equation (1.2.4) corresponds to equation (1.1.4). If the condition $\lambda \gg \omega$ is not satisfied, the presence of the internal degree of freedom may greatly affect the explicit motion. So, for instance, in the limiting case $\lambda \ll \omega$ the explicit motion is described almost exactly by the equation

$$m_1 \ddot{X}_1 = F \sin \omega t \qquad (2.2.5)$$

i.e. it looks as if the mass of the system were reduced to the value m_1.

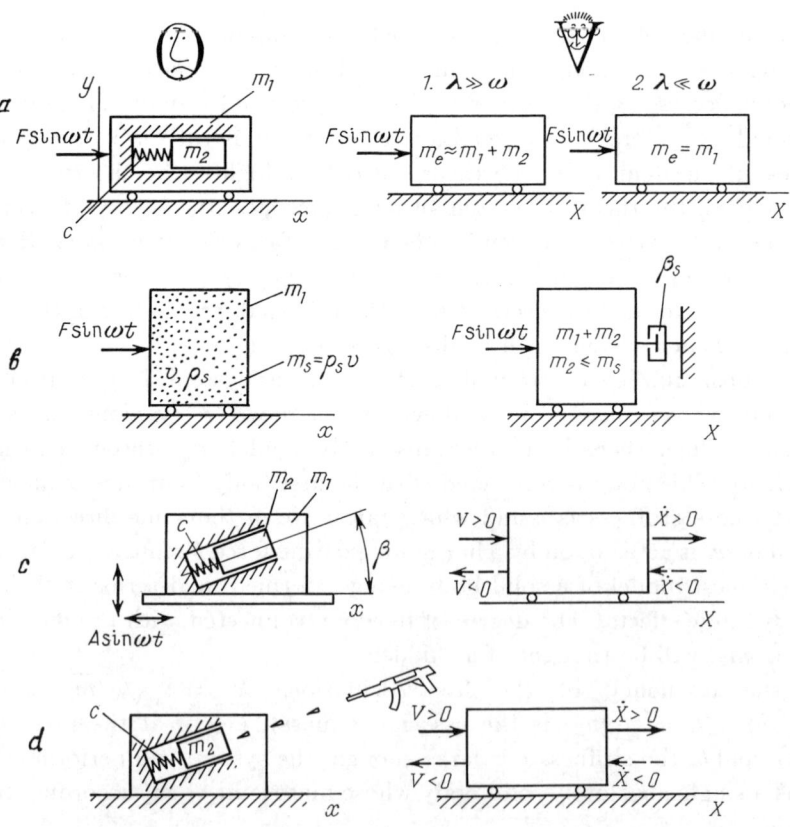

Figure 2.2. A solid body with internal degrees of freedom under periodic actions. On the left there are pictures, seen by the observer **O**, on the right there are those, seen by the observer **V**.

Equation (1.2.5) can also be written in form of equation (1.1.5):

$$(m_1 + m_2)\ddot{X}_1 = F \sin \omega t + V \tag{2.2.6}$$

where the additional force is $V = (m_2/m_1)F \sin \omega t$. Thus according to (1.2.5) the presence of the internal (hidden) degree of freedom of the solid body leads to a seeming change in the mass of the body with respect to the high frequency effects. This circumstance does not contradict the general thesis, formulated in 2.1, since, according to (1.2.6) the mass can be considered unchanged (equal to $m_1 + m_2$) but an additional force V appeared in the equation of the explicit motion. It should be noted that if the exciting force had acted along the axis y , along which the system always moves as a single body, the presence of the hidden degree of freedom would not have been manifested. Neither would it manifest itself if we were interested just in the motion of the center of inertia of the system, i.e. if the value $X = (m_1 X_1 + m_2 X_2)/(m_1 + m_2)$, where x_2 is the coordinate of the mass m_2, would be taken for the coordinate under consideration. Then in accordance with the theorem about the motion of the center of inertia, irrespective of the relationship between λ and ω,the equation

$$(m_1 + m_2)\ddot{X} = F \sin \omega t$$

is valid.

It should be also noted that the question of an expedient form of writing the equations of motion of a solid body with the internal degrees of freedom has been considered in the monograph by Lurie [346].

Henceforward we will more than once come across the effect of a change in the inertial properties of the system, equivalent to the action of certain additional forces. In particular, this effect (as well as the appearance of a certain dissipative force) takes place if the mass m_1 represents a vessel full of suspension (Fig. 2.2,b). If the vessel is still acted upon by an exciting harmonic force $F \sin \omega t$, then its stationary motion will go on in accordance with the equation

$$(m_1 + m_2)\ddot{X} + \text{v}k^*\dot{X} = F \sin \omega t \tag{2.2.7}$$

where $m_2 = \rho^* \text{v}$ and $\text{v}k^* = \beta_s$ are the effective mass and the coefficient of damping the suspension respectively, v is the volume of the vessel, ρ^* and k^* are the effective density and the damping factor respectively, referred to a unit of the volume of suspension (the equations for ρ^* and k^* and also the main suppositions at which equation (1.2.7) is valid are given in 16.2) It is remarkable that ρ^* never exceeds the density of the suspension ρ_s i.e. the mass $m_1 + \rho^* \text{v}$ is never greater than that of the system $m_1 + m_2 = m_1 + \rho_s \text{v}$.

As above, equation (1.2.7) can be quite easily reduced to

$$(m_1 + \rho_s \text{v})\ddot{X} = F \sin \omega t + V(\dot{X}, t), \qquad (2.2.8)$$

corresponding to the form of equation (1.1.5) with the additional force

$$V(\dot{X}, t) = \frac{m_1 + \rho_s \text{v}}{m_1 + \rho^* \text{v}} \left[-\text{v}k^* \dot{X} + \frac{(\rho_s - \rho^*)\text{v}}{m_1 + \rho_s \text{v}} F \sin \omega t \right]. \qquad (2.2.9)$$

Still more peculiar is the behaviour of the system of the type shown in Fig. 2.2,b if the external body m_1 is on a rough surface. In this case we assume that the direction of a possible motion of the internal body of a mass m_2 does not coincide with the direction of the axis x, forming with it a certain acute angle β. Suppose the plane performs either vertical or horizontal oscillations. Then, as we can see (12.2.1), the body will be transported along the plane either to the right or to the left, depending on the frequency of vibration ω. Tilting the plane, we can make the body move upwards along the plane. These conclusions have been proved experimentally.

It is also of interest that endowing the body m_2 with periodic pulses of the same direction, e.g. shooting at it from a gun (the plane in this case is considered to be fixed, see Fig. 2.2,d), we can make the body move in the direction of the gun [274, 275, 391].

The observer who does not know that the system has a hidden degree of freedom (that observer is called by us the observer $\mathbf{V}$ to distinguish him from the observer $\mathbf{O}$, see 5 of the chapter 1) will think that the described effects are the results of the appearance of an additional force $\mathbf{V}$ acting along the plane in one or other direction. A lightminded observer may even say that due to vibration there is "an inversion" — a change in the direction of gravity! Similar mistakes and paradoxes have already been mentioned.

It is easy to guess, that the reaction to the external action of the solid body with a gyroscope inside will also be very peculiar. Of interest are specific features of motion of the artificial satellite of the Earth, possessing the internal degrees of freedom (see 20.6).

The inverse problem, i.e. the problem of identification, is very interesting too, and is of practical importance. With regard to the system under consideration — the body with the internal (hidden) degrees of freedom — this problem may be formulated in the following way. How is it possible, judging by the reactions of the body (the "black box") to various "testing" force effects, to establish the contents of that "box"? Not less interesting and important is the problem of synthesis — how to form the internal structure of the body in accordance with the preassigned character of the reaction of the body to the external actions. But we cannot afford to dwell on the analysis of those problems in this book.

2.2.3   Systems with Cyclic Coordinates

Elements of mechanics of the systems with hidden motions can be found in the classical works by Routh, Thomson, and Tait, referring to the dynamics of systems with cyclic coordinates (e.g. see [371]). Some of their terms have been adopted by us, but we used them here in a wider sense of the word.

By *cycle* we mean the generalized coordinates which are not involved explicitly in the expression for the kinetic energy of the system T, and the generalized forces, corresponding to those coordinates, are equal to zero. All the other generalized coordinates are called *positional* . The positional coordinates $x_1 = X_1, \ldots, x_l = X_l$ and the corresponding motions are regarded as explicit, and the cyclic coordinates $x_{l+1} = \overline{\psi}_{l+1}, \ldots, x_n = \overline{\psi}_n$ are considered to be hidden. The specific feature of this case lies in the fact that the transformations leading to the main equations of type (1.1.6), i.e. the equations, containing only the explicit (positional) coordinates, can be performed up to the end in an elegant analytic form.

In accordance with the definition of cyclic coordinates, the Lagrange equations for the cyclic (hidden) coordinates will look as follows:

$$\frac{d}{dt}\frac{\partial T}{\partial \dot{\overline{\psi}}_s} = 0 \qquad (s = l+1, \ldots, n), \tag{2.2.10}$$

and therefore they allow the following first integrals:

$$\frac{\partial T}{\partial \dot{\overline{\psi}}_s} = C_s \qquad (s = l+1, \ldots, n). \tag{2.2.11}$$

While the equations for the positional (explicit) coordinates will be

$$\frac{d}{dt}\frac{\partial T}{\partial \dot{X}_s} - \frac{\partial T}{\partial X_s} = Q_s(\dot{\mathbf{X}}, \mathbf{X}, \dot{\overline{\mathbf{\Psi}}}, \overline{\mathbf{\Psi}}, t) \qquad (s = 1, \ldots, l) \tag{2.2.12}$$

where Q_s denotes the generalized forces, corresponding to the explicit coordinates. In the general case these forces can depend on both the explicit- and the hidden generalized coordinates and velocities, $\mathbf{X}$ designates the vector of the explicit motions, and $\overline{\mathbf{\Psi}}$ denotes the vector of the hidden motions.

If we restrict our consideration to the systems with holonomic and stationary constraints, the kinetic energy of the system can be presented in the following form:

$$T = \frac{1}{2} \sum_{r=1}^{l} \sum_{s=1}^{l} a_{rs} \dot{X}_r \dot{X}_s + \sum_{s=1}^{l} \sum_{r=l+1}^{n} b_{rs} \dot{X}_s \dot{\overline{\psi}}_r + \sum_{r=l+1}^{n} \sum_{s=l+1}^{n} c_{rs} \dot{\overline{\psi}}_r \dot{\overline{\psi}}_s \quad (2.2.13)$$

where $\mathbf{a} = \| a_{rs} \|$ and $\mathbf{c} = \| c_{rs} \|$ are the positive matrices, with a_{rs}, b_{rs} and c_{rs} depending only on the explicit coordinates $X_1, \ldots, X_l$. Then relations (1.2.11) are the linear equations with respect to $\overline{\psi}_{l+1}, \ldots, \overline{\psi}_n$ from which it is always possible to find the expressions for those values. Substituting these expressions into equations (1.2.13), we can express the kinetic energy of the system via the explicit coordinates and velocities X_s and $\dot{X}_s$ and also by the integration constants $C_{l+1}, \ldots, C_n$. The integration of the expressions for $\overline{\psi}_s$ will lead to the appearance of $n - l$ additional arbitrary constants $C'_{l+1}, \ldots, C'_n$ in the expressions for $\overline{\psi}_s$.

As is shown in the cited classical works, as a result of such transformations, equations (1.2.12) for the explicit motions can be presented in the following form:

$$\frac{d}{dt} \frac{\partial R_2}{\partial \dot{X}_s} - \frac{\partial R_2}{\partial X_s} = Q_s^{(1)}(\dot{\mathbf{X}}, \mathbf{X}, \mathbf{C}^{(1)}, t) + \frac{\partial \overline{R}_0}{\partial X_s} + \sum_{r=1}^{l} g_{rs} \dot{X}_r$$

$$(s = 1, \ldots, l). \quad (2.2.14)$$

Here $\mathbf{C}^{(1)}$ is a vector with the components $C_{l+1}, \ldots, C_n; C'_{l+1}, \ldots, C'_n$,

$$Q_s^{(1)}(\dot{\mathbf{X}}, \mathbf{X}, \mathbf{C}^{(1)}, t) =$$

$$Q_s \left[\dot{\mathbf{X}}, \mathbf{X}, \dot{\overline{\mathbf{\Psi}}}(t, C_{l+1}, \ldots, C_n), \overline{\mathbf{\Psi}}(t, C_{l+1}, \ldots, C_n, C'_{l+1}, \ldots, C'_n) \right],$$

$$\overline{R}_0 = -\frac{1}{2|c|} \sum_{s=l+1}^{n} \sum_{r=l+1}^{n} C_{rs} C_r C_s,$$

$$R_2 = \frac{1}{2} \sum_{r=1}^{l} \sum_{s=1}^{l} (a_{rs} + a'_{rs}) \dot{X}_r, \dot{X}_s, \quad g_{rs} = -g_{sr} = \frac{\partial a_r}{\partial X_s} - \frac{\partial a_s}{\partial X_r},$$

$$a_k = \frac{1}{|c|} \sum_{m=l+1}^{n} \sum_{r=l+1}^{n} C_r C_{mr} b_{mk} \quad (2.2.15)$$

with C_{rs} being a cofactor of the element c_{rs} of the determinant $|c|$ of the matrix $\mathbf{c} = \| c_{rs} \|$, while the matrix with the elements $a_{rs} + a'_{rs}$ where

$$a'_{rs} = -\frac{1}{|c|} \sum_{p=l+1}^{n} \sum_{q=l+1}^{n} C_{qp} b_{ps} b_{qr} \quad (2.2.16)$$

is positive.

Equations (1.2.14) comply with the following remarkable statement: *the explicit motion of the system under consideration proceeds in the way as if the system had the kinetic energy R_2 (i.e. a matrix of the inertial coefficients $\| a_{rs} + a'_{rs} \|$) and as if besides the initial generalized forces it were acted upon by the forces $\Gamma = \sum\limits_{r=1}^{l} g_{rs} \dot{X}_r$ called the gyroscopic forces, and by the conservative forces with the potential energy $-R_0$.*

Thus, when writing down the main equations in form (1.2.14), the ignoring of hidden motions results not only in the appearance in these equations of certain additional forces, but also in the change of the matrix of the inertial coefficients (in respect to the matrix $\| a_{rs} \|$ corresponding to the explicit motions in the absence of the hidden motions). We have already come across such circumstance in 2.2.2. of this section. In this case the matrix a'_{pq} can be called the *matrix of the associated masses*. But the main equations can be also written in the form, corresponding to that of the main equations (1.1.6), when the matrix of the inertial coefficients remains unchanged, and the consideration of the hidden motions, present in that system, leads only to the appearance of additional forces. To this end it is sufficient to present the form R_2 as

$$R_2 = T_x + T' \tag{2.2.17}$$

where

$$T_x = \frac{1}{2} \sum_{r=1}^{l} \sum_{s=1}^{l} a_{rs} \dot{X}_r \dot{X}_s$$

is the kinetic energy of the system, in which the hidden motions are absent, and in the expressions $\dfrac{d}{dt} \dfrac{\partial T'}{\partial \dot{X}_s} - \dfrac{\partial T'}{\partial X_s}$ it is necessary to substitute the second derivatives $\ddot{X}_s$ with their expressions found from the initial equations of motion (1.2.10), (1.2.12), and substitute $\dot{\boldsymbol{\Psi}}$ and $\boldsymbol{\Psi}$ with their expressions via $\dot{\mathbf{X}}$, $\mathbf{X}$ and the constants $\mathbf{C}$. Designating the result of that substitution by $[E_s(T')]$ we will present equation (1.2.14) as

$$\frac{d}{dt} \frac{\partial T_x}{\partial \dot{X}_s} - \frac{\partial T_x}{\partial X_s} = Q_{s0}(\dot{\mathbf{X}}, \mathbf{X}, t) + V_s(\dot{\mathbf{X}}, \mathbf{X}, \mathbf{C}^{(1)}, t) \tag{2.2.18}$$

where the additional force is

$$\mathbf{V}_s(\dot{\mathbf{X}}, \mathbf{X}, \mathbf{C}^{(1)}, t) = -E_s(T') - Q_{s0}(\dot{\mathbf{X}}, \mathbf{X}, t)$$

$$+ Q_s^{(1)}(\dot{\mathbf{X}}, \mathbf{X}, \mathbf{C}^{(1)}, t) + \frac{\partial \overline{R}_0}{\partial X_s} + \sum_{r=1}^{l} g_{rs} \dot{X}_r, \tag{2.2.19}$$

and Q_{s0} has the same meaning as R_0 in the equations (1.1.4).

The form of notation of the main equations (1.2.14) for systems with cyclic coordinates (and in some other cases as well) is more convenient than (1.2.18), so it is notation (1.2.14) that is usually used. Equations (1.2.18) are given here just to show that they fit in the context of the general statements presented in 2.1. Though in the case when $b_{pq} = 0$ and therefore $a_k = 0$, $g_{rs} = 0$ and $T' = 0$, i.e. when the system refers to the class of the so called *gyroscopically unbound systems* (for them $g_{rs} = 0$), both notations of the main equations practically coincide. It is remarkable that the functions $-\overline{R}_0$, reflecting (though not always fully) the effect of the hidden motions upon the explicit motions, and playing in equations (1.2.14) the role of the potential energy, in the case of the gyroscopically unbound systems proves to be nothing but the kinetic energy of hidden motions (see e.g. [371]):

$$-\overline{R}_0 = \frac{1}{2} \sum_{r=l+1}^{n} \sum_{s=l+1}^{n} c_{rs} \dot{\overline{\psi}}_r \dot{\overline{\psi}}_s .$$

It is this circumstance that forms the basis of Hertz's conception [234] that "...every potential energy can be considered to be a kinetic energy of certain hidden motions, which cannot be observed directly".

The presence of the function $\overline{R}_0$ in the equations (1.2.14) leads particularly to the fact that when solving the question of the stability of stationary motions of systems with cyclic coordinates, the role of the potential energy is played by the function $W = \Pi - \overline{R}_0$ where Π is the "ordinary" potential energy of the system. Namely, if in the stationary motion under consideration W has a minimum, this motion is stable in respect to the positional (explicit) coordinates and velocities, at least for the perturbances which do not violate the values of the cyclic integrals (1.2.11) (the Routh theorem, see e.g. [371, 372].)

Running ahead we will remark here that the "potentiality" of the effect of the ignored motions upon the motions taken into account plays the main role when establishing both the "potentially on the average" and the adequate extreme signs of stability of the motion of systems with the ignored fast motions. Such systems play a special role in vibrational mechanics (see chapter 3).

2.2.4   Motion of a Body in a Viscous Incompressible Fluid

The system which presents a solid body moving in a viscous fluid i.e. in a medium with a distributed mass possesses " $\infty + 6$" degrees of freedom. It is natural that in the applied investigations, when it is the motion of the body that is of the main interest, it is expedient to consider the motion of the body alone, trying to reduce the action the fluid exerts on it to certain forces. In other words, an infinite number of coordinates, determining the motion of the fluid, is in this case ignored (hidden).

When solving many applied problems of this class, the action of the fluid upon the body (i.e. the force, mentioned above as additional) is reduced to the force which depends only on the velocity of the body relative to the fluid and linearly - on the corresponding acceleration, which leads to the concept of the associated masses and moments of inertia. As a result, the number of degrees of freedom under consideration coincides with that of the solid body. Under certain conditions (though not always worded) such simplification does not lead to any serious mistakes. In other cases the hidden degrees of freedom of the system require a more careful consideration.

Let us consider the situation using the classical problem about the non-stationary translational motion of a ball in an unlimited volume of a viscous incompressible fluid at Reinolds small numbers. This problem was considered first by Boussinesq then by Ozeen. An elegant solution was found by Lurie [345] who used the operational calculus. The task was reduced to solving a system of differential equations

$$\dot{\mathbf{u}} = -\frac{1}{\rho}\operatorname{grad} p + \nu \Delta \mathbf{u}, \quad \operatorname{div} \mathbf{u} = 0,$$

presenting the Navier-Stokes equations with rejected convective terms and the equation of incompressibility respectively. Here $\mathbf{u}(x, y, t)$ is the velocity of the fluid, $p = p(x, y, z, t)$ is the pressure, ρ is the density of the fluid, $\nu = \mu/\rho$ is the kinematic viscosity coefficient (μ being the viscosity coefficient), and Δ is the Laplas operator. The boundary conditions imply that the velocity of the fluid infinitely far from the center of the ball is equal to zero and that the velocities of the fluid on the surface of the ball S are equal to the velocity of the ball $v(t)$.

$$\mathbf{u}|_{r \to \infty} = 0 \quad \left(r = \sqrt{x^2 + y^2 + z^2} \right); \quad \mathbf{u}|_s = \mathbf{v}(t).$$

These relationships must be considered together with the differential equation of the motion of the ball in the fluid:

$$m_0 \dot{\mathbf{v}}(t) = \mathbf{F} + \mathbf{V}_1 \tag{2.2.20}$$

Here m_0 is the mass of the ball, $\mathbf{F}$ is the external force, and $\mathbf{V}_1$ is the force, exerted on the ball by the fluid. The latter is determined by integrating the stresses, distributed over the surface of the ball. Besides, it is necessary to take into consideration the initial conditions. Let us assume that at the initial moment of time $t = 0$ the system was at rest:

$$\mathbf{u}(x, y, z, 0) = 0, \qquad \mathbf{v}(0) = 0$$

The solution of the task results in the following equation for the force $\mathbf{V}_1$, known as *the Boussinesq equation*:

$$\mathbf{V}_1 = -\frac{1}{2}m_1\dot{\mathbf{v}}(t) - 6\pi\mu a \left[\mathbf{v}(t) + \frac{a}{\sqrt{\pi\nu}}\int\limits_0^t \frac{d\mathbf{v}(\tau)/d\tau}{\sqrt{t - \tau}}\, d\tau\right]. \qquad (2.2.21)$$

Here m_1 is the mass of the fluid within the space of the ball, a is the radius of the ball. It is the equation (1.2.21) that determines the additional force, caused by the presence of an infinite number of hidden degrees of freedom of the system. In this connection it should be noted that expression (1.2.21) is exact: equation (1.2.20) in view of (1.2.21) is absolutely equivalent to the above-described original system under the indicated initial conditions. This equation is not of a second order, as may seem on the face of it, but of an "infinite order", which is connected with the presence of an integral term in expression (1.2.21) and is seen very well when making integration by parts.

Thus, generally speaking, the additional force, is not reduced to the first two summands in equation (1.2.21), determining the associated mass $1/2m_1$ and the Stokes resistance $6\pi\mu av(t)$. The role of the last summand may prove to be quite essential at the initial period of motion, and in some other cases.

It should be noted that the equation of motion (1.2.20) just like the corresponding equtions in 2.2.2. and 2.2.3 can be written in the form

$$m_0\dot{\mathbf{v}}(t) = \mathbf{F} + \mathbf{V}, \qquad (2.2.22)$$

not containing in the left part the added mass of the fluid, i. e. in the form, corresponding to that of equation (1.1.5). Then the additional force is

$$\mathbf{V} = -\frac{m_0}{m_0 + \frac{1}{2}m_1}\left\{6\pi\mu a\left[\mathbf{v}(t) + \frac{a}{\sqrt{\pi\nu}}\int\limits_0^t \frac{d\mathbf{V}(\tau)/d\tau}{\sqrt{t - \tau}}\, dt\right] + \frac{1}{2}\frac{m_1}{m_0}\mathbf{F}\right\}. \qquad (2.2.23)$$

2.2.5   *Vibrational Mechanics — Systems with Hidden Fast Motions*

As was mentioned before, this book is mainly devoted to the class of systems with hidden fast motions. Certain examples were considered in 2.2.2. of this section. For such systems the explicit motions $\mathbf{X}$ in the equalities (1.2.2) correspond to the slow motions, and the hidden motions $\mathbf{\Psi}$ — to the fast motions.

Chapter 3

Basic Statements and Mathematical Apparatus of Vibrational Mechanics

3.1 Vibrational Mechanics as the Mechanics of Systems with Hidden Motions

We will point out again the main prerequisites defining the subject-matter of vibrational mechanics:

1. A great number of phenomena and processes of fundamental and applied interest, taking place under vibration in non-linear mechanical systems [a], are characterized by the imposition of "fast" oscillations of a high frequency on the "slow" motions [b].

2. It is the slow motions that are as a rule of the main interest.

By *vibrational mechanics* we mean the mechanics, describing the slow motions attendant on the vibration in non-linear mechanical systems, or in other words, the mechanics for the observer who "does not see" the fast forces or fast motions. Making use of the point of view and terms of chapter 2, we may also say that the vibrational mechanics is a mechanics of systems with the hidden fast motions. It is natural to hope that such mechanics will prove to be simpler than that, describing the total motion of the system. And in future we will see more than once that it is really so.

In 2.1 it has been established that the mechanics of the systems with hidden

---

[a]Here and further on we mean not only the external vibrational effect upon non-linear systems, but also the effect of the vibration which appears within the system, i.e. the self-excited oscillations. It is remarkable that the " fast" self-excited oscillations can appear in the system in which only the "slow" forces are acting. Classic examples of it can be provided by self-excited oscillations in the systems with dry friction and by turbulent pulsations in the flow of viscous fluid.

[b]the mathematical meaning of the notions "the slow variable" and "the fast variable" will be defined in 3.2.4 of this chapter.

motions is characterized by the necessity to take into account in the equations of explicit motions certain additional forces; in this case, following Kapitsa, we will call them vibrational forces. So one of the main tasks of vibrational mechanics is to find expressions for the vibrational forces and to get equations, at least the approximate ones, of slow motions.

3.2　Method of a Direct Separation of Motions as an Effective General Method of Solving Problems of Vibrational Mechanics

3.2.1　Preliminary Remarks

In this section we will present an effective method of obtaining expressions for the vibrational forces and of working out the main equations of vibrational mechanics. This method can be called *method of direct separation of motions*. It will be regularly used in this book.

The method of direct separation of motions comprises two stages. First, like it is in the mechanics of systems with hidden motions, the initial differential equations of the motion of the system are transformed to a system of integro-differential equations of an order "twice higher" with regard to the explicit (further on "slow") components and the hidden (further on "fast") components, selecting an expression for the additional (further "vibrational") forces. This transformation is performed with due regard for the method of the subsequent approximate solution of the equations, but it is valid irrespective of the rate of change of the components in time.

At the second stage the system obtained is solved by an approximate method with due regard for the difference in the tempe of change of the fast (hidden) and slow (explicit) components.

After presenting the method at the euristic level its substantiation is given in the sense of asymptotic methods.

3.2.2　The Initial Equation and Its Reduction to a System of Integro-differential Equations for the Explicit and Hidden Components of Motions

With rather general suppositions, the differential equations of motion of the systems under consideration can be presented in the following form [c]

---

[c]Below, in 3.6 of this section, we also consider the case when the equations of motion are written in form of the Lagrange equations of the second type. All the variable and constant values, involved in the equations, whenever necessary should be considered dimensionless. So in the inequality $\omega \gg 1$ the frequency ω is believed to be dimensionless : $\omega = \omega_* / \omega_0$, where ω_0 is the dimensional frequency and ω_0 is a certain frequency characteristic of the system, say, the largest frequency of free oscillations of the linearized system.

$$\mathbf{m\ddot{x}} = \mathbf{F}(\mathbf{\dot{x}}, \mathbf{x},t) + \mathbf{\Phi}(\mathbf{\dot{x}}, \mathbf{x},t,\omega t) \tag{3.2.1}$$

where $\mathbf{x}$ is the n-dimensional vector of the generalized coordinates, $\mathbf{m}$ is a positive constant ("mass"), ω is the positive parameter (in future the "large" parameter), $\mathbf{F}$ and $\mathbf{\Phi}$ are the n-dimensional vectors of forces, with $\mathbf{\Phi}$ being an almost periodical function of the argument $\tau = \omega t$ (in particular — the periodic function τ with a period of 2π); in future, i.e. at the second stage the time t will be called *slow time* , $\mathbf{F}$ will be called *slow force* , while τ will be called *fast time* , and $\mathbf{\Phi}$ *fast force* . As for the smoothness of the functions $\mathbf{F}$ and $\mathbf{\Phi}$, the usual conditions are supposed to be satisfied, which provides the existence of all the solutions of differential equations which are considered below. The conditions of the existence of the average values, introduced below, are also believed to be satisfied. In a number of cases these conditions are rendered concrete (see, e.g. p. 75).

In accordance with what has been said in 3.21, we will assume that

$$\mathbf{x} = \mathbf{x}(t, \tau) = \mathbf{X}(t) + \mathbf{\Psi}(t, \tau) \tag{3.2.2}$$

where $\mathbf{X}$ is the explicit (in future — the slow), and $\mathbf{\Psi}$ is the hidden (in future — the fast) component of the vector of the generalized coordinates. Here we will assume that $\mathbf{\Psi}$ is an almost periodical (in particular — periodical) function of τ, assuming for certainty that [d]

$$< \mathbf{\Psi}(t, \tau) >= 0, \tag{3.2.3}$$

i.e. we will assume that the average value of the hidden component with respect to τ with the fixed ("frozen") slow time t is equal to zero. According to (3.2.3) the explicit component $\mathbf{X}$ in accordance with (3.2.2) is the corresponding average value of the coordinate $\mathbf{x}$:

$$\mathbf{X}(t) =< \mathbf{x}(t, \tau) > . \tag{3.2.4}$$

Substituting expressions (3.2.2) into the differential equations (3.2.1), and then adding and subtracting in their right sides the expression

---

[d]the angular brackets here and below indicate the averaging with respect to argument $\tau = \omega t$ which can enter into the expression which is being averaged either explicitly or via the function $\mathbf{\Psi}$, with $< \ldots >= \lim\limits_{T \to \infty} \frac{1}{T} \int\limits_{0}^{T} \ldots d\tau$ in the case of an almost periodic function and $< \ldots >= \int\limits_{0}^{2\pi} \ldots d\tau$ in the case of 2π- periodic function of τ.

$$\tilde{\mathbf{F}}(\dot{\mathbf{X}}, \mathbf{X}, \dot{\boldsymbol{\Psi}}, \boldsymbol{\Psi}, t) - <\boldsymbol{\Phi}(\dot{\mathbf{X}} + \dot{\boldsymbol{\Psi}}, \mathbf{X} + \boldsymbol{\Psi}, t, \tau)> \qquad (3.2.5)$$

where

$$\tilde{\mathbf{F}}(\dot{\mathbf{X}}, \mathbf{X}, \dot{\boldsymbol{\Psi}}, \boldsymbol{\Psi}, t) = \mathbf{F}(\dot{\mathbf{X}} + \dot{\boldsymbol{\Psi}}, \mathbf{X} + \boldsymbol{\Psi}, t) - \mathbf{F}(\dot{\mathbf{X}}, \mathbf{X}, t) \qquad (3.2.6)$$

denotes the function which becomes zero when $\dot{\boldsymbol{\Psi}} = \mathbf{0}, \quad \boldsymbol{\Psi} = \mathbf{0}$.

Now, having the right to choose (instead of one unknown function **x** two functions **X** and $\boldsymbol{\Psi}$ have been introduced), we will demand that a certain group of terms of the relation that has been obtained should also become zero. And namely we will demand that the equation [e]

$$m\ddot{\boldsymbol{\Psi}} = \tilde{\mathbf{F}}(\dot{\mathbf{X}}, \mathbf{X}, \dot{\boldsymbol{\Psi}}, \boldsymbol{\Psi}, t) + \boldsymbol{\Phi}(\dot{\mathbf{X}} + \dot{\boldsymbol{\Psi}}, \mathbf{X} + \boldsymbol{\Psi}, t, \tau)$$
$$- <\tilde{\mathbf{F}}(\dot{\mathbf{X}}, \mathbf{X}, \dot{\boldsymbol{\Psi}}, \boldsymbol{\Psi}, t)> - <\boldsymbol{\Phi}(\dot{\mathbf{X}} + \dot{\boldsymbol{\Psi}}, \mathbf{X} + \boldsymbol{\Psi}, t, \tau)> \qquad (3.2.7)$$

should be satisfied. Then the equation

$$m\ddot{\mathbf{X}} = \mathbf{F}(\dot{\mathbf{X}}, \mathbf{X}, t) + <\tilde{\mathbf{F}}(\dot{\mathbf{X}}, \mathbf{X}, \dot{\boldsymbol{\Psi}}, \boldsymbol{\Psi}, t)> + <\boldsymbol{\Phi}(\dot{\mathbf{X}} + \dot{\boldsymbol{\Psi}}, \mathbf{X} + \boldsymbol{\Psi}, t, \tau)>$$
$$(3.2.8)$$

should also be satisfied, its right side being, as can be easily seen, the result of averaging the right side of the initial equation (3.2.1) with respect to τ. The argument in favour of such a "splitting" of equation (3.2.1) will be given below. Here we will mark that the system of integro-differential equations (3.2.7),(3.2.8) that has been obtained, is equivalent to the initial equation (3.2.1), at least in the sense that if there is a certain solution **X**, $\boldsymbol{\Psi}$ of this system, the function $\mathbf{x} = \mathbf{X} + \boldsymbol{\Psi}$ will be the solution of equation (3.2.1).In other words, for the existence of the solution of the equation (3.2.1) of type (3.2.2) it is sufficient that there should be the corresponding solutions **X**, $\boldsymbol{\Psi}$ of the system (3.2.7), (3.2.8).

The system of equations (3.2.7), (3.2.8) can also be arrived at in the following way. Let us substitute expression (3.2.2) into the initial equation (3.2.1) and average both sides of it with respect to the time τ which enters it both explicitly and via the function $\boldsymbol{\Psi}$. Then, after determining the function $\mathbf{F}(\dot{\mathbf{X}}, \mathbf{X}, t)$, we arrive at the equation (3.2.8). Equation (3.2.7) is obtained by means of subtracting equation (3.2.8) from the initial equation (3.2.1). Looking ahead, we may say the terms of the initial equation which further on will be considered to be fast and slow respectively as if they were "balanced" separately

---

[e]This equation corresponds to the relations (2.1.3) chapter 2.

by equations (3.2.7) and (3.2.8). Note that when obtaining equations (3.2.7) and (3.2.8) that way of reasoning was used by Kapitsa in his above-mentioned publications [257, 258] on the pendulum with a vibrating axis of suspension.

Of special interest for applications are cases when part of the components of the vector in equalities (3.2.2) is identically equal to zero, i.e. part of the generalized coordinates **x** presents the hidden (in future — fast) generalized coordinates. To make it concrete let us assume that

$$x_1 = X_1 + \psi_1, \ldots, \quad x_k = X_k + \psi_k; \quad x_{k+1} = \overline{\psi}_{k+1}, \ldots, \quad x_n = \overline{\psi}_n \quad (3.2.9)$$

so that the number of the hidden generalized coordinates is $r = n - k$. For the hidden coordinates the corresponding components of the functions **F** and of the average $< \boldsymbol{\Phi} >$ are equal to zero. As a result, equations (3.2.7) for those coordinates with due regard for equalities (3.2.2) coincide with the initial equations (3.2.1). Then the number n of equations (3.2.7) for the hidden components $\boldsymbol{\Psi}$ remains the same, and the number of equations (3.2.8) for the explicit components **X** which are to be considered appears to be equal to k, i.e. it is reduced by the number r of the hidden generalized coordinates.

3.2.3   The Case when a Separate Equation for the Explicit Component is Obtained: Theorem 2

As was mentioned, solutions of the system (3.2.7), (3.2.8) in which $\boldsymbol{\Psi}$ are periodical or almost periodical with respect to $\tau = \omega t$ are of a considerable interest. Let us assume that we have managed to find such solutions

$$\boldsymbol{\Psi} = \boldsymbol{\Psi}^*(\dot{\mathbf{X}}, \mathbf{X}, t, \tau) \quad (3.2.10)$$

which are isolated at the given **X** and $\dot{\mathbf{X}}$. Let them also be asymptotically stable with respect to the hidden generalized coordinates $\overline{\boldsymbol{\Psi}}$ i.e. to the coordinates which do not contain any explicit components, and also with respect to the corresponding generalized velocities $\dot{\overline{\boldsymbol{\Psi}}}$ all over the region of the change of other generalized coordinates and velocities. Then for every such solution $\boldsymbol{\Psi} = \boldsymbol{\Psi}^*$ a certain additional force can be obtained:

$$\mathbf{V}(\dot{\mathbf{X}}, \mathbf{X}, t) = < \tilde{\mathbf{F}}(\dot{\mathbf{X}}, \mathbf{X}, \dot{\boldsymbol{\Psi}}^*, \boldsymbol{\Psi}^*, t) > + < \boldsymbol{\Phi}(\dot{\mathbf{X}} + \dot{\boldsymbol{\Psi}}^*, \mathbf{X} + \boldsymbol{\Psi}^*, t, \tau) >, \quad (3.2.11)$$

and equation (3.2.8) can be written as

$$m\ddot{\mathbf{X}} = \mathbf{F}(\dot{\mathbf{X}}, \mathbf{X}, t) + \mathbf{V}(\dot{\mathbf{X}}, \mathbf{X}, t), \quad (3.2.12)$$

which corresponds to the form of equation (2.1.7) of chapter 2.

The fulfillment of this condition of asymptotic stability of the solution $\boldsymbol{\Psi} = \boldsymbol{\Psi}^*$ provides, under rather general assumptions, the definiteness of the expression for the additional force within a certain range of changing the initial conditions for the hidden generalized coordinates $\overline{\boldsymbol{\Psi}}$ at sufficiently large values of t. It also provides a mutual conformity of properties of the stability of motions of the initial system (3.2.1) with respect to $\dot{\mathbf{x}}, \mathbf{x}$ and of system (3.2.12) with respect to $\dot{\mathbf{X}}, \mathbf{X}, \dot{\overline{\boldsymbol{\Psi}}}, \overline{\boldsymbol{\Psi}}$. And namely, if the transformation of the variables $\dot{\mathbf{x}}, \mathbf{x}$ to the variables $\dot{\mathbf{X}}, \mathbf{X}$, defined by the relations

$$x_s = X_s(t) + \psi_s(\dot{\mathbf{X}}, \mathbf{X}, t, \omega t) \quad (s = 1, \ldots, k);$$
$$x_s = \overline{\psi}_s(\dot{\mathbf{X}}, \mathbf{X}, t, \omega, t) \quad (s = k+1, \ldots, k+r = n),$$

and also the inverse transformation possess the properties of uniqueness and continuity in the vicinity of the non-perturbed motions under consideration, then the stable (asymptotically stable) with respect to $\dot{\mathbf{X}}$ and $\mathbf{X}$ solutions of Eq.(3.2.12) are answered by stable (asymptotically stable) solutions $\mathbf{x} = \mathbf{X} + \boldsymbol{\Psi}$ of the initial system (3.2.1). And vice versa, the stable (asymptotically stable) solutions $\mathbf{x}$ of the initial system (3.2.1) are answered by stable (asymptotically stable) solutions $\mathbf{X}$ of the system (3.2.12).

Thus, we have come to the conclusion which can be considered as a specialization of the main thesis of mechanics of the systems with hidden motions as applied to the case under consideration (theorem 1):

T h e o r e m 2. *Let the solutions of the initial differential equation (3.2.1) be present in form of (3.2.2) where the respective explicit and hidden components* $\mathbf{X}$ *and* $\boldsymbol{\Psi}$ *are searched for as solutions of the system of integro-differential equations (3.2.7), (3.2.8). Let then equations (3.2.7) allow, under the given* $\mathbf{X}(t)$ *from the region under consideration, the isolated almost periodical (in particular, the 2π-periodical) by $\tau = \omega t$ solutions of* $\boldsymbol{\Psi} = \boldsymbol{\Psi}^*$, *these solutions being asymptotically stable over all hidden generalized coordinates* $\overline{\boldsymbol{\Psi}}$ *and velocities* $\dot{\overline{\boldsymbol{\Psi}}}$ *with all the other variables in the region under consideration changing. Then the explicit component* $\mathbf{X}$ *satisfies equation (3.2.12) in which the additional force* $\mathbf{V}$ *is found according to equation (3.2.11) and when t is large enough, it is quite a definite function of* $\dot{\mathbf{X}}$, $\mathbf{X}$ *and t all over the region of attraction of each of the given solutions* $\boldsymbol{\Psi}^*$. *Provided the transformation from* $\dot{\mathbf{x}}, \mathbf{x}$ *to* $\dot{\mathbf{X}}, \mathbf{X}, \dot{\overline{\boldsymbol{\Psi}}}, \boldsymbol{\Psi}$ *and backwards is unique and continuous, in the vicinity of the unperturbed motions under consideration there is a full correspondence between the properties of the stability of motions of the initial system (3.2.1) by the variables* $\dot{\mathbf{x}}$ *and* $\mathbf{x}$ *and system (3.2.12) by the variables* $\dot{\mathbf{X}}$ *and* $\mathbf{X}$.

This statement can also be formulated in the following way: if we "do not want to notice" either the component of the motion $\boldsymbol{\Psi}$ or the force $\boldsymbol{\Phi}$, which depend on the argument $\tau = \omega t$, then at the assumed version of "splitting" the initial equation (3.2.1), we must first find from equation (3.2.7) the hidden component $\boldsymbol{\Psi}$, possessing the above- mentioned properties , and, secondly, we must determine the explicit component $\mathbf{X}$ from (3.2.12) in which the force $\boldsymbol{\Phi}$ is not given in its direct form and a certain additional force $\mathbf{F}$ is added to the force $\mathbf{V}$.

Up to now we have not yet made use of the assumption about the value of the parameter ω and, accordingly, about the rate of the change of the forces $\mathbf{F}$ and $\boldsymbol{\Phi}$ as well as components $\mathbf{X}$ and $\boldsymbol{\Psi}$. At the same time without that assumption relations (3.2.7) – (3.2.12) have a formal rather than constructive meaning, since the solution of system (3.2.7), (3.2.8) is, in the general case, not simpler than that of the initial equation (3.2.1).

3.2.4   The Main Assumption of Vibrational Mechanics, Its Formalization and Conditions of its Fulfillment; Theorem 3 and 4

Let us now make the *main assumption of vibrational mechanics* , which implies that the initial equation (3.2.1) may have solutions of type (3.2.2) or that system (3.2.7), (3.2.8) may have adequate solutions $\mathbf{X}$, $\boldsymbol{\Psi}$ with the component $\mathbf{X}$ in those solutions being "actually slow" and the component $\boldsymbol{\Psi}$ being actually "fast" [f]. First, however we should define the mathematical meaning of the notions "slow" and "fast".

For this purpose we will use the symbols X_0 and ψ_0 to denote the scale of the components X and ψ defined in such a way that the values $X_* = X/X_0$ and $\psi_* = \psi/\psi_0$ should be of the order of unity

$$X_* = X/X_0 \sim 1, \quad \psi_* = \psi/\psi_0 \sim 1, \qquad (3.2.13)$$

and let T be the shortest period of time t during which the variable ψ undergoes changes of the order of ψ_0. Then we will assume that the component X is changing very slowly as compared to ψ (or for brevity sake in comparison with ψ it is slow and, accordingly, ψ is fast in comparison with X, provided the following relationship is satisfied[g]:

$$\frac{X|_{t+T} - X|_t}{X_0 T} : \frac{\psi|_{t+T} - \psi|_t}{\psi_0 T} \approx \frac{\dot{X}}{X_0 T} : \frac{\dot{\psi}}{\psi_0 T} \sim \varepsilon.$$

---

[f] Here, as before, the components of the corresponding vectors, coordinates and forces are printed in the ordinary (not bold) type. However, the indices, indicating the number of the component are omitted, which makes the notation simpler.

[g] In this relationship, and also in (3.2.14)–(3.2.19) the differences $X|_{t+T} - X|_t$, $\psi|_{t+T} - \psi|_t$ and the derivatives of X and ψ imply the maxima of their absolute values.

That relationship expresses the demand that the relative speed of changing the component X should be a value of the order of ε as compared to the relative speed of changing ψ, with ε being a small parameter. Having calculated the derivative of the function $\psi(t, \omega t)$ with respect to t, we will write the relationship under consideration as

$$\frac{X}{X_0 T} \bigg/ \frac{\dot{\psi}}{\psi_0 T} = \frac{1}{X_0} \frac{dX}{dt} \bigg/ \frac{1}{\psi_0} \left(\frac{\partial \psi}{\partial t} + \omega \frac{\partial \psi}{\partial \tau} \right) \sim \varepsilon. \qquad (3.2.14)$$

Hence it can be seen that for the validity of the assumption about the rate of changing the components X and ψ, it is sufficient (though not necessary!) to identify the small parameter ε with the value $1/\omega$ (see the footnote on page 38) and demand that the values $(1/\psi_0)(\partial \psi/\partial \tau)$ and $(1/X_0)(\partial X/\partial t)$ should be of the same order, and $(1/\psi_0)(\partial \psi/\partial t)$ be of the same or of a higher order with respect to $\varepsilon = 1/\omega$ in particular it may be $(1/\psi_0)(\partial \psi/\partial t) \equiv 0$. Then

$$\frac{\dot{X}}{\dot{\psi}} \frac{\psi_0}{X_0} \sim \varepsilon = \frac{1}{\omega}. \qquad (3.2.15)$$

Thus we come to the following statement (denoting, as usual, the order of the value x by $O(x)$):

T h e o r e m   3.   *If there exist solutions of the initial differential equation (3.2.1) of type (3.2.2) or the adequate solutions X, ψ of system (3.2.7), (3.2.8), then the conditions*

$$\omega = \frac{1}{\varepsilon} \gg 1,$$

$$O\left(\frac{1}{\psi_0} \frac{\partial \psi}{\partial \tau} \right) = O\left(\frac{1}{X_0} \frac{dX}{dt} \right), \quad O\left(\frac{1}{\psi_0} \frac{\partial \psi}{\partial t} \right) \geq O\left(\frac{1}{\psi_0} \frac{\partial \psi}{\partial \tau} \right) \quad (3.2.16)$$

are sufficient for the validity of the main assumption of vibrational mechanics.

Notions of the fast and slow forces can also be formalized in a similar way, there is however no necessity in it, as will be shown later.

The simplest example of the pair of functions X and ψ, satisfying condition (3.2.16) is provided by $X = X_0 \sin t$ and $\psi = \psi_0 \sin \omega t$. In this connection we must emphasize that condition (3.2.16) and, consequently, the main assumption of vibrational mechanics, as it was formulated above, does not impose any restriction on the ratio of the absolute values of the components X and ψ: the value ψ_0 is not necessarily small in comparison to X_0, it may be comparable to X_0 and even larger than that. In other words, the amplitude of the vibration of high frequency ψ_0 can be of the same order or even much larger than the scale of change of the slow component. We will come across such a case in chapter 20 where some problems of celestial mechanics are discussed.

In much the same way as in equality (3.2.14) we find

$$\frac{\ddot{X}}{X_0 T^2} : \frac{\ddot{\psi}}{\psi_0 T^2} = \frac{1}{X_0}\frac{d^2 X}{dt^2} \Big/ \frac{1}{\psi_0}\left(\frac{\partial^2 \psi}{\partial t^2} + 2\omega\frac{\partial^2 \psi}{\partial t \partial \tau} + \omega^2\frac{\partial^2 \psi}{\partial \tau^2}\right), \qquad (3.2.17)$$

hence, believing the following conditions to be satisfied,

$$O\left(\frac{1}{\psi_0}\frac{\partial^2 \psi}{\partial \tau^2}\right) = O\left(\frac{1}{X_0}\frac{d^2 X}{dt^2}\right), \ O\left(\frac{1}{\psi_0}\frac{\partial^2 \psi}{\partial t^2}, \frac{\varepsilon}{\psi_0}\frac{\partial^2 \psi}{\partial t \partial \tau}\right) \geq O\left(\frac{1}{\psi_0}\frac{\partial^2 \psi}{\partial \tau^2}\right),$$
$$(3.2.18)$$

we obtain

$$\frac{X}{\ddot{\psi}}\frac{\psi_0}{X_0} \sim \varepsilon^2 = \frac{1}{\omega^2}. \qquad (3.2.19)$$

From relations (3.2.15)and (3.2.19) it follows that if

$$\psi_0/X_0 \sim \varepsilon^n \quad (n = \ldots, -1, 0, 1, 2, \ldots), \qquad (3.2.20)$$

i.e. if the fast component ψ is the value of the order n as compared to the slow component X, then

$$\dot{X}/\dot{\psi} \sim \varepsilon^{1-n}, \quad \ddot{X}/\ddot{\psi} \sim \varepsilon^{2-n}. \qquad (3.2.21)$$

Particularly, in an important case when ψ is a small value of the 1-st order as compared to X, we have

$$n = 1, \quad X/\psi \sim \frac{1}{\varepsilon}, \quad \dot{X}/\dot{\psi} \sim 1, \quad \ddot{X}/\ddot{\psi} \sim \varepsilon, \qquad (3.2.22)$$

and in case ψ and X are of the same order,

$$n = 0, \quad X/\psi \sim 1, \quad \dot{X}/\dot{\psi} \sim \varepsilon, \quad \ddot{X}/\ddot{\psi} \sim \varepsilon^2. \qquad (3.2.23)$$

Comparing relation (3.2.21) to equations (3.2.7) and (3.2.8), we conclude that for the validity of the main assumption it is necessary that the right side of equation (3.2.7) should be of the order ε^{n-2} if the order of the right side of equation (3.2.8) is taken to be unity, which, evidently, can always be done. In other words, it is necessary that these equations should be presented as

$$m\ddot{\mathbf{X}} = \mathbf{Q}, \quad m\ddot{\mathbf{\Psi}} = \mathbf{P}/\varepsilon^{2-n} \qquad (3.2.24)$$

where

$$\mathbf{Q} = \mathbf{F} + <\tilde{\mathbf{F}}> + <\mathbf{\Phi}>, \quad \mathbf{P}/\varepsilon^{2-n} = \tilde{\mathbf{F}} - <\tilde{\mathbf{F}}> + \mathbf{\Phi} - <\mathbf{\Phi}>,$$

with $|\mathbf{P}|$ and $|\mathbf{Q}|$ being the values of the same order.

Along with it, should we write down equations (3.2.7), (3.2.8), using the relative variables $X_* = X/X_)$ $\psi_* = \psi/\psi_0$, then according to (3.2.18), the orders of their right sides must always differ by two units, i.e. these equations may be written as

$$\mathbf{m\ddot{X}}_* = \mathbf{Q}_*, \quad \mathbf{m\ddot{\psi}}_* = \mathbf{P}_*/\varepsilon^2 \qquad (3.2.25)$$

where $|\mathbf{P}_*|$ and $|\mathbf{Q}_*|$ are values of the same order, connected with $|\mathbf{P}|$ and $|\mathbf{Q}|$ by the relationships

$$|\mathbf{P}_*| = |\mathbf{P}|\varepsilon^n/\psi_0, \quad |\mathbf{Q}_*| = |\mathbf{Q}|/X_0.$$

It is necessary to bear in mind that when using the variables $\mathbf{X}_*$ and $\mathbf{\Psi}_*$, equality (3.2.2) is written as

$$\mathbf{x} = X_0\mathbf{X}_*(t) + \psi_0\mathbf{\Psi}_*(t, \omega t). \qquad (3.2.26)$$

Thus, the main assumption under consideration leads to a certain demand to the relative order of the right sides of equations (3.2.7) and (3.2.8), depending on the relative order of the components $\mathbf{X}$ and $\mathbf{\Psi}$. Finally, as it should be expected, this imposes certain conditions on the functions $\mathbf{F}$ and $\mathbf{\Phi}$ in the right sides of the initial equation (3.2.1).

What has been said can be formulated as the following statement:

T h e o r e m 4: *For the existence of the solutions of the initial differential equation* (3.2.1) *of type* (3.2.2) *or the corresponding solutions* $\mathbf{X}$, $\mathbf{\Psi}$, *of system* (3.2.7), (3.2.8), *satisfying conditions* (3.2.15), (3.2.19) *and such that* $\psi \sim \varepsilon^n X$ *(n being a whole number or zero), it is necessary that system* (3.2.7), (3.2.8) *should allow the notation in form* (3.2.24) *or* (3.2.25).

3.2.5   The Main Equation of Vibrational Mechanics. Vibrational Forces, Observers **O** *and* **V**

Let the main assumption of vibrational mechanics be satisfied, i. e. let there be a solution of equation (3.2.17) of type (3.2.2) or the corresponding solutions of system (3.2.7), (3.2.8) satisfying conditions (3.2.16). Let us assume further that the conditions of theorem 2 are satisfied. Then we will call equation (3.2.12), containing only the slow component of motion alone, the *main equation of vibrational mechanics* , or the *equation of slow motions* , and the expression $\mathbf{V}(\mathbf{\dot{X}}, \mathbf{X}, t)$ for the additional force — the *vibrational force* [h]. Equation (3.2.8)

---

[h]Note that in [95, 99, 103] we designated the vibrational force by $\mathbf{W}$ which differs in sign from the designation $\mathbf{V}$ introduced here. So $\mathbf{V} = -\mathbf{W}$.

will then be called the *equation of fast motions* . As has been mentioned in
the introduction, vibrational mechanics can be considered to be a mechanics in
which the observer does not notice any fast forces or fast motions, and perceives
only the slow forces and slow motions. Such an observer (observer **V**) can
be contrasted with an "ordinary" observer **O** who perceives both the slow and
fast forces and motions. Further on we will widely use these images, taking,
as may be required, the positions of either the first or the second observer.
In accordance with what has been said, the observer **V**, so as not to come in
conflict with the laws of mechanics, must take into consideration not only the
"ordinary" slow forces **F**, but also the additional slow forces — the vibrational
forces **V** .

Note that according to equations (3.2.6), (3.2.9) and (3.2.11), the vibrational force is obtained by means of averaging the eigenfast force **Φ** and the
fast contribution **F** which is separated from the slow force F. In accordance
with it, we will distinguish *the eigenvibrational force*

$$\mathbf{V}^{(s)} =< \mathbf{\Phi} > \tag{3.2.27}$$

and the *induced vibrational force*

$$\mathbf{V}^{(i)} =< \tilde{\mathbf{F}} > . \tag{3.2.28}$$

Mark that the induced vibrational force may be different from zero even in the
absence of a fast external perturbation — due to fast self-excited oscillations
which may appear in systems with slow forces (see the footnote on p. 37 and
2.2.7 of this section).

One can see (and it will be demonstrated by numerous examples) that
equation (3.2.12) at least in its adequate approximate version is much simpler
than the initial equation (3.2.1); the slow component **X**, whose change this
equation describes, being of utmost interest for researchers.

3.2.6   Method of an Approximate Derivation of the Expression of Vibrational Forces and of Composing the Main Equation of Vibrational Mechanics, its Grounding; Theorem 5

As before, we will consider valid both the main assumption of vibrational mechanics and the conditions of theorem 2. Then it seems natural to use the
following method of an approximate finding of the vibrational force **V** and of
composing the main equation (3.2.12). First we will describe it at an euristic level. We begin by solving the equation of fast motions (3.2.7). Since the
change of the values $\dot{\mathbf{X}}, \mathbf{X}$ and t for the typical period of the fast motion $2\pi/\omega$

is relatively small, in solving Eq.(3.2.7) those values are being regarded as '
'frozen", i.e. as fixed parameters.

Let Eq. (3.2.7), with $\dot{\mathbf{X}}, \mathbf{X}$ and t being frozen, admit either one or several
almost-periodical (particularly 2π-periodical) with respect to $\tau = \omega t$ solutions
$\boldsymbol{\Psi} = \boldsymbol{\Psi}^*(\dot{\mathbf{X}}, \mathbf{X}, t, \tau)$ satisfying condition (3.2.3) and asymptotically stable with
respect to all fast generalized coordinates and velocities, while all the other
fast components all over the region are changing. This assumption is usually
checked up very easily and it is really valid under the conditions of theorem 2.
Besides, Eq.(3.2.7) is such that the necessary condition (3.2.3) of the existence
of almost-periodical (particularly — 2π-periodical with respect to τ) solutions
is automatically fulfilled for it. To make sure of it, it is enough to average
Eq.(3.2.7) with respect to $\tau = \omega t$ ($\dot{\mathbf{X}}, \mathbf{X}$ and t being frozen). What has been
said holds the key to the way adopted by us of "splitting" the initial equation
(3.2.1) into two equations — (3.2.17) and (3.2.8).

Substituting the definite solution $\boldsymbol{\Psi} = \boldsymbol{\Psi}^*(\dot{\mathbf{X}}, \mathbf{X}, t, \tau)$ into equations (3.2.9),
(3.2.11), we will find an approximate expression for the vibrational force
$\mathbf{V}(\dot{\mathbf{X}}, \mathbf{X}, t)$ after which the main equation (3.2.12) can be composed which,
of course, will also be approximate.

We will show that the described approximate method can be substantiated
in the sense of asymptotic methods, at least in the case when we search for
solutions of the type

$$\mathbf{x} = \mathbf{X}(t) + \boldsymbol{\Psi}(t, \omega t) = X_0[\mathbf{X}_*(t) + \varepsilon\boldsymbol{\Psi}_*(t, \tau)], \quad \varepsilon = \frac{1}{\omega} \ll 1, \qquad (3.2.29)$$

i.e. solutions of type (3.2.22). We will start from the differential equations
(3.2.7), (3.2.8), presented in the relative variables $\mathbf{X}_*$ and $\boldsymbol{\Psi}_*$ in form of system
(3.2.25). Let us transfer in this system to the independent variable $\tau = \omega t =
t/\varepsilon$. Marking the full derivative with respect to this variable by a prime, we
will assume that $\mathbf{X}_*' = \varepsilon\mathbf{Y}_*$, $\boldsymbol{\Psi}_*'\boldsymbol{\Phi}_*$. Then the system of equations (3.2.24) will
be written as

$$m\boldsymbol{\Phi}_*' = \mathbf{P}_*, \quad \boldsymbol{\Psi}_*' = \boldsymbol{\Phi}_*, \quad m\mathbf{Y}_*' = \varepsilon\mathbf{Q}_*, \quad \mathbf{X}_*' = \varepsilon\mathbf{Y}_*, \quad t' = \varepsilon \qquad (3.2.30)$$

where according to (3.2.29)

$$\mathbf{P}_* = \mathbf{P}_*(\dot{\mathbf{x}}, \mathbf{x}, t, \tau) = \mathbf{P}_*[X_0(\mathbf{Y}_* + \boldsymbol{\Phi}_*), X_0(\mathbf{X}_* + \varepsilon\boldsymbol{\Psi}_*), \varepsilon\tau, \tau],$$
$$\mathbf{Q}_* = \mathbf{Q}_*(\dot{\mathbf{x}}, \mathbf{x}, t, \tau) = \mathbf{Q}_*[X_0(\mathbf{Y}_* + \boldsymbol{\Phi}_*), X_0(\mathbf{X}_* + \varepsilon\boldsymbol{\Psi}_*), \varepsilon\tau, \tau].$$

The system (3.2.30) refers to the so called *equations with multidimensional
fast motions*. For such systems Volosov [572] elaborated a general scheme of

averaging, based on the ideas of Bogolyubov. As can be easily seen, conditions of the applicability of this scheme are fulfilled if Eq. (3.2.7) at the frozen $\dot{\mathbf{X}}, \mathbf{X}$ and t has almost-periodic (in particular — 2π-periodic) with respect to $\tau = \omega t$ solutions $\boldsymbol{\Psi} = \boldsymbol{\Psi}^*(\dot{\mathbf{X}}, \mathbf{X}, t, \tau)$, asymptotically stable with respect to all fast generalized coordinates and velocities, with the change of all the other fast components and of the frozen $\dot{\mathbf{X}}, \mathbf{X}$, and t all over the region under consideration. It is what had been assumed by us before. We can also show that the scheme of Volosov can also refer to the integro-differential equations that are being considered here.

According to that scheme equations of the first approximation for the slow variables $\mathbf{X}_*$ and $\mathbf{Y}_*$ look in this case as follows

$$\mathbf{X}'_* = \varepsilon \mathbf{Y}_*, \quad m\mathbf{Y}' = \varepsilon \overline{\mathbf{Q}}_1. \tag{3.2.31}$$

Here $\overline{\mathbf{Q}}_1$ designates the expression for $\mathbf{Q}_*$ which is obtained when we substitute an almost-periodical or periodical solution of the first two equations (3.30) for the function $\boldsymbol{\Psi}$. That solution was found under the assumption that slow variables are constant. In other words, according to (3.2.9),(3.2.11),(3.2.23) and (3.2.25).

$$\overline{\mathbf{Q}}_1 = \mathbf{F}(\dot{\mathbf{X}}, \mathbf{X}, t) + \mathbf{V}(\dot{\mathbf{X}}, \mathbf{X}, t).$$

But then equations (3.2.31), after returning to the independent variables t and after excluding $\mathbf{Y}_*$, will coincide with equation (3.2.31), composed by the approximate method stated here. The scheme of Volosov makes it also possible to find other subsequent approximations to the solution of system (3.2.7), (3.2.8), however, there is no necessity in it, at least in all the numerous applied problems considered below.

What has been said enables us to formulate the following statement.

T h e o r e m 5. *Let the main assumptions of vibrational mechanics be fulfilled for the initial differential equation (3.2.1) or system (3.2.7), (3.2.8) and let the motions of type (3.2.9) be searched for, i.e.the motions in which the fast components $\boldsymbol{\Psi}$ are of the order $\varepsilon = 1/\omega$ as compared to the slow components $\mathbf{X}$. Then to obtain the expression of vibrational force (3.2.11) and the equation of slow motions (the main equation of vibrational mechanics) (3.2.12) in the first asymptotic approximation, it is sufficient to find the almost-periodic (in particular 2π-periodic) with respect to $\tau = \omega t$ solutions of the equation of fast motions (3.2.7) at the constant ("frozen") values of the slow variables $\dot{\mathbf{X}}, \mathbf{X}, t$, asymptotically stable with respect to all fast generalized coordinates and velocities, all the other fast variables changing and $\dot{\mathbf{X}}, \mathbf{X}$ and t being frozen all over the region under consideration.*

Equation (3.2.12) composed in that way, must be valid on the one hand with $t > t_0$, where t_0 is the time of the establishment of the hidden fast motions, and on the other hand — with $t < T_0$, where T_0 is the boundary of the interval of validity of the asymptotic approximation (it is naturally assumed that $T_0 \gg t_0$).

3.2.7   An Important Special Case; Theorem 6

Let us consider the case when the initial system (3.2.1) can be presented in the following way (to simplify our reasoning and notation we will first consider x to be a scalar):

$$m\ddot{x} = F(\dot{x}, x, t) + \omega\Phi_1(x, t, \tau), \quad \omega \gg 1 \qquad (3.2.32)$$

where the almost-periodic with respect to τ function Φ has a zero average value with respect to this argument at the fixed x and t:

$$< \Phi(x, t, \tau) >_{x, t = const} = 0. \qquad (3.2.33)$$

Here, like in 2.2.6, we are searching for the solution of the type

$$x = X(t) + \varepsilon\psi_1(t, \tau), \quad \varepsilon = 1/\omega \ll 1, \qquad (3.2.34)$$

i.e. the solution of type (3.2.29). In this case, considered by Malakhova [361], finding the function ψ_1 and the expression for the vibrational force becomes much more simple and the ground of the approximate method stated in 2.2.6 is obtained directly by means of using the first and second theorems of Bogolyubov.

In this case the equation of fast motions (3.2.7) looks as

$$m\varepsilon\ddot{\psi}_1 = F(\dot{X} + \varepsilon\dot{\psi}_1, X + \varepsilon\psi_1, t) - < F(\dot{X} + \varepsilon\dot{\psi}_1, X + \varepsilon\psi_1, t) >$$
$$+ \frac{1}{\varepsilon}[\Phi_1(X + \varepsilon\psi_1, t, \tau) - < \Phi_1(X + \varepsilon\psi_1, t, \tau) >],$$

or, considering (3.2.33) and with an accuracy to the terms of a higher order with respect to ε

$$m\ddot{\psi}_1 = \frac{1}{\varepsilon^2}\Phi_1(X, t, \tau). \qquad (3.2.35)$$

Looking ahead, we will note that the latter equation corresponds to the so called *purely inertial approximation* (see 3.2.8).

Let $\psi_1^* = \psi_1^*(X, t, \tau)$ be a certain almost-periodic solution of equation (3.2.35), found at the frozen X and t. Then, according to the approximate method, stated in 3.2.6, in the main equation of vibrational mechanics (3.2.12), i.e. in the equation

$$m\ddot{X} = F(\dot{X}, X, t) + V(\dot{X}, X, t), \qquad (3.2.36)$$

written with the same accuracy with respect to ε as equation (3.2.35), the expression for the vibrational force (3.2.11) will be

$$V(\dot{X}, X, t) = < F\left[\dot{X} + \frac{\partial \psi_1^*(X, t, \tau)}{\partial \tau}, X, t\right] - F(\dot{X}, X, t) >$$

$$+ < \frac{\partial \Phi_1(X, t, \tau)}{\partial X} \psi_1^*(X, t, \tau) > . \qquad (3.2.37)$$

(Recall that according to what has been said in (3.2.4), the values $\varepsilon\dot{\psi}_1$ and $\dot{X}$ are of the same order, while $\varepsilon\psi_1$ is supposed to be small as compared to X). We will show that equation (3.2.36) in which the vibrational force is determined according to equation (3.2.37), can be obtained by using the first theorem of Bogolyubov. To this end we transform equation (3.2.32) into a standard form by means of changing the variables

$$x = X + \varepsilon\psi_1^*(X, t, \tau), \qquad x' = \varepsilon Y + \varepsilon\frac{\partial \psi_1^*(X, t, \tau)}{\partial \tau}$$

where the full derivative with respect to τ is again marked with a prime. As a result, we obtain the system

$$X' = \varepsilon[Y + \varepsilon A(Y, X, t, \tau, \varepsilon)], \quad Y' = \varepsilon\left\{\frac{1}{m}F[Y + \frac{\partial \psi_1^*(X, t, \tau)}{\partial \tau}, X, t]\right.$$

$$+ \frac{1}{m}\frac{\partial \Phi_1(X, t, \tau)}{\partial \tau}\psi_1^*(X, t, \tau) - \frac{\partial^2 \psi_1^*(X, t, \tau)}{\partial \tau \partial X}Y\frac{\partial^2 \psi_1^*(X, t, \tau)}{\partial \tau \partial t}$$

$$\left. + \varepsilon B(Y, X, t, \tau, \varepsilon)\right\}, \quad t' = \varepsilon \qquad (3.2.38)$$

where A and B are certain functions of the indicated arguments. Let functions F and Φ_1 be such that the right sides of equations (3.2.38) should satisfy the conditions of Bogolyubov's first theorem. Then applying to these equations the principle of averaging, and retaining the same designations for the averaged variables, we will obtain the following equation of the first approximation

$$\dot{X} = \varepsilon Y, \quad Y' = \varepsilon R(Y, X, t), \quad t' = \varepsilon. \qquad (3.2.39)$$

where

$$R(Y, X, t) = \frac{1}{m} < (F + V) >$$

$$= \frac{1}{m} < F\left[Y + \frac{\partial \psi_1^*(X, t, \tau)}{\partial \tau}, X, t\right] + \frac{\partial \Phi_1(X, t, \tau)}{\partial X}\psi_1^*(X, t, \tau) >$$

It is obvious that the latter equations lead to equation (3.2.36) in which $V(\dot{X}, X, t)$ are determined according to (3.2.37); which is precisely what was to be proved. It should be noted that relations of type (3.2.36), (3.2.37) were obtained by Chelomey [152] in another way.

Let us consider the question of stability of the positions of equilibrium $Y = 0$, $X = X_*$ which are determined by equations (3.2.30), i.e. the so called *positions of quasi-equilibrium* for the initial equation (3.2.32) (see 4.3). Here we will assume that functions F and Φ_1 do not depend explicitly on t and consequently $R = R(Y, X)$. The indicated positions of equilibrium, if they exist, are determined as solutions of the equation

$$R(0, X) \equiv \frac{1}{m}[F(0, X) + V(0, X)] = 0. \tag{3.2.40}$$

Then, under more strict requirements to the right sides of equations (3.2.32) it is possible to apply Bogolyubov's second theorem [136]. If, according to this theorem, the roots of the equation

$$\begin{vmatrix} -\lambda & 1 \\ \frac{\partial R}{\partial X}\Big|_{\substack{Y=0 \\ X=X_*}} & \frac{\partial R}{\partial Y}\Big|_{\substack{Y=0 \\ X=X_*}} - \lambda \end{vmatrix} = 0 \tag{3.2.41}$$

have negative real parts, then the stationary solutions $Y = 0$, $X = X_*$ of the equations of the first approximation will be answered, at the sufficiently small ε all over the interval $-\infty < \tau < \infty$ by the asymptotically stable almost-periodic solutions of the initial equations (3.2.32), i.e. by the quasi-equilibrium positions for this equation.

But system (3.2.39) is equivalent to equation (3.2.36), and equation (3.2.41) is equivalent to the characteristic equation for the equation in variations, corresponding to the stationary solution of equation (3.2.36). From here it follows that the asymptotically stable stationary solutions $\dot{X} = 0$, $X = X_*$ of Eq.(3.2.36) are answered by the asymptotically stable almost-periodic solutions of the initial equations (3.2.32), i.e. by the positions of quasi-equilibrium for this equation.

What has been said can be formulated as the following statement.

T h e o r e m 6. *Let us assume that the initial differential equation can be presented in form (3.2.32) where the function $\omega\Phi_1$ ("the fast force") satisfies equality (3.2.33) and the functions F, Φ_1 are such that the right sides of equations (3.2.38) satisfy the conditions of Bogolyubov's first theorem, and the solutions of type (3.2.34) are being considered. Then in the first asymptotic approximation, the slow component X satisfies equation (3.2.36) (the main equation of vibrational mechanics), in which the fast component ψ_1^* is found* *as an almost- periodic solution of the differential equation* (3.2.35) *at the fixed* *("frozen")* X *and* t.

If the functions F and Φ_1 *do not explicitly depend on the time* t, *and the* *right sides of equation* (3.2.38) *satisfy the conditions of Bogolyubov's second the-* *orem, then the asymptotically stable positions of the equilibrium* $\dot{X} = 0, X = X_*$ *of the system, defined by Eq.* (3.2.36), *correspond to the the asymptotically sta-* *ble positions of the quasi-equilibrium* (3.2.34) *of the system, defined by equation* (3.2.32).

Note that in the case of the system with many degrees of freedom, when the functions $F, \varphi_1, x, X, \psi_1$ and V are vectors $\mathbf{F}, \boldsymbol{\Phi}_1, \mathbf{x}, \mathbf{X}, \boldsymbol{\Psi}_1$ and $\mathbf{V}$, the expression in the last angular brackets of equation (3.2.37) for the vibrational force $\mathbf{V}$ should be understood as the product of the matrix $\partial \boldsymbol{\Phi}_1 / \partial \mathbf{X}$ multiplied by the vector $\boldsymbol{\Psi}_1^*$.

We will also note that another approach to the substantiation of the presented approximate method in the special case under consideration was suggested by Kirghetov, on the basis of the concept of μ-*stability*, introduced by him [256, 270].

The author was informed of the important circumstances concerning this item by Fidline in 1998. The specific procedure of averaging suggested by him made it possible not only to obtain very accurately the results stated here, but also strictly and under more general assumptions to consider the case when function Φ_1 in equation (3.2.32) depends on $\dot{x}$. The author hopes that A. Fidline will soon have these results published.

3.2.8 On the Simplifications of Solving Equations for the Fast Component of Motion. Purely Inertial Approximation

Apart from these mentioned above, there are other expedient methods of an approximate solution of the equation for the fast component $\boldsymbol{\Psi}$. From general positions, the rightfulness of such approximate methods follows from what has been said in 2.1 on the equations of systems with hidden motions: the information which is contained in the initial equation (3.2.1) and also in the system (3.2.7), (3.2.8) is excessive if we are interested only in the change of the slow component $\mathbf{X}$. It is natural that then we can restrict ourselves to an approximate finding of the fast component $\boldsymbol{\Psi}$. To be concrete, the possibility of an approximete finding of $\boldsymbol{\Psi}$ follows from the fact that this function appears in equation (3.2.11) for the vibrational force only under the integration sign.

One of the possible simplifications when solving equation (3.2.7) consists in finding $\boldsymbol{\Psi}$ in the form of a sum of a small number of harmonics or a small number of terms of a power series of a small parameter, which need not necessarily coincide with the parameter $\varepsilon = 1/\omega$. We must emphasize that here we mean an approximate calculation of $\boldsymbol{\Psi}$ when $\dot{\mathbf{X}}, \mathbf{X}$ and t are "frozen", i.e. within the frames of the method, given in 3.2.6. Besides, as was already mentioned, $\boldsymbol{\Psi}$ is often considered to be small as compared to $\mathbf{X}$ and it is possible to linearize the expression for the forces $\mathbf{F}$ and $\boldsymbol{\Phi}$ with respect to $\boldsymbol{\Psi}$ (and sometimes to $\dot{\boldsymbol{\Psi}}$). In particular, in many cases it is possible to consider only the linear terms in the series of the function $\tilde{\mathbf{F}}$ by the powers $\dot{\boldsymbol{\Psi}}$ and $\boldsymbol{\Psi}$, believing that according to (3.2.6)

$$\tilde{\mathbf{F}} = \left(\frac{\partial \mathbf{F}}{\partial \dot{\mathbf{x}}}\right) \cdot \dot{\boldsymbol{\Psi}} + \left(\frac{\partial \mathbf{F}}{\partial \mathbf{x}}\right) \cdot \boldsymbol{\Psi} \tag{3.2.42}$$

where the derivatives are calculated when $\dot{\boldsymbol{\Psi}} = \mathbf{0}, \boldsymbol{\Psi} = \mathbf{0}$. It is however necessary to bear in mind that due to (3.2.28) and (3.2.3) there exists only the eigenvibrational force $\mathbf{V}^{(s)}$, while the induced vibrational force $\mathbf{V}^{(i)}$ is absent. In other words, the induced component exists in the case when the slow force $\mathbf{F}$ is non-linear with respect to $\dot{\mathbf{x}}$ and $\mathbf{x}$.

On the other hand, even in case of the absence of the fast force $\boldsymbol{\Phi}$ in the initial equation (3.2.1), the vibrational force can be different from zero on account of its induced component. Such a situation is characteristic of the autonomous systems in which fast oscillations appear in spite of the fact that there are only slow forces in the system (see the footnote on p. 37).

When solving the equations of fast motions (3.2.7), the approximation which may be called *purely intertial approximation* is of special significance. It is based on the assumption that in this equation the "oscillating component" of the fast force is much greater than the "oscillating component" of the slow force, i.e.

$$|\boldsymbol{\Phi} - <\boldsymbol{\Phi}>| \gg |\tilde{\mathbf{F}} - <\tilde{\mathbf{F}}>|, \tag{3.2.43}$$

the second component therefore can be neglected as compared to the first one. In many cases relation (3.2.43) appears in a most natural way from the statement of the problem; it can be formalized by means of introducing a small parameter (the system considered in 3.2.7 can serve as one of the examples).

Quite frequently, when $\boldsymbol{\Psi} = \boldsymbol{\Phi} - <\boldsymbol{\Phi}>$ does not depend on $\dot{\boldsymbol{\Psi}}$ or $\boldsymbol{\Psi}$ and is an almost-periodic function τ, presented as a sum

$$\boldsymbol{\Psi} = \sum_{s=1}^{p} \mathbf{f_s}(\dot{\mathbf{X}}, \mathbf{X}, t, \tau) \tag{3.2.44}$$

where

$$\mathbf{f_s}(\dot{\mathbf{X}}, \mathbf{X}, \mathbf{t}, \tau) = \mathbf{A_s}(\dot{\mathbf{X}}, \mathbf{X}, t) \cos \nu_s t + \mathbf{B_s}(\dot{\mathbf{X}}, \mathbf{X}, t) \sin \nu_s t$$

and $\nu_s > 0$ are certain numbers, the wanted approximate solution of equation (3.2.7) has the form

$$\Psi = -\frac{1}{m} \sum_{s=1}^{p} \frac{\mathbf{f_s}(\dot{\mathbf{X}}, \mathbf{X}, t)}{\nu_s^2}. \qquad (3.2.45)$$

In future we will widely use a purely inertial approximation and, particularly, expressions (3.2.44) and (3.2.45).

3.2.9   Additional Remarks, Certain Generalizations

In connection with the method of an approximate obtaining of the expressions for the vibrational force and for the main equation of vibrational mechanics, stated in 3.2.6-3.2.8 of this section, we will make the following remarks and additions.

1. The main equation (3.2.12) will not change, that is there won't be any mistake if either all the slow forces or just some of them are referred to the fast forces.

2. If the initial equation is written in form of the Lagrange equation of the second type

$$\mathbf{E}(T) = \mathbf{Q} \qquad (3.2.46)$$

where

$$\mathbf{E}(\ ) = \frac{d}{dt}\frac{\partial}{\partial\dot{\mathbf{x}}} - \frac{\partial}{\partial\mathbf{x}} \qquad (3.2.47)$$

is the Euler operator, $\mathbf{Q}$ is the generalized force, and T is the system kinetic energy, then the equation of slow motions can be written in the following way

$$< \mathbf{E}(T_{\mathbf{X}+\mathbf{\Psi}}) >=< \mathbf{Q}_{\mathbf{X}+\mathbf{\Psi}} > \qquad (3.2.48)$$

where $T_{\mathbf{X}+\mathbf{\Psi}}$ and $\mathbf{Q}_{\mathbf{X}+\mathbf{\Psi}}$ denote the result of substituting the values $\mathbf{x} = \mathbf{X} + \mathbf{\Psi}$ and $\dot{\mathbf{x}} = \dot{\mathbf{X}} + \dot{\mathbf{\Psi}}$ into the corresponding expression. Then the equation of fast motions will be

$$\mathbf{E}(T_{\mathbf{X}+\mathbf{\Psi}}) - < \mathbf{E}(T_{\mathbf{X}+\mathbf{\Psi}}) >= \mathbf{Q}_{\mathbf{X}+\mathbf{\Psi}} - < \mathbf{Q}_{\mathbf{X}+\mathbf{\Psi}} > . \qquad (3.2.49)$$

In this case equation (3.2.48) can always be written in the form, corresponding to that of the main equation of vibrational mechanics (3.2.12):

$$\mathbf{E}(T_{\mathbf{X}}) = \mathbf{F} + \mathbf{V}. \qquad (3.2.50)$$

Here the expression for the vibrational force $\mathbf{V} = \mathbf{V}(\dot{\mathbf{X}}, \mathbf{X}, t)$ can be easily obtained by means of comparing equations (3.2.48) and (3.2.50) and taking into account that the initial equation (3.2.46) is always soluble with respect to **x**; the "ordinary" slow force is designated by $\mathbf{F}(\dot{\mathbf{X}}, \mathbf{X}, t)$.

3. The approximate method presented here can also be used in the case when under the 2π-periodic function Φ with respect to τ the equation of fast motions (3.2.7) allows a periodic solution with respect to τ, its period being $T = 2\pi q/p$ (with q and p being relatively prime numbers), and also in the case of the autonomous initial equation (3.2.1). In the first case the corresponding expressions should be subject to 2π averaging and in the second case the period of averaging — T, unknown beforehand, is determined in the process of solving the equation. It should be also noted that in the case when the fast forces do not depend on τ, equations (3.2.7) or (3.2.49) admit the solutions $\Psi \equiv 0$.

Finally, the equation of fast motions can allow solutions of a stochastic character. In that case the necessary condition for the applicability of that method is the existence of the average

$$< \ldots >= \lim_{T \to \infty} \frac{1}{T} \int\limits_0^T (\ldots) d\tau.$$

4. In theorem 4 the case in point is the condition necessary for the existence of solutions of the initial differential equation (3.2.1) which can be presented in form (3.2.2), or for the corresponding solutions **X** and Ψ of system (3.2.7), (3.2.8), satisfying the main assumption of vibrational mechanics about the rate of change of the components **X** and Ψ. The question about the real existence of such solutions can be solved efficiently by means of constructing them actually in every separate case and by checking a posteriori the assumption about the rate of change of the components **X** and Ψ. Such a checking is most desirable since the motions, described by equation (3.2.12), may prove to be fast, despite the fact that the motions, described by the same equation at $\mathbf{V} \equiv 0$, were slow (see e.g.[79] where special attention was paid to that circumstance in conformity to the method of harmonic linearization).

It should be also noted that in case the forces **F** and Φ do not depend on the slow time t, which happens quite often, (hence **V** does not depend on t either), and when it is only the stable stationary solutions of the equations of slow motions $\mathbf{X} = 0$, $\mathbf{X} = \mathbf{X}_*$ that are of interest, as a rule it is possible to state a priori that in a sufficiently close vicinity of such solutions the conditions we are discussing are fulfilled. Indeed, for such solutions the right sides of the equations of slow motions become zero, and in the indicated vicinity they can be arbitrarily small with a proper choice of the vicinity. At the same time the right sides of the equations of fast motions at $\mathbf{X} = 0$, $\mathbf{X} = \mathbf{X}_*$ generally speaking do not become zero.

When evaluating a posteriori the range of the validity of suppositions, it is useful to compare the frequency ω with the frequencies λ_i of the small oscillations of the system taking into account the vibrational forces in the vicinity of the stationary states. Relation

$$\lambda_i \ll \omega \qquad (3.2.51)$$

must be fulfilled, though the consideration of a number of concrete problems shows that it is usually sufficient that the condition $\lambda_i \leq 3\omega$ should be valid.

In case other regions of the change of $\dot{\mathbf{X}}, \mathbf{X}$ are of interest, it is necessary to compare the orders of the right sides of the equations for those regions, and then establish the correlation between the orders of $\mathbf{X}$ and $\boldsymbol{\Psi}$ at which the conditions of the validity of this method will be fulfilled. So, for instance, if the right sides of equations (3.2.7) and (3.2.8) prove to be of the same order, it can be expected that the solutions found will be valid under the initial conditions for $\dot{\mathbf{X}}, \mathbf{X}$ and under the system parameters which provide the fulfillment of the condition $\boldsymbol{\Psi} \sim \varepsilon^2 \mathbf{X}$.

5. Important generalization of the method of direct separation of motions into mechanical systems with constrains has been given by Yudovich [586].

3.2.10   *Summary of the Main Relations and Results*

Let us make a brief summary of the main equations and results of this section, since we are to refer to them often in future.

The initial differential equation is

$$m\ddot{\mathbf{x}} = \mathbf{F}(\dot{\mathbf{x}}, \mathbf{x}, t) + \boldsymbol{\Phi}(\dot{\mathbf{x}}, \mathbf{x}, t, \omega t) \qquad (3.2.52)$$

where $\mathbf{x}$ is the vector of the generalized coordinates of the system, $m = const$, $\omega > 0$ is the parameter, and $\boldsymbol{\Phi}$ is an almost-periodic (in particular, 2π-periodic) function of the argument $\tau = \omega t$. Solutions, sought for this equation, are of the type

$$\mathbf{x} = \mathbf{X}(t) + \boldsymbol{\Psi}(t, \omega t) \qquad (3.2.53)$$

where $\boldsymbol{\Psi}(t, \omega t$ is an almost-periodic (in particular 2π-periodic) function of $\tau = \omega t$, satisfying the condition[i]

$$< \boldsymbol{\Psi}(t, \tau) >= 0. \qquad (3.2.54)$$

---

[i]see the footnote on page 39

The component $\mathbf{X}(t)$ is considered to be explicit, and $\boldsymbol{\Psi}(t, \omega t)$ is considered to be hidden. Part of the components of the vector $\mathbf{X}(t)$ can be equal to zero — the corresponding generalized coordinates are called *hidden*.

The component $\mathbf{X}$ and $\boldsymbol{\Psi}$ obey the system of integro-differential equations

$$m\ddot{\mathbf{X}} = \mathbf{F}(\dot{\mathbf{X}}, \mathbf{X}, t) + <\tilde{\mathbf{F}}(\dot{\mathbf{X}}, \mathbf{X}, \dot{\boldsymbol{\Psi}}, \boldsymbol{\Psi}, t)> + <\boldsymbol{\Phi}(\dot{\mathbf{X}} + \dot{\boldsymbol{\Psi}}, \mathbf{X} + \boldsymbol{\Psi}, t, \tau)>,$$
$$(3.2.55)$$

$$m\ddot{\boldsymbol{\Psi}} = \tilde{\mathbf{F}}(\dot{\mathbf{X}}, \mathbf{X}, \dot{\boldsymbol{\Psi}}, \boldsymbol{\Psi}, t) - <\tilde{\mathbf{F}}(\dot{\mathbf{X}}, \mathbf{X}, \dot{\boldsymbol{\Psi}}, \boldsymbol{\Psi}, t)> + \boldsymbol{\Phi}(\dot{\mathbf{X}} + \dot{\boldsymbol{\Psi}}, \mathbf{X} + \boldsymbol{\Psi}, t, \tau)$$
$$- <\boldsymbol{\Phi}(\dot{\mathbf{X}} + \dot{\boldsymbol{\Psi}}, \mathbf{X} + \boldsymbol{\Psi}, t, \tau)> \qquad (3.2.56)$$

where

$$\tilde{\mathbf{F}}(\dot{\mathbf{X}}, \mathbf{X}, \dot{\boldsymbol{\Psi}}, \boldsymbol{\Psi}, t) = \mathbf{F}(\dot{\mathbf{X}} + \dot{\boldsymbol{\Psi}}, \mathbf{X} + \boldsymbol{\Psi}, t) - \mathbf{F}(\dot{\mathbf{X}}, \mathbf{X}, t). \qquad (3.2.57)$$

If equation (3.2.56) allows almost-periodic (in particular 2π-periodic) with respect to $\tau = \omega t$ solutions $\boldsymbol{\Psi} = \boldsymbol{\Psi}^*$, asymptotically stable with respect to all hidden generalized coordinates and velocities while all other variables in the region under consideration are changing, then when t is large enough, for each of such solutions $\boldsymbol{\Psi} = \boldsymbol{\Psi}^*$, we may assume that

$$\mathbf{V}(\dot{\mathbf{X}}, \mathbf{X}, t) = <\tilde{\mathbf{F}}(\dot{\mathbf{X}}, \mathbf{X}, \dot{\boldsymbol{\Psi}}^*, \boldsymbol{\Psi}^*, t)> + <\boldsymbol{\Phi}(\dot{\mathbf{X}} + \dot{\boldsymbol{\Psi}}^*, \mathbf{X} + \boldsymbol{\Psi}^*, t, \tau)>,$$
$$(3.2.58)$$

and equation (3.2.55) then is written in the form containing only the explicit component

$$m\ddot{\mathbf{X}} = \mathbf{F}(\dot{\mathbf{X}}, \mathbf{X}, t) + \mathbf{V}(\dot{\mathbf{X}}, \mathbf{X}, t). \qquad (3.2.59)$$

The force $\mathbf{V}$ presents the so called additional force which appears in equations of explicit motions due to the presence of hidden motions.

The equations and formulas given here are valid irrespective of the rate of change of the components $\mathbf{X}$ and $\boldsymbol{\Psi}$: they present the general relations of the mechanics of systems with hidden motions.

The main supposition of vibrational mechanics consists in the assumption that the initial equation (3.2.52) has solutions of type (3.2.53) or that system (3.2.55), (3.2.56) has the corresponding solutions in which

$$\frac{\dot{X}}{X_0 T} : \frac{\dot{\psi}}{\psi_0 T} \sim \varepsilon \qquad (3.2.60)$$

where X_0 and ψ_0 are the scales of the components X and ψ, defined in such a way that

$$X_* = X/X_0 \qquad 1, \qquad \psi_* = \psi/\psi_0 \sim 1; \qquad (3.2.61)$$

T is the period of time during which the variable ψ undergoes changes of the order of ψ_0, ε being a small parameter. This supposition is realized if the frequency ω is large enough; in this case we can assume that

$$\omega = 1/\varepsilon \gg 1, \tag{3.2.62}$$

and, provided the following relations with respect to the orders of the derivatives are valid,

$$O\left(\frac{1}{\psi_0}\frac{\partial\psi}{\partial t}\right) \geq O\left(\frac{1}{\psi_0}\frac{\partial\psi}{\partial\tau}\right), \ O\left(\frac{1}{X_0}\frac{\partial X}{\partial t}\right) = O\left(\frac{1}{\psi_0}\frac{\partial X}{\partial\tau}\right) \tag{3.2.63}$$

When all those conditions are fulfilled, the component $\mathbf{\Psi}$ is called slow, and the component $\mathbf{X}$ is called fast, equation (3.2.55) is called the equation of slow motions, and equation (3.2.56) is the equation of fast motions. Equation (3.2.59), containing only the slow component $\mathbf{X}$ is called the main equation of vibrational mechanics, and expression (3.2.58) for the additional force is called the vibrational force.

$$\mathbf{V} = \mathbf{V}^{(s)} + \mathbf{V}^{(i)}, \ \ \mathbf{V}^{(s)} =< \mathbf{\Phi} >, \ \ \mathbf{V}^{(i)} =< \tilde{\mathbf{F}} > \tag{3.2.64}$$

where the component $\mathbf{V}^{(s)}$ is called *properly vibrational force* and the component $\mathbf{V}^{(i)}$ is called *induced vibrational force*.

If

$$\psi_0/X_0 \sim \varepsilon^n \tag{3.2.65}$$

where n is a whole number or a zero, then

$$\dot{X}/\dot{\psi} \sim \varepsilon^{1-n}, \ \ \ddot{X}/\ddot{\psi} \sim \varepsilon^{2-n}. \tag{3.2.66}$$

If we assume here that the order of the right side of equation (3.2.55) is unity, then for existing the solutions, satisfying conditions (3.2.65), (3.2.66), it is necessary that the system (3.2.55), (3.2.56) should be presented in form

$$m\ddot{\mathbf{X}} = \mathbf{Q}, \ \ \ \ m\ddot{\mathbf{\Psi}} = \mathbf{P}/\varepsilon^{2-n} \tag{3.2.67}$$

where

$$\mathbf{Q} = \mathbf{F} =< \tilde{\mathbf{F}} > + < \mathbf{\Phi} >= \mathbf{F} + \mathbf{V},$$
$$\mathbf{P}/\varepsilon^{2-n} = \tilde{\mathbf{F}} - < \tilde{\mathbf{F}} > + \mathbf{\Phi} - < \mathbf{\Phi} >= \tilde{\mathbf{F}} + \mathbf{\Phi} - \mathbf{V}, \tag{3.2.68}$$

with $|\mathbf{P}|$ and $|\mathbf{Q}|$ being the values of the same order. In the relative variables $\mathbf{X}_* = \mathbf{X}/X_0$, $\mathbf{\Psi}_* = \mathbf{\Psi}/\psi_0$ equations (3.2.55), (3.2.56) must allow the form

$$m\ddot{\mathbf{X}}_* = \mathbf{Q}_*, \ \ \ \ m\ddot{\mathbf{\Psi}}_* = \mathbf{P}_*/\varepsilon^2 \tag{3.2.69}$$

where $|\mathbf{P}_*| = |\mathbf{P}|\varepsilon^n/\psi_0$ and $|\mathbf{Q}_*| = |\mathbf{Q}|/X_0$ are also of the same order. In this case to obtain the expressions of the vibrational force $\mathbf{V}$ and the main equation of vibrational mechanics (3.2.59) in the first asymptotic approximation it is necessary to find the almost-periodic (in particular 2π-periodic) with respect to $\tau = \omega t$ solution $\mathbf{\Psi} = \mathbf{\Psi}_*(\dot{\mathbf{X}}, \mathbf{X}, t, \tau)$ of the equation of fast motions (3.2.56) at the "frozen" (constant) $\dot{\mathbf{X}}, \mathbf{X}$, and t, asymptotically stable with respect to all fast generalized coordinates and velocities over the investigated region of the values of other variables. It is this solution that should be used when finding the expression for $\mathbf{V}$ by formula (3.2.58). This approximate method is well-grounded, at least in case the fast component $\mathbf{\Psi}$ can be considered small of the order ε as compared to the slow component $\mathbf{X}$ i.e. when in the relations (3.2.65) – (3.2.67) the number $n = 1$. Then in the special case when the expression for the fast force $\mathbf{\Phi}$ can be given in the form

$$\mathbf{\Phi} = \frac{1}{\varepsilon}\mathbf{\Phi}_1(\mathbf{x}, t, \tau), \quad \varepsilon = 1/\omega \ll 1 \tag{3.2.70}$$

where the function $\mathbf{\Phi}_1$ has a zero average with respect to τ at the fixed $\mathbf{x}$ and t, the expression for the vibrational force looks as (see the note at the end of p. 53)

$$\mathbf{V}(\dot{\mathbf{X}}, \mathbf{X}, t) = \left\langle \mathbf{F}\left[\dot{\mathbf{X}} + ds\frac{\partial\mathbf{\Psi}_1^*(\mathbf{X}, t, \tau)}{\partial\tau}, \mathbf{X}, t\right] - \mathbf{F}(\dot{\mathbf{X}}, \mathbf{X}, t)\right\rangle$$
$$+ \left\langle\frac{\partial\mathbf{\Phi}_1(\mathbf{X}, t, \tau)}{\partial\mathbf{X}} \cdot \mathbf{\Psi}_1^*(\mathbf{X}, t, \tau)\right\rangle \tag{3.2.71}$$

where $\mathbf{\Psi}_1^*(\mathbf{X}, t, \tau)$ is an almost-periodic solution of the equation of fast motions, found at the frozen $\mathbf{X}$ and t. It is written in this case as

$$m\ddot{\mathbf{\Psi}}_1 = \frac{1}{\varepsilon^2}\mathbf{\Phi}_1(\mathbf{X}, t, \tau). \tag{3.2.72}$$

Taking into account that the function $\mathbf{\Psi}$ enters into equation (3.2.58) under the sign of averaging, along with the method shown here, it is possible to use other approximate methods of finding the function $\mathbf{\Psi}$ which lead to the expression for the vibrational force $\mathbf{V}$ with an acceptable accuracy.

The method presented here can be easily used in the case when the initial equation is written in form of the Lagrange equation of the 2-nd type (see note 2 in 3.2.9 of this section).

In the case when part of the generalized coordinates are fast (to be specific, let us assume that these coordinates are $x_{k+1} = \overline{\psi}_{k+1}, \ldots, x_n = \overline{\psi}_n$), the number of the main equations of vibrational mechanics which are to be considered (3.2.59), written in the scalar form, is equal to k. In other words, the order

of the system (3.2.59) is by $2(n - k)$ units less than that of the initial system
(3.2.52).

3.2.11   On the Procedure of the Practical Use of the Method

The use of any method, including even the strict mathematical method or
result, when solving an applied problem comprises as a rule a number of stages
which do not belong to those mathematically strict, but can be referred to
the methods realized at the so called rational level of rigour[j]. This certainly
refers to the method of direct separation of motions as well. The procedure
of using that method, at least in the form stated above, comprises a number
of rational elements, including the euristic elements. We will describe this
procedure schematically. We will emphasize that it allows various modifications
and improvements, in future however we will try to keep to it when solving
concrete problems.

1. A preliminary conclusion , based on the experimental data and/or on
the euristic grounds, deals with the question whether the motion of the system
under consideration presents an imposing of fast oscillations upon the slow
motion. In other words, a prognosis is made about the fulfillment of the main
assumption of vibrational mechanics.

2. A system of equations is composed (3.2.55), (3.2.56), based on the initial
equations of motion.

3. A supposition is made about the relative order of the value of the fast
and slow components $\mathbf{X}$ and $\boldsymbol{\Psi}$, i.e. the supposition about the n in the relation
$\boldsymbol{\Psi} \sim \varepsilon^n \mathbf{X}$, and the values of the parameters and of groups of summands are estimated, and large and small parameters and groups of summands are selected.
On account of that supposition and those estimations, due simplifications are
made in the equations.

4. Almost-periodic (in particular 2π-periodic) with respect to $\tau = \omega t$ solu-
tions $\boldsymbol{\Psi} = \boldsymbol{\Psi}^*(\dot{\mathbf{X}}, \mathbf{X}, t, \omega, t)$ of the equation of fast motions at the frozen (fixed)
$\dot{\mathbf{X}}, \mathbf{X}$ and t are searched for as a rule approximately. Among them such solutions
are selected which are asymptotically stable with respect to all fast generalized
coordinates and velocities all over that region of values of the variables. These
solutions are used while finding approximate expressions for the vibrational
force by equation (3.2.58), after which the corresponding main equations of
vibrational mechanics (3.2.59) are composed.

This approximate method is well-grounded, at least for the case when

---

[j] This circumstance and some other specific features of applied mathematical research are
discussed in detail in the book [119].

$\mathbf{\Psi} \sim \varepsilon\mathbf{X}$, i.e. when $n = 1$. As that takes place, if the expression for the fast force $\mathbf{\Phi}$ can be presented as (3.2.70), the finding of the function $\mathbf{\Psi}$ and of the vibrational force is essentially simplified.

5. Solutions of equations (3.2.59) which are of interest are being studied.

6. In questionable cases the suppositions made are checked a posteriori (taking into account remark 4 in 3.2.9).

3.2.12 *Short Bibliographic Review*

The idea of dividing the motions into fast and slow plays an important role in the modern theory of non-linear oscillations. This fruitful idea was best embodied and rigidly grounded in the method of averaging, whose development is connected with the names of Krylov, Bogolyubov, Mitropolsky, Zubarev, Volosov, Arnold and with their numerous disciples and followers [24, 133, 134, 135, 136, 137, 572, 385, 386, 387, 388]. Among other important publications we will mark [189, 196, 601, 257, 258, 279, 280, 390, 433, 442, 462, 471, 523, 235].

The method of direct separation of motions, as it is treated in this section, has been developed and grounded in the works of the author [93, 95, 99] (see also [103, 114, 564, vol. 2]); here the presentation of the method has been essentially refined and supplemented. The primary source of this approach was the work by Kapitsa [257, 258] in which by using the euristic method he considered the problem of the behaviour of the pendulum with a vibrating axis of suspension and suggested the notion of vibrational torque. Afterwards this method was used by a number of authors to solve various applied problems [190, 191, 192, 170, 239, 274, 449, 593, 199, 200, 594]; those investigations will be discussed in the corresponding sections of the book. In the final equations of a number of those publications expressions are used which correspond to the vibrational forces and moments, i.e. those equations being in fact equations of vibrational mechanics, though they were obtained by other methods (see 3.3).

It should be noted that certain elements of such an approach to the solution of problems of non-linear mechanics and of the theory of non-linear oscillations can also be found in the publications, preceding the appearance of Kapitsa's work on the pendulum, such elements can be seen in the investigations devoted to the statistical physics, to the theory of turbulence, to non-linear acoustics, (the theory of acoustic flows and radiational pressure), to radio-electronics (the theory of detecting signals), and also in later publications on the oscillations in non-linear systems, especially in the systems of control [442].

3.3   On the Main Peculiarities and Advantages of the Approaches of Vibrational Mechanics and of the Method of Direct Separation of Motions as Compared to Other Similar Methods

3.3.1   *Peculiarities and Limitations*

1. The approach, called vibrational mechanics and the method of direct separation of motions have been adapted to investigate processes presentable in the form of the sum of the main "slow" motion $X(t)$ and the fast (not necessarily small) oscillations $\psi(t, \omega t)$ i.e. in the form

$$x = X(t) + \psi(t, \omega t) \qquad (3.3.1)$$

Here lies its distinction from, e.g., the classical version of Van der Pol's method when the solution sought has the form

$$x = A(t) sin[\omega(t)t + \beta(t)] \qquad (3.3.2)$$

where $A(t)$, $\omega(t)$ and $\beta(t)$ are the slowly changing functions of time.

As can be seen from the content of this book, cases when solutions of the type of (3.3.1) that are of interest are quite numerous and highly diversified.

2. When using in practice the method of direct separation of motions, the way of obtaining an approximate solution of the equation of fast motions (3.2.6) is of special importance. The main limitation in using this method lies in the fact that the component X changes slowly enough as compared to the component ψ , i.e. that the frequency ω is sufficiently large. That fact imposes certain constraints on the relative orders of the right sides of equations (3.2.55) and (3.2.56) (of 3.2.10). Usually it is sufficient that the typical period of changing the component X should be at least three times larger than that of the component ψ.

Another limitation lies in the requirement that the equations of fast motions (3.2.56) should have solutions asymptotically stable by the corresponding variables (i.3.2.10). Though for the separate equation for the variable X (not necessarily slow!) to be valid, it is sufficient to satisfy only the second of the above mentioned limitations.

3.3.2   *Advantages*

1. The main advantage of this method and this approach lies in the fact that the main equations (the equations of slow motions) are obtained in the form of equations of dynamics. The result of the action of vibration on the nonlinear

system is expressed in these equations by means of adding to the "ordinary" slow forces the so called vibrational forces V. The appearance of these forces, as well as the process of obtaining them, have in this case a distinct physical meaning.

The indicated advantage of this method is illustrated by the content of the book. In particular it plays an essential part when considering the following problems:

1) When revealing the so called potential on the average dynamic systems — systems for which vibrational forces have a potential. Slow motions are described in this case by the potential system, while the initial system is essentially non-potential (a more exact definition is given in chapter 4 of the book where several classes of the potential on the average dynamic systems are considered).

2) When solving complicated problems in case it is difficult even to compose the initial differential equations of motion of the system. In some cases expressions for vibrational forces, found as a result of solving simple "model" problems, can be used with certain modifications to solve much more complicated problems. Problems on slow flows of the granular material, caused by vibration (see chapter 15 of the book) are the examples.

3) When solving problems on the action of vibration on systems with dry friction and impacts. In these cases, instead of the initial non-smooth or discontinuous system for the description of slow motions one gets a system with "smooth" vibrational forces. Such systems are considered in chapters 9 and 13–15 of the book).

4) When solving problems of controlling the properties of non-linear mechanical systems, including vibro-rheological properties of non-linear media, by means of vibration (see chapter 18).

5) When solving problems of the inventors. For the creators of new machines and technologies, it is very convenient and expedient to take the position of the observer V and add to all the ordinary forces the additional force V. It gives an opportunity "to neglect" the laws of mechanics which prevent achieving the aim. Thus the inventor can obtain quite unexpected positive effects (see 1.4). This approach — the temporary ignoring in the process of inventing the physical laws — is known under the name of "fantastic analogy" and it proved rather effective [251].

2. The order of the system of slow motion can be much less than the order of the initial system, e.g. in the problem of 6.3 the order of the initial system is $2(n + 1)$, $(n = 1, 2, \ldots)$ and the order of the equation of slow motion is 1.

3. The method of direct separation of motions is "not very sensitive" to the

accuracy of finding the fast component of motion ψ when obtaining the equation
for the slow component X (the main equation of vibrational mechanics). This
is connected with the fact that the fast component enters into the expression
for the vibrational force only under the sign of integration. Due to that to
obtain an acceptable result it is often sufficient to find a crudely approximate
solution of the equation of fast motions. In particular, we can confine ourselves
to the so called purely inertional approach (see 3.8 and e.g. the solution of
certain problems in 9.4 and 10.1).

4. As a result, instead of the difficulties of solving the initial differential
equations, we face the difficulties, connected with solving separately the equations of fast motions and those of slow motions. As follows from what has been
said, these latter difficulties are far lesser than the first.

5. As has been shown in chapter 2, vibrational mechanics can be considered
within the frames of a much more general conception - the conception of a
system with the ignored (hidden) motions.. Hence vibrational mechanics can
be regarded as quite well grounded. The method of direct separation of motions
is also well grounded, at least with regard to the majority of problems, met in
applications.

3.3.3   Final Remarks

The contents of the book show that the enumerated peculiarities and limitations
of the method and approach, discussed above, do not prevent one from solving
numerous applied problems on the action of vibration on nonlinear systems. At
the same time their merits make it possible to get considerable advantages when
solving such problems. Though, a question arises: whether it is possible to get
the same results by other methods. For most of the problems the answer will
be definitely positive. In particular some results, stated in the book, were first
obtained by the author by means of other methods. For instance the solution of
the problems on self-synchronization of mechanical vibro-exciters (chapter 7)
was obtained by Poincare-Lyapunov's method of small parameter, and the
solution of the problems of the theory of vibrational displacement (chapters 9
and 10) was got by exact methods. The reader, just like the author, can also
make sure that by method of direct separation of motions the solution can
be obtained in a much simpler- and often in a much more general form; the
reasons, causing this circumstance, are discussed in i.i. 3 and 4 of 3.3.2. As an
example we refer to the solution of the problem concerning the transportation
of a body up a vertical vibrating pipe (10.1). At present there are a number
of complicated problems of the theory of vibrational displacement and of the

theory of synchronization of vibro-exciters whose solution was obtained by the method of direct separation of motions and which has not yet been obtained (and apparently cannot be obtained) by any other method. Such problems were mentioned above in i. 3.3.2).

3.4   On Other Methods of Obtaining Expressions for the Vibrational Forces and the Main Equations of Vibrational Mechanics

The method of direct separation of motions, presented in 3.2, is just one of the possible methods of obtaining the expressions for vibrational forces and equations of vibrational mechanics. Relations, containing the vibrational forces or moments in the form of summands can in some cases be obtained solving problems by other methods, such as the Poincare method of a small parameter, asymptotic methods, the method of a harmonic balance. It is natural that such relationships present in fact equations of vibrational mechanics. We will also mention the method of successive approximations, proposed by Kolovsky for solving problems of the dynamics of machine aggregates, but which can also be used to solve problems of other types (see 6.4)

Two more methods, in a sense diametrically opposite, should be mentioned, with regard to which the method of direct separation of motions takes an intermediate position. The first of the methods refers to those rare cases when the solution of the initial equation (3.2.52) — either the exact, or the approximate one — is known. Then the finding of the vibrational force $\mathbf{V}$ is reduced to selecting the summands $\mathbf{X}$ and $\boldsymbol{\Psi}$ in that solution taking into account equality (3.2.54), and to the averaging according to formula (3.2.58). Such calculation is not useless, particularly in connection with the fact that the expression obtained for $\mathbf{V}$ can be used as an element when solving more complicated problems in an approximate way. We will make sure of what has been said here when we study the theory of vibrational displacement and of vibrorheology (see 9.4 and 15.1).

Expressions for the vibrational forces can also be obtained by computer, solving numerically the corresponding equations and selecting a proper approximating formula, based on the series of calculations that have been performed.

The second method refers to the opposite case, when the problem is so intricate that neither the initial equation (3.2.52), nor equation (3.2.56) for the fast component $\boldsymbol{\Psi}$ can be solved (and sometimes even composed) in a satisfactory way. Then we must resort to the hypotheses about the type of the function $\mathbf{V}(\dot{\mathbf{X}}, \mathbf{X}, t)$ based mostly on the experimental data and on the ideas, following from the theory of dimensionality and similarity. Examples of such an approach are the *semi-empirical theories of turbulence* [340], the first publications on the theory of vibratory pile driving, in which the ground, which is under the action of vibration, is regarded as a fluid whose viscosity coefficient depends on the parameters of vibration [46], and also a number of investigations on vibratory creep and those which study the effect of vibration upon the substances of the type of concrete mixtures [13, 475]. In spite of a kind of dissatisfaction a theoretician begins to feel at such an approach, there is no denial that such investigations are of practical benefit and they should be considered acceptable, at least until the problem has been elaborated more thoroughly. It should be also noted that the general statements of the mechanics of systems with the ignored motions and of vibrational mechanics, given here, are in fact a basis of a possibility to present the "main" results of the action of vibration upon the system as those caused by the action of vibrational forces. So both approaches to find these forces — the phenomenological and the semi-empirical — can in principal be considered acceptable.

We will also emphasize that the "macroscopic" equation (3.2.59), describing only the slow component of motions, which is of utmost practical interest, requires for its composition far less information than the initial "micro-equation" (3.2.52) or the system that corresponds to it (3.2.55), (3.2.56).

Finally, in a number of cases the vibrational force can be found by direct or indirect experiments (see, say, 10.1).

Chapter 4

Potential on the Average Dynamic Systems and Extremal Signs of Stability of Certain Motions

4.1 On Potentiality on the Average and on the Extremal Signs of Stability

"Potentiality on the average" is a remarkable property of some non-linear systems liable to the action of vibration, either external, or appearing autonomously. It consists in the fact that despite a possible essential non-conservatism of the initial system (3.2.52) or (3.2.46), the equations of slow motions of that system have the form of the equations of motion of the conservative system, that is the main equations of vibrational mechanics (3.2.59) or (3.2.50) allow the notation

$$\frac{d}{dt}\frac{\partial T_X}{\partial \dot{X}_s} - \frac{\partial T_X}{\partial \dot{X}_s} = -\frac{\partial D}{\partial \dot{X}_s} \qquad (s = 1, \ldots, k) \qquad (4.1.1)$$

Here, just like in (3.2.50), T_X denotes the potential energy of the system, corresponding to the slow motions, and $D = D(X_1, \ldots, X_k)$ denotes the function which will be called by us *potential function* . T_X is supposed to be a positive definite quadratic form $\dot{X}$ with the coefficients which may depend on X.

First we will assume that the vibrational forces V_s and also the slow forces F_s depend only on the slow coordinates $X_1, \ldots, X_k$. In this case for the existence of the potential function it is sufficient that both the initial ("ordinary") slow forces F_s and the vibrational forces V_s should be potential, that is there should exist the functions Π_F and D_V, satisfying the relations

$$F_s = -\frac{\partial \Pi_F}{\partial X_s}, \quad V_s = -\frac{\partial D_V}{\partial X_s} \qquad (s = 1, \ldots, k), \quad D = \Pi_F + D_V \qquad (4.1.2)$$

Systems of the type under consideration, allowing the notation in form (4.1.1), i.e. possessing the potential function D, will be called by us *potential on the average dynamic systems*. This name is justified by the fact that both the function D and the expression for the vibrational forces V_s are obtained as a result of the operation of averaging. In this case the function Π_F is the potential energy of the initial slow forces; the function D_V will be called by us *potential energy of vibrational forces*.

Thus the potential function D is the sum of the potential energy of both the original slow forces Π_F and the vibrational forces D_V.

When there is a potential function for the system under consideration (generally speaking, non-conservative!), then an analog holds with the Lagrange-Dirichlet classical theorem on the stability of the positions of equilibrium [106]: points of the minimum of that function may be answered (and under certain additional conditions in fact are answered) by the stable motions of the initial system. It is also essential that both the variables $X_1, \ldots, X_k$ and the potential function can be distinctly interpreted in terms of the characteristic of the system and motions under consideration in all the cases discussed below the function D or its "main part" present an averaged in a certain manner Lagrangian, Hamiltonian, a force function of the system or of its parts. These circumstances offer great advantages in investigations which will be illustrated more than once in the subsequent chapters.

For the existence of the function D with the above-described properties it is not necessary at all for the system to be potential on the average. So, it follows from the Thomson -Tait -Chetayev theorems (see e.g. [372]) that the role of the function D is preserved if the equation of slow motions along with the potential forces also contains the dissipative forces, i.e. they allow the notation in form

$$\frac{d}{dt}\frac{\partial T_x}{\partial \dot{X}_s} - \frac{\partial T_x}{\partial X_s} = -\frac{\partial \Phi}{\partial \dot{X}_s} - \frac{\partial D}{\partial X_s} \qquad (4.1.3)$$

where Φ is the Rayleigh dissipative function which is a positive function of X. Moreover, the existence of the function D with the above-mentioned properties is not necessarily caused by the presence of the equations of type (4.1.1) or (4.1.3). In this connection it should be noted that the above-mentioned way of obtaining the function D by using the method of a direct separation of motions is not the only one possible: as will be shown later, the function with similar properties can be arrived at when solving problems by the Poincare-Lyapunov method and by various asymptotic methods as well as by variational methods. In this case it appears that for certain systems and for some classes of their motions the stable motions can answer the points of both minimum and maximum of the corresponding function; and the meaning of the variables $X_1, \ldots, X_k$ may also appear to be different.

Therefore, generalizing the situation stated above, we will call the signs of stability of the motion of the system, based on the existence of a sufficiently smooth function D of certain parameters of the motion of the system and on the analysis of the character of the extremums of this function with respect to the parameter mentioned, the *extremal signs of stability* of the corresponding motion [a]. The function D will be called by us the potential function also in this general case as well.

We will give a schematic description of obtaining the potential function by using the Poincare method. It was in that way that certain extremal signs of stability were first found. As is known (see, say [90, 363]), while using the Poincare method in the case when the *generative system* allows a set of periodic or almost-periodic solutions, depending on the parameters $\alpha_1, \ldots, \alpha_k$, the corresponding solutions of the initial system can answer the values of those parameters, satisfying a certain system of equations

$$P_s(\alpha_1, \ldots, \alpha_k) = 0 \qquad (s = 1, \ldots, k) \tag{4.1.4}$$

where the functions P_s are expressed by the right sides of the initial equations and via the generating solution. Further on, under certain assumptions about the character of the solutions of the equations in variations, corresponding to the generating system and the generating solution, it is proved [72, 73, 90, 103, 363] that a certain solution of the equations (4.1.4) is really answered by the asymptotically stable solution of the initial system if all the roots κ of the algebraic equation of the k-th degree (δ_{sj} being the Kronecker symbol)

$$\left| \frac{\partial P_i}{\partial \alpha_j} - \delta_{sj} \kappa \right| \qquad (i, j = 1, \ldots, k) \tag{4.1.5}$$

have the negative real parts. If there is at least one root with a positive real part, the corresponding solution is unstable, and the case of the zero- or purely imaginary roots requires an additional investigation.

Now let us assume that there is a certain function $D(\alpha_1, \ldots, \alpha_k)$, continuous together with its derivatives up to the second order inclusively, that the

---

[a]The term *extremal signs of stability* was suggested by Beletsky [57, 58, 59]

relations

$$\frac{\partial D}{\partial \alpha_j} = -P_i(\alpha_i, \ldots, \alpha_k) \qquad (i = 1, \ldots, k) \tag{4.1.6}$$

are fulfilled.

On account of what has been said, it follows from equation (4.1.5) that it is D which is the potential function. It should be noted that the parameters α_s do not represent in this case the slow variables.

Pay attention to the case when the equations for the slow variables $a_1, \ldots, a_l; b_1, \ldots, b_l$ $(l \leq n)$ are written in the canonical form

$$\dot{a}_i = -\frac{\partial H}{\partial b_i}, \quad \dot{b}_i = \frac{\partial H}{\partial a_i}, \quad (i = 1, \ldots, l). \tag{4.1.7}$$

Here the role of the potential function D is played by the Hamilton function $H = H(a_1, \ldots, a_l; b_1, \ldots, b_l)$ and the stable motions may in this case answer both — the points of strict minimum and those of strict maximum of the function D.

Apparently the first result, concerning the extremal signs of stability, follows from the classic work by Poincare [441, item 42 and 79] who studied however the conservative systems [b]. Poincare considered the equations of the type

$$\dot{x}_i = \frac{\partial H}{\partial y_i}, \quad \dot{y}_i = -\frac{\partial H}{\partial x_i}, \quad (i = 1, \ldots, n) \tag{4.1.8}$$

where

$$H = H_0(x_1, \ldots, x_n) + \mu H_1(x_1, \ldots, x_n; y_1, \ldots, y_n)$$
$$+ \mu^2 H_2(x_1, \ldots, x_n; y_1, \ldots, y_n)$$

is the periodic function of the period of 2π of variables $y_1, \ldots, y_n;$ $\mu > 0$ -being a small parameter. At $\mu = 0$ equations (4.1.8) have the solution

$$x_i = a_i, \qquad y_i = \omega_i t + \alpha_i \tag{4.1.9}$$

where a_i and α_i are the constants of integration, ω_i are the functions of $a_1, \ldots, a_n$. Let us assume that under certain values a_i the frequencies ω_i are multiple of $\omega = 2\pi/T$ (below such (4.1.9) solutions are called *synchronous with the frequency* ω — see 4.2), and let $\omega_1 = -\partial H_0/\partial x_1 \neq 0$. Let us also assume that under the corresponding values of a_i the Hessian $|\partial^2 H_0/\partial x^2|$ is different from zero. Since the system under consideration is autonomous, we can assume

---

[b]The author is thankful to Kozlov who drew his attention to this fragment of the work by Poincare

that $\alpha_1 = 0$ [363, 441]. Then, as has been shown by Poincare, if at certain values $\alpha_2 - \alpha_2^*, \ldots, \alpha_n = \alpha_n^*$ the derivatives $\partial \overline{H}_1 / \partial \alpha_i = 0$ $(i = 2, \ldots, n)$ and $|\partial^2 \overline{H}_1 / \partial \alpha^2| = 0$, where [c]

$$\overline{H}_1(a_1, \ldots, a_n; \quad \alpha_1, \ldots \alpha_n) = < [H_1] >,$$

then at the sufficiently small $\mu \neq 0$, the initial system (4.1.8) will have the solution of the period T which at $\mu = 0$ turns into a generating solution (4.1.9) with the parameters $\alpha_1 = 0, \alpha_2^* = \alpha_2, \ldots, \alpha_n = \alpha_n^*$. Investigating the stability of that periodic solution, Poincare showed that the characteristic indexes of that solution can be presented in form of power series expansion $\sqrt{\mu} : \lambda = \lambda_1 \sqrt{\mu} + \lambda_2 \mu + \ldots$. Two characteristic indices are always equal to zero.

Having considered a special case $n = 2$, Poincare obtained for the non-zero indices the following equality for determining λ_1:

$$\omega_1^2 \lambda_1^2 = - \left| \frac{\partial^2 \overline{H}_1}{\partial \alpha_2^2} \right|_{\alpha_2 = \alpha_2^*} \left(\omega_1^2 \frac{\partial^2 H_0}{\partial x_2^2} - 2 \omega_1 \omega_2 \frac{\partial^2 H_0}{\partial x_1 \partial x_2} + \omega_2^2 \frac{\partial^2 H_0}{\partial x_1^2} \right) . \quad (4.1.10)$$

From Poincare's arguments, based on the consideration of expression (4.1.10), it follows that the stable in the first approximation motions will be answered either by the points of minimum or by the points of maximum of the function $\overline{H}_1$.

A rather long space of time separated the fragment of Poincare's classic work stated here and the extremal signs of stability, proposed much later. It was only in late 50-ies that publications began to appear in which the extremal signs of stability are formulated in connection with the motion of more complicated, mostly non-conservative systems. It is essential that for many such systems the extremal signs prove to be "stronger" — they express both necessary and sufficient conditions of stability according to Lyapunov, i.e. in fact they present *criteria of stability*. However for the Hamiltonian systems the corresponding signs express, generally speaking, only the sufficient *conditions of stability in the first approximation with respect to the small parameter*, which enters into the equations in variations. The motion in such cases can be stable according to Lyapunov only provided certain additional conditions are fulfilled. Such a

---

[c]Here and below the brackets [·] indicate that the expression enclosed in them is calculated on the generating solution; remember also that $< \ldots > = \frac{1}{T} \int_0^T \ldots dt$ in the case of periodic functions and $< \ldots > = \lim_{T \to \infty} \frac{1}{T} \int_0^T \ldots dt$ in the case of almost-periodic functions.

situation takes place particularly in Poincare's fragment, stated above. Though the extremal signs do not lose their importance in the latter cases either, since they determine the selection of the constants $\alpha_1, \ldots, \alpha_k$ (or $X_1, \ldots, X_k$) which may be answered by the stable motions.

We must recall that speaking about the *stability of motion according to the first approximation* (in Lyapunov's meaning) one implies a situation when on the basis of the analysis of equations in variations (assuming that they describe the perturbed motion quite exactly), the asymptotic stability of this motion is established. As was shown by Lyapunov, in this case the registration of the non-linear terms in the equations of perturbed motion cannot change the inference of stability.

Speaking about *stability of motion on the first approximation* one means that on the basis of the above analysis only non-asymptotic stability is established. The registration of the non-linear terms can in this case affect considerably the inference of stability according to Lyapunov — the motion may appear to be either stable or unstable.

Speaking about *stability of motion in the first approximation with respect to the small parameter* μ, one implies a situation when the consideration of the first depending on μ members of the decomposition of the characteristic indices of the equations in variations makes it possible to establish only the non-asymptotic stability of motion, assuming those approximations to the characteristic indices give the exact values of those indices. The consideration of the subsequent approximations to the characteristic indices may lead to the conclusion about either stability or instability according to the first approximation. For more detail and more rigorous treatment see e.g. [353, 372].

As could be partially perceived from the previous text, the conditions of existence of the potential function are sufficiently rigid: it is not always possible to establish its presence. Along with that, sections 4.2–4.4 consider three relatively wide classes of systems, important for application, for which by now the function D has been found. In this case the main results are formulated in the form of mathematical theses, which, as a rule, was not done by their authors. The presentation of the results, obtained by the Poincare-Lyapunov method and by asymptotic methods, is given here as a kind of a review. This survey mainly follows the report [111] and publication [120].

4.2   Systems with Synchronized Objects. Integral Signs of Stability (Extremal Properties) of Synchronous Motions

4.2.1   *Systems with Almost Uniform Rotations, Synchronization of Vibroexciters; Theorems 7 and 8*

The integral sign of stability was formulated by Lavrov and the author for the systems with self-synchronized mechanical vibro-exciters [77], (see also chapter 7) and was proved by the author by means of the Poincare-Lyapunov method of small parameter [78] (1960). In the book [90] this sign is generalized for the systems with almost uniform rotations. We will formulate the corresponding result without dwelling on its proof, after which we will get practically the same result by method of direct separation of motions.

Let equations of motion of the system with the generalized coordinates φ_s $(s = 1, \ldots, k)$ and u_r $(r = 1, \ldots, \nu)$ and with the Lagrange function L and with the non-conservative forces Q_{φ_s} and Q_{u_r} be presented in form

$$I_s \ddot{\varphi}_s + k_s(\dot{\varphi}_s - \sigma_s n_s \omega_s) = \mu \Phi_s \qquad (s = 1, \ldots, k), \tag{4.2.1}$$

$$E_{u_r}(L) = Q_{u_r} \qquad (r = 1, \ldots, \nu) \tag{4.2.2}$$

where

$$E_q = \frac{d}{dt} \frac{\partial}{\partial \dot{q}} - \frac{\partial}{\partial q} \tag{4.2.3}$$

is the Euler operator, corresponding to the generalized coordinate q; I_s, k_s and ω are the positive constants, $\sigma_s = \pm 1$, $\mu > 0$ is a small parameter and the functions Φ are determined from the condition of the identity of equations (4.2.1) to the corresponding group of the Lagrange equations .

$$\mu \Phi_s = I_s \ddot{\varphi}_s + k_s(\dot{\varphi}_s - \sigma_s n_s) - E_{\varphi_s}(L) + Q_{\varphi_s} \tag{4.2.4}$$

The numbers σ_s are introduced for convenience sake when writing down the relations which refer to the problems with the synchronization of rotating bodies (see e.g. 6.1). The functions L, Q_{φ_s}, Q_{u_r} and Φ_s may depend on the generalized coordinates and velocities of the system and on the time t; these functions are 2π- periodic with respect to φ_s and $2\pi/\omega$ - periodic with respect to t, the functions Q_{φ_s}, Q_{u_r} and Φ_s may also depend on the small parameter μ.

Concerning the character and smoothness of the functions which enter into the differential equations considered in this chapter, for the validity of every result stated here it is sufficient to assume that these equations can be presented in form $\dot{x}_s = X_s(x_1, \ldots, x_n, t) + \mu f_s(x_1, \ldots, x_n, t, \mu)$ $(s = 1, \ldots, n)$ where the right sides are determined at every real value of t, at the values μ lying on a certain segment $[0, \mu_0]$, and at the values $x_1, \ldots, x_n$, lying in a certain closed region G of the space of these variables. In the indicated region the right sides of the equations are continuous with respect to t, the functions f_s allowing the continuous partial derivatives of the first order with respect to the variables $x_1, \ldots, x_n, \mu$, while X_s do not depend on μ and allow the continuous derivatives of the second order with respect to $x_1, \ldots, x_n$. In some case these requirements may be weakened. With respect to t the right sides can be either periodic with a certain period T, or almost-periodic, or not depending explicitly on that variable. In the case of almost-periodic equations it is believed that at any fixed μ from the segment $[0, \mu_0]$ and any almost-periodic $x_1, \ldots, x_n$, belonging to the region G, the function $X_s(x_1, \ldots, x_n, t)$ and $f_s(x_1, \ldots, x_n, t, \mu)$ are also almost-periodic with respect to t. All the variables and parameters are considered to be dimensionless .

We will call the coordinates φ_s *rotational coordinates* , and the coordinates u_r — *oscillatory coordinates*.

The described systems belong to the class of systems with *almost uniform rotations*. The generating equations ($\mu = 0$) corresponding to equations (4.2.1) allow a set of solutions

$$\varphi_s^0 = \sigma_s(n_s \omega t + \alpha_s) \qquad (4.2.5)$$

depending on k arbitrary parameters $\alpha_1, \ldots, \alpha_k$. Let the generating equations, corresponding to Eq.(4.2.2) and solution (4.2.5), allow at any α_s the asymptotically stable $2\pi/\omega$-periodic solution u_r^0.Then the following theorem is valid:

T h e o r e m 7. *Solutions of system* (4.2.1), (4.2.2) *of the form*

$$\varphi_s = \sigma_s[(n_s \omega t + \alpha_s) + \psi_s(t, \mu)],$$
$$u_r = u_r^0(t) + v_r(t, \mu) \qquad (4.2.6)$$

where ψ_s *and* v_r *are* 2π- *periodic functions* t, *becoming zero at* μ *(such solutions and the motions corresponding to them are called synchronous motions), i.e. the solutions of the initial system which at* $\mu = 0$ *turn into the generating solution* φ_s^0 *and* u_r^0, *can answer the values of the constants* $\alpha_1, \ldots, \alpha_k$, *satisfying the equation*

$$P_s(\alpha_1, \ldots, \alpha_k) \equiv \tfrac{\sigma_s}{k_s} < [\mu \Phi_s] > \equiv \tfrac{\sigma_s}{k_s} < [-E_{\varphi_s}(L) + Q_{\varphi_s}] > = 0$$
$$(s = 1, \ldots, k). \qquad (4.2.7)$$

Every solution of these equations, for which all the roots of the algebraic equation of the k- *degree*

$$\left| \frac{\partial P_s}{\partial \alpha_j} - \delta_{sj} \kappa \right| = 0 \qquad (s, j = 1, \ldots, k) \qquad (4.2.8)$$

have negative real parts, is really answered by the asymptotically stable syn-
chronous solution (4.2.6); if there is at least one root with a positive real part,
the corresponding solution is unstable; cases when there are zero- or purely
imaginary roots require an additional investigation.

As follows from (4.2.7), to obtain the expressions $P_s(\alpha_1, \ldots, \alpha_k)$ playing the
main role in the theorem stated above, it is sufficient to average the right sides
of equations (4.2.1) on the generating solution φ_s^0 and u_r^0. These expressions
can be transformed into a more convenient form, after which it will be not
difficult to obtain the formula for the potential function D and formulate the
theorem in form of the extremal criterion of stability.

We will introduce the function

$$\Lambda = \Lambda(\alpha_1, \ldots, \alpha_k) = <[L]> \qquad (4.2.9)$$

representing the Lagrange function of the system, calculated for the generating
solution and averaged for the period $2\pi/\omega$. Calculating the derivative of that
function with respect to α_j, as a result of quite simple transformations, including the integration by parts taking into account equalities (4.2.2), (4.2.5) and
the periodicity of the corresponding functions, we obtain

$$\frac{\partial \Lambda}{\partial \alpha_j} = \frac{\partial <[L]>}{\partial \alpha_j} = \sum_{s=1}^{k} \left\langle \left[\frac{\partial L}{\partial \dot{\varphi}_s}\right] \frac{\partial \dot{\varphi}_s^0}{\partial \alpha_j} + \left[\frac{\partial L}{\partial \varphi_s}\right] \frac{\partial \varphi_s^0}{\partial \alpha_j} \right\rangle$$

$$+ \sum_{r=1}^{\nu} \left\langle \left[\frac{\partial L}{\partial \dot{u}_r}\right] \frac{\partial \dot{u}_r^0}{\partial \alpha_j} + \left[\frac{\partial L}{\partial u_r}\right] \frac{\partial u_r^0}{\partial \alpha_j} \right\rangle = - \sum_{s=1}^{k} < [E_{\varphi_s}(L)] \frac{\partial \varphi_s^0}{\partial \alpha_j} >$$

$$- \sum_{r=1}^{\nu} < [E_{u_r}(L)] \frac{\partial u_r^0}{\partial \alpha_j} = -\sigma_j < [E_{\varphi_j}(L)] > - \sum_{r=1}^{\nu} < [Q_{u_r}] \frac{\partial u_r^0}{\partial \alpha_j} >. \qquad (4.2.10)$$

The use of relation (4.2.10) enables us to present Eq.(4.2.7) in the following
form:

$$P_s(\alpha_1, \ldots, \alpha_k) \equiv \frac{1}{k_s} \left\{ \frac{\partial \Lambda}{\partial \alpha_s} + \sigma_s < [Q_{\varphi_s}] > + \sum_{r=1}^{\nu} < \left[Q_{u_r} \frac{\partial u_r^0}{\partial \alpha_s} \right] > \right\} = 0$$
$$\qquad (4.2.11)$$

Then, let there exist a function $B = B(\alpha_1, \ldots, \alpha_k)$ called the *potential of
the averaged non-conservative generalized forces*, such that

$$\frac{\partial B}{\partial \alpha_s} = \sigma_s < [Q_{\varphi_s}] > + \sum_{r=1}^{\nu} < \left[Q_{u_r} \frac{\partial u_r^0}{\partial \alpha_s} \right] > . \qquad (4.2.12)$$

Then, denoting

$$D = D(\alpha_1, \ldots, \alpha_k) = -(\Lambda + B), \qquad (4.2.13)$$

we will present equations (4.2.7), (4.2.11) and (4.2.8) in the following form:

$$P_s(\alpha_1, \ldots, \alpha_k) \equiv -\frac{1}{k_s}\frac{\partial D}{\partial \alpha_s} = 0 \qquad (s = 1, \ldots, k), \qquad (4.2.14)$$

$$\left|\frac{1}{k_s}\frac{\partial^2 D}{\partial \alpha_s, \partial \alpha_j} + \delta_{sj}\kappa\right| = 0 \qquad (s, j = 1, \ldots, k). \qquad (4.2.15)$$

The relations obtained enable us to formulate the above theorem in the following form:

   T h e o r e m 8. *The integral criterion (the extremal sign) of stability* [d]. *Every point of the rough minimum* [e] *of the function* $D = D(\alpha_1, \ldots, \alpha_k) = (\Lambda + B)$ *at the sufficiently small values of* μ *is answered by the only asymptotic stable solution (4.2.6) of the initial system (4.2.1), (4.2.2) which at* $\mu = 0$ *is turned into the generating solution. The absence of the minimum, discovered by the analysis of the terms of the 2-nd order in the power series expansion of the function* D *in terms of* α_s *in the vicinity of the stationary point shows that the corresponding synchronous solution is unstable. Other cases require an additional investigation.*

   In other words the function D for the class of systems and motions under consideration plays the role of the potential function. As that takes place, the parameters $\alpha_1, \ldots, \alpha_k$ do not play the role, at least explicitly, of slow variables.

   We will add some supplementary remarks to the theorems stated here.

1) If $L, Q_{\varphi_s}, Q_{u_r}$ and Φ_s are the analytic functions of the generalized coordinates and velocities, and the functions $L, Q_{\varphi_s}, Q_{u_r}$ and Φ_s depend analytically on the small parameter μ as well , then solution (4.2.6) will be analytic with respect to μ, i.e. it can be presented at the sufficiently small μ in the form of convergent series

$$\varphi_s = \sigma_s[(n_s\omega t + \alpha_s) + \mu\varphi_s^{(1)}(t) + \mu^2\varphi_s^{(2)}(t) + \ldots],$$
$$u_r = u_r^0(t) + \mu u_r^{(1)}(t) + \mu^2 u_r^{(2)}(t) + \ldots \qquad (4.2.16)$$

where $\varphi_s^{(1)}, \varphi_s^{(2)}, \ldots, ; u_r^{(1)}, u_r^{(2)}, \ldots$ are $2\pi/\omega$-periodic functions of t.

2) In the case of the autonomous system, the function D depends only on the differences $\alpha_1 - \alpha_k, \ldots, \alpha_{k-1} - \alpha_k$ and the statement given here refers

---

[d]The fact that Eqs. (4.2.14), (4.2.15) contain the positive coefficients k_s which are absent in equalities (4.1.6) do not contradict the statements contained in the theorem (see, say, [90], 8 chapter 5)

[e]The word *rough* implies here the strict extremum of the function, discovered by analyzing the terms of the 2-nd order in the expansion of that function in the vicinity of the stationary point.

to the minima with respect to those differences while we are dealing with the *asymptotic orbital stability*.

3) At $\partial B/\partial\alpha_s \ll \partial\Lambda/\partial\alpha_s$ particularly at $B = $ const, we can assume that

$$D = -\Lambda, \tag{4.2.17}$$

i.e. the potential function is the averaged Lagrangian of the system, calculated for the generating solution and taken with the opposite sign.

4) As is obvious from the given theorem, conditions of the rough minimum of the function D are, under the stated assumptions, not only sufficient, but also "roughly necessary" in the sense that the absence of the minimum, discovered by the analysis of the terms of the 2-nd order in the expansion of the function D near the stationary point, indicates the instability of the synchronous solution under consideration.

5) If the fact of the asymptotic stability of the generating solution u_r^0 cannot be established on the basis of the analysis of the equations in variations for system (4.2.2) at $\mu = 0$, then in order to obtain the sufficient conditions of stability, it is necessary to add to the conditions of a rough minimum of the function D certain supplementary relations, obtained by the analysis of the subsequent approximations. The same situation takes place in case system (4.2.1) is quasiconservative. Along with that the conditions, following from the requirement of a rough minimum of D are basic in this case too, since it is they that determine the selection of the constants $\alpha_1, \ldots, \alpha_k$ which can be answered by stable solutions.

6) Let the Lagrange function of the system be presented in the form

$$L = L^* + L^{(I)} + L^{(II)} \tag{4.2.18}$$

where

$$L^* = \sum_{s=1}^{k} L_s(\dot\varphi_s, \varphi_s) + \sum_{r=1}^{v} f_r(\dot\varphi_1, \ldots, \dot\varphi_k; \varphi_1, \ldots, \varphi_k)\dot u_r + \sum_{s=1}^{k} F_s(\varphi_s),$$

$$L^{(I)} = \frac{1}{2}\sum_{r=1}^{v}\sum_{j=1}^{v} a_{rj}\dot u_r \dot u_j - \frac{1}{2}\sum_{r=1}^{v}\sum_{j=1}^{v} b_{rj}\dot u_r \dot u_j,$$

$$L^{(II)} = \Psi(\dot\varphi_1, \ldots, \dot\varphi_k; \varphi_1, \ldots, \varphi_k). \tag{4.2.19}$$

Here a_{rj} and b_{rj} are constants, and $L_s, f_r, F_r,$ and Ψ are the functions of the listed variables, with L_s, f_r, F_r and F_r being periodic with respect to φ_s with a period 2π; besides, let $[Q_{u_r}] = 0$, i.e. the non-conservative

forces with respect to the coordinates u_r in the generating approximation are absent. Thus the corresponding systems are quasilinear and
quasiconservative with respect to the oscillatory coordinates. Then the
following relations are valid

$$\frac{\partial \Lambda_s}{\partial \alpha_s} = 0, \quad \frac{\partial \Lambda}{\partial \alpha_s} = \frac{\partial(\Lambda^{(II)} - \Lambda^{(I)})}{\partial \alpha_s} \qquad (4.2.20)$$

where

$$\Lambda_s < [L_s] >, \ \Lambda^{(I)} =< [L^{(I)}] >, \ \Lambda^{(II)} =< [L^{(II)}] >, \ \Lambda =< [L] >, \qquad (4.2.21)$$

and the potential function can be presented in the form

$$D = \Lambda^{(I)} - \Lambda^{(II)} - B \qquad (4.2.22)$$

In the work by Nagaev [395] which will be discussed later on in
4.2.2 the expressions $L_s, L^{(I)}$ and $L^{(II)}$ are called respectively the *eigen-
proper Lagrangians of the autosynchronized objects*, and *Lagrangians of
the systems of the carrying and carried connections between the objects*.
It is remarkable that the values $\Lambda^{(I)}$ and $\Lambda^{(II)}$ enter into the expressions
for D with the opposite signs.

In the problem on synchronization of vibroexciters (see chapter 7)
L_s— are the Lagrangians of the rotors not bound with each other, $L^{(I)}$
is a Lagrangian of an elastically supported solid body or of elastically
bound system of solid bodies, on which rotors are placed and the summand $L^{(II)}$ is conditioned by the presence of direct connections between
the rotors in the form of elastic and damping elements. If $L^{(II)} = 0$,
$B = $ const or $\partial \Lambda^{(I)}/\partial \alpha_s \gg \partial B/\partial \alpha_s$, then we can assume that

$$D = \Lambda^{(I)} =< \Lambda^{(I)} >, \qquad (4.2.23)$$

i.e. *the potential function in this case presents the averaged over the
period Lagrangian of an elastically supported solid body or of a system of
solid bodies carrying the rotors, the function $L^{(I)}$ being calculated in the
generating approximation φ_s^0 and u_r^0.* This result essentially simplifies
the investigation of systems with autosynchronized unbalanced rotors
and has very important applications (see chapter 7 and 8)
7) In some cases, including an important case of the problem about multiple
synchronization of mechanical vibro-exciters, connected with the linear
oscillatory system (see 7.3.3), the function D calculated in the indicated

approximation, proves to be independent either of some or of all α_s (or $\alpha_s - \alpha_k$). Then the minima of the function D over those parameters, determined in that way, cannot be strict, and from the conditions of the minima those α_s (or $\alpha_s - \alpha_k$) cannot be determined. The additional investigation, mentioned in the theorem, is necessary in particular in the indicated cases.

The obtaining of the main result, formulated as the first theorem, among given above, by means of the Poincare-Lyapunov method is rather cumbersome. It is based on the use of the specially proved theorems about the existence and stability of the periodic solutions of differential equations with a small parameter in the case when the generating system allows a set of periodic solutions [72, 73, 363].

It is possible to arrive at all those results, particularly at expression (4.2.8) for the potential function D, in a much more simple way, using the method of direct separation of motions [95, 103]; in that case equations of slow motions (the main equations of vibrational mechanics) are also obtained simultaneously. Though when using that method one has in addition to postulate the smallness of the parameter $\varepsilon = 1/\omega$ and of the corresponding terms in the equations of motion.

Turning to that method, we will search for the solutions of system (4.1.1), (4.1.2) in form

$$\varphi_s = \sigma_s[n_s\omega t + \alpha_s(t) + \psi_s(t, \omega t)], \quad u_r = u_r(t, \omega t), \qquad (4.2.24)$$

regarding the functions $\alpha_s(t)$ as "slow"- and ψ_s and u_r as "fast" components and considering that ψ_s and u_r are the π-periodic functions of the "fast time" ωt, satisfying the conditions

$$< \psi_s(t, \omega t) >= 0, \quad < u_r(t, \omega t) >= 0. \qquad (4.2.25)$$

Now, taking into account (4.2.24), we will pass from system (4.2.1), (4.2.2) to the equations for the slow and fast motions respectively, (see equations (3.2.55) and (3.2.56) chapter 3):

$$I_s\ddot{\alpha}_s = -k_s\dot{\alpha}_s + \mu\sigma_s < \Phi_s >, \qquad (4.2.26)$$

$$I_s\ddot{\psi}_s = -k_s\dot{\psi}_s + \mu\sigma_s(\Phi_s - < \Phi_s >), \qquad (4.2.27)$$

$$E_{u_r}(L) = Q_r. \qquad (4.2.28)$$

By virtue of one of the main assumptions of the method of direct separation of motions, the right sides of equations (4.2.27) must be of the order of $1/\varepsilon^2 = \omega^2$, provided the order of the right sides of equations (4.2.26) is equal to unity and α_s and ψ_s are values of the same order; but if ψ_s is of the order of ε as compared to α_s, then the orders of the right sides of the indicated equations must differ only by $1/\varepsilon$, and so on. Since in the problems under consideration we are interested, as a rule , in the stable stationary regimes $\alpha_s = \text{const}$ and in their neighborhoods, then, according to remark 4 in 3.2.9, these conditions can be considered fulfilled.

In the above supposition in order to obtain equations of the slow motion in the first approximation, it is sufficient to find an approximate asymptotically stable periodical solution of equations (4.2.27), (4.2.28) at the fixed ("frozen") $\dot{\alpha}_s, \alpha_s$ and t, and use it when calculating the average value in the right sides of equations (4.2.26). Besides, taking advantage of the smallness of the parameter μ, we will restrict ourselves to solving the equations of fast motions (4.2.27) and (4.2.28) at $\mu = 0$, and the equations of slow motions (4.2.26) will be solved with an accuracy to the terms, containing μ. As will be clear from what follows, the indicated approximations will prove to be sufficient to obtain all the results stated above, established by the Poincare-Lyapunov method. When solving equations (4.2.27) and (4.2.28) further approximations will be necessary only in case of some special problems on synchronization when an additional investigation is necessary just like it is when using the Poincare-Lyapunov method (see note 7 of this section)

In the indicated supposition the periodic solution of equations (4.2.27), satisfying condition (4.2.25) will be $\psi_s = \psi_s^0 = 0$. We will designate the corresponding solution of equations (4.2.28) by $u_{r_*}^0$ and will recall that the latter value should be found at the "frozen" $\dot{\alpha}_s$ and α_s. It should be emphasized that the solution $u_{r_*}^0$ differs from the solution u_r^0 of the same equations, discussed above, which was considered to be corresponding to $\alpha_s = \text{const}$ and $\dot{\alpha}_s = 0$.

Substituting the functions $\psi_s = \psi_s^0 = 0$ and $u_r^0 = u_{r_*}^0$ into the right side of equation (4.2.26), taking into account equalities (4.2.4)(2.4) and (4.2.24), we will obtain the following approximate (in the sense of the method of direct separation of motions and with an accuracy to the terms linear with respect to μ) equation of slow motions:

$$I_s \ddot{\alpha}_s + k_s \dot{\alpha}_s = \mu \sigma_s < [\Phi_s]_* > \quad (s = 1, \ldots, k) \qquad (4.2.29)$$

where

$$\mu \sigma_s < [\Phi_s]_* > = \frac{\partial \Lambda_*}{\partial \alpha_s} + \sigma_s < [Q_{\varphi_s}]_* > + \sum_{r=1}^{\nu} < [Q_{u_r}]_* \frac{\partial u_{r_*}^0}{\partial \alpha_s} > . \qquad (4.2.30)$$

Functions with the index * are calculated for the solution $\psi_s = \psi_s^0 = 0$, $u_{r_*}^0$ but not for $\psi_s^0 = 0$, u_r^0 which are designated without that index.

In expression (4.2.30) the terms $I_s \ddot{\alpha}_s + k_s \dot{\alpha}_s$ are omitted, since if we are confined to the indicated approximation with respect to μ, then, according to equation (4.2.29), these terms in the right side of the equation can be considered equal to zero. Besides, when obtaining expression (4.2.30), it was taken into account that relation (4.2.10) remains valid also for the solution ψ_s^0, $u_{r_*}^0$, (i.e. for the values marked by the index *) since as was specified beforehand, the operation of averaging (unlike that of differentiation!) is performed only with respect to the fast time $\tau = \omega t$.

Though, the main result of this investigation remains valid even in case the indicated simplifications of expression (4.2.30) are not used. This result consists in the fact that the extremal sign of the stability of synchronous motions and the corresponding expression (2.13) for the potential function D, formulated above as a theorem, follows quite readily from Eq. (4.2.29). Indeed, the stationary solutions $\alpha_s = \text{const}$ of equations (4.2.29) must satisfy the equations

$$\mu \sigma_s < [\Phi_s] > \equiv \frac{\partial \Lambda}{\partial \alpha_s} + \sigma_s < [Q_{\varphi_s}] > + \sum_{r=1}^{\nu} < [Q_{u_r}] \frac{\partial u_r^0}{\partial \alpha_s} > = 0$$

$$(s = 1, \ldots, k) \qquad (4.2.31)$$

where when designating the functions, the index * is omitted, since the corresponding functions are calculated here at $\dot{\alpha}_s$, therefore they exactly coincide with the functions introduced before without that index.

Let $\alpha_1 = \alpha_1^0, \ldots, \alpha_k = \alpha_k^0$ be a certain solution of equations (4.2.31); let us consider the question of its stability. Assuming that $\alpha_s = \alpha_s^0 + x_s$, we will compose linerized equations of the perturbed motion for equations (4.2.26) and for the indicated solution

$$I_s \ddot{x}_s + k_s \dot{x}_s = \mu \sigma_s \sum_{j=1}^{k} \left\{ \left(\frac{\partial < [\Phi_s]_* >}{\partial \dot{\alpha}_j} \right) \dot{x}_j + \left(\frac{\partial < [\Phi_s]_* >}{\partial \alpha_j} \right) x_j \right\}$$

$$(s = 1, \ldots, k) \qquad (4.2.32)$$

where the round brackets, in which the derivatives of the functions $< [\Phi_s] >$ are enclosed, indicate that these derivatives are calculated at $\alpha_j = \alpha_j^0 = \text{const}$. At $\mu = 0$ the characteristic equation which corresponds to equations (4.2.32), has k negative roots $\lambda_s^0 = -k_s/I_s$ $(s = 1, \ldots, k)$ and k zero roots $\lambda_{k+s}^0 = 0$, therefore the resolution concerning the question of stability will depend on the subsequent approximations to those latter roots. Assuming in the indicated characteristic equation $\lambda = \lambda^0_{k+s} + \mu\kappa' = \mu\kappa'$, we will arrive at the following algebraic equation of the k-degree in order to determine the values $\kappa = \mu/\kappa$:

$$\left| \frac{\sigma_s}{k_s} \frac{\partial < \mu[\Phi_s] >}{\partial\alpha_s} - \delta_{sj}\kappa \right| = 0 \quad (s, j = 1, \ldots, k). \tag{4.2.33}$$

Here it has been taken into account that according to the designations that have been adopted

$$\left(\frac{\partial < [\Phi_s]_* >}{\partial\alpha_j} \right) = \frac{\partial < [\Phi_s] >}{\partial\alpha_j}.$$

If all the roots of equation (4.2.33) have negative real parts, then the motion under consideration is stable. If there is at least one root with a positive real part, it is unstable. Cases of the zero- or purely imaginary roots require an additional investigation. But the equations (4.2.31) and (4.2.33) coincide with equations (4.2.7) and (4.2.8) respectively, which means that the results, obtained by both methods, also coincide; and the supplementary remarks, given above, are also correct.

If there exists a potential of the averaged non-conservative forces, i.e. if there is a function $B = B(\alpha_1, \ldots, \alpha_k)$, satisfying relation (4.2.12), then the potential function D is, as before, defined by expressions (4.2.13). It is easy to see that if there exists a function B_* (not necessarily depending only on $\alpha_1, \ldots, \alpha_k$) such that

$$\frac{\partial B_*}{\partial\alpha_s} = \sigma_s < [Q_{\varphi_s}]_* > + \sum_{r=1}^{\nu} < [Q_{u_r}]_* \frac{\partial u_r^0}{\partial\alpha_s} >, \tag{4.2.34}$$

then the equations of slow motions can be written as

$$I_s\ddot{\alpha}_s + k_s\dot{\alpha}_s = -\frac{\partial D_*}{\partial\alpha_s} \quad (s = 1, \ldots, k) \tag{4.2.35}$$

where

$$D_* = -(\Lambda_* + B_*). \tag{4.2.36}$$

We will note that D_* is not a potential function, since generally speaking it depends not only on $\alpha_1, \ldots, \alpha_k$ but also on $\dot{\alpha}_1, \ldots, \dot{\alpha}_k$. It is also necessary to remember that the right sides of equations (4.2.35) were assumed to be sufficiently small.

Along with that if we bear in mind that according to (4.2.24), the functions $\dot{\alpha}_s$ are contained in all the expressions of the right sides of the equations only in the combination $n_s\omega + \dot{\alpha}_s$, then when the relation $\dot{\alpha}_s \ll \omega$ is satisfied, we can assume that

$$u^0_{r_*} \approx u^0_r,\ \Lambda_* \approx \Lambda,\ B_* \approx B,\ [Q_{\varphi_s}]_* \approx [Q_{\varphi_s}],\ [Q_{u_r}]_* \approx [Q_{u_r}], \qquad (4.2.37)$$

and then the equations of slow motions (4.2.29) and (4.2.35) will be written as

$$I_s\ddot{\alpha}_s + k_s\dot{\alpha}_s = \frac{\partial\Lambda}{\partial\alpha_s} + <\sigma_s[Q_{\varphi_s}]> + \sum_{r=1}^{\nu} <[Q_{u_r}]\frac{\partial u^0_r}{\partial\alpha_s}>, \qquad (4.2.38)$$

$$I_s\ddot{\alpha}_s + k_s\dot{\alpha}_s = -\frac{\partial D}{\partial\alpha_s} \qquad (4.2.39)$$

where $D = D_*|_{\dot{\alpha}_1=\dot{\alpha}_2=...=\dot{\alpha}_k=0} = D(\alpha_1,\ldots,\alpha_k)$ is the potential function, defined as before by expression (4.2.13).

In conclusion it should be noted that the proof of the integral sign of the stability of synchronous motions of the systems under consideration, based on the use of the variational relation, was given by Lurie [347].

In connection with problems on the synchronization of mechanical vibro-exciters in quasilinear oscillatory systems, Hodzhaev and Sperling [265, 506, 507] proposed expressions for the averaged Lagrangian via *harmonic coefficients of influence* (see also [90, 103]) which are quite convenient for practical use. A geometrical form of the integral criterion of stability, also convenient when solving the indicated applied problems, was suggested by Lavrov [321]. Some results of the cited publications are considered in chapter 7.

4.2.2   *Systems of Quasiconservative Objects, Canonical Systems; Theorem 9*

In the work by Nagaev [395] (1965) the integral sign of stability was obtained by the Poincare-Lyapunov method for the system of weakly connected quasi-conservative objects; it was shown that the expression for the potential function has the form (see also [90, 103])

$$D = -(\Lambda + B)\sigma \qquad (4.2.40)$$

where

$$\sigma = \text{sgn}\, e_s(\omega_s),\quad e_s(\omega_s) = \frac{1}{\omega_s}\frac{dh_s}{d\omega_s}; \qquad (4.2.41)$$

ω_s is the frequency of the isolated purely conservative s-th object, $h_s(\omega_s)$ is its energy constant. It is remarkable that the sign in expression (4.2.40) is determined by the sign of the values $e_s(\omega)$. Depending on that sign Nagaev distinguishes the following: the *rigidly anisochronous objects* ($e_s > 0,\ \sigma = 1$);

as an example we can mention the rotating rotors - see chapter 7); the *softly anisochronous objects* ($e_s < 0$, $\sigma = -1$ as an example we can mention point masses, revolving round a fixed center under the action of gravitation — see chapter 20); the *isochronous objects* ($e_s = \infty$ as an example we can mention the linear oscillators, for which ω_s does not depend on h_s, so that $d\omega_s/dh_s = 0$). The expression (4.2.40) is valid provided the type of anisochronism of all the objects is the same.

For systems with the quasilinear carrying bonds under rather general assumptions about the type of the carried bonds, expression (4.2.40) can be presented as

$$D = (\Lambda^{(I)} - \Lambda^{(II)} - B)\sigma. \qquad (4.2.42)$$

Here $\Lambda^{(I)}$ and $\Lambda^{(II)}$, just like in (4.2.22), are the averaged Lagrangians of systems of the carrying and carried bonds respectively, calculated in the generating approximation.

In the problem on the synchronization of quasiconservative objects, conditions of stability, expressed by the integral sign, are but roughly necessary (see remark 4 in subsection 4.2.1). Sufficient conditions have been obtained by Nagaev and Khodzhaev [395, 397, 398, 265].

By using the asymptotic methods and the method of integral manifolds the results of the work by Nagaev [395] were generalized by Gurtovnik and Neimark in [228] for the case of the so called incomplete synchronism (the corresponding notion is formulated in the same article).

Formulas (4.2.40) and (4.2.42) refer to the case of essentially anisochronous objects. As was shown by Nagaev in the same work [395], the corresponding integral sign of stability can also be formulated in the case of almost isochronous objects, which requires a special consideration. For the same case of almost isochronous objects — quasilinear oscillators — the integral sign was also proved by Valeev and Ganiev in an article [555]. The authors believe that the Lagrangian of the system has the following form

$$L(q, \dot{q}, t, \mu) = \frac{1}{2} \sum_{j=1}^{n} (\dot{q}_j^2 - \omega_j^2 q_j^2) + \mu l_1(q, \dot{q}, t, \mu) \qquad (4.2.43)$$

where $\mu > 0$ is a small parameter; l_1 is a periodic function of time t with a period $T_1 = 2\pi/\omega$. The motion which is near to that of the resonance is being considered here: $\omega_j^2 - \nu_j^2 = O(\mu)$, $\nu_j = \omega p_j/N$ $(j = 1, \ldots, n)$ where $p_j (j = 1, \ldots, n)$ and N are positive integers. Equations of motion have the following form:

$$\ddot{q}_r + \nu_r^2 q_r = -\mu \left(\frac{d}{dt} \frac{\partial l_2}{\partial \dot{q}_r} - \frac{\partial l_2}{\partial q_r} \right),$$

$$l_2(\dot{q}, q, t, \mu) = l_1(\dot{q}, q, t, \mu) + \frac{1}{2\mu} \sum_{j=1}^{n} q_j^2 (\nu_j^2 - \omega_j^2). \qquad (4.2.44)$$

The generating solution of equations (4.2.44)

$$q_r^0 = a_r \cos \nu_r t + \frac{b_r}{\nu_r} \sin \nu_r t \quad (r = 1, \ldots, n) \qquad (4.2.45)$$

(a_r and b_r being the initial values of q_r^0 and $\dot{q}_r^0$ respectively) is periodical in respect to t with a period $T = 2\pi N/\omega$.

On introducing the function $\Lambda(a, b)$ representing the mean value of the Lagrangian (4.2.43) along the generating periodic solution (4.2.45):

$$\Lambda(a, b) = \frac{1}{T} \int\limits_0^T L[\dot{q}^0(t), q^0(t), t, \mu] dt, \qquad (4.2.46)$$

the authors come to the following statement.

T h e o r e m 9. *If the function $\Lambda(a, b)$ has in the point $a_1 = a_1^0, \ldots, a_n = a_n^0; \quad b_1 = b_1^0, \ldots, b_n = b_n^0$ a minimum or a maximum, then that point determines a stable according to the first approximation periodic solution; other stationary points of the function $\Lambda(a, b)$ require a special consideration.*

That statement has been obtained in a very original way, by the use of the finite-difference equations.

It should be noted that according to the proof given in [555] the words "stable according to the first approximation" imply here in accordance with the adopted terminology (see section 4.1) only *the stability in the first approximation with respect to the small parameter*; to solve the question of stability according to the first approximation in the usual sense of the term it is necessary to investigate the roots of the characteristic equation with an accuracy to higher degrees of μ.

It was shown by Malakhova [111, 361] that the sign of Valeyev and Ganiyev could also be obtained by the use of the theorem of Malkin [363], based on the Poincare-Lyapunov methods. But unlike the formulation given above, here an additional requirement appears that the extrema of the function $\Lambda(a, b)$ should be rough, which is conditoned by the use of the Malkin theorem.

Without giving any proof, the authors of [555] state that the sign of stability, considered here, refers also to the general case of canonical systems with the Hamilton function which is an almost-periodic function of the time t.

Beletsky [55] brought forth, as a hypothesis, a *principle of the extremality of resonance-* (according to the above terminology) — *synchronous motions* in the problem on the planar rotation of a celestial body in respect to its center of mass, moving in the elliptical orbit around a fixed center. He also introduced the term *extremal properties of resonance motions*. This principle consists in the following.

Let $U(\theta, t)$ be a force function of the system, with θ being the angle of deviation of the axis of inertia of the body from the radius-vector of the orbit. The average, according to time, value of the force function is introduced

$$< U >= \overline{U}(\dot{\theta}_0, \theta_0) = \lim_{t \to \infty} \frac{1}{t} \int_0^t U[\theta(\dot{\theta}_0, \theta_0, t), t] dt. \qquad (4.2.47)$$

According to the indicated hypothesis, the limit (4.2.47) exists and reaches its maximum on the set of the initial data $\dot{\theta}_0$, θ_0, answering the stable resonance motions. As an evidence in favor of the principle suggested in [55] the author gives the results of the numerical experiment.

Later a theorem was proved by Beletsky and Kasatkin which confirmed in a general form the idea of the existence of extremal properties in the synchronous (resonance) motions [57] (1980). The authors considered a periodic system, somewhat more general than the canonical one (the system with the phase volume retained). They showed that for the presence of the stable periodic or synchronous motions it is necessary and sufficient that there exist the function $K(\mathbf{x}_0)$ of the initial values $\mathbf{x}_0$ of the phase variables $\mathbf{x}$, having a strict maximum or minimum with respect to these initial values. Then the indicated function is connected with a certain function of the phase coordinates and time $\kappa(\mathbf{x}, t)$ by an integral relationship of type (4.2.47). But the question of the way of finding the functions κ and K remains in this case open. It should be noted that according to the above stated hypothesis of Beletsky, these functions are respectively U and $< U >$.

It is also necessary to remark that the theorem of Beletsky and Kasatkin [57] is in good agreement with the results, obtained by Hapayev and Shinkin [235, 493].

In the work of Kozlov [292] it is mentioned that the averaging according to (4.2.47) along the exact non-degenerate solutions in the general case differs from the averaging along the degenerate (that is depending on a certain number of parameters) solutions, discussed in the other cases, considered above. The result of averaging along the degenerate solutions is affected by the choice of the function which is being averaged; while on the non-generate stable in the linear approximation periodic solutions for any periodical continuously differentiated function

$$\lim_{T \to \infty} \frac{\partial}{\partial \mathbf{x}_0} \left\{ \frac{1}{T} \int_0^T \kappa[\mathbf{x}(t, \mathbf{x}_0), t] dt \right\} = 0;$$

in particular, if $K(\mathbf{x}_0)$ exists and is a continuously differentiated function, then $K'_{\mathbf{x}_0} = 0$ that is $\mathbf{x}_0$ is a stationary point of the function $K(\mathbf{x}_0)$. On the other hand, this statement, proved by Kozlov, does not contradict the theorem of Beletsky and Kasatkin [57] which speaks about the existence of the function $\kappa(\mathbf{x}, t)$ and according to which $\mathbf{x}_0$ is not only a stationary point of the function $K(\mathbf{x}_0)$ but also represents a point of the minimum or maximum of this function, that theorem providing both the necessary and sufficient conditions of the existence of stable, periodic solutions.

4.2.3   Systems with Quasicyclic Coordinates; Theorem 10

In [266] Khodzhaev obtained an integral sign of stability for the systems with the quasicyclic coordinates [f]

$$\dot{p}_r = Q_r \qquad (r = 1, \ldots, m), \tag{4.2.48}$$

$$E_{m+r}(L) = Q_{m+r} \qquad (r = 1, \ldots, n - m) \tag{4.2.49}$$

where, just like before, $E_s(\cdot)$ is the Eiler operator, corresponding to the generalized coordinate q_s, the function

$$L = T(q_{m+1}, \ldots, q_n; \dot{q}_1, \ldots, \dot{q}_n) - \Pi(q_{m+1}, \ldots, q_n)$$

is the Lagrange function; μ is a small parameter; $p_r = \partial T/\partial \dot{q}_r$ $(r = 1, \ldots, m)$ are the quasicyclic impulses; $Q_{m+r}(q_{m+1}, \ldots, q_n; \dot{q}_1, \ldots, \dot{q}_n)$ $(r = 1, \ldots, n - m)$ are the generalized non-conservative forces answering the positional coordinates. The generalized non-conservative forces, corresponding to the quasicyclic coordinates, are supposed to be presented as

$$Q_r = U_r(t) + \mu f_r - \mu h_r \dot{q}_r, \quad U_r(t + \frac{2\pi}{\omega}) = U_r(t), \ < U_r >= 0 \ (r = 1, \ldots, m)$$

---

[f] The definition of the quasicyclic coordinates is given in 2.2.3. The term *quasicyclic coordinates* denotes the coordinates which do not enter explicitly into the expression for the kinetic energy of the system, but which are answered by the generalized forces, different from zero.

with f_r, h_r being constants ($h_r > 0$).

The generating equations, answering equations (4.2.48), have a set of $2\pi/\omega$-periodic solutions which depends on m arbitrary constants $\alpha_1, \ldots, \alpha_m$

$$p_r^0 = \alpha_r + V_r(t) \quad (r = 1, \ldots, m) \tag{4.2.50}$$

where $\dot{V}_r = U_r$, $< V_r > = 0$. Let us assume that equations (4.2.49), corresponding to solution (4.2.50), (see the footnote on p. 73)

$$[E_{m+r}(L_R) - Q_{m+r}] = 0 \quad (r = 1, \ldots, n - m) \tag{4.2.51}$$

where

$$L_R = (T - \sum_{r=1}^{m} p_r \dot{q}_r)|_{\dot{q}_r = \dot{q}_r(p_1, \ldots, p_m; q_{m+1}, \ldots, q_n; \dot{q}_{m+1}, \ldots, \dot{q}_n)} - \Pi$$

is the kinetic potential of Rauss, have at any $\alpha_1, \ldots, \alpha_m$ the asymptotically stable $2\pi/\omega$-periodic isolated solution

$$q_{m+r} = q_{m+r}^0(\alpha_1, \ldots, \alpha_m) \quad (r = 1, \ldots, n - m). \tag{4.2.52}$$

Further on let us assume that the following relations are satisfied

$$\sum_{s=1}^{n-m} \left\langle \left[Q_{m+s} \frac{\partial q_{m+s}}{\partial \alpha_r} \right] \right\rangle = 0 \quad (r = 1, \ldots, m). \tag{4.2.53}$$

Under the formulated assumptions the following statement is valid:

T h e o r e m 10. *At the sufficiently small values of μ every point of rough minimum of the potential function*

$$D = D(\alpha_1, \ldots, \alpha_m) = - < [L_R] > - \sum_{r=1}^{m} \frac{f_r}{h_r} \sigma_r \tag{4.2.54}$$

is answered by the only asymptotically stable periodic[g] solution of system (4.2.48), (4.2.49) *which at $\mu = 0$ becomes a generating solution* (4.2.50), (4.2.52).

This extremal sign of stability is used by Khodzhaev in the theory of electromechanical systems [266, 267].

---

[g]Periodical are here all the generalized velocities and positional (but not quasicyclic) coordinates.

4.3   Systems with Kinematic Excitation of Vibration (the Minimax Sign of Stability); Theorem 11

The minimax sign of stability, suggested by Strizhak [522], can be formulated in the following way (here this formulation is given with a certain modification; see below). Let

$$T = \frac{1}{2} \sum_{m=1}^{n} \sum_{j=1}^{n} a_{1mj}(\mathbf{q}, \mathbf{u})\dot{q}_m \dot{q}_j + \frac{1}{2} \sum_{m=1}^{s} \sum_{j=1}^{s} a_{2mj}(\mathbf{q}, \mathbf{u})\dot{u}_m \dot{u}_j +$$

$$\sum_{m=1}^{n} \sum_{j=1}^{s} a_{3mj}(\mathbf{q}, \mathbf{u})\dot{q}_m \dot{u}_j \qquad (4.3.1)$$

be the kinetic energy, and let

$$\Pi(\mathbf{q}, \mathbf{u}) \qquad (4.3.2)$$

be the potential energy of the system, described by the $n + s$ generalized coordinates $q_1, \ldots, q_n$; $u_1, \ldots, u_s$, the latter being given as the final sums

$$u_j = \mu \sum_k u_{j,k}(q_1, \ldots, q_n, \mu)e^{i\nu_k \omega t} \qquad (j = 1, \ldots, s) \qquad (4.3.3)$$

where $\nu_k \neq 0$, $\nu_{-k} = -\nu_k$, $u_{j,-k} = -u_{j,k}$, $\omega = 1/\mu$, (μ being a small parameter). We will call systems of such type *systems with a kinematic excitation of vibration*.

Thus by the kinematic excitation of vibration we imply here such excitation of the system at which the law of oscillations of part of the generalized coordinates can be considered to be given. This seems to be more preferable than the way of introducing vibration, adopted in [522], though it does not lead to any change in the result.

Let us assume that the forces of viscous friction R_r $(r = 1, \ldots, n)$, answering the variables $q_1, \ldots, q_n$, have the form

$$R_r = \sum_{j=1}^{n} \beta_{rj} \dot{q}_j$$

where the matrix of the coefficients β_{rj} is positive [h]. It should be noted that in the system, considered in [522], the non-potential forces are absent; the case in point here is the formal stability of the positions of quasiequilibrium. The introduction of the dissipative forces makes it possible to formulate the sign of asymptotic stability.

---

[h] see the footnote on p. 21

By *quasiequilibria* of the system we will imply the almost-periodic motions of the type $q_j = q_j^0 + \mu\psi_j(\omega t)$ $(j = 1, \ldots, n)$ where q_j^0 are constants , and $< \psi_j(\omega t) > = 0$, that is motions, representing small high-frequency oscillations near the position $q_j = q_j^0$. We can also say that the quasiequilibria correspond to the positions of equilibrium for the slow components of motion (see below).

Then the following theorem is valid:

T h e o r e m 11. *The minimax sign of stability. If in a certain point* $q_1 = q_1^0, \ldots, q_n = q_n^0$ *the function* $< \min\limits_{\dot{q}} L(\dot{q}, q, t) >$ *where* $L(\dot{q}, q, t)$ *is a Lagrangian of the system, made up in view of expressions (4.3.3), has a rough maximum, then this point at sufficiently small values of* μ *is answered by the asymptotically stable quasiequilibrium of the system.*

The minimax sign of stability was established by Strizhak by means of the asymptotic method for the canonical systems [522, 523] (1981). Later on this sign was proved by Malakhova in two other ways — by the Poincare-Lyapunov method (by means of using Malkin's theorem [363], mentioned in 4.2.2) and by method of direct separation of motions, when the minimax sign of stability acts as a criterion of stability of a certain potential on the average system. We will give here the latter proof, omitted in [120].

Substituting expressions (4.3.3) into the expressions for the kinetic and potential energy (4.3.1) and (4.3.2), we obtain

$$L(\mathbf{q}, \mathbf{q}, t) = T(\mathbf{q}, \mathbf{q}, t) - \Pi(\mathbf{q}, t) = \frac{1}{2}\sum_{m=1}^{n}\sum_{j=1}^{n}[a_{mj}(\mathbf{q})$$

$$+\mu\sum_{k} a_{mj,k}(\mathbf{q})e^{i\nu_k\omega t}]\dot{q}_m\dot{q}_j + \sum_{m=1}^{n}\sum_{k}b_{m,k}(\mathbf{q})e^{i\nu_k\omega t}\dot{q}_m$$

$$-\Pi(\mathbf{q},0) + C_0(\mathbf{q}) + \sum_{\substack{k \\ (k \neq -p)}}\sum_{p}C_{kp}(\mathbf{q})e^{i(\nu_k+\nu_p)\omega t} + \mu R \qquad (4.3.4)$$

where

$$a_{mj} = a_{1mj}(\mathbf{q},0), \quad a_{mj,k} = \sum_{v=1}^{s}\frac{\partial a_{1mj}}{\partial u_v}\bigg|_{u=0}u_{v,k},$$

$$b_{m,k} = \sum_{j=1}^{s}a_{3mj}(\mathbf{q},0)u_{j,k}i\nu_k, \qquad (4.3.5)$$

$$C_0 = \frac{1}{2}\sum_{m=1}^{s}\sum_{j=1}^{s}a_{2mj}(\mathbf{q},0)\sum_{k}u_{m,k}u_{j,-k}\nu_k^2,$$

$$C_{kp} = -\frac{1}{2}\sum_{m=1}^{s}\sum_{j=1}^{s}a_{2mj}(\mathbf{q},0)u_{m,k}u_{j,p}\nu_k\nu_p,$$

$$R = \frac{1}{2}\sum_{m=1}^{s}\sum_{j=1}^{s}\sum_{k}e^{i\nu_k\omega t}\sum_{v=1}^{s}\frac{\partial a_{2mj}}{\partial u_v}\bigg|_{u=0}u_{v,k}\sum_{p}u_{m,p}i\nu_p e^{i\nu_p\omega t}$$

$$\times\sum_{w}u_{j,w}i\nu_w e^{i\nu_w\omega t} - \sum_{m=1}^{s}\frac{\partial\Pi}{\partial u_m}\bigg|_{u=0}\sum_{k}u_{m,k}$$

$$\times e^{i\nu_k\omega t} + \sum_{m=1}^{n}\sum_{j=1}^{s}\sum_{k}e^{i\nu_k\omega t}\sum_{\nu=1}^{s}\frac{\partial a_{3mj}}{\partial u_\nu}\bigg|_{u=0}u_{\nu,k}$$

$$\times\sum_{p}u_{j,p}i\nu_p e^{i\nu_p\omega t}\dot{q}_m + O(\mu). \tag{4.3.6}$$

Expression (4.3.4) is written down provided the values $u_{j,k}$ do not depend on the variables $q_1,\ldots,q_n$ or the parameter μ. In case $u_{j,k} = u_{j,k}(q,\mu)$ the structure of the Lagrangian (4.3.4) is reserved, only the expressions for $a_{mj,k}$ and R will change, which, as will be seen later, will not affect the final result.

The equations of motion of the system can also be presented as

$$\sum_{j=1}^{n}(a_{rj} + \mu\sum_{k}a_{rj,k}e^{i\nu_k\tau})q_j'' + \mu\sum_{j=1}^{n}\sum_{k}a_{rj,k}i\nu_k e^{i\nu_k\tau}q_j'$$

$$+\mu\sum_{k}b_{r,k}i\nu_k e^{i\nu_k\tau} + \mu\sum_{j=1}^{n}q_j'\sum_{k}e^{i\nu_k\tau}\left(\frac{\partial b_{r,k}}{\partial q_j} - \frac{\partial b_{j,k}}{\partial q_r}\right)$$

$$+\sum_{j=1}^{n}\sum_{m=1}^{n}\left(\frac{\partial a_{r,j}}{\partial q_m} - \frac{1}{2}\frac{\partial a_{m,j}}{\partial q_r}\right)q_j'q_m' + \mu^2\frac{\partial\Pi(\mathbf{q},0)}{\partial q_r} - \mu^2\frac{\partial C_0}{\partial q_r} + \mu^2 H_r^{(0)}$$

$$+\mu^2\sum_{j=1}^{n}(H_{jr}^{(1)}q_j' + H_{jr}^{(2)}q_j'') + \sum_{j=1}^{n}\sum_{m=1}^{n}(\mu U_{jmr} + \mu^2 V_{jmr})q_j'q_m'$$

$$+\mu\sum_{j=1}^{n}\beta_{rj}q_j' + O(\mu^3) = 0 \quad (r = 1,\ldots,n). \tag{4.3.7}$$

Here $H_r^{(0)}$, $H_{jr}^{(1)}$, $H_{jr}^{(2)}$, U_{jmr}, V_{jmr} are the known functions of the fast time $\tau = \omega t$ and of the generalized coordinates $q_1,\ldots,q_n$, $\ <H_r^{(0)}> \ = 0$; the differentiation in respect to τ is marked by a prime. Following the method of direct separation of motions, we will seek the solution of system (4.3.7) in the form

$$q_j = Q(\mu\tau) + \mu\psi_j(\tau) \tag{4.3.8}$$

where $Q_j = Q_j(\mu\tau) = Q_j(t)$ are the slow components of motion, and $\mu\psi_j(\tau)$ are the fast components, satisfying the condition $< \psi_j(\tau) >= 0$. With an accuracy to the members of the first order of smallness with respect to μ, which corresponds to the *purely inertial approximation* (see 3.2.5), the following equation of the respectively fast and slow motions can be obtained:

$$\sum_{j=1}^{n} a_{rj}(\mathbf{Q})\psi_j'' + \sum_k b_{r,k} i\nu_k e^{i\nu_k\tau} = 0 \quad (r = 1, \ldots, n), \qquad (4.3.9)$$

$$\sum_{j=1}^{n} a_{rj}(\mathbf{Q})Q_j'' + \sum_{j=1}^{n}\sum_{m=1}^{n}\left(\frac{\partial a_{rj}}{\partial Q_m} - \frac{1}{2}\frac{\partial a_{mj}}{\partial Q_r}\right)Q_j'Q_m' + \mu\sum_{j=1}^{n} b_{rj}Q_j'$$

$$+\mu^2\left\{\left\langle\left(\sum_{j=1}^{n}\sum_{m=1}^{n}\frac{\partial a_{rj}}{\partial Q_m}\psi_m\psi_j'' + \sum_{j=1}^{n}\sum_k a_{rj,k}(\mathbf{Q})e^{i\nu_k\tau}\psi_j''\right.\right.$$

$$+\sum_{j=1}^{n}\sum_k a_{rj,k}(\mathbf{Q})i\nu_k e^{i\nu_k\tau}\psi_j' + \sum_{j=1}^{n}\sum_k \frac{\partial b_{r,k}}{\partial Q_j}i\nu_k e^{i\nu_k\tau}\psi_j$$

$$+\sum_{j=1}^{n}\sum_k \psi_j' e^{i\nu_k\tau}\left(\frac{\partial b_{r,k}}{\partial Q_j} - \frac{\partial b_{j,k}}{\partial Q_r}\right) + \sum_{j=1}^{n}\sum_{m=1}^{n}\left(\frac{\partial a_{rj}}{\partial Q_m} - \frac{1}{2}\frac{\partial a_{mj}}{\partial Q_r}\right)\psi_j^i\psi_m^i\right\rangle$$

$$+\frac{\partial\Pi(\mathbf{Q},0)}{\partial Q_r} - \frac{\partial C_0}{\partial Q_r}\right\} = 0(r = 1, \ldots, n). \qquad (4.3.10)$$

It should be noted that the matrix of the coefficients $a_{rj}(\mathbf{Q})$ ($\mathbf{Q}$ being the vector of the variables $Q_1, \ldots, Q_n$) is positive. From equations (4.3.9) fast motions are found

$$\psi_r = -\sum_{j=1}^{n} a_{rj}^{-1}(\mathbf{Q})\sum_k \frac{b_{j,k}(\mathbf{Q})}{i\nu_k}e^{i\nu_k\tau} \quad (r = 1, \ldots, n) \qquad (4.3.11)$$

which, evidently, satisfy the relations

$$< \psi_j''\psi_m + \psi_j'\psi_m' >= 0, \quad < i\nu_k e^{i\nu_k\tau}\psi_j' + e^{i\nu_k\tau}\psi j'' >= 0$$

$$< i\nu_k e^{i\nu_k\tau}\psi_j + e^{i\nu_k\tau}\psi j' >= 0$$

$$< \sum_{j=1}^{n}\psi_j'\sum_k e^{i\nu_k\tau}\frac{\partial b_{jk}}{\partial Q_r} + \frac{1}{2}\sum_{j=1}^{n}\sum_{m=1}^{n}\frac{\partial a_{mj}}{\partial Q_r}\psi_j'\psi_m' >$$

$$= -\frac{\partial}{\partial Q_r}\sum_{j=1}^{n}\sum_{m=1}^{n}a_{jm}^{-1}(Q)\sum_k b_{j,k}b_{m,-k}. \qquad (4.3.12)$$

$(a_{rj}^{-1}$ are the elements of the matrix, inverse to that of the coefficients a_{rj}). After the transfer to the slow time t the equations of slow motions (4.3.10) (the main equation of vibrational mechanics) will be written in the form, corresponding to equations (4.1.3):

$$\sum_{j=1}^{n} a_{rj}(\mathbf{Q})\ddot{Q}_j + \sum_{j=1}^{n}\sum_{m=1}^{n} \left(\frac{\partial a_{rj}}{\partial Q_m} - \frac{1}{2}\frac{\partial a_{mj}}{\partial Q_r} \right) \dot{Q}_j\dot{Q}_m$$

$$= -\frac{\partial D}{\partial Q_r} - \sum_{j=1}^{n}\beta_{rj}(\dot{Q}_j) \qquad (r = 1,\ldots,n). \qquad (4.3.13)$$

Here

$$D = \Pi(\mathbf{Q},0) - C_0(\mathbf{Q}) + \sum_{j=1}^{n}\sum_{m=1}^{n} a_{jm}^{-1}(\mathbf{Q})\sum_{k} b_{j,k}(\mathbf{Q})b_{m,-k}(\mathbf{Q}). \qquad (4.3.14)$$

It is easy to make sure [361, 522] that expression (4.3.14) for the potential function presents an averaged minimum with respect to the variables $\dot{Q}_1,\ldots,\dot{Q}_n$ of the Lagrangian (4.3.4), calculated with an accuracy to the values of the order of μ, taken with the opposite sign, that is

$$D = D(\mathbf{Q}) = - < \min_{\mathbf{Q}} L(\dot{\mathbf{Q}}, \mathbf{Q}, t) > . \qquad (4.3.15)$$

In the points of the maximum of the function $\varphi(\mathbf{Q}) = < \min_{\mathbf{Q}} L(\dot{\mathbf{Q}}, \mathbf{Q}, t) >$ the potential function $D(\mathbf{Q})$ has a minimum; consequently these points are answered by the stable positions of the quasiequilibrium of the initial system, as it should be according to the criterion of Strizhak.

The criterion of stability under consideration has been formulated above as sufficient. Mind that if the coordinates $u_1,\ldots,u_s$ are periodic with respect to t with a certain period $T = O(\mu)$ and have with respect to t the continuous derivatives of the second order, then the conditions, following from the minimax sign, are also necessary in the sense that in case of the absence in the point $Q_1 = Q_1^0,\ldots,Q_n = Q_n^0$ of the minimum of the function D and provided the characteristic equation of the system describing the slow motions near that point has no roots with the zero real parts, the corresponding periodic solution of the initial system is unstable [363].

It is quite remarkable that to find the positions of quasiequilibrium by means of the sign under consideration it is not necessary to make up and solve the equations of motion of the system; it is enough just to make up an expression for the Lagrangian L (see also 5.1 and 5.2).

4.4   Systems with Dynamic Excitation of Vibration ; Theorem 12

By *systems with dynamic excitation of vibration* we mean the systems whose motion is described by the equations

$$\sum_{j=1}^{n} a_{rj}(\mathbf{q})\ddot{q}_j + \sum_{j=1}^{n}\sum_{m=1}^{n}\left(\frac{\partial a_{rj}}{\partial q_m} - \frac{1}{2}\frac{\partial a_{mj}}{\partial q_r}\right)\dot{q}_j\dot{q}_m$$

$$+\sum_{j=1}^{n}\beta_{rj}\dot{q}_j = -\frac{\partial\Pi}{\partial q_r} + \frac{1}{\mu}\sum_k f_{r,k}(\mathbf{q},\mu)e^{i\nu_k\omega t} \quad (r=1,\dots,n) \quad (4.4.1)$$

where $\nu_k \neq 0$, $\nu_{-k} - \nu_k f_{r,-k} = \overline{f}_{r,k}$; $\Pi(\mathbf{q},\mu)$ is the potential energy of the system; $\omega = 1/\mu$ (μ being a small parameter); the matrix of the inertial coefficients $a_{rj(q)}$ is assumed to be positively definite; $\sum_{j=1}^{n}\beta_{rj}\dot{q}_j$ $(r=1,\dots,n)$ are forces of viscous friction with a full dissipation.

We assume that the following relations are satisfied

$$\left.\frac{\partial f_{m,k}}{\partial q_r}\right|_{\mu=0} = \left.\frac{\partial f_{r,k}}{\partial q_m}\right|_{\mu=0} \quad (m=1,\dots,n; \ r=1,\dots,n). \quad (4.4.2)$$

Under the formulated conditions the following theorem is correct

T h e o r e m 12. *If in a certain point $q_1 = q_1^0, \dots, q_n = q_n^0$ the function*

$$D = \Pi|_{\mu=0} + \Pi_V \quad (4.4.3)$$

where

$$\Pi_V = \frac{1}{2}\sum_{j=1}^{n}\sum_{m=1}^{n}a_{jm}^{-1}(\mathbf{q})\sum_k\frac{f_{j,k}(q,0)f_{m,-k}(q,0)}{\nu_k^2} \quad (4.4.4)$$

has a rough minimum, then at the sufficiently small values of μ this point is answered by the asymptotically stable quasiequilibrium of the system (4.4.1).

Here the elements of the matrix inverse to that of the coefficients a_{jm} are designated by a_{jm}^{-1}.

Just like in the case of kinematic excitation, Malakhova considered the question of existence and stability of the quasiequilibrium system (4.4.1) in two ways — by means of the Malkin theorem and by method of direct separation of motions [120, 361] (1990). Below we give the proof by means of the latter method, omitted in [120].

System (4.4.1) can be written as (about the character and smoothness of the functions, considered below, see 4.2.1)

$$\sum_{j=1}^{n}a_{rj}(\mathbf{q})q_j'' + \sum_{j=1}^{n}\sum_{m=1}^{n}\left(\frac{\partial a_{rj}}{\partial q_m} - \frac{1}{2}\frac{\partial a_{mj}}{\partial q_r}\right)q_j'q_m'$$

$$= \mu \sum_{k} f_{r,k}(\mathbf{q}, \mu) e^{i\nu_k \tau} - \mu \sum_{j=1}^{n} \beta_{rj} q'_j - \mu^2 \frac{\partial \Pi}{\partial q_r} \quad (r = 1, \ldots, n) \quad (4.4.5)$$

where the differentiation with regard to the fast time $\tau = \omega t$ is marked by a prime. In accordance with the method of direct separation of motions we will assume that the solution of system (4.4.5) looks as (4.3.8) and will pass to the equations for the fast and slow components of motion. In a purely inertial approximation when solving the equations of fast motions, we will obtain

$$\psi_r = -\sum_{j=1}^{n} a_{rj}^{-1}(\mathbf{Q}) \sum_{k} \frac{f_{j,k}(\mathbf{Q}, 0)}{\nu_k^2} e^{i\nu_k \tau}, \quad (4.4.6)$$

and then the equations of slow motions will have the form

$$\sum_{j=1}^{n} a_{rj}(\mathbf{Q}) Q'' + \sum_{j=1}^{n} \sum_{m=1}^{n} \left(\frac{\partial a_{rj}}{\partial Q_m} - \frac{1}{2} \frac{\partial a_{mj}}{\partial Q_r} \right) Q'_j Q'_m$$

$$+ \mu \sum_{j=1}^{n} \beta_{rj} Q'_j + \mu^2 \left(-V_r + \frac{\partial \Pi}{\partial Q_r} \Big|_{\mu=0} \right) = 0 \quad (r = 1, \ldots, n) \quad (4.4.7)$$

where

$$V_r = -\frac{\partial}{\partial Q_r} \left(\frac{1}{2} \sum_{j=1}^{n} \sum_{m=1}^{n} a_{jm}^{-1}(\mathbf{Q}) \sum_{k} \frac{f_{j,k}(\mathbf{Q}, 0) f_{m,-k}(\mathbf{Q}, 0)}{\nu_k^2} \right)$$

$$+ \left\langle \sum_{j=1}^{n} \sum_{k} \left(\frac{\partial f_{m,k}}{\partial Q_r} \Big|_{\mu=0} - \frac{\partial f_{r,k}}{\partial Q_m} \Big|_{\mu=0} \right) \psi_m e^{i\nu_k \tau} \right\rangle \quad (r = 1, \ldots, n), \quad (4.4.8)$$

just as before, are vibrational forces. In a general case vibrational forces (4.4.8) may have no potential. However, if conditions (4.4.2) are satisfied, the potential energy Π_V of vibrational forces (4.4.8) exists, and expression (4.4.4) is valid for it.

After the transfer to the slow time t the equations of slow motions (4.4.7) (the main equations of vibrational mechanics) will take the form

$$\sum_{j=1}^{n} a_{rj}(\mathbf{Q}) \ddot{Q}_j + \sum_{j=1}^{n} \sum_{m=1}^{n} \left(\frac{\partial a_{rj}}{\partial Q_m} - \frac{1}{2} \frac{\partial a_{mj}}{\partial Q_r} \right) \dot{Q}_j \dot{Q}_m$$

$$+ \sum_{j=1}^{n} \beta_{rj} \dot{Q}_j = -\frac{\partial D}{\partial Q_r} \quad (r = 1, \ldots, n) \quad (4.4.9)$$

where the function D is defined by relation (4.4.3). Thus, provided conditions (4.4.2) are satisfied, then just as before, the points of rough maximum of the potential function D will be answered by the stable positions of the quasiequilibrium of the system.

If in the equations (4.4.1) instead of the final sums in the right sides there are certain functions $f_r(q, \mu, t)$, periodic with respect to t with a period $T = O(\mu)$, then, provided the characteristic equation of the system, describing the slow motions near the points of quasiequilibrium, has no roots, with the zero real parts, this sign gives not only the sufficient, but also the necessary conditions of the asymptotic stability of the quasiequilibriums of the initial system.

It should be noted that the expression for the potential function, analogous to (4.4.3) was found by Landau and Lifshits in their well known course [313] (1958) in the process of solving problems on the behaviour of a particle in a one-dimensional fast-oscillating field. However, the generalization of the equation for the case of a system with many degrees of freedom, proposed by them, is valid only in case the additional conditions (4.4.2), providing the existence of the potential of vibrational forces, are satisfied. These conditions are most essential, for they require the realization of equalities with respect to the amplitudes of the harmonic components of the field. That circumstance was mentioned by Khodzhaev and Shatalov [491]. Practically simultaneously with the book [313], articles by Gaponov and Miller were published [193, 194], containing the expression obtained by them for the potential function in the problem on the motion of a charged particle in a three-dimensional fast-oscillating field.

Remarkable is the existence of the potential function, discovered by Vorovich [577] (1964) — "the potential energy of the amplitudes of stationary oscillations"— in the process of solving the problem on the oscillations of a round plate under the action of a random load by using the asymptotic methods.

It was shown by Beletsky and Golubitskaya [60] (1987) that in the model of a two-leg walking, investigated by them, the synchronous (resonance) regimes are answered by the minima of the functional, characterizing the energy input in the system. On that basis a number of essential periodic regimes of walking has been discovered.

Part II

Vibrational Mechanics of Machines, Mechanisms and Pendulum Devices

Chapter 5

Devices of Pendulum Type

5.1 Pendulum with a Vibrating Axis of Suspension

5.1.1 Brief Bibliographic Review

As is known, the vertical vibration of the axis of the pendulum may destabilize — make unstable — its lower position of equilibrium; it is also known that the vibrational effect may cause a loss of stability of the rectilinear elastic rod if the value of the longitudinal load is essentially less than the Euler critical value (see, say, [428, 584]). The publications [257, 258] (1951) which, as was mentioned, stimulated the development of the approach, presented in this book, Kapitsa obtained the condition at which there is an opposite effect: the upper position of the pendulum is stabilized by the vertical vibration of its axis. Kapitsa also refers to the previous work [247] (1950) where the solution was obtained in another, more complicated way. Though, apparently, the first investigation of the problem was published by Stephenson as far back as 1908 [515, 516, 41]; see also the work by Erdelyi [175] (1934). In the well known book by Bogolyubov [134] (1950) the problem is considered by asymptotic method with the help of the appropriate transformation of the variables.

Later this problem was used by a number of authors as a standard example when illustrating the efficiency of different methods of non-linear mechanics (see, say, [134, 137, 572, 390, 395]), and also it played the role of a model problem when studying the behavior of certain mechanisms under vibration [239, 273, 449, 499]. The method of Kapitsa was successfully used in the book by Ragulskis [449] (1963) for the investigation of such systems.

The solution of the problem on the behavior of the pendulum with a vibrating axis and of a number of similar systems is given in the books by Strizhak [521, 522, 523] (1981–1982) who used the asymptotic methods and also applied the mini-max sign of stability, proposed by her (see 4.3). Results of a thorough analytical and numerical investigation of different regimes (including the chaotic regimes) of motion of the pendulum are stated in the articles by Batalova, Belyakova, and Bukhalova (see, say, [51, 52, 53, 144]).

Besides the cited publications, extensive literature has been devoted to the investigation of the behavior of the pendulum with a vibrating axis. That is explained by both the fundamental and applied importance of this problem. The corresponding references can be found, say, in [52, 53, 557, 595, 521, 584, 585, 249, 542]. Nevertheless the problem is far from being exhausted: new interesting data are constantly being published (see, e.g.[176, 326]).

In this section we consider, under certain limitations, the problem on the behavior of a pendulum with an axis, vibrating in two mutually perpendicular directions according to the harmonic law with a certain frequency ω so that the trajectory of the oscillations is ellipse. Here we used the method of direct separation of motions in the form, stated in 3.2, and the mini-max sign of stability (3.3). Due to the peculiarities of those methods one has not been able to investigate the behavior of the pendulum all over the range of change of its parameters and explain certain effects. It has been possible, however, as will be shown later, to consider in a simple and obvious form most of the regularities of great practical importance.

5.1.2   Equation of Motion

Let the axis of suspension of a pendulum perform oscillations according to the law (Fig. 5.1)

$$x = H \sin \omega t, \quad y = G \cos(\omega t + \theta) \tag{5.1.1}$$

where G and H are the amplitudes of oscillations in the vertical and horizontal directions respectively, ω is the frequency, θ is the phase shift. Then the motion of the pendulum is described by the differential equation

$$I\ddot{\varphi} + h\dot{\varphi} + mgl \sin \varphi + ml\omega^2[H \cos \varphi \sin \omega t - G \sin \varphi \cos(\omega t + \theta)] = 0 \tag{5.1.2}$$

where φ is the angle of deviation of the pendulum from its lower vertical position, I, m and l are respectively the moment of inertia, the mass and the distance from the center of gravity C to the axis of suspension of the pendulum O, g is the free fall acceleration, and h is the coefficient of viscous resistance.

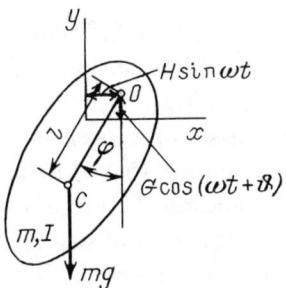

Figure 5.1. Pendulum with a vibrating axis of suspension.

5.1.3 *Solving the Problem by the Method of Direct Separation of Motions*

In accordance with what has been said in 3.2 we assume that the law of motion
of the pendulum under vibration has the form

$$\varphi = \alpha(t) + \psi(t, \omega t) \tag{5.1.3}$$

where $\alpha(t)$ is the main component, and $\psi(t, \omega t)$ is the fast 2π-periodic component in respect to ωt, the latter satisfying the condition

$$< \psi(t, \omega t) >= 0 \tag{5.1.4}$$

and in this case it is considered small as compared to α. Solving the problem
that interests us, we consider the moment of forces of the viscous resistance $k\dot{\varphi}$
and the moment of force of gravity $m\,g\,l \sin \varphi$ to be the slow components, while
the moment of the force of inertia, conditioned by the vibration of the axis of
the pendulum, is considered to be the fast component.

In this case equations (3.2.55), (3.2.56) have the form

$$I\ddot{\alpha} = -h\dot{\alpha} - m l \sin \alpha + V, \tag{5.1.5}$$

$$I\ddot{\psi} = \mathbf{\Psi}, \tag{5.1.6}$$

with

$$\mathbf{\Psi} = \mathbf{\Psi}(\alpha, \dot{\psi}, \psi, \omega t) = -h\dot{\psi} - m\,g\,l\,[\sin(\alpha + \psi) - < \sin(\alpha + \psi) >]$$

$$-m\,l\,\omega^2[H\cos(\alpha+\psi)\sin\omega t - G\sin(\alpha+\psi)\cos(\omega t+\theta)$$
$$- < H\cos(\alpha+\psi)\sin\omega t - G\sin(\alpha+\psi)\cos(\omega t+\theta) >],$$
$$V = m\,g\,l\,[\sin\alpha - <\sin(\alpha+\psi)>] - m\,l\,\omega^2 < [H\cos(\alpha+\psi)\sin\omega t$$
$$-G\sin(\alpha+\psi)\cos(\omega t+\theta)] > . \tag{5.1.7}$$

As was mentioned in chapter 3, this system is equivalent to the initial equation (5.1.2), at least in the sense that if its solution α, ψ is found, then the function $\varphi = \alpha + \psi$ will satisfy equation (5.1.2).

Let us now make the main assumption concerning the rate of changing the functions α and ψ and solve equation (5.1.6) in a crude way, considering also in the process of solving the equation the value α to be constant (frozen); the question of the admissibility in this case of such a method will be discussed by us later, when performing an a posteriori check of the assumptions made by us. Let us make the following supposition about the smallness of dimensionless parameters:

$$\frac{p_0}{\omega} \sim \varepsilon, \qquad \frac{m\,l\,\sqrt{G^2+H^2}}{I} \sim \varepsilon, \qquad \frac{h}{I\omega} \sim \varepsilon. \tag{5.1.8}$$

Here $p_0 = \sqrt{m\,g\,l/I}$ is the frequency of the small free oscillations of the pendulum, and $\varphi \geq 0$ is a small parameter. Then equation (5.1.6) can be written as

$$I\ddot{\psi} = \varepsilon\mathbf{\Psi}_1, \tag{5.1.9}$$

with

$$\varepsilon\mathbf{\Psi}_1 = \mathbf{\Psi} = \mathbf{\Psi}(\alpha, \dot{\psi}, \psi, \omega t). \tag{5.1.10}$$

In the initial approximation, i.e. at $\varepsilon = 0$ the solution of the equation (5.1.9), satisfying the condition (5.1.4), is $\psi = \psi_0 = 0$ and the first approximation is defined as a periodic solution $\psi = \psi_1$ of the equation

$$I\ddot{\psi}_1 = -ml\omega^2[H\cos\alpha\sin\omega t - G\sin\alpha\cos(\omega t+\theta)]. \tag{5.1.11}$$

Hence, taking into consideration the conditions (5.1.4) we can easily find

$$\psi = \psi_1 = \frac{ml}{I}[H\cos\alpha\sin\omega t - G\sin\alpha\cos(\omega t+\theta)]. \tag{5.1.12}$$

We will restrict ourselves to this approximation which belongs to those, called by us purely inertial (see 3.2.8). Now we will make use of the supposition concerning the smallness of the function ψ as compared to α and will linearize the expression for the vibrating moment (5.1.7) with respect to ψ. Bearing in mind that $<\sin\omega t> = <\cos\omega t> = 0$ we obtain

$$V = ml\omega^2[H\sin\alpha <\psi\sin\omega t> + G\cos\alpha <\psi\cos(\omega t+\theta)>]. \tag{5.1.13}$$

Substituting here expression (5.1.12) and taking into account the equalities

$$< \sin^2 \omega t > = < \cos^2 \omega t > = \frac{1}{2}, \quad < \sin \omega t \cos \omega t > = 0, \tag{5.1.14}$$

we will find the following expression for the vibrational moment:

$$V = V(\alpha) = -\frac{(ml\omega)^2}{4I}[(G^2 - H^2)\sin 2\alpha + 2GH \cos 2\alpha \sin \theta]. \tag{5.1.15}$$

If we introduce the designations

$$\frac{G}{\sqrt{G^2 + H^2}} = \cos \gamma, \quad \frac{H}{\sqrt{G^2 + H^2}} = \sin \gamma, \quad V_0 = \frac{(ml\omega\sqrt{G^2 + H^2})^2}{4I}, \tag{5.1.16}$$

we can make expression (5.1.15) take the following form

$$V = V(\alpha) = -V_0(\sin 2\alpha \cos 2\gamma + \cos 2\alpha \sin 2\gamma \sin \theta). \tag{5.1.17}$$

As a result, the equation of the slow motions (5.1.5) (the main equation of vibrational mechanics) will be presented in the following form:

$$I\ddot{\alpha} + h\dot{\alpha} + mgl\sin \alpha - V(\alpha) = 0. \tag{5.1.18}$$

If we introduce the potential energy of slow forces Π_F, the potential energy of vibrational forces Π_V and the potential function D according to the formulas

$$\Pi_F = -mgl\cos \alpha,$$
$$\Pi_V = -\frac{1}{2}V_0(\cos 2\alpha \cos 2\gamma - \sin 2\alpha \sin 2\gamma \sin \theta)$$
$$\equiv -\frac{(ml\omega)^2}{8I}[(G^2 - H^2)\cos 2\alpha - 2GH \sin 2\alpha \sin \theta],$$
$$D = \Pi_F + \Pi_V, \tag{5.1.19}$$

then the equation (5.1.18) will be written as

$$I\ddot{\alpha} + h\dot{\alpha} = -\frac{dD}{d\alpha}. \tag{5.1.20}$$

Thus, the system under consideration belongs to the class of systems, discussed in 4.1. Despite its essentially non-conservative character, the corresponding equation of vibrational mechanics allows a notation in the form, characteristic of the conservative system which is under the action of a dissipative force.

According to the equation (5.1.18) the positions of the quasi-equilibrium of the pendulum, i.e. positions of equilibrium for the slow component of motion $\alpha = \alpha_*$ will be found from the equation

$$mgl\sin\alpha_* - V(\alpha_*) = 0, \tag{5.1.21}$$

the definite position of equilibrium being asymptotically stable, provided the following condition is satisfied:

$$mgl\cos\alpha_* - V'(\alpha_*) > 0, \tag{5.1.22}$$

which, taking into consideration equalities (5.1.15) and (5.1.17), can be presented in one of the following forms

$$mgl\cos\alpha_* + \frac{(ml\omega)^2}{2I}[(G^2 - H^2)\cos 2\alpha_* - 2GH\sin 2\alpha_* \sin\theta] > 0, \tag{5.1.23}$$

$$mgl\cos\alpha_* + 2V_0(\cos 2\alpha_* \cos 2\gamma_* - \sin 2\alpha_* \sin 2\gamma \sin\theta) > 0, \tag{5.1.24}$$

It is easy to see that relations (5.1.21)–(5.1.24) are also obtained from the conditions of the minimum of the potential function D.

Let us now make a posteriori check of the validity of the assumptions, made by us, including the assumptions about the rate of changing the components α and ψ. As an example, we will dwell here on that check in more detail than below and even in more detail than it is necessary to do when considering remark 4 in 3.2.8. We will study the solutions of equation (5.1.18), corresponding to the positions of stable quasi-equilibrium of the pendulum $\alpha = \alpha_*$, and the oscillations of the pendulum $\alpha(t) = \alpha_* + \beta(t)$ in the vicinity of those positions. In the process, for the value β from (5.1.18) a linear differential equation will be obtained:

$$I\ddot{\beta} + h\dot{\beta} + p_V^2\beta = 0$$

a general solution of which is

$$\beta = \beta_0 e^{-\frac{h}{I}t}\sin(p_V t + \rho)$$

where β_0 and ρ are constants and

$$p_V = \sqrt{\frac{mgl\cos\alpha_*}{I} - \frac{V'(\alpha_*)}{I}} \tag{5.1.25}$$

is the frequency of oscillations under consideration. Now, using equation (5.1.12) we can evaluate the relation $\dot{\alpha}/\dot{\psi}$. Designating•the scale of change of the component ψ by $\psi_0 = ml(G + H)/I$ and taking the value β_0 for the scale of change of the component α we will have

$$\frac{\dot{\alpha}}{\dot{\psi}} = \frac{\alpha_0 e^{\frac{h}{I}t}[-\frac{h}{I}\sin(p_V t + \rho) + p_V \cos(p_V t + \rho)]}{\psi_0 \omega \left[\frac{H}{G+H}\cos\alpha_* \sin\omega t - \frac{G}{G+H}\sin\alpha_* \cos(\omega t + \theta)\right]}$$

$$\sim \frac{\alpha_0}{\psi_0}\left|\frac{h}{I\omega} + \frac{p_V}{\omega}\right|. \tag{5.1.26}$$

According to the formulas (5.1.16), (5.1.17) and (5.1.26)

$$p_V < \sqrt{\frac{mgl}{I} + \frac{2V_0}{I}} = p_0\sqrt{1 + \frac{\omega^2}{p_0^2}\frac{m^2 l^2(G^2 + H^2)}{2I^2}}$$

$$\approx p_0\sqrt{1 + \frac{1}{2}\frac{\omega^2}{p_0^2}\frac{G^2 + H^2}{l^2}}. \tag{5.1.27}$$

In accordance with the assumptions (5.1.8) the second term under the radical sign in the last expression presents the ratio of two small values of the same order. Therefore we can consider that $p_V \sim p_0$ and from the same assumptions (5.1.8)

$$\frac{\dot{\alpha}}{\dot{\psi}}\frac{\psi_0}{\alpha_0} \sim \varepsilon, \tag{5.1.28}$$

just as it should be according to the relations (3.2.60) of chapter 3 for the main assumption of vibrational mechanics to be satisfied. Since we had assumed that $\psi_0/\alpha_0 \sim \varepsilon$, then in our case

$$\dot{\alpha}/\dot{\psi} \sim 1, \quad \ddot{\alpha}/\ddot{\psi} \sim \varepsilon, \tag{5.1.29}$$

and the ratio of the right hand sides of the equations (5.1.5) and (5.1.6) must also be of the order of ε. It is easy to see that the latter condition is also satisfied due to the relations (5.1.8) and (5.1.26)–(5.1.29). Thus, we can say that the obtained solution of the problem satisfies all the assumptions that have been made.

It should be noted that the conclusion about the validity of the assumptions made, at least for the motions, lying in a sufficiently close vicinity of the stationary regimes $\alpha = \alpha_*$, could be made without resorting to the computations (and it is such regimes that are of the main interest).We must recall (see note 4 in 3.2.8) that in such vicinity the right hand side of equation (5.1.5) remains preassigned small as compared to the right hand side of equation (5.1.6).

It should be also noted that the above obtained relations (5.1.15)–(5.1.24) can be also arrived at by means of reducing the initial equations (5.1.2) to that considered in 3.2.7 as a special case. To this aim we will admit along with the relations (5.1.8) that in the motions under consideration

$$h\dot{\varphi}/I\omega^2 = h\varphi_1/I\omega \sim \varepsilon^2 ; \qquad (5.1.30)$$

the validity of this supposition can be checked a posteriori. Then for the ratio of the groups of terms of the equation (5.1.2) we will obtain

$$\frac{ml\omega^2[H\cos\varphi\sin\omega t - G\sin\varphi\cos(\omega t + \theta)]}{h\dot{\varphi} + mgl\sin\varphi}$$

$$= \frac{\dfrac{ml}{I}[H\cos\varphi\sin\omega t - G\sin\varphi\cos(\omega t + \theta)]}{h\dot{\varphi}/I\omega^2 + p_0^2/\omega^2} \sim \frac{1}{\varepsilon},$$

and it is possible to write down that equation in the form

$$I\ddot{\varphi} = F(\dot{\varphi}, \varphi) + \omega\Phi_1(\varphi, \tau), \qquad (5.1.31)$$

corresponding to the equation (3.2.32) of chapter 3, with

$$F(\dot{\varphi}, \varphi) = -h\dot{\varphi} - mgl\sin\varphi,$$
$$\Phi_1(\varphi, \tau) = -ml\omega[H\cos\varphi\sin\tau - G\sin\varphi\cos(\tau + \theta)],$$
$$\tau = \omega t, \quad <\Phi_1>_{\varphi=const} = 0, \qquad (5.1.32)$$

and as before, solutions of the following type are sought:

$$\varphi = \alpha(t) + \varepsilon\psi_1(t, \tau), \quad \varepsilon = 1/\omega. \qquad (5.1.33)$$

The equation of fast motions (3.2.35)(chapter 3) now looks as

$$I\ddot{\psi}_1 = \frac{1}{\varepsilon^2}\Phi_1(\alpha, \tau), \qquad (5.1.34)$$

and its periodic solution at $\alpha = $ const will be

$$\psi_1^* = \frac{ml\omega}{I}[H\cos\alpha\sin\tau - G\sin\alpha\cos(\tau + \theta)]. \qquad (5.1.35)$$

Using this solution, we obtain, according to equation (3.2.37) (chapter 3), the expression for the vibrational moment

$$V(\alpha) = F[\dot{\alpha} + \frac{\partial\psi_1^*(\alpha, \tau)}{\partial\tau}, \alpha] - F(\dot{\alpha}, \alpha)$$

$$- \left\langle \frac{\partial\Phi_1(\alpha, \tau)}{\partial\alpha}\psi_1^*(\alpha, \tau) \right\rangle = \frac{(ml\omega)^2}{I}\left\langle [H\sin\alpha\sin\tau \right.$$

$$\left. + G\cos\alpha\cos(\tau + \theta)] \cdot [H\cos\alpha\sin\tau - G\sin\alpha\cos(\tau + \theta)] \right\rangle$$

$$= -\frac{(ml\omega)^2}{4I}[(G^2 - H^2)\sin 2\alpha + 2GH\cos 2\alpha\sin\theta], \qquad (5.1.36)$$

coinciding with the expression (5.1.15), found before. Evaluating the ratio $h\dot{\varphi}/I\omega^2$ analogously to how it was done above for the values $\dot{\alpha}/\dot{\psi}$ and $\ddot{\alpha}/\ddot{\psi}$, we come to the conclusion that the assumption (5.1.30) was valid.

5.1.4   The Use of the Mini-Max Criterion of Stability

Solving this problem by the Poincare-Lyapunov method also leads to the results, stated in item 5.1.3, and the relation (5.1.21)–(5.1.24) can be easily obtained from the mini-max criterion of Strizhak (see 4.3). For the use of that criterion we will make up an expression for the kinetic and potential energy of the system

$$T = \frac{1}{2}J\dot{\varphi}^2 + \frac{1}{2}m(\dot{x}_c^2 + \dot{y}_c^2), \qquad \Pi = m\,g\,y_c. \qquad (5.1.37)$$

Here

$$x_c = x - l\sin\varphi, \qquad y_c = y - l\cos\varphi \qquad (5.1.38)$$

are the coordinates of the center of gravity of the pendulum C, J is the moment of inertia, with respect to the center of gravity, and the coordinates of the axes of suspension x and y are determined by the formulas (5.1.1). In view of equalities (5.1.38) we obtain the following expression for the Lagrange function of the system:

$$L = T - \Pi = \frac{1}{2}I\dot{\varphi}^2 + \frac{1}{2}m[\dot{x}^2 + \dot{y}^2 - 2l\dot{\varphi}(\dot{x}\cos\varphi - \dot{y}\sin\varphi)]$$
$$-m\,g\,(y - l\cos\varphi), \qquad (5.1.39)$$

where $I = J + ml^2$ just like before is the moment of inertia of the pendulum in respect to the axis of suspension. Hence we can easily find

$$\min_{\dot{\varphi}} L = -\frac{m^2l^2}{2I}(\dot{x}\cos\varphi - \dot{y}\sin\varphi)^2 + m\,g\,l\cos\varphi + \frac{1}{2}m(\dot{x}^2 + \dot{y}^2) - m\,g\,y. \quad (5.1.40)$$

Now, taking into account (5.1.1), performing the averaging for the 2π-period over ωt and changing the angle φ for its slow component α we obtain:

$$< \min_{\dot{\alpha}} L > = -\frac{ml\omega^2}{8I}[(G^2 - H^2)\cos 2\alpha - 2GH\sin 2\alpha \sin\theta]$$
$$+m\,g\,l\cos\alpha + C, \qquad (5.1.41)$$

or, taking into account the designations (5.1.16),

$$< \min_{\dot{\alpha}} L > = \frac{1}{2}V_0(\cos 2\alpha \cos 2\gamma - \sin 2\alpha \sin 2\gamma \sin\theta) + m\,g\,l\cos\alpha + C. \quad (5.1.42)$$

Here the symbol C designates the summand which does not depend on α and which is unessential for the subsequent consideration. In accordance with the sign of Strizhak and with (4.3.15) (chapter 4) the potential function is

$$D = - < \min_{\dot{\alpha}} L >= -\frac{1}{2}V_0(\cos 2\alpha \cos 2\gamma - \sin 2\alpha \sin 2\gamma \sin \theta) - m\,g\,l\cos \alpha.$$
$$(5.1.43)$$

This expression coincides with expression (5.1.19), found by the method of direct separation of motions, which is what we wished to prove. It is remarkable that when using the mini-max sign it was not necessary to compose and use the differential equation of motion of the system (5.1.2).

It should be noted that it is not difficult to consider the cases when the oscillations of the axis of suspension of the pendulum along the directions x and y are combinations of harmonics or the arbitrary enough $2\pi/\omega$-periodic functions of the time t. One of such cases was considered by Kapitsa in the above cited articles [257, 258].

5.1.5   Behavior of the Pendulum Depending on the Type of Vibration

Let us now consider the main regularities of the behavior of the pendulum under the action of vibration of its axis of suspension. We will dwell upon studying a number of specific cases

1.Vertical vibration of the axis($G = A$, $H = 0$, A is the amplitude of vibration). In this case according to (5.1.15) equation (5.1.21) acquires the form

$$m\,g\,l\sin \alpha_* + \frac{(m\,l\,A\omega)^2}{4I}\sin 2\alpha_* = 0 \tag{5.1.44}$$

and from that equation four positions of the quasi-equilibrium of the pendulum are found:

$$\alpha_*^{(1)} = 0, \quad \alpha_*^{(2)} = \pi, \quad \alpha_*^{(3-4)} = \pm \arccos\left[-\frac{2\,g\,I}{m\,l\,(A\omega)^2}\right] \tag{5.1.45}$$

The isolated third and fourth positions exist, provided the following condition is satisfied

$$\frac{m\,l\,(A\omega)^2}{2\,g\,I} > 1. \tag{5.1.46}$$

The condition of stability (5.1.23) of the quasi-equilibrium positions that have been found in the case under consideration are reduced to the inequality

$$\cos \alpha_* + \frac{m\,l\,(A\omega)^2}{2\,g\,I}\cos 2\alpha_* > 1. \tag{5.1.47}$$

Hence it follows that the lower equilibrium position of the pendulum $\alpha = \alpha_*^{(1)} = 0$ is always stable, and the upper (upside-down) position $\alpha_*^{(2)} = \pi$ is stable provided condition (5.1.46) is satisfied. This conclusion is the classic result we spoke about in 5.1 of this chapter. It is also easy to see that the positions of quasi-equilibrium $\alpha = \alpha_*^{(3-4)}$, existing when equality (5.1.34) is satisfied, are unstable.

The same conclusions can be easily drawn from the considerations of extremums of the potential function D which in the case under consideration looks as

$$D = -m\,g\,l\,\cos\alpha_* - \frac{(m\,l\,A\omega)^2}{8I}\cos 2\alpha. \qquad (5.1.48)$$

It should be noted that for the mathematical pendulum, when $I = m\,l^2$, the condition (5.1.46) can be given one of the following forms:

$$A\omega > \sqrt{2\,g\,l}, \qquad \frac{m(A\omega)^2}{2} > m\,g\,l,$$

that is to express as a demand that the amplitude of speed of vibration should exceed the speed of the free fall of the body from the height, equal to the length of the pendulum, or that the kinetic energy of vibration should exceed the potential energy, acquired by the pendulum when rising to a height l [257, 258].

The energy regularities, established above, which follow from the presence in this problem of the potential function $D = \Pi + \Pi_V = - < \min_{\dot{\alpha}_*}(T - \Pi) > $ can be considered as a generalization of the elegant interpretations given above.

It must be remarked that the conclusion about the stability of the lower position of the pendulum under the vertical vibration of the axis of suspension remains valid if the parameters of the system satisfy relations (5.1.8), i.e. in the region where this investigation is valid. As is well known (this is also a classical result), in certain regions of the change of parameters, the lower position of the pendulum becomes unstable. We mean here the phenomenon of the *parametric instability of the pendulum*, described by the Ince-Strutt well known diagram. Fig. 5.2 shows a fragment of this diagram, with the ratio $a = 4p_0^2/\omega^2$ traditionally plotted on the abscissa, and the ratio $b = 2mlA/I$ plotted on the axis of ordinates. In this case the positive values of a are answered by the lower position of the pendulum, and the negative values are answered by the upper position of the pendulum. The regions of stability of the equilibrium positions are shaded. The diagram is built for the case when the damping is absent ($h = 0$), which is in this case of no principal importance. The curve $b^2/2a = m\,l(A\omega)^2/2g\,I = 1$ corresponds to the boundary of validity of inequality (5.1.46).

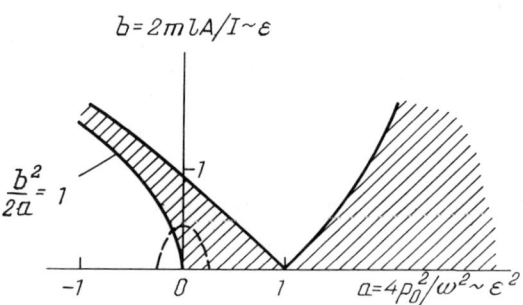

Figure 5.2. Stability region of the upper and lower positions of the pendulum with a vertically vibrating axis of suspension (fragment of the diagram of Ince-Strutt).

The diagram shows that, say, at $a = 0.2$, $b = 1.0$ the lower position of the pendulum is unstable, which does not follow from the investigation, given above. Neither follows from it the possibility of a rather complicated chaotic behavior of the pendulum in certain regions of the change of the parameters (see, say, [52, 53]). It should be noted, however, that both the parameteric resonance and the chaotic oscillations take place at such values of the amplitudes and frequencies of the point of suspension which cannot be referred to vibration in the sense of the definition that has been accepted above(see the footnote on p. 3).

Along with it in the region $|a| = 2p_0^2/\omega^2 \sim \varepsilon^2 \ll 1$, $b = 2mlA/I \sim \varepsilon \ll 1$ where the assumptions made are fulfilled and where the parameters of oscillations of the point of suspension answer the indicated definition (this area in Fig. 5.2 is conventionally limited by a semi-oval), exactly the same conclusion follows from the diagram: the condition of stability of the upper position of the pendulum is the relation $b^2/2a > 1$ which coincides with inequality (5.1.46).

2. **Rectilinear harmonic vibration at the angle γ to the vertical line** ($\theta = -\frac{1}{2}\pi$, $A = \sqrt{G^2 + H^2}$).

First let us consider this case assuming that the acceleration of the free fall g is negligibly small as compared to the acceleration of vibration $A\omega^2$ or that the pendulum oscillates in a horizontal plane. The quasi-equilibrium positions according to equalities (5.1.17) and (5.1.21) will be determined from the equation

$$\sin 2(\alpha_* - \gamma) = 0, \qquad (5.1.49)$$

and the condition of stability (1.22) is reduced to the inequality

$$\cos 2(\alpha_* - \gamma) > 0. \qquad (5.1.50)$$

Thus, in this case there are also four positions of quasi-equilibrium:

$$\alpha_*^{(1)} = \gamma, \quad \alpha_*^{(2)} = \gamma + \pi, \quad \alpha_*^{(3-4)} = \gamma \pm \frac{1}{2}\pi, \qquad (5.1.51)$$

the first two of which are stable, while the third and the fourth are unstable. At $\gamma = 0$ and $g = 0$ the result obtained coincides, just as it must, with that, found above for the case of vertical vibration. The equation of slow motions of the pendulum (5.1.18) in this case will look as

$$I\ddot{\alpha} + h\dot{\alpha} + V_0 \sin 2(\alpha - \gamma) = 0. \qquad (5.1.52)$$

If we introduce the angle $\beta = \alpha - \gamma$ which is the angle of deviation of the pendulum from the position $\alpha = \gamma$ and linearize the vibrational moment near the positions of stable equilibrium $\alpha = \alpha_*^{(1)} = \gamma$ (i.e. $\beta = \beta_*^{(1)} = 0$) and $\alpha = \alpha_*^{(2)} = \gamma + \pi$ (i.e. $\beta = \beta_*^{(2)} = \pi$), then this equation will acquire the form

$$I\ddot{\beta} + h\dot{\beta} + 2V_0\beta = 0. \qquad (5.1.53)$$

Hence it follows that the action of vibration is reduced to the fact that the pendulum seems to be "attracted " to the positions $\alpha = \alpha_*^{(1)} = \gamma$ and $\alpha = \alpha_*^{(2)} = \gamma + \pi$ by a spring of the rigidity $c_V = 2V_0/a_0^2 = (ml\,A\omega)^2/2Ia_0^2$ (a_0 being the distance from the point of fixing the spring to the axis of the pendulum — see Fig. 5.3). That is how the situation can be explained by the observer **V**. He also explains the so called "drift" (the displacement from the correct position) of the needles of any device on a vibrating base, for instance of a compass, by the attraction to the indicated positions.

In case of gravity the slanting vibration will cause the shift of the lower stable position of equilibrium of the pendulum towards the shortest turn to the direction of vibration and, moreover, it may stabilize the pendulum in the vicinity of the upper unstable position.

The observer **V** explains the remarkable fact (which Kapitsa paid attention to [257, 258] that the pendulum clock under the effect of vertical vibration is always fast due to the action of the invisible springs with a rigidity c_V. Indeed, in this case the equation of slow motions of the pendulum (5.1.18) near the lower stable position of equilibrium will have the form

$$I\ddot{\alpha} + h\dot{\alpha} + (m\,g\,l + c_V\,a_0^2)\alpha = 0, \tag{5.1.54}$$

that is the frequency of free oscillations (without taking account of damping) will be

$$p_V = \sqrt{(m\,g\,l + c_V\,a_0^2)/I},$$

while without vibration the corresponding frequency is

$$p_0 = \sqrt{m\,g\,l/I} < p_V.$$

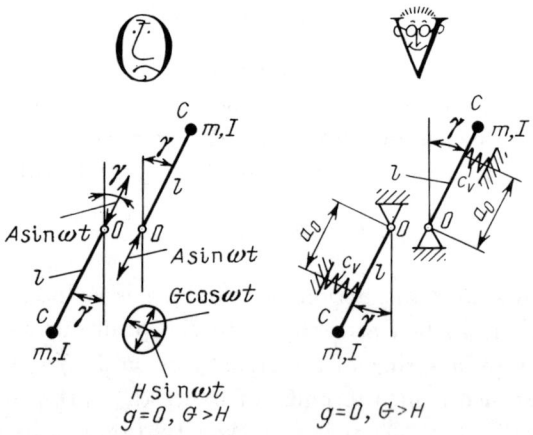

Figure 5.3. In case of rectilinear vibration in the absence of gravity the pendulum is stabilized in the direction of vibration, in case of an elliptical vibration — along the major axis of the elliptical trajectory. In case of gravity the pendulum is "attracted" towards the indicated directions.

3. Vibration in the elliptical trajectory.

Just as before, we will first consider the case when the pendulum oscillates in a horizontal plane or when the acceleration of a free fall is negligibly small as compared to the acceleration of vibration. Then, evidently, without breaking the generality, we can assume that $\theta = 0$, while the amplitudes G and H will be the semi-axes of the elliptical trajectory. Equation (5.1.21) for the determination of the quasi-equilibrium positions $\alpha = \alpha_*$ will in this case have the form

$$\sin 2\alpha_* = 0, \tag{5.1.55}$$

and the condition of stability of these positions (5.1.23) is reduced to the inequality

$$(G^2 - H^2)\cos 2\alpha_* > 0. \tag{5.1.56}$$

Hence it follows that from the four quasi-equilibrium positions

$$\alpha_*^{(1)} = 0, \quad \alpha_*^{(2)} = \pi, \quad \alpha_*^{(3-4)} = \pm\frac{1}{2}\pi \tag{5.1.57}$$

with $G > H$, those positions of the pendulum are stable at which it is located along one of the semi-axes G, and with $H > G$ — along one of the semi-axes H. In other words, in this case when in the absence of vibration any position of the pendulum is a position of equilibrium, in case of vibration the pendulum tends to get located along the major semi-axis of the elliptical trajectory, that is it has only two stable positions of quasi-equilibrium.

The situation is more complicated in the case of the inclined axis of suspension of the pendulum and when the effect of gravity is appreciable. Then the described effect manifests itself just as a tendency — the pendulum seems to be attracted to the directions of the major semi-axes of the elliptical trajectory of vibration. As before, vibration may in this case cause the shift of equilibrium positions, stable without vibration, and the stabilization (perhaps also with a shift) of the unstable positions. All those effects can be explained in a simple way in the context of vibrational mechanics as a result of the action of the corresponding positional vibrational moments (see 5.3).

5.1.6   *Multilink Pendulums*

Vibrational excitation of the axis of suspension can stabilize the upside-down equilibrium-position of the pendulum composed of two or more (n) rigid rods connected to each other by hinges. The publications [3, 521, 149, 419, 4] are devoted to the experimental and theoretical consideration of such systems. The possibility of performing the Indian rope-trick $(n \to \infty)$ has also been discussed [419, 4]

5.2   Chelomei's Pendulum with a Vibrating Axis of Suspension

5.2.1   Brief Bibliographical Review

By *Chelomei's pendulum* we mean a system, consisting of a rod, able to revert round a certain axis ("the axis of suspension"), and of a solid body ("the washer") able to move along the rod (Fig. 5.4).

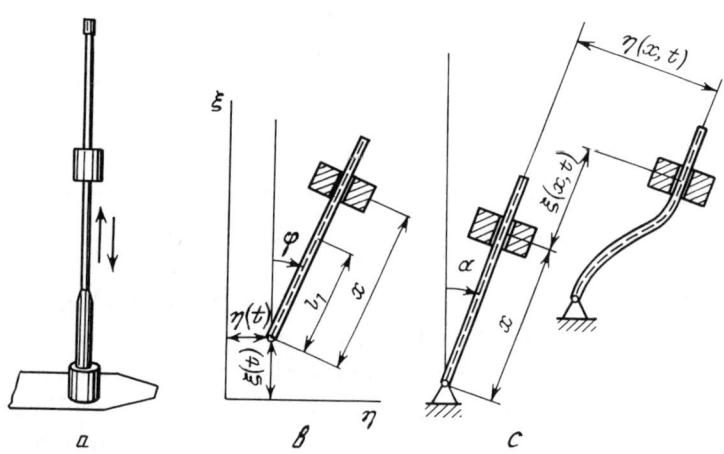

Figure 5.4. Chelomey's pendulum with a vibrating axis of suspension: a) scheme of a laboratory installation [149]; b) model with a deformable rod, performing the given oscillations.

It was experimentally found by Chelomei [149] (1983) that due to the vertical vibration of the axis of suspension, the upper ("upside-down") position of the rod under certain conditions proves to be stable, with the washer on the rod taking a certain fixed position (Fig. 5.4,*a*).

The article by Chelomei at once attracted the attention of a several researchers, who studied the described system independently and who published their results almost simultaneously. In the work by Menyailov and Movchan [370] (1984) the behavior of Chelomei's pendulum was studied by method of averaging, under the assumption that the vibration of the rod was both horizontal and vertical. In the article by the author and Malakhova [112] (1986) under the same assumption the behavior of the washer was considered on both — an absolutely rigid rod and an elastic rod, vibrating in the preassigned manner in the standing wave regime. The authors used the method of direct separation of motions and also the mini-max sign of stability. The article by Kurbatov,

Chelomei junior and Hromushkin [305] (1986) contains the results of the numerical integrating of the equations of motion of the system in the case of a purely vertical vibration of the axis of suspension. The behavior of the washer on a vibrating elastic rod was also studied by Ragulskis and Naghinyavichus [451] (1986). By use of another analytic method the conditions of stability of the washer in the upper position of the rod were obtained by Kirghetov [269] (1986).

Later Astashev, Babitsky, Vepric and Krupenin [30] (1989) used the method of direct separation of motions to consider the behavior of the washer on the string or on an elastic rod when exciting their oscillations by means of the given distributed load. On the basis of the results of their own investigations and on the work of their predecessors the authors advanced an idea of a possibility of using a mobile washer as an effective damper of oscillations. That idea was checked experimentally.

The work [112] is quoted here in view of the supplements made by Malakhova in her thesis [361] (1990).

5.2.2   Pendulum with an Absolutely Rigid Rod

First we will suppose that the rod, along which the washer is being displaced, is rectilinear and cannot be deformed (Fig. 5.4,b). Let its axis of suspension oscillate according to the law

$$\xi(t) = a sin\omega t, \quad \eta(t) = bsin(\omega t + \theta) \tag{5.2.1}$$

where a and b are the amplitudes of the vertical and horizontal oscillations respectively, ω is the frequency, and θ is the phase shift. Expressions for the kinetic and potential energy of the system have the following form;

$$T = \frac{1}{2}(J_1 + J + mx^2)\dot{\varphi}^2 + \frac{1}{2}m\dot{x}^2 + (m_1 l_1 + mx)(\dot{\eta}\cos\varphi - \dot{\xi}\sin\varphi)\dot{\varphi}$$

$$+m(\dot{\eta}\sin\varphi + \dot{\xi}\cos\varphi)\dot{x} + \frac{1}{2}(m_1 + m)(\dot{\xi}^2 + \dot{\eta}^2), \tag{5.2.2}$$

$$\Pi = (m_1 l_1 + mx)g\cos\varphi + (m_1 + m)g\xi. \tag{5.2.3}$$

Here φ is the angle of rotation of the rod; m_1, J_1 and l_1 are respectively its mass, its moment of inertia in respect to the axis of suspension and the distance from the axis of suspension to the center of gravity; x is the coordinate of the washer, counted along the axis of the rod from the axis of suspension; m and J are the mass and the moment of inertia of the washer; g is a free fall acceleration.

Assuming that the resistance to the oscillations of the rod with a washer and the resistance to the displacement of the washer along the rod are of a nature of viscous friction with the coefficients h_1 and h_2 respectively, the equations of motion will be written in the following form

$$(J + J_1 + mx^2)\ddot{\varphi} + 2mx\dot{x}\dot{\varphi} + h_1\dot{\varphi} - (m_1 l_1 + mx)(g - a\omega^2 \sin \omega t) \sin \varphi$$
$$-(m_1 l_1 + mx)b\omega^2 \sin(\omega t + \theta) \cos \varphi = 0,$$
$$\ddot{x} - x\dot{\varphi}^2 + h_2\dot{x} + (g - a\omega^2 \sin \omega t) \cos \varphi - b\omega^2 \sin(\omega t + \theta) \sin \varphi = 0. \quad (5.2.4)$$

Considering the frequency of oscillations to be high as compared to the typical frequencies of free oscillations,

$$\lambda_1 = \sqrt{(m_1 + m)l_1 g/(J_1 + J + ml^2)}, \quad \lambda_2 = \sqrt{g/l_1}, \quad (5.2.5)$$

we will assume that

$$\frac{\lambda_1}{\omega}, \quad \frac{\lambda_2}{\omega}, \quad \frac{a}{l_1}, \quad \frac{b}{l_1}, \quad \frac{h_1}{(J_1 + J + ml_1^2)\omega}, \quad \frac{h_2}{m\omega} \sim \varepsilon, \quad (5.2.6)$$

that is we will consider the indicated relations to be small values of the order of ε.

Then equations of motion (5.2.4) can be presented in the following form

$$\ddot{\varphi} = F_\varphi(\dot{\varphi}, \varphi, \dot{x}, x) + \omega\Phi_{1\varphi}(\varphi, x, \tau),$$
$$\ddot{x} = F_x(\dot{\varphi}, \varphi, \dot{x}, x) + \omega\Phi_{1x}(\varphi, x, \tau), \quad (5.2.7)$$

where

$$F_\varphi = [-2mx\dot{x}\dot{\varphi} - h_1\dot{\varphi} + (m_1 l_1 + mx)g \sin \varphi]/(J_1 + J + mx^2),$$
$$F_x = x\dot{\varphi}^2 - h_2\dot{x} - g \cos \varphi,$$
$$\Phi_{1\varphi} = (m_1 l_1 + mx)[b\omega \cos \varphi \sin(\tau + \theta) - a\omega \sin \tau]/(J_1 + J + mx^2),$$
$$\Phi_{1x} = a\omega \cos \varphi \sin \tau + b\omega \sin \varphi \sin(\tau + \theta), \quad \tau = \omega t. \quad (5.2.8)$$

Since we are interested in the solutions of equations (5.2.8) of the form

$$\varphi = \alpha(t) + \varepsilon\psi_{1\varphi}(t, \tau), \quad x = X(t) + \varepsilon\psi_{1x}(t, \tau) \quad (5.2.9)$$

where α and X are the slow components, and $\psi_{1\varphi}$ and ψ_{1x} are the fast components, and since

$$< \Phi_{1\varphi} >|_{\substack{\varphi=const=0 \\ x=const}} = 0, \quad < \Phi_{1x} >|_{\substack{\varphi=const=0 \\ x=const}} = 0, \quad (5.2.10)$$

the problem under consideration corresponds to a special case which has been discussed in 3.2.7. As that takes place, the equations of fast motions will be written in the form

$$\ddot{\psi}_{1\varphi} = \frac{1}{\varepsilon^2}\Phi_{1\varphi}(\alpha, X, \tau), \quad \ddot{\psi}_{1x} = \frac{1}{\varepsilon^2}\Phi_{1x}(\alpha, X, \tau), \tag{5.2.11}$$

corresponding to the purely inertial approximation. According to (5.2.8), the periodic solution of these equations at the frozen α and X will be

$$\psi_{1\varphi}^* = -\frac{m_1 l_1 + mX}{J_1 + J + mX^2}[b\omega \cos\alpha \sin(\tau + \theta) - a\omega \sin\alpha \sin\tau],$$
$$\psi_{1x}^* = -a\omega \cos\alpha \sin\tau - b\omega \sin\alpha \sin(\tau + \theta). \tag{5.2.12}$$

In the case under consideration (two degrees of freedom) expressions (3.2.37) (chapter 3) for the vibrational forces will look as

$$V_\alpha =< F_\varphi(\dot{\alpha} + \frac{\partial\psi_{1\varphi}^*}{\partial\tau}, \alpha, \dot{X} + \frac{\partial\psi_{1x}^*}{\partial\tau}, X) - F_\varphi(\dot{\alpha}, \alpha, \dot{X}, X) >$$
$$+ < \frac{\partial\Phi_{1\varphi}}{\partial\alpha}\psi_{1\varphi}^*, + \frac{\partial\Phi_{1\varphi}}{\partial X}\psi_{1x}^* >,$$
$$V_x =< F_x(\dot{\alpha} + \frac{\partial\psi_{1\varphi}^*}{\partial\tau}, \alpha, \dot{X} + \frac{\partial\psi_{1x}^*}{\partial\tau}, X) - F_x(\dot{\alpha}, \alpha, \dot{X}, X) >$$
$$+ < \frac{\partial\Phi_{1x}}{\partial\alpha}\psi_{1\varphi}^*, + \frac{\partial\Phi_{1x}}{\partial X}\psi_{1x}^* > . \tag{5.2.13}$$

Performing calculations according to these formulas and in view of expressions (5.2.8), (5.2.12) and of the relations of type (5.1.14), we arrive at the following expressions for the generalized vibrational forces:

$$V_\alpha = V_\alpha(\alpha, X) = \frac{\omega^2}{4}\left[\frac{(m_1 l_1 + mX)^2}{J_1 + J + mX^2} - m\right]$$
$$\times[2ab\cos 2\alpha \cos\theta - (a^2 - b^2)\sin 2\alpha],$$
$$V_x = V_x(\alpha, X) = \frac{\omega^2}{2}m\left[X\left(\frac{m_1 l_1 + mX}{J_1 + J + mX^2}\right)^2 - \frac{m_1 l_1 + mX}{J_1 + J + mX^2}\right]$$
$$\times(a^2 \sin^2\alpha + b^2 \cos^2\alpha - ab\sin 2\alpha \cos\theta). \tag{5.2.14}$$

As a result, the equations of slow motions (the main equations of vibrational mechanics) will be presented as

$$(J_1 + J + mX^2)\ddot{\alpha} + 2mX\dot{X}\alpha + h_1\dot{\alpha} = (m_1 l_1 + mX)g\sin\alpha + V_\alpha,$$
$$m\ddot{X} - mX\dot{\alpha}^2 + h_2\dot{X} = -mg\cos\alpha + V_x. \tag{5.2.15}$$

It is easy to see that the vibrational forces (5.2.14) are answered by the "potential energy"

$$\Pi_V = \left[\frac{(m_1 \, l_1 + mX)^2}{J + J_1 + mX^2} - m\right]\frac{\omega^2}{4}(a^2 \sin^2 \alpha + b^2 \cos^2 \alpha - ab \sin 2\alpha \cos \theta),$$

(5.2.16)

while the potential energy, corresponding to the force of gravity, is

$$\Pi = (m_1 \, l_1 + mX)g \cos \alpha. \tag{5.2.17}$$

So the positions of the stable quasi-equilibrium of the system can be found as the points of the minimum of the potential function

$$D = D(\alpha, X) = \Pi(\alpha, X) + \Pi_V(\alpha, X), \tag{5.2.18}$$

and the equations of slow motions (5.2.15) can be presented in the form

$$(J_1 + J + mX^2)\ddot{\alpha} + 2mX\dot{X}\alpha + h_1\dot{\alpha} = -\partial D/\partial \alpha,$$
$$m\ddot{X} - mX\dot{\alpha}^2 + h_2\dot{X} = -\partial D/\partial X, \tag{5.2.19}$$

It should be noted that the same equation for the potential function can be easily obtained by means of using the mini-max sign of Strizhak (see 4.3). Indeed, according to formulas (5.2.2) and (5.2.3)

$$\min_{\dot{\alpha},\dot{X}} L(\dot{\alpha}, \alpha, \dot{X}, X) = \min_{\dot{\alpha},\dot{X}}(T - \Pi)$$
$$= -\frac{1}{2}\frac{(m_1 \, l_1 + mX)^2(\dot{\eta}\cos \alpha - \dot{\xi}\sin \alpha)^2}{J_1 + J + mX^2} - \frac{1}{2}m(\dot{\eta}\sin \alpha + \dot{\xi}\cos \alpha)^2$$
$$+\frac{1}{2}(m_1 + m)(\dot{\xi}^2 + \dot{\eta}^2) - (m_1 \, l_1 + mX)g\cos\alpha + (m_1 + m)g\xi, \quad (5.2.20)$$

and in accordance with expression (4.3.15) (chapter 4)

$$D = D(\alpha, X) = - < \min_{\dot{\alpha},\dot{X}} L(\dot{\alpha}, \alpha, \dot{X}, X) > .$$

Performing the averaging in view of formulas (5.2.1), we arrive at the expression for the function D, coinciding with (5.2.18) to an accuracy of an unessential constant. Just like above, when using the mini-max sign, it was not necessary to consider the differential equations of motion, which, as was already emphasized, is a typical peculiarity of that sign.

We will not dwell here on an a posteriori check of the suppositions about the ratio of the orders of terms in the right hand sides of equations (5.2.7) since it is done similarly to that made in section 3 and confirms the suppositions made (see also note 4 in 3.2.8).

Let us turn to the analysis of the relations we have obtained. We are interested in the stability of the quasi-equilibria of the system $\alpha = \alpha_*$, $X = X_*$ corresponding to the vertical (upper and lower) positions of the rod when $\sin \alpha_* = 0$. Such positions are possible in case $a \cos \theta = 0$. In that case X_* is the solution of the equation

$$\pm g - \frac{b^2 \omega^2}{2} \frac{(m_1 l_1 + mX)(J_1 + J - m_1 l_1 X)}{(J_1 + J + mX^2)^2} = 0. \tag{5.2.21}$$

Here the upper sign answers the value $\alpha_* = \pi$ and the lower sign is $\alpha_* = 0$. The conditions of stability of the quasi-equilibria under consideration are reduced to the inequalities

$$\left. \frac{\partial^2 D}{\partial \alpha^2} \right|_{\substack{\alpha = \alpha_* \\ X = X_*}} = \pm (m_1 l_1 + mX_*)g + \left[\frac{(m_1 l_1 + mX)^2}{J_1 + J + mX_*^2} - m \right] \frac{\omega^2}{2}(a^2 - b^2) > 0, \tag{5.2.22}$$

$$\frac{1}{m} \left. \frac{\partial^2 D}{\partial X^2} \right|_{\substack{\alpha = \alpha_* \\ X = X_*}} = \frac{b^2 \omega^2}{2} \{[m(J_1 + J) - 2mm_1 l_1 X_* - m_1^2 l_1^2](J_1 + J + mX_*^2)$$
$$-4mX_*(m_1 l_1 + mX_*)(J_1 + J - m_1 l_1 X_*)\}(J_1 + J + mX_*^2)^{-3} > 0. \tag{5.2.23}$$

Let $m_1 l_1 \neq 0$. Then, by introducing a dimensionless variable $q = \dfrac{m_1 l_1 X}{(J_1 + J)}$ and the dimensionless parameters

$$a_1 = \frac{a^2 \omega^2 m_1 l_1}{2g(J_1 + J)}, \qquad b_1 = \frac{b^2 \omega^2 m_1 l_1}{2g(J_1 + J)}, \qquad c = \frac{m(J_1 + J)}{m_1^2 l_1^2},$$

equations (5.2.21) and inequalities (5.2.22), (5.2.23) can be presented in the form

$$\pm 1 - b_1 \frac{(1 + cq)(1 - q)}{(1 + cq^2)^2} = 0, \tag{5.2.24}$$

$$\pm (1 + cq) + \frac{1 + 2cq - c}{1 + cq^2}(a_1 - b_1) > 0, \tag{5.2.25}$$

$$b_1 c[(1 + cq^2)(c - 1 - 2cq) - 4cq(1 - q)(1 + cq)](1 + cq^2)^{-3} > 0. \tag{5.2.26}$$

It is easy to see that when the vibration is purely horizontal ($a_1 = 0$), the upper position of the pendulum ($a_* = 0$) is unstable, since in that case $q > 1$ due to equation (5.2.24) and, consequently,

$$-(1 + cq) + b_1 \frac{c - 1 - 2cq}{1 + cq^2} < 0,$$

that is inequality (5.2.25) is not satisfied. We must note that that was to be expected due to the results of 5.1.5.

Let us now assume that $b_1 c \neq 0$ and consider the lower vertical position of the rod when the vibration of the axis of suspension is purely horizontal. Transforming inequality (5.2.26) to the form

$$c^2 q^2 (3 - 2q) - c(1 - 6q + 3q^2) + 1 < 0$$

and considering that $q < 1$, we may obtain

$$q < 1/9, \qquad c_1(q) < c < c_2(q),$$

where

$$c_1(q) = \frac{1 - 6q + 3q^2 - (1 - q)\sqrt{(1 - q)(1 - 9q)}}{2q^2(3 - 2q)} \qquad (5.2.27)$$

$$c_2(q) = \frac{1 - 6q + 3q^2 + (1 - q)\sqrt{(1 - q)(1 - 9q)}}{2q^2(3 - 2q)}, \qquad (5.2.28)$$

relation (5.2.25) being also fulfilled.

Fig. 5.5, a in the plane of the parameters c, b_1 depicts the region of the existence of a stable lower vertical position of the quasi-equilibrium of Chelomei's pendulum ($\alpha_* = \pi$, $X = X_*$ where $0 < X_* < (J_1 + J)/9m_1 l_1$) with a purely horizontal vibration of the axis of suspension. On the segment $1 < c \leq 5.4$ this region is bounded below by the curve $b_1 = b_1^{(1)}(c)$, and on the interval $5.4 < c < \infty$ — by the curve $b_1 = b_1^{(2)}(c)$. Here

$$b_1^{(1)}(c) = b_1(c, q)|_{q=q_1(c)}, \qquad b_1^{(2)}(c) = b_1(c, q)|_{q=q_2(c)},$$

$b_1(c, q) = (1 + cq^2)^2/(1 + cq)(1 - q)$, $q = q_1(c)$ and $q = q_2(c)$ are the functions, reverse to the functions $c = c_1(q)$ and $c = c_2(q)$ respectively, defined by formulas (5.2.27) and (5.2.28). The conditions of the existence and stability of the lower vertical quasi-equilibrium position in the case of a weightless rod were obtained in [478].

Now let us assume that the vibration of the axis of suspension is both vertical and horizontal, with $\cos \theta = 0$, so that the trajectory of vibration of the axis is an ellipse with the semi-axes a and b. Let us consider the quasi-equilibria of the system at which the rod takes the upper position, and the washer is localized at a certain place on the rod. For the upper position of the rod $q > 1$ and, consequently, $(1 + 2cq - c)/(1 + cq^2) > 0$; therefore the inequality (5.2.25) can be satisfied if we choose the parameter a_1 from the condition

$$a_1 > b_1 + \frac{(1 + cq)(1 + cq^2)}{1 + 2cq - c}$$

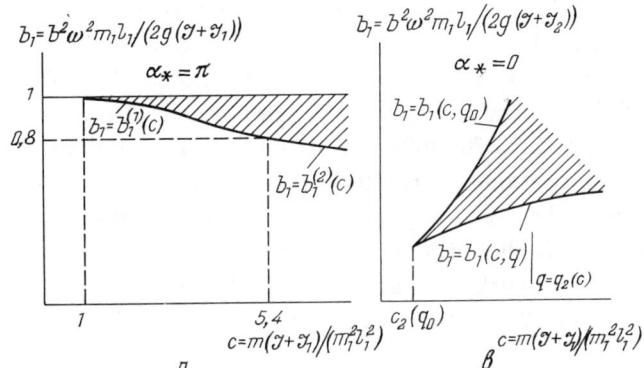

Figure 5.5. Quasi-equilibrium stability regions of Chelomey's pendulum, corresponding to the vertical positions of the rod: a) the lower position ($a_* = \pi$); b) the upper ("upside-down") position ($a_8 = 0$).

Thus, for the stability of the upper vertical position of Chelomei's pendulum it is necessary that the amplitude of the vertical vibration of the point of support of the rod should exceed the amplitude of the horizontal vibration; this conclusion is in agreement with what was stated in 5.1.5. It should be also noted that according to (5.2.23) a stable stabilization of the washer on a vertically erected rod is possible only if there is a horizontal component of the vibration of the support (this conclusion refers, however, only to the model and type of motion under consideration and to the corresponding region of the change of parameters of the system).

In future we will assume that the fulfillment of condition (5.2.25) is provided by the choice of the corresponding value of the amplitude of the vertical vibration. From inequality (5.2.26) we obtain $q > 3/2$, $c > c_2(q)$ where

$$c_2(q) = \frac{-(1 - 6q + 3q^2) + (q - 1)\sqrt{(q - 1)(9q - 1)}}{2q^2(2q - 3)} \tag{5.2.29}$$

Since for the rod whose length is l the inequality $X_* < l$ must be complied with, the region of the existence of a stable upper vertical quasi-equilibrium of the pendulum is determined by the relations

$$3/2 < q < q_0, \quad c > c_2(q), \quad b_1(c, q) = (1 + cq^2)^2/(1 + cq)(1 - q),$$

where $q_0 = m_1 l_1 l/(J_1 + J)$. In the plane of the parameters c, b (Fig. 5.5,b) the indicated region is limited by the curves $b_1 = b_1(c, q_0)$ and $b_1 = b_1(c, q)|_{q=q_2(c)}$ where $q = q_2(c)$ is the function, inverse to function (5.2.29).

It should be noted that in case the rod is homogeneous, from the condition $q > 3/2$ it follows that $X_* > l$, i.e. there are no stable positions of quasi-equilibrium of the washer on a vertically erected rod. This result also follows from the investigations [370, 269], conducted by other methods. Article [305], as was mentioned, contains the results of the numerical integration of the equations of motion of the system in the case of a purely vertical vibration of the support of the rod. It should be noted, however, that the region of the change of parameters in this work does not correspond to the assumptions about the smallness of ratios (5.2.6) (also see the note at the end of 5.2.3).

5.2.3   *The Behavior of the Washer on a Vibrating Deformable Rod*

In connection with the study of the behavior of Chelomei's pendulum under vibration, and also in connection with certain applications it seems of interest to consider another model, different from that given in 5.2.2, — a washer, set on a rectilinear deformable rod and performing the prescribed oscillations (vibration) in the regime of a standing wave (Fig. 5.4,c).

Let $\xi(x, t)$ and $\eta(x, t)$ be respectively the longitudinal and transverse displacements of the point of the axis of the rod with a coordinate x, counted from the point of support along the non-deformed position of that axis, at which it is tilted at a certain angle α to the vertical. The coordinate x plays also the role of a generalized coordinate of the washer.

Expressions for the kinetic and potential energy of the washer have the form

$$T = \frac{1}{2}m\left\{\left[\dot{x} + \frac{\partial\xi(x,t)}{\partial x}\dot{x} + \frac{\partial\xi(x,t)}{\partial t}\right]^2 + \left[\frac{\partial\eta(x,t)}{\partial x}\dot{x} + \frac{\partial\eta(x,t)}{\partial t}\right]^2\right\}$$

$$+ \frac{1}{2}J\left[\frac{\partial^2\eta(x,t)}{\partial x^2}\dot{x} + \frac{\partial^2\eta(x,t)}{\partial x\partial t}\right]^2, \tag{5.2.30}$$

$$\Pi = mg\{[x + \xi(x,t)]\cos\alpha - \eta(x,t)\sin\alpha\}. \tag{5.2.31}$$

The angle of rotation of the washer with an accuracy to values of a higher order of smallness with respect to $\partial\xi/\partial x$, $\partial\eta/\partial x$ was assumed to be equal to $\partial\eta/\partial x$. The equation of motion of the washer, corresponding to equations (5.2.30), (5.2.31), has the form (like in 5.2.2, we assume that the resistance to the motion of the washer along the rod is of a nature of viscous friction)

$$\ddot{x}\left\{m\left[\left(1 + \frac{\partial\xi}{\partial x}\right)^2 + \left(\frac{\partial\eta}{\partial x}\right)^2\right] + J\left(\frac{\partial^2\eta}{\partial x^2}\right)^2\right\} + h_2\dot{x}$$

$$+2\dot{x}\left\{m\left[\left(1+\frac{\partial\xi}{\partial x}\right)\frac{\partial^2\xi}{\partial x\partial t}+\frac{\partial\eta}{\partial x}\frac{\partial^2\eta}{\partial x\partial t}\right]+J\frac{\partial^2\eta}{\partial x^2}\frac{\partial^3\eta}{\partial x^2\partial t}\right\}$$

$$+\dot{x}^2\left\{m\left[\left(1+\frac{\partial\xi}{\partial x}\right)\frac{\partial^2\xi}{\partial x^2}+\frac{\partial^2\eta}{\partial x^2}\frac{\partial\eta}{\partial x}\right]+J\frac{\partial^2\eta}{\partial x^2}\frac{\partial^3\eta}{\partial x^3}\right\}$$

$$+m\left[\frac{\partial^2\xi}{\partial t^2}\left(1+\frac{\partial\xi}{\partial x}\right)+\frac{\partial^2\eta}{\partial t^2}\frac{\partial\eta}{\partial x}\right]+J\frac{\partial^3\eta}{\partial x\partial t^2}\frac{\partial^2\eta}{\partial x^2}$$

$$+mg\left[\left(1+\frac{\partial\xi}{\partial x}\right)\cos\alpha-\frac{\partial\eta}{\partial x}\sin\alpha\right]=0. \qquad (5.2.32)$$

Let, as was mentioned, the shifts ξ and η be of the nature of standing waves, that is they are given as

$$\xi=\xi(x,t)=a(x)\sin\omega t,\quad \eta=\eta(x,t)=b(x)\sin(\omega t+\theta) \qquad (5.2.33)$$

where ω is the frequency, θ is the phase shift, and the value $a_0=\sup[a(x),b(x)]$ is small as compared to the length of the half-wave L_0, so that the ratio a_0/L_0 can be considered to be of the order of a small parameter ε. Similarly to the assumptions in 5.2.2, we also assume that

$$\frac{h_2}{m\omega}\sim\varepsilon,\quad \frac{1}{\omega}\sqrt{\frac{g}{L_0}}\sim\varepsilon \qquad (5.2.34)$$

and that solutions are sought of equation (5.2.32) of the form

$$x=X(t)+\varepsilon\psi_1(t,\tau),\quad \tau=\omega t. \qquad (5.2.35)$$

With the formulated assumptions, the use of the method of direct separation of motions leads to the reasoning and calculations also quite similar to those given in 5.2.2. As a result, we come to the following equation of slow motions (the main equation of vibrational mechanics)

$$m\ddot{X}+h_2\dot{X}=-mg\cos\alpha+V(X) \qquad (5.2.36)$$

where

$$V(X)=\frac{m\omega^2}{2}b(X)\frac{db(X)}{dX}+\frac{\omega^2}{2}\frac{db(X)}{dX}\frac{d^2b(X)}{dX^2} \qquad (5.2.37)$$

It is remarkable that in this approximation the vibrational force V does not depend on the amplitude of the longitudinal component of the vibration of the rod $a(x)$, and is determined by the transverse component alone: when the motion is slow, the washer "does not perceive" the longitudinal vibration. So, when the vibration is purely longitudinal, there are no positions of quasi-equilibrium of the washer on the rod; though it is necessary to make a certain reservation here, similar to that made in 5.2.2 (see p. 117). When the vibration is transverse, the positions of the quasi-equilibrium $X = X_*$ are determined from the equation

$$m\, g \cos \alpha = V(X_*). \tag{5.2.38}$$

The vibrational force V is answered by the "potential energy"

$$\Pi_V = -\frac{m\omega^2}{4}b^2(X) - \frac{J\omega^2}{4}\left[\frac{db(X)}{dX}\right]^2. \tag{5.2.39}$$

Consequently, one can judge about the stability of a possible position of quasi-equilibrium by the presence of the minimum of the potential function

$$D = \Pi + \Pi_V = mgX\,\cos\alpha - \frac{m\omega^2}{4}b^2(X) - \frac{J\omega^2}{4}\left[\frac{db(X)}{dX}\right]^2 \tag{5.2.40}$$

where $\Pi = m\,g\,X\cos\alpha$ is the "usual" potential energy, corresponding to the gravitational force.

Let

$$b(X) = b_0 \sin\frac{\pi X}{L_0}, \tag{5.2.41}$$

Then equation (5.2.38) takes the form

$$g\cos\alpha = \frac{\pi b_0^2\omega^2}{4L_0}\left(1 - \frac{\pi^2 J}{mL_0^2}\right)\sin\frac{2\pi X_*}{L_0}, \tag{5.2.42}$$

and the quasi-equilibrium positions of the washer will exist provided the following conditions are satisfied:

$$\frac{mb_0^2\omega^2}{4L_0}\left|1 - \frac{\pi^2 J}{mL_0^2}\right| > g\cos\alpha. \tag{5.2.43}$$

It is easy to conclude that when the following inequality is valid

$$mL_0^2 > \pi^2 J, \tag{5.2.44}$$

those quasi-equilibria will be stable which are located nearer to antinodes of the form of the transverse oscillations $b(X) = b_0\sin\pi X/L_0$ and when the opposite inequality is satisfied — those which are nearer to the nodes.

If the position of the rod is horizontal ($\alpha = \pm\pi/2$), then in the first case the stable positions of quasi-equilibrium will be answered exactly by the points of antinodes, while in the second case they will be answered quite exactly by the nodes. It should be noted that the corresponding result for the case $J = 0$ was also obtained by Ragulskis and Naghinyavichus, cited above [451].

It should be also noted that like in 5.2.2, expression (5.2.40) for the potential function D, and, consequently, the results of the investigation of the stability of quasi-equilibrium positions of the system can also be obtained by means of the mini-max sign of stability of Strizhak (see 4.3). Indeed, according to expressions (5.2.30) and (5.2.31),

$$\min_{\dot{X}} L(\dot{X}, X) = \min_{\dot{X}}(T - \Pi)$$

$$= -\frac{\left\{ m\left[\left(1 + \frac{\partial \xi}{\partial X}\right)\frac{\partial \xi}{\partial t} + \frac{\partial \eta}{\partial X}\frac{\partial \eta}{\partial t}\right] + J\frac{\partial^2 \eta}{\partial X^2}\frac{\partial^2 \eta}{\partial X \partial t}\right\}^2}{2\left\{ m\left[\left(1 + \frac{\partial \xi}{\partial X}\right)^2 + \left(\frac{\partial \eta}{\partial X}\right)^2\right] + J\left(\frac{\partial^2 \eta}{\partial X^2}\right)^2\right\}}$$

$$+\frac{1}{2}m\left[\left(\frac{\partial \xi}{\partial t}\right)^2 + \left(\frac{\partial \eta}{\partial t}\right)^2\right] + \frac{1}{2}J\left(\frac{\partial^2 \eta}{\partial X \partial t}\right)^2 - mg[(X + \xi)\cos\alpha - \eta\sin\alpha],$$

where $\xi = \xi(X, t)$ and $\eta = \eta(X, t)$. Performing the averaging in accordance with formula (4.3.15) (chapter 4) and using expressions (5.2.33), we obtain

$$D = -\lim_{\varepsilon \to 0} < \min_{\dot{X}} L(\dot{X}, X) >= -\frac{m\omega^2}{4}b^2(X) - \frac{\omega^2}{4}\left[\frac{db(X)}{DX}\right]^2$$

$$+mgX\cos\alpha \qquad (5.2.45)$$

where $\xi = \xi(X, t)$ and $\eta = \eta(X, t)$. Performing the averaging in accordance with formula (4.3.15) (chapter 4) and using expressions (5.2.33), we obtain

$$D = -\lim_{\varepsilon \to 0} < \min_{\dot{X}} L(\dot{X}, X) >= -\frac{m\omega^2}{4}b^2(X) - \frac{\omega^2}{4}\left[\frac{db(X)}{DX}\right]^2$$

$$+mgX\cos\alpha \qquad (5.2.46)$$

which coincides with formula (5.2.40).

Summing up, we must admit that Chelomei's experiment (Fig. 5.4,a) so far has not yet been fully explained. The theoretical analysis performed almost simultaneously and independently by different researchers [112, 269, 370], failed to find a way of stabilizing the washer when the vibration of the axis of suspension of the rod is purely vertical. In order to do it within the frames of that analysis it proved to be necessary that there should also be a horizontal vibration, which is beyond question because a purely longitudinal vibration of the rod is not "perceived" by the washer. It is not clear, however, whether the stabilization of the washer in Chelomei's experiment was the result of his overlooking the transverse vibration of the axis of suspension of the rod, or that due to the longitudinal vibration there appeared the transverse either elastic or "solid-body" oscillations of the rod itself. The latter point of view is disputed in [305] where the numerical investigation has been performed in a wider range of parameter changes than in (5.2.6). It seems, however, that this question requires additional investigations.

5.3   The Follower-loaded Double Pendulum with a Vibrating Axis of Suspension

The partially follower-loaded double pendulum with elastic and damping elements in the hinges (sometimes it is called Ziegler's pendulum) is the simplest model to study important effects connected with the loss of stability of the rectilinear form of the equilibrium of the rod. Jensen investigated the behavior of such a pendulum under the action of the longitudinal vibration of its axis of suspension and showed that in such a way it is possible under certain conditions to essentially broaden the region of stability of the rectilinear form of equilibrium of the system [248, 249]. The result agrees very well with that of V. N. Chelomey and S. V. Chelomey [148, 149, 150, 151, 152] (also see 5.4).

Figure 5.6 shows the scheme of the system which has been considered. It consists of two rigid massless rods of the same length l. The rods are linked with each other and with the support by means of hinges with the linear torsional elastic and damping elements with the coefficients c and h respectively. The rods carry the masses $2m$ and m positioned at the ends of the first and second rods respectively. This mass distribution most closely corresponds to a continuous beam. The angles of the rods in respect to the straight position are given as θ_1 and θ_2. The system is subject to a partially follower-load P acting at the free rod end. The load arrangement is characterized by the parameter α where $\alpha = 1$ corresponds to the pure tangential loading and $\alpha = 0$ corresponds to the pure conservative loading.

The axis of suspension is subjected to a rectilinear harmonic displacement given as $G \sin \omega t$ where G is the amplitude and ω is the frequency of the displacement.

The small amplitude of the resonant excitations which has been considered implies that $G \ll l$ and $\omega \gg \lambda_2$ where λ_2 is the highest natural frequency of the pendulum.

Having used for the solution of the problem the approach of vibrational mechanics and the method of direct separation of motions Jensen obtained the differential equations of the slow motions of the system.

The investigation of stability of a rectilinear form of the equilibrium of the system, based on those equations, has led to a result given in Fig. 5.7. That figure, taken from [249] just like the description given below, shows the regions

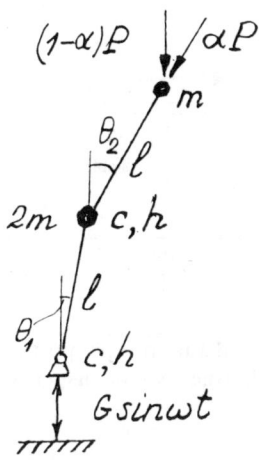

Figure 5.6. The elastic partially follower-loaded double pendulum with an oscillating axis of suspension.

of stability of the indicated form of equilibrium in the plane of the parameters α and $p = P\dfrac{l}{c}$ for the four different values of the parameter $\nu = \dfrac{3}{4}(G\omega_1)^2$ characterizing the intensity of vibration $\omega_1 = \omega\sqrt{ml^2/c}$; the curves are built for the values of the parameters $h_1 = h\sqrt{cml^2} = 0.1$ and $\omega_1 = 25$. Figure 5.7 also indicates whether the stability of the upright position is lost due to divergence (the dashed lines) or due to the flutter (the solid lines).

With the addition of vibrational forcing ($\nu \neq 0$) the region of stability is broadened. Stability boarders are moved up (down) for $p > 0$ ($p < 0$) respectively, especially for the upper right-hand divergence region, found for $\alpha >\approx 1.4$, $p > 0$. In this region the stability of the upright position is improved significantly with the addition of the excitation of the axis. However, as an exception, around $\alpha = 1.3$ the presence of the vibrational forcing may destabilize the pendulum. The stability boarder for the flutter moves to the right with the growing values of ν. This implies that the loads which previously did not affect the upright position may now cause the flutter of the pendulum.

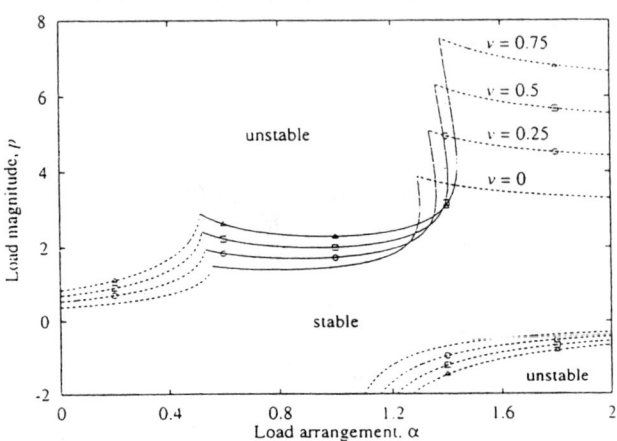

Figure 5.7. Regions of stability of the upright position of a partially follower-loaded double pendulum with an oscillating axis of suspension.

The markers in Fig. 5.7 represent the stability boarders, computed by numerical integration of the initial system of the equation by means of the Runge-Kutta standard algorithm. A very good agreement is noted, which shows that the method of direct separation of motions accurately captures the full effect of the excitation of the axis of the pendulum in respect to stability.

Jensen has also considered the case of bi-directional oscillations of the axis of suspension of the pendulum [249].

The effect of stabilization of the system under consideration has been investigated in [7].

5.4   Some   Generalizations   and   Applications.
On Vibrational Stabilization and Destabilization, Vibrational Shift (Drift)

One of the most glowing facts, illustrated in this chapter by the example of pendulums, is, undoubtedly, the vibrational stabilization of the positions of equilibrium of the mechanical systems which in the absence of vibration are unstable. On the other hand, as was marked, the equilibrium positions, stable in the absence of vibration, may prove to be unstable under vibration — a phenomenon of a parametrical resonance being an example of it.

In the absence of vibration the system may have no positions of equilibrium at all, or it may have a set of equilibrium positions, distributed continuously and under the effect of vibration it may acquire either one or several isolated quasi-equilibrium or equilibrium positions. A simplest example of it is a pendulum with a vertically placed axis of suspension, vibrating along the elliptical trajectory. In the absence of vibration all the positions of such a "pendulum" are those of equilibrium, and in case of vibration, as shown in 5.1.3, the pendulum is located along one of the major semi-axes of the elliptical trajectory of oscillations, i.e., it has only two positions of stable quasi-equilibrium. If such a pendulum were acted upon by a certain constant moment, then in the absence of vibration it would rotate about its own axis, i.e., it would not have any equilibrium positions at all, while an intensive vibration can stabilize the pendulum near the directions of major semi-axes of the elliptical trajectory of vibration.

The article by Yudovich [587] is devoted to the investigation of the quasi-equilibrium positions of the particle on a vibrating smooth surface.

Vibration may cause a shift of the stable equilibrium position of the system, which can be easily demonstrated by a pendulum with a vibrating axis of suspension. That effect is often called *vibrational shift*. It is manifested by the deviation of a compass needle from the right direction or by the regular mistakes in the readings of other devices caused by vibration [239, 282, 358, 499, 153]. The effect of vibrational displacement (drift) allows in the context of vibrational mechanics the consideration in a general form. Let the behavior of the system be described in the absence of vibration by the equation

$$m\ddot{x} = F(\dot{x}) + T(x) \tag{5.4.1}$$

where m is the mass, $T(x)$ is a certain slow positional force, and $F(\dot{x})$ is the resistance force which is supposed to be of the nature of viscous friction, i.e., we will assume that $F(0) = 0$, $F'(0) < 0$; in that case $F(\dot{x})$ is also slow. Then the equilibrium positions of the system $x = x_*$ are defined by the equation

$$T(x_*) = 0. \tag{5.4.2}$$

In case of vibration, as was shown above, under rather general assumptions the slow component of motion of the system is described by the equation

$$m\ddot{X} = T(X) + F(\dot{X}) + V(\dot{X}, X) \tag{5.4.3}$$

where $V(\dot{X}, X)$ is the vibrational force. As a result, the positions of equilibrium $X = X_*$ (positions of quasi-equilibrium) will be determined from the equation

$$T(X_*) + V(0, X_*) = 0, \tag{5.4.4}$$

different from (5.4.2); therefore, if $V(0, X_*) \neq 0$, which can be very rarely otherwise, then the values X_* will, generally speaking, differ from x_*, i.e. a vibrational drift will take place $\Delta = X_* - x_*$ there. Here we mean, of course, the stable positions of equilibrium, i.e. the values x_* and X_*, satisfying the conditions

$$T'(x_*) < 0, \quad T'(X_*) + V'_X(0, X_*) < 0, \quad F'(0) + V'_{\dot{X}}(0, X_*) < 0 \qquad (5.4.5)$$

where V'_X and $V'_{\dot{X}}$ denote the corresponding partial derivatives of the function $V(\dot{X}, X)$. If the vibrational force is small as compared to $T(x)$, then the value of the drift Δ can be determined by the equation

$$\Delta = -V(0, x_*)/T'(x_*). \qquad (5.4.6)$$

The simple examination that has been made can be readily generalized for the case of the system with many degrees of freedom.

All the other regularities, discussed here, characteristic of the effect of vibration on non-linear mechanical systems, can also be easily interpreted in the general form from positions of vibrational mechanics. And namely, let us assume that the mechanical system is such that its positions of equilibrium and their stability in the absence of vibration are determined by the potential energy Π. In case of vibration, the positions of quasi-equilibrium and their stability, as was shown in chapter 4, are in many cases determined by the potential function

$$D(X) = \Pi(X) + \Pi_V(X) \qquad (5.4.7)$$

which differs from $\Pi(X)$ by the term $\Pi_V(X)$, presenting the potential energy of vibrational forces. It is natural that the addition of the function $\Pi_V(X)$ may result in the fact that $D(X)$ will have new, as compared to $\Pi(X)$, "potential pits", answering the positions of stable quasi-equilibrium, and also in the shifting, or even in the absolute disappearance, of the potential pits, etc.

The systems, considered in this chapter, are but the simplest systems in which the remarkable regularities, we have been discussing here, can be manifested . We will make a brief review of such regularities, found in more complicated systems, including those which are considered in the subsequent chapters of this book.

Chelomei paid attention to the circumstance that, similarly to the situation with the pendulum , it is possible to increase the stability of the parametrically excited elastic systems with regard to the constant or slowly changing forces — the so called *static stability* [148] (see also [564, vol.2]). Among other things it was established by him that such stability can be achieved even when the static loads, acting on the vibrating system, exceed the Eiler critical forces. Chelomei Junior continued these investigations [150, 151, 152] and studied the cases when vibration caused the loss of the stability of the rod under the subcritical loads. Dynamic stabilization of unstable systems is the subject matter of investigations made by Valeev [556], and Zevin and Filonenko [596]. Vibrational stabilization of beams has been considered in the publications by Agafonov, Jensen, Krylov, Sorokin, Tchernyak, Thomsen [7, 250, 249, 533, 300, 535, 532].

Besides the publications, cited above, which refer to systems of pendulum type, a great number of investigations have been devoted to drifts, caused by the vibration of various gyroscopic devices. One of the first of such investigations belongs to Ishlinsky who by simple reasoning and by calculations explained the appearance of the precession of a gyroscope, caused by the vibration of its base with the pliability of the elements of suspension [240]. Among other investigations of that cycle we will mention [291, 153, 496].

Very interesting regularities of the behavior of the inertially-excited non-linear oscillatory systems, investigated by father and son Ragulskis [452], can be considered to be a distinctive manifestation of the effect of vibrational drift.

A certain peculiarity is characteristic of the effects of vibrational drift in systems with dry friction. The equilibrium positions in such systems, without vibration, continuously fill up a certain segment, often called *dead zone*. The vibrational force in such cases essentially depends on the slow component of velocity $\dot{X}$ and does not become zero when $\dot{X} = 0$, which leads to a drift. Such systems are considered in detail in 11.2 and in 15.3, particularly in connection with the specific features of the behavior of the granular materials in the vibrating vessels.

Most interesting from both the conceptual and applied points of view are the effects, appearing under the action of vibration, in systems "fluid – gas – solid particles". The description of some experiments with such systems is given in the work by Chelomei, cited above [149]. In these systems vibration may also cause both the stabilization and the destabilization of the states of equilibrium. It may also lead to a change in the form of the free surface of the fluid or in the surface of the "fluid – gas" interface in different fields of force, for instance in the gravity field. It may lead to the separation and stable localization of the liquid, solid and gaseous phases in certain zones of the vibrating volume. In some cases such effects are of a paradoxical character, particularly in the situations when the heavier components of the system are located higher than the lighter ones. Original results and detailed reviews of such investigations can be found in the books by Ganiyev and Ukrainsky and their fellow-workers [190, 191, 192] and also in publications by Bezdenezhnykh, Briskman, Cherepanov and Sharov

[66], Lukovsky and Timokha [344], Lyubimov and Cherepanov [351].

Of considerable applied interest are, in particular, the investigations of the behavior of solid particles and of bubbles of air in the vibrational fields of the type of standing wave; one of such publications is the article by Dukhin [170] which was already mentioned above. It is remarkable that, depending on the circumstances, the particles drift either to the nodes or to the antinodes of the wave as to the stable quasi-equilibrium positions, just like it happens in the case of a washer on a vibrating rod ([112, 149]; also see 5.2.3). The discovered regularities of the behavior of the particles and bubbles can, undoubtedly, be used for organizing processes in chemical technology, in enrichment and in various industries. As for the specific features of the behavior of the washer on a vibrating rod or string, their potentialities to cancel the oscillations were discussed in [30] which was already mentioned in 5.1.

The effects of stabilization and localization of the systems under vibration are extended not only to the states of equilibrium but to the motions as well. One of the examples of such an effect is the synchronization of rotating bodies, particularly of unbalanced rotors, which is discussed in detail in chapter 7. Another striking example is the stabilization and control of the motion of charged particles in the electromagnetic fields. It is known that such particles cannot have equilibrium positions in a static electromagnetic field. That makes the content of the corresponding classical theorem. But the imposition of a high frequency pulsating field results in the appearance of stable quasi-equilibrium positions, which in the case of the "moving standing wave" can be used for the acceleration of particles. This idea was advanced and grounded in the above mentioned publications by Gaponov and Miller [193, 194]. Making use of the conception of potential on the average dynamic systems, we can say that in this case due to the high frequency pulsation of the field, moving potential pits are created for the particle, and the latter is drawn in by them.

Linkov and Urman [332, 333, 334] in connection with the theory of a non-contact magnetic suspension considered the behavior of the conducting magnetic bodies in a magnetic field. Particularly, they showed that while in the constant field only the fast rotation of a dynamically symmetrical body (the gyroscope wheel) about the axis with a larger moment of inertia is stable, in the alternating field the rotation round the axis with a smaller moment of inertia, as well as round the axes, not coinciding with the main axes of inertia, can also be stable.

It goes without saying that these effects prove to be possible due to the non-linearity of equations, describing the behavior of the charged particle and the indicated bodies in the electromagnetic fields. They will be considered in some general form in chapter 19.

Summing up what has been said, we must emphasize that all the regularities, considered here, whose simplest analogs are the peculiarities of the behavior of a pendulum with a vibrating axis of suspension, allow a simple interpretation and mathematical description within the frames of vibrational mechanics and the conception of potentiality on the average, which in some cases has been done by the authors of the corresponding investigations. And still it seems that the abilities and advantages of the indicated concepts have not yet been fully utilized when describing the phenomena under consideration.

Chapter 6

Rotor Mechanisms. Machine Aggregates

6.1 Unbalanced Rotor on a Vibrating Base — Vibrational Maintenance of Rotation

6.1.1 Brief Bibliography

The vibration of the axis of an unbalanced rotor with a certain frequency ω can steadily maintain its stationary rotation with a mean angular velocity $< \dot{\varphi} >= \pm(p/q)\omega$ where p and q are whole positive numbers. It is characteristic of that wonderful phenomenon that the power, "taken away" from the source of oscillations and transformed into rotation, can in fact be sufficiently large, especially in the main regime, when the mean angular velocity of rotation is equal in absolute value to the frequency of vibration $\omega(p = q, < \dot{\varphi} >= \omega)$. Numerous important applications were based on that circumstance.

The condition under which it is possible to maintain by vibration the rotation of the unbalanced rotor in the indicated main regime and under the rectilinear harmonic oscillations of the axis was obtained by Bogolyubov [134, 136] (1950) by asymptotic method.

The author [69] (1954) has considered a more general case when the axis performs phase shifted harmonic oscillations in two, mutually perpendicular, directions. Conditions of the existence and stability of the main regime were obtained by the Poincare-Lyapunov method of small parameter.

Barkan and Shekhter observed and later Shekhter investigated the case when $< \dot{\varphi} >= \omega/2$ [492] (1959). The same case was examined by Biryukov by the Poincare-Lyapunov method (1959)(see [90]). Koughey investigated the effect as applied to the theory of the game-exercise "hula-hoop" [163] (1960) (also see 6.2)

In the monograph by Ragulskis [449] (1963) the regimes $< \dot{\varphi} >= \omega$, 2ω and

3ω were considered by the method of direct separation of motions; those regimes were also observed by him in the laboratory experiment and were reproduced on an electronic simulating installation.

Akulenko, Volosov, Moisseyev and also Morgunov and Chernousko (1963 and later) considered the rotational motions of a *generalized pendulum* — the system, described by the equation of the type

$$\ddot{z} + f(x) = \mu\varphi(\dot{z}, z, t)$$

where $f(z)$ is a periodic function of z, φ is a periodic function of t and μ is a small parameter. In some of those publications, whose main results are reflected in the book by Moiseev [390], examples have been also considered which refer to several special cases of the problem of the eigenpendulum (rotor) with a harmonically oscillating axis.

An important question about the regions of attraction of certain stationary regimes of the rotation of a rotor with a vibrating axis was considered in the publications of Agranovskaya [8] (1963), Mitulis [389] (1965–1966), Batalova [51] (1967), Kumpikas and Ragulskis [304] (1966) in which the analytic investigation is combined with a numerical analysis. Most fully in this respect the problem was considered in publications of Batalova, Belyakova and Bukhalova (see, particularly, [51, 52, 53, 144]) where they also discovered quite complicated ("chaotic") regimes of motion.

Questions of the theory of the phenomenon, results of experiments and some applications are considered in the publications by Inoue, Araki, Hayahsi, Miyaura, Matsushita, and Okada [238](1966–1974).

Bykhovsky has studied the vibrational maintenance of the rotation of an unbalanced rotor with an eccentrically attached pendulum [146] (1971).

The brief review given here is far from being complete. The presentation and generalization of the main results of the investigations of the phenomenon which were obtained for the corresponding period can be found in [90, 103, 114, 564, vol. 2].

In this section the solution of the problem is obtained by the method of direct separation of motions in the form, stated in the first part of the book. The work by Jaroshevich and the author [121] was also used here as well as the results of the analysis of the solution, stated in the above cited books by the author.

Now a few words about the title of the section. By *mechanisms* one usually implies devices, transforming the motion. From this point of view neither rotors, nor pendulums by themselves are mechanisms. Here, however, they can be referred to mechanisms, for they act as a peculiar kind of the transformers

of an oscillatory motion: in one case into a rotational motion, in another case
into that of quasi-equilibrium.

6.1.2   Equation of Motion, the Main Equation of Vibrational Mechanics

Like in the case of a pendulum (chapter 5), let the horizontally located axis
of the unbalanced rotor O_1 perform harmonic oscillations in two mutually perpendicular directions, according to the law

$$x = H\sin\omega t, \quad y = G\cos(\omega t + \theta) \tag{6.1.1}$$

where H, G are the amplitudes, ω is the frequency, and θ is the angle, characterizing the phase shift between the components of oscillations (Fig. 6.1,a).
Then the equation of motion of the rotor will have the form

$$I\ddot{\varphi} = mg\,\varepsilon_1\cos\varphi - m\,\varepsilon_1\omega^2[H\sin\omega t\sin\varphi + G\cos\varphi\cos(\omega t + \theta)]$$
$$+L(\dot{\varphi}) - R(\dot{\varphi}). \tag{6.1.2}$$

Here φ is the angle of rotation of the rotor, counted clockwise, m, I and ε_1
are the mass, the moment of inertia in respect to the axis of rotation and the
eccentricity of the rotor respectively, g is the free fall acceleration;

$$R(\dot{\varphi}) = R^0(|\dot{\varphi}|)\operatorname{sgn}\dot{\varphi} \tag{6.1.3}$$

is the anti-torque moment which can consider not only the resistance in the
bearing, but also the useful load, with $R^0(|\dot{\varphi}|) > 0$ being the modulus of that
moment. For generality it is assumed that the rotor can be acted upon by the
torque $L(\dot{\varphi})$ corresponding, say, to the static characteristic of an electric engine
of asynchronous type [a].

We will consider the rotation of a rotor with the frequency $\dot{\varphi}$ close to the
frequency of a uniform rotation $\sigma p\omega/q$ where $\sigma = 1$ or -1, depending on the
direction of rotation of the rotor in the regime under study, and p and q are
rather small mutually simple natural numbers, characterizing the multiplicity
of the regime. And namely, let us study the motion of the form

$$\varphi = \sigma\left[\frac{p\omega t}{q} + \alpha(t) + \psi(t, \omega t)\right] \tag{6.1.4}$$

---

[a] As usual, when solving problems, considered in this chapter, as well as problems on the
synchronization of vibroexciters (chapter 7), we will restrict ourselves to the static characteristics of the engines.

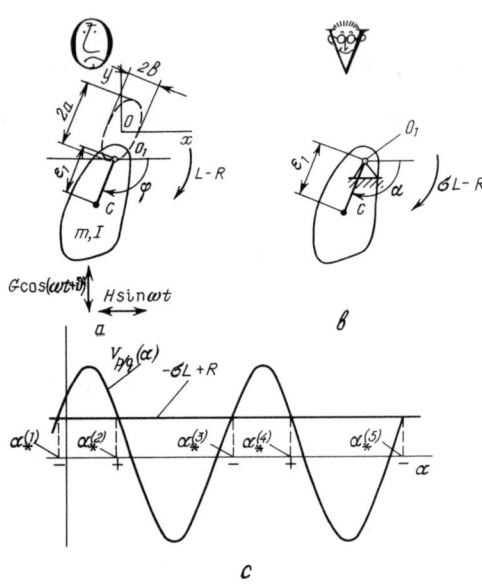

Figure 6.1. Unbalanced rotor with a vibrating axis.

where α is the main slow component, and ψ is the small fast $2\pi q$-periodic component with respect to the fast time ωt with

$$\frac{1}{2\pi q}\int\limits_0^{2\pi q}\psi(t,\tau)d\tau = <\psi(t,\tau)> = 0.$$

The corresponding motions will be called by us *regimes of the type* p/q.

Using the method of direct separation of motions, we will refer the moments $L(\dot\varphi)$ and $R(\dot\varphi)$ to the slow forces, and all the others to the fast forces and will pass from equation (6.1.2) to the system of two equations for the functions α and ψ (see equations (2.2.55) and (2.2.56) chapter 3).

$$I\ddot\alpha + k\dot\alpha = V_{p/q}(\alpha) + \sigma L(\sigma p\omega/q) - R^0(p\omega/q), \tag{6.1.5}$$

$$I\ddot\psi + k\dot\psi = \Psi(\alpha,\psi,\omega t). \tag{6.1.6}$$

Here we have

$$V_{p/q}(\alpha) = <\Phi(\alpha,\psi,\omega t)>,$$

$$\Phi(\alpha, \psi, \omega t) = \sigma \, m \, g\varepsilon_1[\cos(p\omega t/q + \alpha) - \psi \sin(p\omega t/q + \alpha)]$$
$$-m\,\varepsilon_1\omega^2\{H\sin\omega t[\sin(p\omega t/q + \alpha) + \psi\cos(p\omega t/q + \alpha)] +$$
$$\sigma G\cos(\omega t + \theta)[\cos(p\omega t/q + \alpha) - \psi\sin(p\omega t/q + \alpha)]\},$$
$$\Psi(\alpha, \psi, \omega t) = \Phi(\alpha, \psi, \omega t) - <\Phi(\alpha, \psi, \omega t)>. \qquad (6.1.7)$$

When obtaining equations (6.1.5)-(6.1.7), the right-hand side of the equation (6.1.2) was linearized over ψ and also the expression $L - R$ near the point $\dot{\varphi} = \sigma p\omega/q$ and the following designation was made: $k = -d(L-R)/\dot{\varphi}|_{\dot{\varphi}=\sigma p\omega/q} > 0$. As for the applicability of the method of direct separation of motions in the case under consideration, which consists in the analysis of the relative orders of the right-hand sides of equations (6.1.5) and (6.1.6) by the parameter $\varepsilon = 1/\omega$, we will speak about it later.

The expression $V_{p/q}(\alpha)$ in the equation of slow motion (6.1.5) is a vibrational torque, corresponding to the regime of a p/q type. As was mentioned, within the frames of this method, in order to determine this moment accurately enough, the equation of fast motion (6.1.6) can be solved in a crude way, assuming the slow variable α to be fixed ("frozen"). To this end we will write equation (6.1.6) as

$$I\ddot{\Psi} = \mu\Psi_1 \qquad (\mu\Psi_1 = \Psi - k\dot{\psi}), \qquad (6.1.8)$$

considering the value $\mu > 0$ as a small parameter; it should be noted that in this case it is not necessary to identify it with the small parameter $\varepsilon = 1/\omega$. It is quite easy to write out the conditions, imposed by this assumption upon the parameters of the system; they are analogous to (5.1.8) in chapter 5.

In the initial ("zero") approximation, the solution of equation (6.1.8), satisfying condition (6.1.4) will be $\psi = \psi_0 = 0$. In that case, according to equations (6.1.7), it is easy to obtain the expression for the vibrational torque

$$V_{p/q}^{(0)}(\alpha) = \begin{cases} -\dfrac{m\,\varepsilon_1\omega^2}{2}[H\cos\alpha + \sigma G\cos(\alpha - \theta)] & \text{at} \quad p = q = 1 \\ 0 & \text{at} \quad p \neq q \end{cases} \qquad (6.1.9)$$

It is remarkable that even in a very rough approximation, when determining the function ψ, the same expression of the vibrational torque for the main regime $p = q = 1$ was obtained by the Poincare method [69, 90]. Though, when using the zero approximation, the vibrational torque could be found only for that regime alone.

On calculating the first and second approximations for the function ψ after some rather simple, though somewhat clumsy calculations, we will obtain by equation (6.1.7) the following expressions for the vibrational torques [b].

$$V_{1/1} = -\frac{m\,\varepsilon_1\omega^2}{2}[H\cos\alpha + \sigma G\cos(\alpha - \theta)]$$

$$+\frac{3}{32}\frac{(m\,g\,\varepsilon_1)^2 m\varepsilon_1}{I^2\omega^2}[H\cos\alpha - \sigma G\cos(\alpha - \theta)]$$

$$+\frac{(m\,\varepsilon_1)^3\omega^2}{512I^2}[H\cos\alpha + \sigma G\cos(\alpha - \theta)][H^2 + G^2 - 2\sigma GH\cos\theta]$$

$$-\frac{k(m\,\varepsilon_1)^2}{I^2\omega^3}\left[\frac{1}{2}g^2 + \frac{\omega^4}{64}(H^2 + G^2)\right],$$

$$V_{1/2} = -\frac{2(m\,\varepsilon_1)^2 g}{I}[\sigma H\sin 2\alpha + G\sin(2\alpha - \theta)],$$

$$V_{2/1} = \frac{\sigma m\,g\,\varepsilon_1(m\,\varepsilon_1)^2}{576I^2}[5H^2\cos\alpha + 22G^2\cos(\alpha - 2\theta)]$$

$$+18\sigma GH\cos(\alpha - \theta)] - \frac{k(m\varepsilon_1)^2}{I^2\omega^3}$$

$$\times\left\{\frac{1}{16}g^2 + \frac{\omega^4}{108}[14(H^2 + G^2) + 13\sigma GH\cos\theta]\right\},$$

$$V_{1/3} = \frac{729(m\,g\,\varepsilon_1)^2 m\,\varepsilon_1}{32I^2\omega^2}[H\cos 3\alpha + \sigma G\cos(3\alpha - \theta)]$$

$$-\frac{729(m\,\varepsilon_1)^3\omega^2}{2048I^2}[H^3\cos 3\alpha - \sigma G^3\cos(3\alpha - \theta)]$$

$$+GH[2\sigma H\cos(3\alpha - \theta) + G\cos(3\alpha - 2\theta) - 2G\cos 3\alpha] - \sigma GH^2\cos(3\alpha + \theta)$$

$$-\frac{k(m\,\varepsilon_1)^2}{I^2\omega^2}\left\{\frac{27}{2}g^2 - \frac{\omega^4}{512}[7(H^2 + G^2) + 18\sigma GH\cos\theta]\right\},$$

$$V_{3/1} = \frac{(m\,\varepsilon_1)^3\omega^2}{2048I^2}[H^3\cos\alpha + \sigma GH^2\cos(\alpha - \theta)]$$

$$-G^2 H\cos(\alpha - 2\theta) - \sigma G^2\cos(\alpha - 3\theta)]$$

$$-\frac{k(m\,\varepsilon_1)^2}{I^2\omega^3}\left[\frac{1}{54}g^2 + \frac{\omega^4}{512}(H^2 + G^2 + 14\sigma GH\cos\theta)\right],$$

$$V_{2/3} = -\frac{729\,\sigma m\,g\,\varepsilon_1(m\,\varepsilon_1)^2}{64}\frac{1}{I^2}[H^2\cos 3\alpha + G^2\cos(3\alpha - 2\theta) + 2\sigma GH\cos(3\alpha - \theta)]$$

$$-\frac{27k(m\,\varepsilon_1)^2}{I^2\omega^3}\left\{\frac{1}{16}g^2 - \frac{\omega^4}{250}[31(H^2 + G^2) + 63\sigma GH\cos\theta]\right\}. \qquad (6.1.10)$$

Thus, taking into account two approximations for the function ψ , it is possible to obtain the expressions of the vibrational torques in the multiple regimes

---

[b] These equations were obtained by Malakhova who has refined the corresponding formulas of the article [117] and corrected some errors.

$p/q = 1/2, 2/1, 1/3, 3/1, 2/3$ and specify the expression for the vibrational moment in the main regime $p/q = 1$. It will be hardly expedient to calculate the subsequent approximations, the corresponding corrections to values of the vibrational torques being small, and the regimes with the large values of p and q difficult to be realized (see below).

6.1.3 Stationary Regimes of Rotation of the Rotor and Their Stability

Having expression (6.1.10), from the equation of slow motions (6.1.5) — the main equation of vibrational mechanics — we obtain the following relation in order to find the values of the angle α, corresponding to possible regimes of the stationary rotation of the rotor:

$$V_{p/q}(\alpha) + \sigma L(\sigma p\omega/q) = R^0(p\omega/q). \tag{6.1.11}$$

The condition of the presence of real solutions with respect to α will be the condition of the existence of the corresponding regime of rotation of equation (6.1.11). A certain value $\alpha = \alpha_*$, found from equation (6.1.11), will answer the stable motion if

$$\left.\frac{dV_{p/q}}{d\alpha}\right|_{\alpha=\alpha_*} < 0; \tag{6.1.12}$$

In case the opposite inequality is satisfied, the motion under consideration is unstable. We come to this conclusion if we compose an equation in variations for equation (6.1.5) and for the solution $\alpha = \alpha_*$.

Fig. 6.1,c shows a graphic interpretation of what has been stated: the stationary regimes correspond to the points of intersection of the curve $y = V_{p/q}(\alpha)$ and the horizontal line $y = -\sigma L + R^0$. The sign "plus" marks the points, corresponding to the stable regimes, and the sign "minus" to those unstable. If the line $y = -\sigma L + R^0$ does not intersect the curve $y = V_{p/q}(\alpha)$, then the regime under study is impossible. Since in cases of the regimes of type $p/q = 1/1, 2/1 and 3/1$ the expression for $V_{p/q}(\alpha)$ contains only the $\sin\alpha$ and $\cos\alpha$, there can be in those cases only one stable stationary regime, essentially different from any other (with the values α_* differing by $2\pi n$ where $n = \pm 1, \pm 2, \ldots$). In case $p/q = 1/2$, there can be two such stable motions, and in case $p/q = 1/3$ and $p/q = 2/3$ there can be three.

It is not difficult to build up the potential functions, corresponding to the expressions (6.1.10) and equality (6.1.11), that is such functions $D_{p/q}$ that

$$dD_{p/q}/d\alpha = -[V_{p/q}(\alpha) + \sigma L(\sigma p\omega/q) - R^0(p\omega/q)];$$

The stable motions of the type under consideration will correspond to the points of the minimum of those functions.

The a posteriori check of the validity of the main suppositions of the method of direct separation of motions is performed in this case in the way analogous to that in 5.2. However, there is no special necessity in it, we being interested only in the vicinities of the stable stationary regimes of rotation of the rotor, corresponding to $\dot{\alpha} = 0$ and $\alpha = \alpha_*$ (see note 4 in 3.2.8). As for the other regions of the values $\dot{\alpha}$ and α, the main suppositions are valid for them provided $\psi \approx \varepsilon^2 \alpha$. It is also easy to see that for the main regime the frequency of oscillations of the rotor λ near the stationary value $\alpha = \alpha_*$ is of the order of $\omega \sqrt{m \varepsilon_1 \sqrt{G^2 + H^2/I}}$ that is $\lambda \ll \omega$ due to the assumptions made.

6.1.4   *Vibrational Maintenance of Rotation of the Rotor*

Let us consider the most interesting case when the rotating moment is absent, that is when $L(\dot{\varphi}) = 0$. Then equation (6.1.11) will take the form

$$V_{p/q}(\alpha) = R^0(p\omega/q) \qquad (6.1.13)$$

that is it expresses a condition that the moment of the resistance forces should be compensated by the vibrational moment. In case this equation has real solutions $\alpha = \alpha_*$, there exist stationary regimes of the rotation of the rotor with a frequency $p\omega/q$. If inequality (6.1.12) is satisfied, the corresponding motions are stable. In its turn, for the equation (6.1.13) to have real solutions with respect to α it is necessary and sufficient that the inequality

$$(V_{p/q})_{\max} > R^0(p\omega/q) \qquad (6.1.14)$$

should be satisfied; $(V_{p/q})_{\max}$ is the greatest value with respect to α of the vibrational torque (*modulus of the vibrational torque*) . According to (6.1.13) this value determines the maximal moment of resistance which can be overcome by the rotor in the regime under consideration, and also the maximal power

$$(N_{p/q})_{\max} = (V_{p/q})_{\max}\omega, \qquad (6.1.15)$$

which can be transferred by vibration to the shaft of the rotor.

Thus, inequality (6.1.14) presents the main condition for the ability of a vibrational maintenance of the rotation of an unbalanced rotor. The values $(V_{p/q})_{\max}$ involved in this condition are, according to (6.1.10), defined by the formulas:.

$$(V_{1/1})_{\max} = m \varepsilon_1 \omega^2 A_{1/1}, \qquad (V_{1/2})_{\max} = \frac{4(m \varepsilon_1)^2 g A_{1/2}}{I},$$

$$(V_{2/1})_{\max} = \frac{(m\varepsilon_1)^3 g A_{2/1}}{576 I^2}, \quad (V_{1/3})_{\max} = \frac{729(m\varepsilon_1)^3 A_{1/3}^3}{2048 I^2},$$

$$(V_{1/3})_{\max} = \frac{(m\varepsilon_1)^3 \omega^2 A_{3/1}^3}{2048 I^2}, \quad (V_{2/3})_{\max} = \frac{729(m\varepsilon_1)^3 g A_{2/3}^2}{64 I^2}. \qquad (6.1.16)$$

(Here we neglected the terms, proportional to the damping factor k which are usually not large, as compared to others; though those terms can be easily taken into account by adding them to the moment of the resistance forces $R^0(p\omega/q)$; $A_{1/1}, A_{1/2}, \ldots, A_{2/3}$ denote the positive values with the dimension of the amplitude, and they are defined by the relations

$$A_{1/1}\cos\chi_{1/1} = \frac{1}{2}H\left[1 - \nu\left(1 + \frac{2C^2}{B^2}\right)\right] + \frac{1}{2}\sigma G\cos\theta\left[1 + \nu\left(1 - \frac{2C^2}{B^2}\right)\right],$$

$$A_{1/1}\sin\chi_{1/1} = \frac{1}{2}\sigma G\sin\theta\left[1 - \nu\left(1 - \frac{2C^2}{B^2}\right)\right];$$

$$A_{1/2}\cos\chi_{1/2} = \frac{1}{2}(H + \sigma\cos\theta), \quad A_{1/2}\sin\chi_{1/2} = \frac{1}{2}\sigma G\sin\theta;$$

$$A_{2/1}^2\cos\chi_{2/1} = 5H^2 + 18\sigma GH\cos\theta + 22G^2\cos 2\theta,$$

$$A_{2/1}^2\sin\chi_{2/1} = 18\sigma GH\sin\theta + 22G^2\sin 2\theta,$$

$$A_{1/3}^3\cos\chi_{1/3} = \frac{2}{3}B^2(H + \sigma G\cos\theta) + \sigma G\cos\theta(G^2 - H^2)$$
$$- H^3 + G^2 H(2 - \cos 2\theta),$$

$$A_{1/3}^3\sin\chi_{1/3} = \sigma G\sin\theta\left(\frac{2}{3}B^2 + G^2 - 3H^2 - 2\sigma GH\cos\theta\right);$$

$$A_{3/1}^3\cos\chi_{3/1} = H^2(H + \sigma G\cos\theta) - G^2(H\cos 2\theta + \sigma G\cos 3\theta),$$

$$A_{3/1}^3\sin\chi_{3/1} = \sigma G(H^2\sin\theta - \sigma GH\sin 2\theta - G^2\sin 3\theta);$$

$$A_{2/3}^2\cos\chi_{2/3} = H^2 + G^2\cos 2\theta + 2\sigma GH\cos\theta,$$

$$A_{2/3}\sin\chi_{2/3} = G^2\sin 2\theta + 2\sigma GH\sin\theta,$$

$$(\nu = 3(m\varepsilon_1 g)^2/16 I^2\omega^4,$$

$$B^2 = 96g^2/\omega^4, C^2 = G^2 + H^2 - 2\sigma GH\cos\theta). \qquad (6.1.17)$$

Expressions (6.1.16) are essentially simplified in case of vibration in a circular trajectory of the radius R when $G = H = R$, $\theta = 0$ or $\theta = \pi$:

$$(V_{1/1})_{\max} = \frac{m\varepsilon_1\omega^2 R}{2}(1 + \sigma^*) - \frac{3}{32}\frac{(m\varepsilon_1)^3 g^2}{I^2\omega^2}(1 - \sigma^*),$$

$$(V_{1/2})_{\max} = \frac{2(m\varepsilon_1)^2 g R}{I}(1 + \sigma^*), (V_{2/1})_{\max} = \frac{(m\varepsilon_1)^3 g R^2}{64 I^2}(3 + 2\sigma^*),$$

$$(V_{1/3})_{\max} = \frac{729}{32}\frac{(m\,\varepsilon_1)^3 g^2 R}{I^2 \omega^2}(1+\sigma^*), \quad (V_{3/1})_{\max} = 0,$$

$$(V_{2/3})_{\max} = \frac{729}{32}\frac{(m\,\varepsilon_1)^3 g R^2}{I^2}(1+\sigma^*), \qquad\qquad (6.1.18)$$

$$\sigma^* = \sigma\mathrm{sgn}(\cos\theta) = \pm 1 \qquad\qquad (6.1.19)$$

It is easy to see that the value σ^* introduced in that way is equal to 1, if we consider the rotation of the rotor in the direction, coinciding with that of the motion of axis of the rotor along the elliptic trajectory (6.1.1) and $\sigma^* = -1$ in case the indicated directions do not coincide.

From equations (6.1.17) and (6.1.18) it follows that the condition of maintaining the rotation in the direction of motion of its axis along the elliptic trajectory is "softer" for every regime than the corresponding condition for the case when the indicated directions are opposite (in the first case the vibrational torques being larger than in the second). When the axis oscillates along a circular trajectory, the rotation of the rotor in the opposite direction is for the majority of multiple regimes absolutely impossible. The regime of the rotation of the rotor of type 3/1 in the case of circular oscillations is, under the assumptions made, absolutely impossible either, irrespective of the direction of rotation.

Let us consider more thoroughly the case of the main regime $p = q = 1$, which is most interesting for applications. Should we neglect the value ν, usually small as compared to 1, then, according to equations (6.1.17), we will have

$$A_{1/1} \approx A = \frac{1}{2}\sqrt{G^2 + H^2 + 2\sigma GH \cos\theta} \qquad\qquad (6.1.20)$$

and then

$$(V_{1/1})_{\max} = m\,\varepsilon_1\,\omega^2 A = FA \qquad\qquad (6.1.21)$$

where $F = m\varepsilon_1\omega^2$ denotes a centrifugal force, developed under the rotation of the rotor if its axis is fixed. Making use of the well known formulas of analytic geometry, it is easy to express the semi-axes of the elliptic trajectory of vibration a and b via the parameters G, H, and θ:

$$a = \frac{\sqrt{2}GH|\cos\theta|}{\sqrt{G^2 + H^2 - \sqrt{(G^2 + H^2)^2 - 4G^2 H^2 \cos^2\theta}}}$$

$$b = \frac{\sqrt{2}GH|\cos\theta|}{\sqrt{G^2 + H^2 + \sqrt{(G^2 + H^2)^2 - 4G^2 H^2 \cos^2\theta}}} \qquad (6.1.22)$$

and give formula (6.1.17) the following simple form

$$A_{1/1} \approx A = \frac{1}{2}(a + \sigma^* b). \tag{6.1.23}$$

The value A can be called the *effective amplitude of oscillations* of the axis of the rotor. In accordance with (6.1.23) this amplitude is equal to half the sum of the semi-axes of the ellipse (6.1.1) if we consider the rotation of the rotor in the direction of the motion of its axis along the elliptic trajectory ($\sigma^* = 1$) and it is equal to the half-difference of the semi-axes provided the indicated directions do not coincide ($\sigma^* = -1$).

Thus, formula (6.1.21) allows the following quite simple interpretation: the modulus of the vibrational torque in the main regimes approximates the product of the centrifugal force $F = m\varepsilon_1\omega^2$ by the effective amplitude of vibration of the axis of the rotor. In case of a rectilinear vibration of the axis $b = 0$, and $A = 1/2a$, that is the effective amplitude is merely equal to half the amplitude of the oscillations of the axis. When the axis of the rotor vibrates in a circular trajectory of the radius $R = a = b$, the effective amplitude is $A = R$ provided the rotation of the rotor is in the same direction as the motion of the axis along the circumference. If the indicated directions are opposite, both the effective amplitude and the rough absolute value of the vibrational torque ($V_{1/1}$) become zero.

From formula (6.1.23) it follows that the conditions of the rotation of the rotor in the direction of motion of the rotor axis along the elliptic trajectory ($\sigma^* = 1$) are more favorable than those in the opposite direction ($\sigma^* = -1$). The only exception is the case of a rectilinear vibration $b = 0$, when both directions of the rotation "have the same rights".

Formula (6.1.21) may serve for a crude estimation of the vibrational torque, transferred to the unbalanced rotor with a vibrating axis, and also in more complicated cases, for instance when investigating self-synchronization (see 7.2 and 7.4). It should be also noted that the condition of a possibility of a vibrational support of rotation (6.1.14) for the main regime acquires an especially simple form in the case when the resistance to the rotation of the rotor is conditioned mostly by the resistance in the bearings. So for the rolling bearings $R^0(\omega) = \frac{1}{2}fFd$ where f is the friction coefficient in the bearer and d is the diameter of the inner ring. As a result, in view of (6.1.21), inequality (6.1.14) is reduced to the requirement

$$2a/d > f. \tag{6.1.24}$$

Taking into consideration that usually $10^{-3} \leq f \leq 10^{-2}$, it is clear that the rotation of the rotor can be supported by the oscillations whose amplitude is far smaller than the size of its bearings.

A circumstance, rather important for applications, should also be noted: the magnitude of the maximal power $(N_{1/1})_{max}$ which can be transferred by vibration to the shaft of the rotor in the main regime at the parameters that can be virtually realized is rather great. So, for instance, at $m\varepsilon_1 = 10kg \cdot m$, $A = 0.25 \cdot 10^{-2}m$ and $\omega = 314s^{-1}$ we obtain

$$(N_{1/1})_{\max} = (V_{1/1})_{\max}\omega = m\varepsilon_1 A\omega^3$$
$$= 10 \cdot 0.25 \cdot 10^{-2} \cdot 314^3 \approx 0.8 \cdot 10^6 N \cdot m/s \approx 800kW!$$

In view of relationships (6.1.10), (6.1.20) and (6.1.21), equation (6.1.13) for the main regime will take the following form (bearing in mind that we neglect the value $\nu = 3(m\varepsilon_1)^2 g^2/16I^2\omega^4$ as compared to unity):

$$- FA\cos(\alpha - \chi) = R^0(\omega) \qquad (6.1.25)$$

where the angle χ is defined by the equalities (6.1.17). If condition (6.1.14) is satisfied, that is the inequalities $FA > R^0(\omega)$ (or in case of rolling bearings — inequalities (6.1.24)) are satisfied, then that equation has two essentially different solutions:

$$\alpha_*^{(1)} = \chi + \delta, \qquad \alpha_*^{(2)} = \chi - \delta, \qquad (6.1.26)$$

where

$$\frac{1}{2}\pi < \delta = \arccos\left[-\frac{R^0(\omega)}{FA}\right] < \pi, \qquad (6.1.27)$$

and according to (6.1.12) the condition of stability is reduced to the inequality $\sin(\alpha - \chi) < 0$ so that the solution $\alpha_*^{(1)}$ is answered by the unstable motion and the solution $\alpha_*^{(2)}$ is answered by the stable motion.

Let us now turn to other ("multiple") regimes. From formulas (6.1.16) – (6.1.18) it follows that the regimes of type $1/2, 2/1$ and $2/3$ are impossible when there is no free fall acceleration g, that is in the case when the axis of the rotor is vertical; while all the other regimes under consideration can exist even at $g = 0$. The numerical analysis of expressions (6.1.16) shows that the absolute value of the vibrational torque and of the maximal power transferred for the main regime considerably exceeds those values for the multiple regimes. So in case of a circular vibration when the direction of rotation of the rotor coincides with it ($\sigma^* = 1$) and at the values of the parameters $m\varepsilon_1 = 10kg \cdot m$, $R = 0.25 \cdot 10^{-2}m$, $I = 1kg \cdot m^2$, $\omega = 147s^{-1}$ according to formulas (6.1.18) we obtain

$$(N_{1/1})_{\max} = (V_{1/1})_{\max}\omega \approx 79kW, \quad (N_{1/2})_{\max} = (V_{1/2})_{\max}\frac{\omega}{2} \approx 0.7kW,$$
$$(N_{1/3})_{\max} = (V_{1/2})_{\max}\frac{\omega}{3} \approx 2.5 \cdot 10^{-2}kW,$$
$$(N_{2/1})_{\max} = (V_{2/1})_{\max}2\omega \approx 1.4 \cdot 10^{-3}kW, \quad (N_{3/1})_{\max} = 0$$

Therefore while the main regime finds numerous practical applications, the realization of the multiple regimes, at least without using any additional devices, involves great difficulties.

One of the ideas is to apply to the rotor a moment which periodically depends on the angle of rotation of the rotor φ, for instance by using a spring, eccentrically connected to the rotor [92, 121]. With such a spring (Fig. 6.2) in the corresponding formulas for $V_{p/q}$ instead of the value $m\, g\, \varepsilon_1$ there will be the value $m\, g\, \varepsilon_1 + T_0 l$ where T_0 is the strain of the spring, and l is the distance from the point of the hinge fixing of the spring to the axis of the rotor. So for instance if we assume that $T_0 l = 60 m\, g\, \varepsilon_1$, then all the other conditions being the same, we obtain $(N_{1/2})_{\max} = 43\text{kW}$, $(N_{1/3})_{\max} = 93\text{kW}$, $(N_{2/1})_{\max} = 0.085\text{kW}$.

Other ways of increasing the values $(N_{p/q})_{\max}$ at $p/q \neq 1$ are also possible. They are based on the increase of the degree of non-uniformity of rotation of the rotor in the stationary regimes.

In conclusion it should be noted that for the rotors installed in the ordinary rolling bearings a "rigid" excitation of the considered regimes is characteristic — they appear and are steadily maintained provided the rotor has been previously accelerated up to the value of the frequency of rotation, relatively close to that of the stationary rotation.

It should be also noted that the unbalanced rotor on a vibrating base can be considered to be a kind of a peculiar synchronous engine, not requiring any supply of electric energy: the vibrating base, on which the rotor is installed, serving as a channel, transmitting the energy. Such an engine can be called *vibro-engine of a synchronous type* (vibro-engines of an asynchronous type are

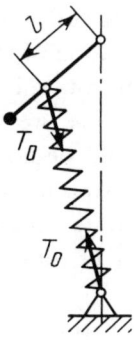

Figure 6.2. One of the ways of improving the conditions of a vibrational support of the multiple regimes of rotation of the unbalanced rotor.

discussed in 10.4).

6.1.5   *Vibrational Capture of the Rotation of an Unbalanced Rotor of an Electric Engine. How the Efficiency of an Engine Can Be More than Unit*

Considering the main regime, let us write down equation (6.1.11) in the form

$$\sigma L(\sigma\omega) - R^0(\omega) + FA\cos(\alpha - \chi) = 0. \tag{6.1.28}$$

The condition of the existence of real roots of this equation is the inequality

$$|\,\sigma L(\sigma\omega) - R^0(\omega)\,| < FA = (V_{1/1})_{\max}. \tag{6.1.29}$$

Provided this condition is satisfied, equation (6.1.28) allows two essentially different solutions: one of which is stable, while the other is unstable (see also Fig. 6.1,c). Inequality (6.1.29) is in fact that condition under which the rotation of the rotor of the electric engine is captured by the external frequency ω. In case there is no vibration, the rotor rotates, generally speaking, with some other frequency.

To give to this condition, from the standpoint of physics, a more distinct form we will introduce here the so called *partial angular velocity* of the rotation of the rotor of the electric engine ω_0, implying by it the frequency of the rotation of the rotor on a fixed base counted, unlike φ, (see Fig. 6.1,a) in the direction of the rotation of the rotor in the motion under consideration. It is evident that the velocity ω_0, defined in that way, satisfies the equation

$$\sigma L(\sigma\omega_0) = R^0(\omega_0)\,\mathrm{sgn}\,\omega_0. \tag{6.1.30}$$

Mark that the vibration of the base on which the engine with the unbalanced rotor is installed may cause the rotation of the rotor in the direction opposite to that in which the engine tends to rotate it. In this case, answering the work of the engine in the generating regime, we have $\sigma\dot{\varphi}_0 = \omega_0 < 0$, that is the partial velocity is negative. It should be noted that the notion of the partial velocity can be considered as an extension of the notion of the frequency of self-oscillations to the rotating objects. An essential difference, however, lies in the fact that the partial velocity of rotation can be both either positive or negative.

Let it be however that $\omega_0 > 0$ that is the engine works in the normal regime. Then in case the frequency of vibration ω coincides with ω_0, then according to (1.30) $\sigma L(\sigma\omega) - R^0(\omega) = \sigma L(\sigma\omega_0) - R^0(\omega_0)$ and condition (6.1.29) is certain to be fulfilled. In other words, if in the absence of the vibration of the base of the rotor the rotor rotated in the stationary regime with the frequency $\dot{\varphi}_0 = \sigma \omega_0$, then in the presence of vibration with the frequency $\omega = \omega_0$ the rotor will also be able to rotate with the same frequency. However, the regime under consideration will also exist in the case when the oscillation frequency is different from the partial velocity ω_0 but does not differ from it too much, so that the vibrational moment can compensate the "excessive moment" $\sigma L(\sigma \omega) - R^0(\omega)$. Thus, there will be, generally speaking, an interval of the change in the vibration frequency.

$$- \Delta_1 < \omega - \omega_0 < \Delta_2 \qquad (\Delta_1 > 0, \quad \Delta_2 > 0) \qquad (6.1.31)$$

within which the rotation of the rotor is captured by the external frequency. The width of that interval is called *capture band*. In case Δ_1 and Δ_2 are relatively small, on linearizing the relationships $\sigma L(\sigma \omega) - R^0(\omega)$ and $F = m\,\varepsilon_1\,\omega^2$ near $\omega = \omega_0$, from inequality (6.1.29) we obtain

$$\Delta_1 = \omega_0 - \omega_1 = \frac{m\,\varepsilon_1\,\omega_0^2}{k + 2m\,\varepsilon_1\,\omega_0}, \quad \Delta_2 = \omega_2 - \omega_0 = \frac{m\,\varepsilon_1\,\omega_0^2}{k - 2m\,\varepsilon_1\,\omega_0} \qquad (6.1.32)$$

where

$$k = - \left. \frac{d[\sigma L(\sigma \omega) - R^0(\omega)]}{d\omega} \right|_{\omega = \omega_0}, \qquad (6.1.33)$$

usually $k > 2m\,\varepsilon_1\,\omega_0 > 0$.

From formulas (6.1.32) it follows that there is an absence of the *threshold of capture* — i. e. such value of the effective amplitude of vibration A at which the capture band disappears. For the ordinary simplest self-oscillating system the analogous fact was established by Andronov and Vitt [20].

It should be noted that the capture band for the system under consideration can be sufficiently wide. In particular, from what has been said in 6.1.4 it follows that the partial velocity ω_0 can be equal to zero (the engine being absent or turned off from the circuit), and despite that the rotation of the rotor can be captured by vibration. Moreover, apart from the interval (6.1.31), containing the frequency ω_0, there can also be other intervals of the frequency change ω in which capture takes place. These regions can be revealed by constructing graphs of the functions $\sigma L(\sigma \omega)$, $R^0(\omega)$ and $m\,\varepsilon_1\,\omega^2 A$ [90].

From these relationships it is clear that at $\omega < \omega_0$ the vibrational torque retards the rotor and at $\omega > \omega_0$ it pushes it ahead. In the latter case the power greater than that supplied is taken off from the shaft of the engine. As a result, the researcher who did not notice the vibration of the base or the fact that the rotor was unbalanced, i.e. the observer **W** (see the Preface) will arrive at the wrong conclusion that the efficiency of the engine exceeds a unit. The observer **V** will easily explain this paradox by the presence of the vibrational torque.

Mark that in the case of multiple regimes, the investigation of the effect of capture is carried out in a similar way, and the results are also similar, with the only difference that the effect, generally speaking, appears to be manifested much weaker due to smaller values of moduluses of vibrational torques (see 6.1.4).

6.2   Devices "A Roller (Ball) in a Vibrating Cavity" and "A Ring on a Vibrating Rod (Hula-Hoop)". Vibrational Maintenance of the Planetary Motion

6.2.1   *About the Devices under Consideration: a Brief Bibliographic Reference*

The unbalanced rotor, rotating in the bearings, which was considered in 6.1. has two other related devices, presented in Fig. 6.3. The first device (Fig. 6.3,a) presents a circular cylindrical roller or a ball, lying with a clearance in a cylindrical cavity.] A second device (Fig. 6.3,b) consists of a circular ring, also with a clearance, set on a cylindrical rod. In the first case it is the vibration of the body in which the cavity is located, in the second case it is the

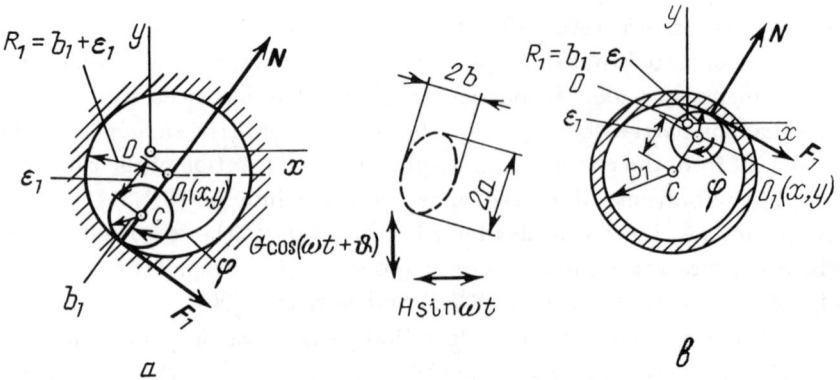

Figure 6.3. Vibrational maintenance of planetary motion. a) a roller (ball) in vibrating cavity; b) a ring on a vibrating rod ("hula-hoop").

vibration of the rod with the frequency ω that under certain conditions causes and steadily maintains the revolving of the roller or the ball in the cavity and of the ring around the rod. The centers of gravity of bodies C being rolled moving in a circle around the centers O_1 of the cavity or of the rod with the average frequency $<\dot{\varphi}> = \pm p\omega/q$ where just like above p and q are certain positive integers. In other words, in this case there is an *effect of vibrational maintenance of the planetary rotation*. A well known example of the second device is the game-exercise "hula-hoop".

The effect of vibrational maintenance of the stationary planetary motion found an important application in technology, particularly in new machines for crushing and grinding the ore and other solid materials — *in vibro-inertial crushers and mills* [86, 456, 238]. The process of crushing or grinding the material in such machines takes place in the clearance between the surfaces of the cavity and the roller, ball or any other body of revolution that is being rolled. These applications are based first of all on the ability to transfer by means of vibration to the body that is being rolled a considerable power, especially in the main regime ($p = q = 1$), just like it is for the unbalanced rotor, (see 6.1.4), and secondly on the fact that in this case the stationary planetary motion in the indicated regime is softly excited — it very soon becomes stationary under practically any initial conditions corresponding to the start of the machine.

The effect of maintaining the planetary motion of the roller in the cavity, performing two-component harmonic oscillations, is considered by the Poincare-Lyapunov method of small parameter in the work of the author [69] (1954). As applied to the theory of crushing and grinding machines, this effect was investigated by Dubrovin, Rudin, Rundquist and the author [80, 81, 168, 468] (1960-61). An experimental investigation of the regimes of establishing a planetary motion of the roller under vibration was performed by the author by method of a high-speed filming [90] (1960). Those investigations were presented and generalized in books [90, 103, 114]. The theory of the game-exercise "hulla-hoop" was considered in the work by Coughey [163] (1960). The investigation of the planetary motion of a roller in a vibrating cavity was performed by Inoue, Araki, Hayashi, Matsushita, Miyaura and Okada [238] (1966-1967).

In this section the problem is considered by method of direct separation of motions, and the results obtained coincide with those obtained before by other methods.

6.2.2 *Equation of Motion, the Main Equation of Vibrational Mechanics*

We will assume that in the motions that are being considered the roller or the ball does not leave the surface of the cavity, and the ring does not leave the surface of the rod, both bodies running about the corresponding surfaces without slipping. The conditions under which these assumptions are fulfilled will be considered below. Then the motion of the roller (ball) and of the ring can be considered as the motion of the system with one degree of freedom. We assume that the oscillations of the body in which the cavity is located and those of the rod are given by formulas (6.1.1) just like in 6.1. As a result, the motion of the bodies under consideration will be described by the equation

$$I\ddot{\varphi} = mg\,\varepsilon_1\cos\varphi - m\,\varepsilon_1\,\omega^2[H\sin\omega t\sin\varphi + G\cos(\omega t + \theta)\cos\varphi]$$
$$-R(\dot{\varphi}, \varphi, \omega t). \tag{6.2.1}$$

Here φ is the angle between the lines O_1C connecting the centers of the cavity (rod) with those of the roller or the ball (ring) and the axis Ox of the fixed axes of the coordinates xOy (Fig. 6.3); m is the mass of the roller or the ball (ring); b_1 is the radius of the roller (the internal radius of the ring); $\varepsilon_1 = O_1C$ is the eccentricity, in the case of the roller or the ball $R_1 = b_1 - \varepsilon_1$ being the radius of the cavity, and in the case of a ring $R_1 = b_1 - \varepsilon_1$ being the radius of the rod; $I = m\varepsilon_1^2\left(1 + \frac{I_C}{mb_1^2}\right)$ is the reduced moment of inertia of the roller or of the ball (ring) with I_c being a moment of inertia of the corresponding body with respect to the axis passing through its center of gravity. The moment of resistance to the rerolling of the roller (ring) is denoted by R:

$$R = R(\dot{\varphi}, \varphi, \omega t) = f'\varepsilon_1 N \operatorname{sgn}\dot{\varphi}, \tag{6.2.2}$$

where f' is the *coefficient of resistance to rerolling*, and N is the normal reaction between the contacting bodies, defined by the formula

$$N = m(\varepsilon_1\dot{\varphi}^2 + g\sin\varphi - \ddot{x}\cos\varphi + \ddot{y}\sin\varphi), \tag{6.2.3}$$

in which x and y are given by equalities (6.1.1). Formula (6.2.3) is obtained from the equation of the motion of the center of gravity C of the roller or of the ball (ring) projected on the normal to its trajectory, considered in the mobile axes, parallel to those of the fixed system xOy.

If equalities (6.2.2) and (6.2.3) are taken into account, the differential equation (6.2.1) takes the following final form:

$$I\ddot{\varphi} = mg\,\varepsilon_1\cos\varphi - m\varepsilon_1\omega^2[H\sin\omega t\sin\varphi + G\cos(\omega t + \theta)\cos\varphi]$$
$$-mf'\,\varepsilon_1[H\omega^2\sin\omega t\cos\varphi - G\omega^2\cos(\omega t + \theta)\sin\varphi + \varepsilon\dot{\varphi}^2 + g\sin\varphi]\operatorname{sgn}\dot{\varphi} \tag{6.2.4}$$

It is of the same type as the equation of motion of the rotor (6.1.2). As before, we will seek its solution of the form (6.1.3), restricting ourselves to the main regime ($p = q = 1$). Therefore we will write out the main equation of vibrational mechanics for the problem under consideration without dwelling on the process of obtaining it by method of direct separation of motions:

$$I\ddot{\alpha} + k\dot{\alpha} = V_{1/1}(\alpha) - f'm\,\varepsilon_1^2\omega^2. \tag{6.2.5}$$

Here $k = 2f'm\,\varepsilon_1^2\omega > 0$ is the damping factor, and

$$V_{1/1}(\alpha) = -\frac{1}{2}m\,\varepsilon_1\omega^2\{H(\cos\alpha - f'\sin\alpha)$$
$$+\sigma G[\cos(\alpha - \theta) - f'\sin(\alpha - \theta)]\} \tag{6.2.6}$$

is the vibrational torque which considers (in the averaged form) the presence of the quickly changing part of the moment of resistance $R(\dot{\varphi}, \varphi, \omega t)$ and has therefore a more complicated structure than the corresponding expression (6.1.10), made   up   with   the   same   accuracy   (assuming   that $\nu = 3(m\,\varepsilon_1)^2g^2/16I^2\omega^4 \approx 0$).

Making use of the designations (6.1.19) – (6.1.23), and also introducing the angle ρ' according to the equality

$$\tan\rho' = f', \tag{6.2.7}$$

we may reduce formula (6.2.6) to the following simple form:

$$V_{1/1}(\alpha) = -\frac{m\,\varepsilon_1 A\omega^2}{\cos\rho'}\cos(\alpha + \rho' - \chi). \tag{6.2.8}$$

The values of the angle $\alpha = \alpha_*$, corresponding to the stationary regimes of the planetary motion of bodies, are thus found from the equation

$$\cos(\alpha_* + \rho' - \chi) = -\frac{\varepsilon_1}{A}\sin\rho' \tag{6.2.9}$$

which is obtained from equation (6.2.5) at $\alpha = \alpha_* = \text{const}$ and in view of designations (6.2.7). Equation (6.2.9) has real solutions $\varepsilon_1\sin\rho'/A < 1$ provided the condition is satisfied. This inequality is the condition of the possibility of the existence of a stationary regime of the planetary motion of the roller, the ball, or the ring, maintained by vibration. Taking into account designations (6.2.7), that condition can be reduced to the following simple form:

$$\frac{\varepsilon_1}{A} < \frac{\sqrt{1 + (f')^2}}{f'} \tag{6.2.10}$$

The latter inequality being complied with, equation (6.2.9) has two essentially different solutions in respect to α_*; according to the condition of form (6.2.12), one of them corresponds to the stable motion, the other — to the unstable motion. For the case under consideration it is not difficult, just like in 6.1, to build up the potential function $D(\alpha)$, i.e. such function that

$$dD/d\alpha = -V_{1/1}(\alpha) + f' m \,\varepsilon_1 \omega^2.$$

It remains for us to consider the conditions under which the roller, the ball or the ring run along the corresponding surfaces without leaving them and without sliding. These conditions are expressed by the inequalities

$$N > 0, \qquad |F_1|, | < f_1 N \tag{6.2.11}$$

where F_1 is the friction force at the point of contact between the bodies, f_1 is the coefficient of static friction. The expression for the frictional force, obtained in the way similar to expression (6.2.3) for the force N, looks as

$$F_1 = m(g \cos \varphi + \ddot{x} \sin \varphi + \ddot{y} \cos \varphi). \tag{6.2.12}$$

Taking into account expressions (6.1.1), it is easy to conclude that for the expressions (6.2.11) to be satisfied, it is enough to satisfy the inequality

$$\frac{1 + f_1}{f_1} \left(\frac{g}{\varepsilon_1 \omega^2} + \frac{G + H}{\varepsilon_1} \right) < 1. \tag{6.2.13}$$

This condition is usually fulfilled with a considerable safety margin.

6.2.3   The Discussion of Results, their Applications.   The Critical Slot of Vibro-Inertial Crushing and Grinding Machines

Let us consider the condition of a possibility of motions under consideration (6.2.10) more thoroughly. The meaning of that condition lies in the fact that the eccentricity must not exceed the effective amplitude of oscillations A more than $\sqrt{1 + (f')^2}\big/ f'$ times. Taking into consideration that according to the experimental data (see, say, [90, p. 678]) the coefficient f' varies for the cylindrical roller within the limits $^c 0.1 < f' < 0.4$, we obtain that the value $\eta^* = \sqrt{1 + (f')^2}\big/ f'$ is within the limits $10 > \eta^* > 2.7$, with the values $f' = 0.1$, $\eta^* = 10$ corresponding to the rolling motion of the cylindrical

---

^{c}The range of changes f shown here is somewhat wider than that given in the book [90]. That is done in accordance with the new data obtained at the "Mechanobr" Institute.

roller over the smooth metallic surface, and the values $f' = 0.4$, $\eta^* = 2.7$ corresponding to the rolling motion of the roller over the layer of the material being destroyed which is located in the cavity.

The results presented here lead to the idea of a *critical slot* of vibro-inertial crushing and grinding machines [80, 81, 168], which consists in the following. The *slot S* in such a machine implies a double value of the eccentricity ε_1 i. e. the value $S = 2\varepsilon_1$, and by the *critical value* of that slot S^* we imply its value, maximum allowable by condition (6.2.10). In other words

$$S^* = 2\varepsilon_1^* = \frac{\sqrt{1 + (f')^2}}{f'} A = \eta^* A, \qquad (6.2.14)$$

and according to (6.2.10) for the normal work of the machine in the regime under consideration it is necessary that the inequality

$$S < S^* = \eta_* A. \qquad (6.2.15)$$

should be satisfied. Taking into consideration the real limits of the change of the value η^* presented above, we come to the conclusion that $S^* = (2.7 \div 10)A$ i. e. the critical slot cannot practically exceed the ten-fold value of the effective amplitude of vibration, but can be close to $2.7A$.

The notion of a critical slot plays an essential role when calculating and designing the vibro-inertial crushing and grinding machines. It should be noted that condition (6.2.14) can be easily given an energy form, i.e. it must be expressed as a condition that the power, spent on the process of crushing or grinding, should not exceed a certain limiting (critical) value.

Specification of the notions of the critical slot and the critical power as applied to the concrete vibro-inertial machines can be found in the publications, cited above. Ibid., in the presentation [456] and in the book [114] data are given about the machines of that type, which make it possible to look forward to an essential progress in the technique and technology of crushing and grinding. It should be emphasized that when creating such machines it is very important to establish the regime of a planetary motion of bodies (the rolling motion), considered above, at the initial conditions corresponding to the start of the machine, i.e. to provide a soft excitation of that regime [90].

In conclusion, another remarkable circumstance should be noted. When studying the motion of a ball, placed into a tube, sealed at the ends, performing harmonic motions with the amplitude A (Fig. 6.4), one obtains the following conditions of the existence of a stationary regime with two successive co-impacts of the ball against the opposite ends of the tube at every period of its oscillations (see, for example, [90, 273, 274]):

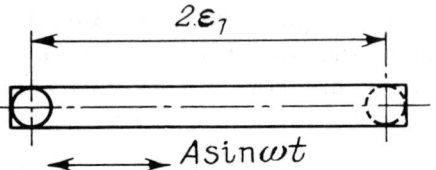

Figure 6.4. A ball in a vibrating tube.

$$\frac{\varepsilon_1}{A} < \frac{\sqrt{1 + \rho^2}}{\rho}, \quad \rho = \frac{2}{\pi}\frac{1 - R}{1 + R} \qquad (6.2.16)$$

where R is the restitution coefficient at the impact and ε_1 is the half length of the stroke of the ball inside the tube. In other words, the analogue of the equation (6.2.10) is obtained, the value ρ playing the role of the rerolling resistance coefficient to rolling. Moreover, when the restitution coefficient R changes within its widest limits $0.2 < R < 0.55$, which are given for the systems, similar to that considered in the reference literature, the values ρ lie in the range of $0.2 < \rho < 0.4$, comparatively close to the range of change of the coefficient f', indicated above. When designing some vibro-striking crushing and grinding machines, inequality (6.2.16) plays the same role as inequalities (6.2.10) and (6.2.15) used when designing the vibro-inertia cone crushers and mills.

Thus we have here a remarkable agreement between the results, obtained for the two-dimensional "smooth" systems and for the one-dimensional vibro-striking system. This agreement has not yet received its proper interpretation, we belive it is not just a casual coincidence.

6.3   Unbalanced Rotor (Mechanical Unbalanced Vibro-Exciter) in the Oscillatory System — Vibrational Retardation of Rotation, Zommerfield's Effect

6.3.1   *On the Effects under Consideration: Brief Bibliographic Reference*

In the majority of the existing vibrational machines and devices oscillations are excited by mechanical unbalanced vibro-exciters. Such exciters, installed on one of the bodies of the system, are the unbalanced rotors, rotated by electric engines of an asynchronous type (Fig. 6.5,*a–f*). The behavior of the oscillatory system with a mechanical vibro-exciter is characterized by a number of remarkable regularities, having a great practical significance.

1. If the exciter whose axis of rotation performs the assigned periodical oscillations (see 6.1.5) either gives the energy to the source of vibration (vibration retards the rotation), or takes away the energy from that source (vibration maintaines the rotation), then in the case under consideration there is always a vibrational retarding of the rotor. This effect is easy to understand: in case of a dissipation in the oscillatory system, the engine must spend its energy not only to overcome the moment of resistance in the bearing of the rotor, but also to maintain the excited oscillations. It is also natural that the vibrational retardation of the rotor grows up abruptly near the resonances of the oscillatory part of the system (first we will assume that the oscillatory part of the system is linear): the frequency of vibration is stabilized near one of the eigenvalues of p that is it hardly changes at all with the increase of the power supplied to the rotor, remaining less than p (Fig. 6.5,*b*).

2. When the power, supplied to the rotor, reaches a certain critical value N^*, and the frequency ω reaches the value very close to p, the oscillations are disrupted, and the frequency of rotation of the rotor increases abruptly up to the value ω^*, the amplitude of oscillations falling abruptly down. After that when the power, supplied to the engine, either decreases or increases, the frequency changes smoothly enough. The right hand slope of the peak of the resonant curve proves to be unrealizable at that method of exciting the oscillations (Fig. 6.5,*e*).

The second of the mentioned regularities which cannot be explained within the frames of the linear theory of forced oscillations was apparently first described by Zommerfeld as far back as 1902 [502], and later by Timoshenko [546] (1934). It is often called *Zommerfeld's effect*. The experimental investigation of that effect was performed by Kalishuk [255] (1939) and by Martyshkin [367] (1940). They also made the first attempts of a theoretical investigation of that question. They failed however to obtain a full explanation of the regularities they observed because they did not in fact consider the question of the stability of motion.

Later Rocard [458] (1949) stated the problem of investigating Zommerfeld's effect as a problem of the interaction of the exciter of oscillations with the oscillatory system, but due to a mistake, made by him, come to the wrong conclusion concerning the region of the stability of motion.

A scrupulous investigation of that effect by the Poincare method was per-

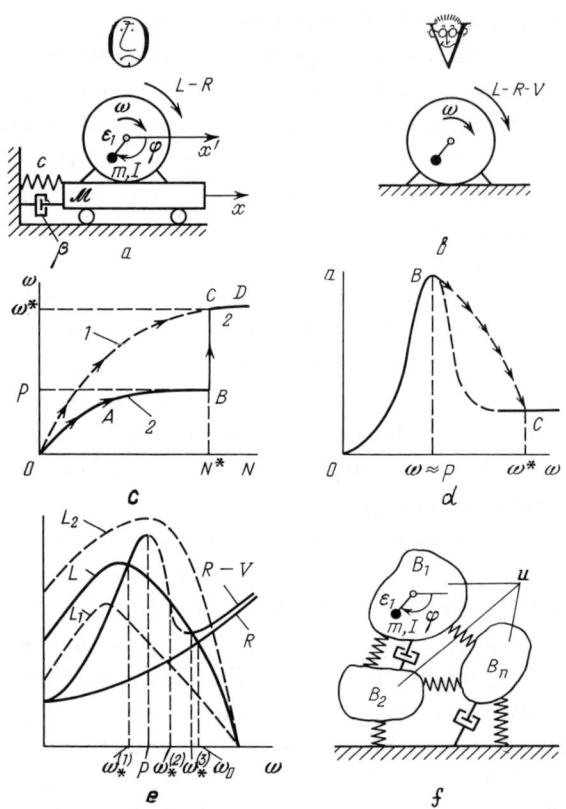

Figure 6.5. Unbalanced rotor in an oscillatory system.

formed by the author in [68] (1953) and as applied to the relative system — to the Boisse-Sarda regulator — in the joint work of Dzhanelidze and the author [71] (1955). The generalization of those results was given in the book [90]. This effect was also investigated in the monograph by Maset [369] (1955) and was given a thorough mathematical study in the article by Bolshakov, Zeldin, Mints and Fufaev [139] (1965).

The theory of the corresponding systems (*oscillatory systems with a limited excitation*) received a comprehensive development in the works by Kononenko [288], Frolov and their disciples and followers (1958 and later). Most detailed information about the publications of that line of investigations is given in the monograph by Alifov and Frolov [12] and also in the book [90] and in the reference book [564, vol. 2]. A brief discussion of those effects is given in the books by Panovko and Gubanova [426] and by the author [114]. Among the latest publications the article by Petchenev should be mentioned [434] containing essential additions to the results of the investigations made before (also see 6.3.3).

This section shows that the regularities described can be easily obtained and interpreted from the position of vibrational mechanics. From the point of view of the observer **V** all of them are explained by the appearance of the vibrational torque, acting on the shaft of the rotor (Fig. 6.5,*b*). The advantages of such an approach are in this case especially appreciable since the consideration of the complicated nonlinear system with $n+1$ degree of freedom (n being the number of the degrees of freedom of the oscillatory part of the system) is reduced to the consideration of one autonomous equation of slow motions, having the first order, that is corresponding to the system with "1/2 degree of freedom".

Recently several publications [366, 365, 373, 512, 376] have been devoted to a thorough theoretical, experimental and numerical investigation of Zommerfield's effect.

In some of those publications the approach of vibrational mechanics is successfully used.

6.3.2   *Simplest Case: the Oscillatory Part of the System is Linear and has One Degree of Freedom*

The simplest system in which the described effects are fully displayed is shown in Fig. 6.5,*a*. An unbalanced rotor, set in rotation by an electric engine of an asynchronous type, is installed on a platform of mass $\mathcal{M}$, connected to a fixed base by means of a linear elastic element of a rigidity c and a damping element with a viscous friction coefficient β. The equation of the motion of such system has the form

$$I\ddot{\varphi} = L(\dot{\varphi}) - R(\dot{\varphi}) + m\,\varepsilon_1(\ddot{x}\sin\varphi + g\cos\varphi),$$
$$M\ddot{x} + \beta\dot{x} + cx = m\,\varepsilon_1(\dot{\varphi}^2\cos\varphi + \ddot{\varphi}\sin\varphi). \qquad (6.3.1)$$

Here φ is the angle of rotation of the rotor, counted from the horizontal direction; x is the shift of the platform from the position, corresponding to the non-deformed elastic element; m, I and ε_1 are the mass, the moment of inertia and the eccentricity of the rotor respectively ; $M = \mathcal{M} + m$ is the mass of the system, $L(\dot{\varphi})$ is the torque, transferred to the rotor from the electric engine,

$R(\dot{\varphi})$ is the moment of the resistance to the rotation of the rotor, which may include the resistance in the bearings and the disposable load, just like in 6.1.

Using the method of direct separation of motions, we will consider the solutions of equations (6.3.1) of the type

$$\dot{\varphi} = \omega(t) + \psi(t, \omega t), \quad x = x(t, \omega t) \tag{6.3.2}$$

where ω is a slowly changing function of time, while ψ and x are the fast-changing functions of time, with ψ and x being 2π-periodic with respect to $\tau = \omega t$ and their average values for the period with respect to τ are equal to zero:

$$< \psi(t, \tau) >= 0, \quad < x(t, \tau) >= 0. \tag{6.3.3}$$

We also assume that $\psi \ll \omega$. Like in 6.1, we linearize the difference $L - R$ close to $\dot{\varphi} = \dot{\varphi}_0 = \omega$:

$$L(\dot{\varphi}) - R(\dot{\varphi}) = L(\omega) - R(\omega) - k(\dot{\varphi} - \omega), \quad k = k(\omega) > 0. \tag{6.3.4}$$

Then equations (3.2.55), (3.2.56) of chapter 3 will be presented as

$$I\dot{\omega} = L(\omega) - R(\omega) + V_1(\omega), \tag{6.3.5}$$

$$I\dot{\psi} = -k\psi + \Psi(\ddot{x}, \varphi), \tag{6.3.6}$$

$$M\ddot{x} + \beta\dot{x} + cx = m\,\varepsilon_1(\dot{\varphi}^2 \cos\varphi + \ddot{\varphi}\sin\varphi) \tag{6.3.7}$$

where we have

$$V_1(\omega) = m\,\varepsilon_1 < \ddot{x}\sin\varphi + g\cos\varphi >,$$
$$\Psi(\ddot{x}, \varphi) = m\,\varepsilon_1[\ddot{x}\sin\varphi + g\cos\varphi - < \ddot{x}\sin\varphi + g\cos\varphi >], \tag{6.3.8}$$

and we bear in mind that the expressions, corresponding to equality (6.3.2) were substituted for φ, $\dot{\varphi}$ and $\ddot{\varphi}$.

Assuming that according to what was said in 3.2, $\omega(t)$ is large enough, we will seek solutions of system (6.3.6), (6.3.7) with the frozen t, $\omega(t)$ and $\dot{\omega}(t)$; besides, we will suppose that the expression $\Psi(\ddot{x}, \varphi)$ is small and we can assume that

$$\Psi(\ddot{x}, \varphi) = \mu\Psi_1(\ddot{x}, \varphi) \tag{6.3.9}$$

where $\mu > 0$ is a small parameter. In the first approximation, the solution of equation (6.3.6), satisfying condition (6.3.3), will be $\psi = \psi^0 = 0$. Then, according to (6.3.2), we will have $\dot\varphi^0 = w(t)$ and we can assume that

$$\varphi = \varphi^0 = \int_0^t w(t)dt = w(t)t + \alpha(t) \tag{6.3.10}$$

where $\alpha(t)$ is a certain function t. We substitute expression (6.3.10) into equation (6.3.7), taking into account that according to our assumption $\ddot\varphi = \dot w(t) \ll w^2(t) = \dot\varphi^2$; therefore in this approximation we can neglect the second term in the right hand part of that equation as compared to the first term. Then the asymptotically stable at μ sufficiently small 2π-periodic with respect to wt solution of equation (6.3.7) at the frozen t, satisfying condition (6.3.3), will be

$$x = x^0(t, wt) = \frac{m\,\varepsilon_1}{M\Delta}\cos(wt + \alpha + \gamma) \tag{6.3.11}$$

where

$$\Delta = \sqrt{(1 - \lambda^2)^2 + 4n^2}, \quad \lambda = p/w, \quad p = \sqrt{c/M}, \quad n = \beta/2Mw,$$
$$\sin\gamma = -2n/\Delta, \qquad \cos\gamma = -(1 - \lambda^2)/\Delta. \tag{6.3.12}$$

Substituting expressions (6.3.10) and (6.3.11) into (6.3.8), and performing the averaging, we obtain the following approximate expression for the vibrational torque:

$$V_1(w) \approx V(w) = -\frac{(m\,\varepsilon_1 w)^2}{M}\frac{n}{(1 - \lambda^2)^2 + 4n^2}. \tag{6.3.13}$$

So, the main equation of vibrational mechanics in this approximation has the form

$$I\dot w = L(w) - R(w) + V(w), \tag{6.3.14}$$

that is it corresponds to the system with "1/2 degree of freedom", while the initial system (6.3.1) had two degrees of freedom. We will also mark that this equation differs from the classic *equation of the machine assembly* [15, 124] by the presence of the term $V(w)$, which determines the peculiar behavior of the system.

The expression for the vibrational torque (6.3.13) can be presented in the form

$$V = -\frac{1}{2}Fa\sin\gamma = -V_{\max}\sin\gamma \tag{6.3.15}$$

where $F = m\varepsilon_1\omega^2$ is the amplitude of the exciting force, developed by the rotor, in the case of a fixed platform, and

$$a = m\varepsilon_1/M[(1 - \lambda^2)^2 + 4n^2]^{1/2}$$

is the amplitude of the oscillations of the platform in case of the stationary motion. This formula agrees very well with expression (6.1.21) since according to (6.1.23) in case of rectilinear oscillations the effective amplitude is $A = \dfrac{1}{2}a$.

We have quite deliberately obtained here the main equation (6.3.14), omitting for brevity sake some arguments, substantiating the conclusion. If necessary, they can be reproduced quite readily. We will just mention that the obtained result agrees very well with those found by other methods. particularly, regarding the stationary regimes, by the Poincare-Lyapunov method of small parameter [90].

Let us dwell on the consideration of such regimes. The frequency of rotation of the rotor in these regimes is determined from the equation

$$L(\omega) = R(\omega) - V(\omega). \tag{6.3.16}$$

Fig. 6.5,e shows the dependences $R(\omega)$, $R(\omega) - V(\omega)$ and $L(\omega)$, with $L(\omega)$ roughly corresponding to the static characteristic of an asynchronous electric engine. The figure indicates three such characteristics: L, L_1 and L_2. From the figure as well as from the equations (6.3.12) and (6.3.15) one can see that first, the vibrational torque is always retarding, as it had to be expected, and, secondly, that its dependence on the frequency of rotation is of a resonant character: $|V(\omega)|$ has a maximum near $\omega = p$. It is essential (and on the face of it paradoxical) that the smaller is the dimensionless coefficient of damping n, the greater is the maximum, that is the weaker the damping in the oscillatory part of the system, the stronger the retarding action of vibration. However, this regularity is quite understandable on the basis of equation (6.3.15): the resonant amplitude a_{max} also increases with the decrease of n.

According to (6.3.16), the points of intersection of the curves $L(\omega)$ and $R(\omega)$ in Fig. 6.5,e correspond to the stationary regimes. With the type of curves under consideration there can be either one or three such points. On setting up the equation in variations for the definite stationary solution $\omega = \omega_*$ of equation (6.3.14), we come to the conclusion that this solution will correspond to the stable motion, provided

$$R'(\omega_*) - V'(\omega_*) > L'(\omega_*). \tag{6.3.17}$$

Hence it follows that the points of the type of $\omega_*^{(1)}$ and $\omega_*^{(3)}$ are answered by the stable motions, while the points of the type of $\omega_*^{(2)}$ are answered by the unstable motions. As a result, it is the right hand slope of the peak of the resonant curve that proves to be unrealizable (Fig. 6.5,*d*).

Should the rotor be established on a fixed base, then the frequency of its stationary rotation ω_0 (*the partial angular velocity* — see 6.1.5) would be determined by the equation

$$L(\omega) = R(\omega)$$

One can see from Fig. 6,5,*e*, that $\omega_0 > \omega_*^{(s)}$ ($s = 1, 2, 3$); this corresponds to the above conclusion that in the system under consideration the retardation of the rotor always takes place.

Imagine that we are smoothly increasing the torque of the engine L and consequently the supplied power $N = L\omega$, seeking to increase the frequency of rotation of the rotor ω; in the context of Fig. 6,5,*e* this corresponds to the transition from the curve L_1 to the curves L and L_2. If the rotor is installed on a fixed base, then ω increases quite smoothly (curve 1 in Fig. 6.5,*c*). But if the rotor is connected with the oscillatory system (curve 2), then a regularity is observed which at the beginning of the section was spoken about as *Zommerfeld's effect*. At the initial segment OA when the vibrational torque is relatively small, the frequency ω increases approximately as much as when the rotor is rotating on a fixed base. Then, when the frequency ω approaches that of the free oscillations p, the increase is going on very slowly, despite a considerable growth of the supplied power (the segment AB), which is accompanied by a growth of the amplitude of oscillations a. Finally, at a certain value $N = N^*$ there is a spasmodic growth of the frequency up to a certain post-resonant value $\omega = \omega^*$; the amplitude of oscillations falling abruptly. At the subsequent smooth increase of N the frequency again changes quite smoothly (the segment CD).

Thus, all the regularities of the behavior of the system get their explanation based on equation (6.3.14) and on formula (6.3.12) for the vibrational torque, playing the role of a kind of an additional load. These regularities were mentioned in 6.3.1.

6.3.3   *General Case: the Oscillatory Part of the System is Non-linear and has Several Degrees of Freedom*

The results of 6.3.2 are naturally generalized for the case of a more complicated system, shown in Fig. 6.5,*f*. The oscillatory part of the system in this case presents several solid bodies $B_1, \ldots, B_n$, connected with each other and with the fixed base by means of elastic and damping elements. This part of the system can have an infinite number of degrees of freedom and can be non-linear. The main equation of vibrational mechanics (6.3.14) in this case retains its form, that is it still corresponds to the system with "1/2 degree of freedom", only the expression for the vibrational moment $V(\omega)$ has a more complicated structure. The algorithm of obtaining this relation by method of direct separation of motions remains the same, only the technical difficulties increase. We will present here some relations of a general character.

Equations of motion of the system can be presented in the form

$$I\ddot{\varphi} = L(\dot{\varphi}) - R(\dot{\varphi}) + M(\varphi, \ddot{\mathbf{u}}_1), \quad D\mathbf{u} = \mathbf{F}(\ddot{\varphi}, \dot{\varphi}, \varphi). \qquad (6.3.18)$$

Here the symbols $\varphi, I, L(\dot{\varphi})$ and $R(\dot{\varphi})$ have the same meaning as in equations (6.3.1); $\mathbf{u}$ is the vector of the generalized coordinates of the oscillatory part of the system; $M(\varphi, \ddot{\mathbf{u}}_1)$ is the moment of forces, acting upon the rotor due to the oscillations of the body on which it is installed ($\mathbf{u}_1$ is the vector of the generalized coordinates of that body; its components are part of the components of the vector $\mathbf{u}$); D is a certain differential operator, not necessarily linear; $\mathbf{F}$ is the vector of a generalized exciting force, acting upon the oscillatory system on the part of the rotor. Then the equation for the vibrational torque in the main equation (6.3.14) has the form

$$V(\omega) = < M(\omega t, \mathbf{u}_1^0) > \qquad (6.3.19)$$

where $\mathbf{u}_1^0$ is the value of the vector $\mathbf{u}_1$, corresponding to the asymptotically stable periodic solution of the equation

$$D\mathbf{u}^0 = \mathbf{F}(0, \omega, \omega t), \qquad (6.3.20)$$

found at the frozen $\omega = \omega(t)$ and t; in case of the non-linear oscillatory part of the system, there may be several such solutions, each of them having its own expression for $V(\omega)$.

In conclusion it should be noted that the simplest equation (6.3.14) which makes it possible to explain the main regularities of the behavior of the system, but does not still give answers to some questions connected with its behavior. A detailed investigation, made by Pechenev in his article [434] cited above, helped to discover a more complicated hierarchy of that system — there are not only fast and slow motions there, but also "half-slow" motions as well. Besides, it proved possible to build a domain of attraction of stable near-resonant regime and describe the process of "swinging" of the rotor near that regime.

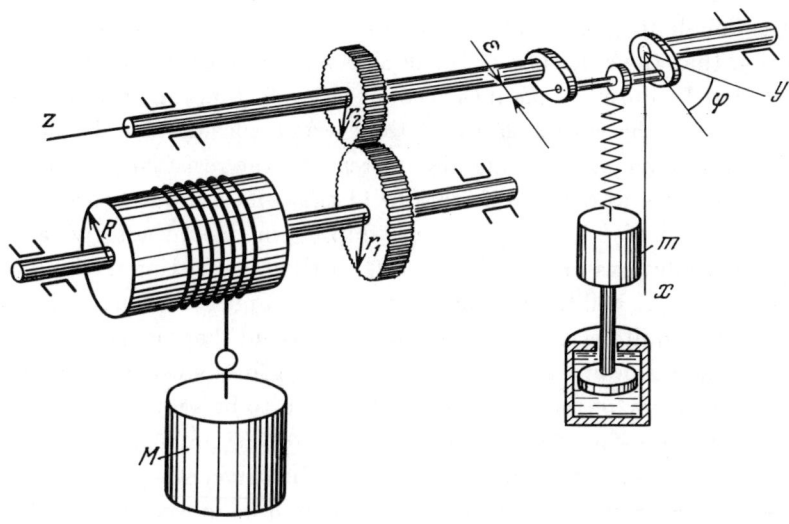

Figure 6.6. Vibrational stabilizer of the speed of lowering the load (the Boisse-Sarda regulator).

6.3.4   On Certain Applications

The effects we have discussed are essential for quite a number of applications. We will dwell briefly on some of them.

One of the most important applications is connected with a method of exciting vibration, common enough in vibrational technology, that is by using the unbalanced rotating rotors (mechanical inertial vibro-exciters). A considerable unbalance of the rotor is formed in this case intentionally, therefore the useful effect of frequency stabilization and the most undesirable effect of a break down of the resonance regime at the fluctuations of the load is quite essential here [90, 564, vol. 4].

A system different on the face of it in which these effects also play the main role is the *Boisse-Sarda regulator* (Fig. 6.6).This regulator provides a constant speed of lowering the main load Mg due to the stabilization of the frequency of oscillations of the "small"load mg near the resonance value $p = \sqrt{c/m}$ [71].

Another group of applications is connected with a harmful effect of vibrational retardation of the unbalanced rotor or of another source of vibration in the oscillatory system as was mentioned above, in the case, considered in 6.2, the observer **W** (see the Preface) who did not notice that the rotor of the engine was unbalanced and that its axis was vibrating, could come to the wrong conclusion that the efficiency of the engine was more than a unit. In our case the fact that he did not take into account those factors might lead him to the conclusion (also the wrong one) that there was a considerable decrease in the power of the engine or a considerable increase in the resistance to the rotation of the rotor. Interesting manifestations of that effect have been described in literature. So, a case is known when a ship engine would not give a nominal number of revolutions n_0, in spite of the fact that the friction in the shaft line, measured in static conditions, was quite normal and low enough. The reason was discovered quite accidentally. It has been found that the power which was supplied was withdrawn ("sucked off") by a length of steel rope, lying on the deck; the frequency of its free oscillations proved to be somewhat less than n_0, and on approaching it, it would begin to vibrate very intensively.

Cases are known when the locomotive engine driver was unable to increase the speed of the train to the wanted value though he moved the handles of the controller. And only when the power supplied greatly exceeded the required level, the speed would grow up spasmodically. The reason lay in the fact that near a certain value of the velocity the frequency of oscillations of the cars, excited by the wheels passing the rail joints, proved to be close to, say, the frequency of oscillations of the fluid in the cisterns that were not quite full. This case is interesting in particular by the fact that the oscillations, "sucking off" the energy, are conditioned not by the unbalance of the rotor, but by someother reason, though the mechanism of the phenomenon remains the same. This peculiar manifestation of the effect of vibrational retardation was marked by Minkin and Volfson who considered its theory as well [571].

6.4   Machine Aggregates

The work of the aggregates in both stationary and non-stationary regimes can be greatly affected by the dependence of the reduced moment of inertia and the moment of resistance forces in the machine aggregate on the angle of rotation of the shaft, by the elasticity of the links of the mechanism and by not ideal characteristic of the engine as well as by some other factors [25, 446, 562, 273, 284, 285, 286]. In the stationary regimes these factors lead to the frequency fluctuations of the rotation of the aggregate, to a certain change in the mean value of that frequency regarding the nominal value, and at the start to the situation quite similar to *Zommerfeld's effect*, described in 6.3.

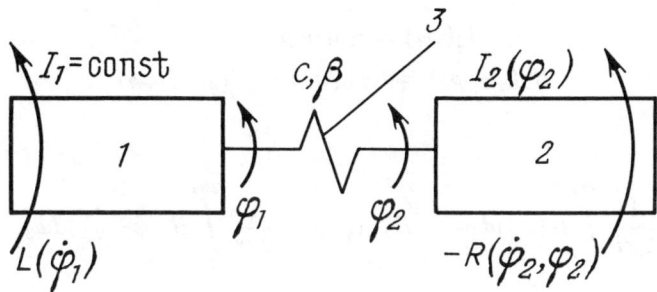

Figure 6.7. Scheme of the machine aggregate.

Considering those two effects, Kolovsky marked that they could be interpreted as a result of the action of the vibrational torque, the expression for which was obtained by him by the Poincare method of small parameter [284]. Kolovsky drew the attention of the author to the expediency of using the conception and the apparatus of vibrational mechanics to solve a number of problems on the dynamics of machine aggregates and robotics. To illustrate the adequate possibilities let us consider the system discussed in [284] by the method of direct separation of motions.

The scheme of the system is presented in Fig. 6.7. The rotor engine of an asynchronous type *1* is connected with the mechanical system of machine *2* by an elastic-damping element (shaft) *3*. The moment of inertia of the engine of the rotor I_1 is assumed to be constant, and the moment of inertia of the machine, reduced to the shaft of the engine, $I_2(\varphi_2) - 2\pi i$ is assumed to be periodically dependent on the reduced (multiplied by i) angle of rotation of the main shaft φ_2 (i being the transmitting relation. The rotating moment of the engine L is considered, as before, to be given by its static characteristic $L(\dot\varphi_1)$ and the reduced moment of the resistance forces $R_2(\dot\varphi_2, \varphi_2)$ is assumed to be the $2\pi i$-periodic function of the angle φ_2. The elastic damping element is considered to be non-inertial; its rigidity and resistance coefficient are designated by c and β respectively.

The equations of motion of the system described look as (4.1)

$$I_1\ddot\varphi_1 + \beta(\dot\varphi_1 - \dot\varphi_2) + c(\varphi_1 - \varphi_2) = L(\dot\varphi_1),$$

$$I_2(\varphi_2)\ddot\varphi_2 + \frac{1}{2}\frac{dI_2(\varphi_2)}{d\varphi_2}\dot\varphi_2^2 + \beta(\dot\varphi_2 - \dot\varphi_1) + c(\varphi_2 - \varphi_1) = -R_2(\dot\varphi_2, \varphi_2).(6.4.1)$$

The periodic functions I_2 and R_2 are presented as

$$I_2(\varphi_2) = I_{2,0} + \mu\tilde{I}_2(\varphi_2),$$
$$R_2(\dot{\varphi}_2, \varphi_2) = R_{2,0}(\dot{\varphi}_2) + \mu\tilde{R}_2(\dot{\varphi}_2, \varphi_2) \tag{6.4.2}$$

where

$$I_{2,0} = \frac{1}{2\pi i}\int\limits_0^{2\pi i} I_2(\varphi_2)d\varphi_2, \quad R_{2,0}(\dot{\varphi}_2) = \frac{1}{2\pi i}\int\limits_0^{2\pi i} R_2(\dot{\varphi}_2, \varphi_2)d\varphi_2; \tag{6.4.3}$$

$$\tilde{I}_2(\varphi_2) = \sum_{n=1}^{\infty} I_{2,n}\sin(n\varphi_2 + \gamma_n),$$
$$\tilde{R}_2(\dot{\varphi}_2, \varphi_2) = \sum_{n=1}^{\infty} R_{2,n}(\dot{\varphi}_2)\sin(n\varphi_2 + \delta_n). \tag{6.4.4}$$

We will consider the value $\mu \geq 0$ to be a small parameter, thus assuming that the deflection of the functions I_2 and R_2 from their average for the period values $I_{2,0}$ and $R_{2,0}(\dot{\varphi}_2)$ are quite small. Besides, similarly to the relations (6.3.4), we will change the moments $L(\dot{\varphi}_1)$ and $R_{2,0}(\dot{\varphi}_2)$ close to their "operating" value of the frequency of rotation $\dot{\varphi}_{1,0} = \dot{\varphi}_{2,0} = \omega$, unknown beforehand due to the autonomy of the system (6.4.1), for their linear expressions

$$L(\dot{\varphi}_1) = L(\omega) - k^*(\dot{\varphi}_1 - \omega), \quad R_{2,0}(\dot{\varphi}_2) = R_{2,0}(\omega) + k^0(\dot{\varphi}_2 - \omega) \tag{6.4.5}$$

where $k^* > 0$ and $k^0 > 0$. Equalities (6.4.2) and (6.4.5) being taken into account, the equations of motion (6.4.1) will be written in the form

$$I_1\ddot{\varphi}_1 + \beta(\dot{\varphi}_1 - \dot{\varphi}_2) + c(\varphi_1 - \varphi_2) + k^*(\dot{\varphi}_1 - \omega) = L(\omega),$$
$$I_{2,0}\ddot{\varphi}_2 + \beta(\dot{\varphi}_2 - \dot{\varphi}_1) + c(\varphi_2 - \varphi_1) + k^0(\dot{\varphi}_2 - \omega) =$$
$$-R_{2,0}(\omega) + \mu\Phi(\ddot{\varphi}_2, \dot{\varphi}_2, \varphi_2) \tag{6.4.6}$$

where we have

$$\Phi(\ddot{\varphi}_2, \dot{\varphi}_2, \varphi_2) = -[I_2(\varphi_2)\ddot{\varphi}_2 + \frac{1}{2}\frac{d\tilde{I}_2(\varphi_2)}{d\varphi_2}\dot{\varphi}^2 + \tilde{R}_2(\dot{\varphi}_2, \varphi_2)]. \tag{6.4.7}$$

We are interested in the solutions of equations (6.4.6) of the form

$$\varphi_s = \omega t + \alpha_s(t) + \psi_s(t, \omega t) \quad (s = 1, 2) \tag{6.4.8}$$

where α_1 and α_2 are the slowly changing functions of time while ψ_1 and ψ_2 are the fast changing functions of time, ψ_1 and ψ_2 having the $2\pi i$ period with respect to ωt and satisfying the equalities

$$< \psi_1 >= 0 \qquad < \psi_2 >= 0 \tag{6.4.9}$$

in which (as well as everywhere below) the averaging over ωt is assumed to be taken for the same period, and the assumption about the slowness of changing α_1 and α_2 is understood in the sense that $\dot{\alpha}_1 \ll \omega$ and $\dot{\alpha}_2 \ll \omega$. Using the method of direct separation of motions in order to find solutions (6.4.8), we will consider all the moments except $\mu\Phi$ in the equations (6.4.6) to be slow and will write down equations (3.2.55) and (3.2.56) of chapter 3 in the following form:

$$I_1\ddot{\alpha}_1 + \beta(\dot{\alpha}_1 - \dot{\alpha}_2) + c(\alpha_1 - \alpha_2) + k^*\dot{\alpha}_1 = L(\omega),$$
$$I_{2,0}\ddot{\alpha}_2 + \beta(\dot{\alpha}_2 - \dot{\alpha}_1) + c(\alpha_2 - \alpha_1) + k^0\dot{\alpha}_2 = -R_{2,0}(\omega) + \mu < \Phi >; \tag{6.4.10}$$

$$I_1\ddot{\psi}_1 + \beta(\dot{\psi}_1 - \dot{\psi}_2) + c(\psi_1 - \psi_2) + k^*\dot{\psi}_1 = 0,$$
$$I_{2,0}\ddot{\psi}_2 + \beta(\dot{\psi}_2 - \dot{\psi}_1) + c(\psi_2 - \psi_1) + k^0\dot{\psi}_2 = \mu(\Phi - < \Phi >). \tag{6.4.11}$$

Seeking the $2\pi i$-periodic with respect to ωt solutions of equations (6.4.11) at the frozen $\dot{\alpha}_1, \dot{\alpha}_2, \alpha_1, \alpha_2$ and t in the form of power series of small parameter

$$\psi_s = \psi_s^{(0)} + \mu\psi_s^{(1)} + \ldots \quad (s = 1, 2), \tag{6.4.12}$$

we obtain

$$\psi_s^{(0)} = 0,$$

$$\psi_s^{(1)} = -\sum_{n=1}^{\infty}\{\ddot{\alpha}_2 I_{2,n} M_s(n\omega) \sin[n(\omega t + \alpha_2) + \gamma_n + \varepsilon_s(n\omega)]$$

$$+ \frac{1}{2}(\omega + \dot{\alpha}_2)^2 n I_{2n} M_s(n\omega) \cos[n(\omega t + \alpha_2) + \gamma_n + \varepsilon_s(n\omega)]$$

$$+ R_{2,n}(\omega + \dot{\alpha}_2) M_s(n\omega) \sin[n(\omega t + \alpha_2) + \delta_n + \varepsilon_s(n\omega)]\}, \tag{6.4.13}$$

where $M_s(n\omega)$ and $\varepsilon_s(n\omega)$ denote respectively the amplitude-frequency- and the phase-frequency characteristics of the linear part of the system, that is the expressions involved in the periodic solutions $y_s = A M_s(\omega) \sin[\omega t + \varepsilon_s(\omega)]$ of the system of equations (A — being a constant)

$$I_1\ddot{y}_1 + \beta(\dot{y}_1 - \dot{y}_2) + c(y_1 - y_2) + k^*\dot{y}_1 = 0,$$
$$I_{2,0}\ddot{y}_2 + \beta(\dot{y}_2 - \dot{y}_1) + c(y_2 - y_1) + k^0\dot{y}_2 = A\sin\omega t. \tag{6.4.14}$$

The use of formulas (6.4.13) in view of relations (6.4.6)–(6.4.8) leads to the expression for the vibrational torque of the following type (with an accuracy to the terms of the order of μ^2):

$$V(\ddot{\alpha}_2, \dot{\alpha}_2, \omega) = \mu^2 V_2(\ddot{\alpha}_2, \dot{\alpha}_2, \omega) = \mu < \Phi >$$

$$= \sum_{n=1}^{\infty} M_2(n\omega) F_n(\ddot{\alpha}_2, \dot{\alpha}_2). \tag{6.4.15}$$

Here $F_n(\ddot{\alpha}_2, \dot{\alpha}_2)$ denotes the functions $\ddot{\alpha}_2$ and $\dot{\alpha}_2$, obtained as a result of simple though rather unwieldy calculations at which we will not dwell here.

In view of formula (6.4.15), the equations of slow motions (6.4.10) — the main equations of vibrational mechanics for the problem under consideration — acquire the form

$$I_1 \ddot{\alpha}_1 + \beta(\dot{\alpha}_1 - \dot{\alpha}_2) + c(\alpha_1 - \alpha_2) + k^* \dot{\alpha}_1 = L(\omega),$$
$$I_{2,0} \ddot{\alpha}_2 + \beta(\dot{\alpha}_2 - \dot{\alpha}_1) + c(\alpha_2 - \alpha_1) + k^0 \dot{\alpha}_2$$
$$= -R_{2,0}(\omega) + V(\ddot{\alpha}_2, \dot{\alpha}_2, \omega). \tag{6.4.16}$$

It should be noted that the applicability of the method of direct separation of motions, according to note 4 in 3.2.9, is guaranteed, provided ω are large enough, at least near the stationary regimes $\alpha_1 = \alpha_1^* = \text{const}$, $\alpha_2 = \alpha_2^* = \text{const}$ which we will consider below. For such regimes from equations (6.4.16) we obtain

$$c(\alpha_1 - \alpha_2) = L(\omega), \quad c(\alpha_2 - \alpha_1) = -R_{2,0}(\omega) + V(0, 0, \omega). \tag{6.4.17}$$

Combining those equalities, we arrive at the following equation for determining the frequency of rotation $\omega = \omega_*$ in the stationary regimes:

$$L(\omega) = R_{2,0}(\omega) - V(0, 0, \omega). \tag{6.4.18}$$

This equation differs from the corresponding equation $L(\omega) = R_{2,0}(\omega)$, provided the reduced moment of inertia I_2 and the moment of the resistance forces R_2 do not depend on the angle of rotation φ_2. This difference is due to the presence of the vibrational torque V. Accordingly, the values of the frequency of rotation, defined by them, ω_* and ω_0 also prove to be different. By its meaning the equations (6.4.18) as well as the analogous equations (6.1.11), (6.1.28) and (6.3.16) of this chapter are equations of the equilibrium of the average moments or (after multiplying by ω) — equations of the energy balance in the system.

Taking into account the " peak character" of the dependence (6.4.15) of the moment V on ω, one can draw the conclusion that it is quite possible that the Zommerfeld effect, described in 6.3, should take place in the system under consideration. This conclusion follows from the consideration of the behaviour of the curves $L(\omega)$ and $R_{2,0} - V(0,0,\omega)$ whose abscissas of intersection points ω_* correspond to the possible stationary regimes. These curves are quite similar to those shown in Fig. 6.5, e (though somewhat different designations are used there). Just like it is in problem in 6.3, in the general case there may be three intersection points $\omega_*^{(1)}$, $\omega_*^{(2)}$ and $\omega_*^{(3)}$, corresponding to three different regimes, the stable regimes answering the first and last points $\omega_*^{(1)}$ and $\omega_*^{(3)}$. That can be established by means of a more detailed solution of that problem (taking into account the transient process of establishment with regard to the variable ω). However, we will not dwell upon it here.

It is easy to see that these results comply very well with those, obtained by Kolovsky in the above cited work [284]. Later Kolovsky investigated the dynamics of the machine aggregate in a much more general form, embracing the peculiarity of the problems of robotics [285, 286]. To solve non-linear problems a method was suggested by him at which the equations of motion of the aggregate are realized by computer by means of algorithms, based on using the method of kinetostatics or the Lagrange equations, after which the equations are integrated by method of successive approximations. The law of the programmed motion of the aggregate is taken to be the initial approximation. The generalized forces are calculated for that law and the law of motion is found in the first approximation, and so on. The convergence of the process of successive approximations is provided by digitization of the moments of time, for which solutions are sought, and by the use of spline-interpolation at the intervals between those discrete moments.

The method of Kolovsky makes it possible after the first or higher approximations have been obtained, to find vibrational torques, i.e. to obtain the main equations of vibrational mechanics for this class of systems.

Chapter 7

Self-synchronization of Mechanical Vibro-exciters

7.1 On the Phenomenon of Synchronization of Unbalanced Rotors (Mechanical Vibro-exciters). A Brief Review of Investigations

Self-synchronization of unbalanced rotors is one of the most remarkable phenomena, caused by vibration. It consists in the fact that two or more rotors, not connected with each other either kinematically or electrically, installed on a mobile base and set into motion by independent asynchronous engines, rotate synchronously, i.e. with the same or with multiple average angular velocities and with definite mutual phases. As that takes place, the coordination of rotation of the rotors appears despite the difference between their *partial angular velocities* , that is those velocities at which they are rotating when installed on a fixed base (see 6.1.5). The tendency to synchronous rotation proves in many cases to be so strong that even should one or several rotors be switched off, it will not break the synchronism: the rotors with switched off engines can keep on rotating indefinitely long. The energy, necessary to maintain their rotation, is transferred from the engines turned on, due to the vibration of the base on which the rotors are installed. This vibration can be hardly noticeable; the observer **V** might think there were elastic shafts or springs between the rotors (Fig. 7.1). It should be noted that the simplest manifestation of that effect (we may also say - its simplest idealization as well) was considered in 6.1. where we discussed the maintenance of rotation of an unbalanced rotor under the action of the preassigned vibration of its axis.

It was an accidental observation of that effect that gave impetus to discover the phenomenon of self-synchronization of unbalanced rotors. On long trials of a vibrational machine with two mechanical vibro-exciters, that is with the

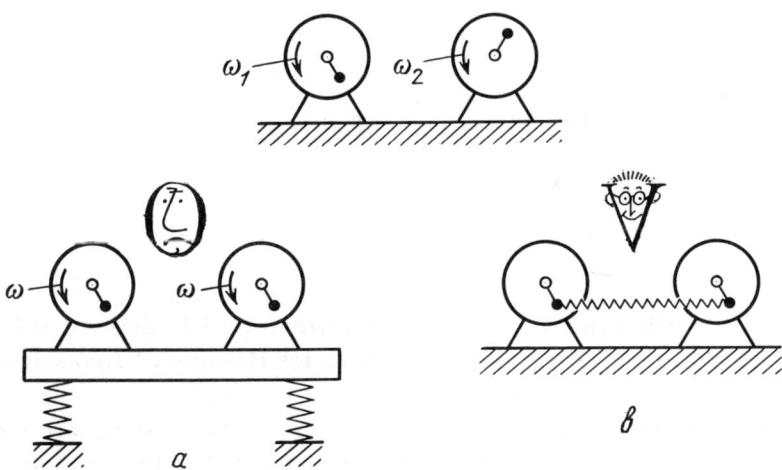

Figure 7.1. Self-synchronization of mechanical vibro-exciters. a) Two or more unbalanced rotors, set on a common mobile base and activated by the asynchronous
engines, rotate with the same average velocity ω, while the same rotors on a fixed
base have different rotational velocities ω_1 and ω_2 (the upper picture). The synchronism of rotation can be retained even if one of the rotors is switched off ($\omega_1 = 0$ or
$\omega_2 = 0$); b) the observer **V** has an illusion as if the rotors were connected with a
spring.

unbalanced rotors, activated by the asynchronous electric motors, the wire,
feeding one of the engines was broken, but the open circuit was revealed only
several hours later for the device kept on working in a normal way.

That observation was registered in Russia by Pliss and Abramovich in the
report of the "Mekhanobr" Institute [438] (1948). The first publication on
self-synchronization of vibro-exciters, as far as we know, appeared later — in
1950 as a patent on the simplest concrete vibrational devices given out to the
Swedish inventor Signul [497, 498]; the claim for it was made in Sweden in
1946, but no material was published on it until the first patent was issued in
(1950)[498].

In 1953 the author gave a physical explanation and the mathematical description of the phenomenon of self-synchronization of mechanical vibro-exciters [68], making use of the Poincare-Lyapunov methods of small parameter and of the theory of the stability of motion. Then the results of that article were developed and supplemented in the articles of the author and also in the publications of other researchers both in Russia and in other countries. [76, 77, 78, 87, 88, 91, 97, 102, 99, 104, 106, 117, 120, 31, 5, 47, 48, 49, 205, 217, 172, 600, 589, 590, 591, 290, 304, 319, 320, 321, 361, 362, 354, 395, 396, 397, 398, 399, 449, 450, 179, 265, 161, 138, 238, 360, 380, 506, 507, 252, 359, 374, 375, 377, 378, 379, 383, 431, 480, 489, 508, 509, 510, 511, 530]. The review and presentation of the main results of the theory and applications of self-synchronization of vibro-exciters were given in the books of the author [90, 103, 114], Nagaev and Guzev [403, 404] in the reference book [564, vol. 4], and in the dissertation of Merten [376].

Some results with the adequate reference will be considered in this chapter in the course of exposition or will be mentioned in the review in 7.3.3.

In the theory and applications of self-synchronization of mechanical vibro-exciters, the integral sign of stability (extremal property) of synchronous motions is of great significance. The review and account of the corresponding investigations are given in 4.2.

The majority of theoretical investigations as well as experimental designs and elaborations were made at the "Mekhanobr" Institute in St.Petersburg, Russia. That Institute and its employees were given a certificate for the scientific discovery of the phenomenon of synchronization of the rotating bodies (rotors) [1].

The discovery and the theoretical explanation of the phenomenon of self-synchronization of unbalanced rotors has opened up fresh opportunities in the vibrational technology. It is connected with the fact that more than one mechanical exciter [a] is used in a great number of vibrational machines and installations. So the use of many relatively low-powered exciters instead of one of a high power makes it possible to disperse the exciting force over the "non-rigorous" vibrating organ of a considerable size and thus keep it oscillating as an absolutely solid body. In other cases the installation of several vibro-exciters is conditioned by a limited efficiency of the rolling bearings, used in industry. This problem is to be faced when designing heavy vibrational machines.

The application of two unbalanced rotors, rotating in the opposite directions with the same frequency makes it possible to obtain a force of constant direction (Fig. 7.2, a, b); and with the adequate directions of the rotation it is also possible

---

[a] As was already mentioned, such vibro-exciters , often called *unbalanced vibro-exciters* , are used nowadays in most vibrational devices.

to obtain circular, angular and spiral oscillations of a body. Finally, in some vibrational machines, particularly in those, meant for crushing and grinding solid materials, besides the ordinary vibro-exciters, actuated by the engines, rolling bodies can also be used (rollers, cones, rings) which, being devoid of engines, must revolve synchronously with the rotation of the rotors of ordinary exciters (Fig. 7.2,*c*); such devices have already been considered in 6.2.

Before the discovery of the effect of self-synchronization of the unbalanced rotors the only widespread way of synchronizing their rotation in vibrational machines was installing the rigid kinematic bonds between the rotors in form of gears, chain drives, etc., (see, say, Fig. 7.2,*a*). The disadvantage of that method is a considerable noise and a quick wear of the gear wheel or chain gear due to presence of the pulsating loads or those of alternating sign. Besides, the kinematic way of synchronization retarded the development of vibrational technology, since its use could not be expedient in many cases of practical importance, for instance when the distance between the vibro-exciters was considerably large. When using self-synchronization, the kinematic links between the rotors prove to be unnecessary (Fig. 7.2,*b*). The discovery of the phenomenon of self-synchronization of the unbalanced rotors and the subsequent elaboration of the theory and methods of designing devices with self-synchronized vibro-exciters led to the creation of a new class of vibrational machines — conveyers, feeders, screens, crushers, mills, concentration tables, special stands, etc., By now about three hundred inventions have been registered, (mostly in the former USSR and in Russia), based on the use of the effect of self-synchronization. Many of those inventions could not have been made without the theoretical investigations or methods of calculation. The thing is, as will be seen later, that a number of peculiarities of self-synchronization of vibro-exciters can hardly be predicted on the basis of the purely intuitive consideration or discovered by means of pointless experiments.

At the same time the resources of using the self-synchronization of vibro-exciters are far from being exhausted: new ideas keep on emerging, facilities are being elaborated, some devices are being perfected.

The discovery and development of the theory of self-synchronization of unbalanced rotors helped us to understand the tendency towards synchronization as a common property of material objects of different nature — the tendency to self-organizing, opposite to that of a chaotic behavior. In fact, a new section of modern nonlinear mechanics has appeared — the *theory of synchronization of dynamic systems* . In particular, it has been established that the integral criterion of stability (extremal property) of synchronous motions describes the tendency to synchronization and the regularities of motion not only of vibro-

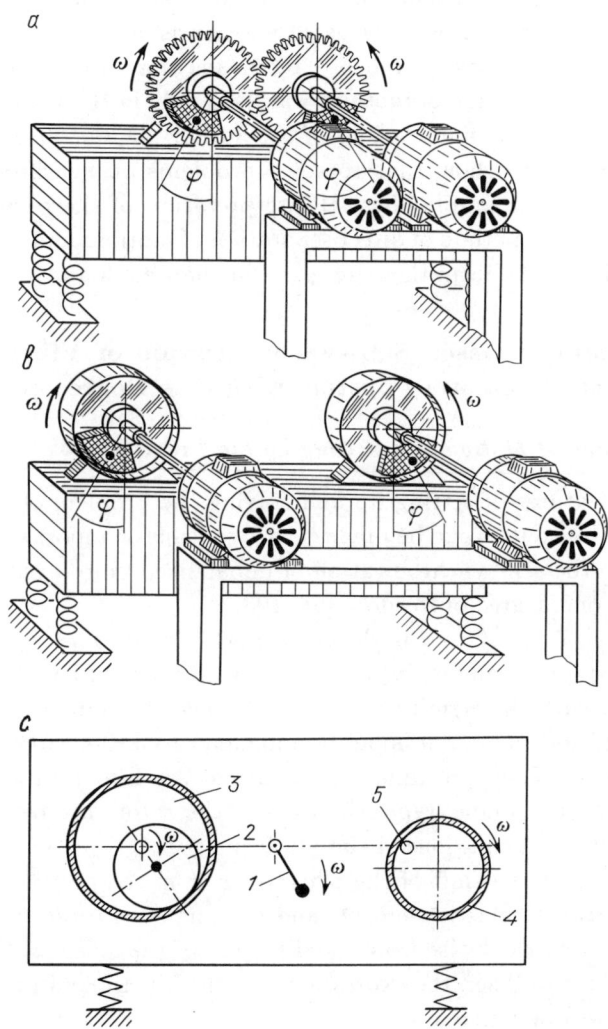

Figure 7.2. Potentialities of the use of self-synchronization of rotors. a) The "old" way of coordinating the rotation of the rotors of vibro-exciters for obtaining the directed oscillations; b) the new way for achieving the same object (the gear wheel being unnecessary); ensuring synchronous rolling of the body of revolution (1 — the unbalanced vibro-exciter; 2 — the roller, rolling in cylindrical cavity 3; 4 —the ring, rolling on axis 5).

exciters, but also of the orbital motions of bodies of the Solar system. These investigations are also discussed in 7.8. and in 20.1–20.5.

A greater part of the results of the theory of synchronization of the mechanical vibro-exciters that was discussed above was obtained by the Poincare-Lyapunov method, a smaller number — by means of asymptotic methods. The method of direct separation of motions was used only in the book by Ragulskis [449] and later on in the publications of the author [93, 95, 99, 103].

In this chapter we show that to solve problems on synchronizing vibro-exciters and to give a simple physical interpretation of the results, good use could be made of the method of direct separation of motions and of the extremal signs of stability in the form they are given in chapters 3 and 4 of this book.

7.2   The Simplest Case: Self-synchronization of Vibro-exciters in the Linear Oscillatory System with One Degree of Freedom

7.2.1   *Equations of Motion and Setting up the Problem*

Most specific features of setting up and solving the problem of synchronizing mechanical vibro-exciters can be made clear by means of a most simple example which refers to the self-synchronization of unbalanced vibro-exciters on a rigid platform with one degree of freedom [68, 103].

The scheme of the system is shown in Fig. 7.3, a, b. The rigid platform B (the carrying or supporting body) is connected with a fixed base by means of an elastic element with the rigidity c and a non-linear damping element with the resistance coefficient β. A k number of unbalanced vibro-exciters are installed on the platform. They are unbalanced rotors, set into rotation by electric engines of an asynchronous type; the axes of the rotors are perpendicular to the direction of the oscillations of the platform.

We will consider the shift of the platform x from the position, corresponding to the unstrained elastic element, and the angles of rotation of the rotors $\varphi_1, \ldots, \varphi_k$, counted clockwise from the direction of the axis Ox, to be the generalized coordinates. Then the expressions for the kinetic and potential energy of the system will be written as

$$T = \frac{1}{2}\mathcal{M}\dot{x}^2 + \frac{1}{2}\sum_{s=1}^{k} J_{C_s}\dot{\varphi}_s^2 + \frac{1}{2}\sum_{s=1}^{k} m_s(\dot{x}_{C_s}^2 + \dot{y}_{C_s}^2), \tag{7.2.1}$$

$$\Pi = \frac{1}{2}cx^2 + \sum_{s=1}^{k} m_s\varepsilon_s g(1 - \sin\varphi_s). \tag{7.2.2}$$

Here $\mathcal{M}$ is the mass of the platform, m_s and J_{C_s} are respectively the mass and the moment of inertia of the rotor of the s-th vibro-exciter with respect to the axis, passing through its center of gravity C_s, g is the free fall acceleration, ε_s is the eccentricity, and

$$x_{C_s} = u_s + x + \varepsilon_s \cos \varphi_s, \quad y_{C_s} = v_0 - \varepsilon_s \sin \varphi_s \qquad (7.2.3)$$

are the coordinates of the center of gravity of the s-th rotor in the system of the fixed axes xOy (uO_1v being the axes, rigidly connected with the platform and coinciding with the axes xOy with $x = 0$; u_s and v_0 being constants, presenting the coordinates of the axes of rotation of the rotors v_s in the axes uO_1v).

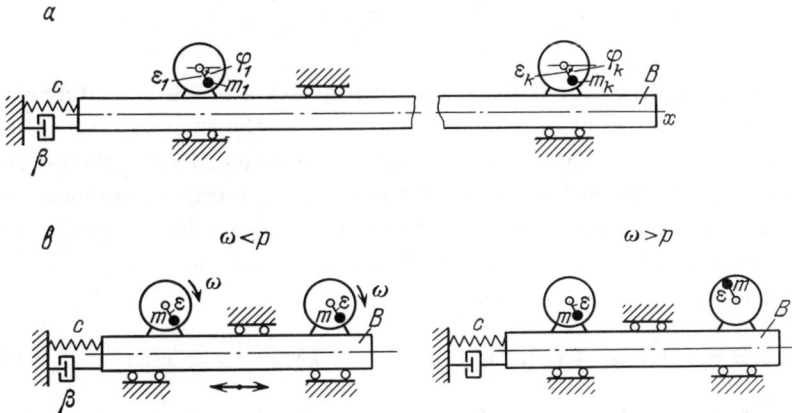

Figure 7.3. Self-synchronization of vibro-exciters in linear oscillatory system with one degree of freedom.

In view of formulas (7.2.3) the expression for the kinetic energy is transformed into

$$T = \frac{1}{2}M\dot{x}^2 + \frac{1}{2}\sum_{s=1}^{k} I_s \dot{\varphi}_s^2 - \dot{x} \sum_{s=1}^{k} m_s \varepsilon_s \dot{\varphi}_s \sin \varphi_s, \qquad (7.2.4)$$

where

$$M = \mathcal{M} + \sum_{s=1}^{k} m_s, \quad I_s = J_{C_s} + m_s \varepsilon_s^2.$$

The generalized non-conservative forces in this case are

$$Q_x = -\beta \dot{x}, \; Q_s = L_s(\dot{\varphi}_s) - R_s(\dot{\varphi}_s), \qquad (7.2.5)$$

where, like in 6.1 and 6.3, $L_s(\dot{\varphi}_s)$ is the driving torque of the asynchronous electric engine (its static characteristic [b]), and $R_s(\dot{\varphi}_s) = R_s^0(|\dot{\varphi}_s|)\mathrm{sgn}\,\dot{\varphi}_s$ is the moment of forces, resistant to the rotation, which, as a rule, is conditioned by the resistance in the bearings. Taking into account relations (7.2.2), (7.2.4) and (7.2.5), we set up the Lagrange equation and arrive at the following differential equations of motion of the system under consideration:

$$I_s\ddot{\varphi}_s = L_s(\dot{\varphi}_s) - R(\dot{\varphi}_s) + m_s\varepsilon_s(\ddot{x}\sin\varphi_s + g\cos\varphi_s) \quad (s = 1, \ldots, k), \quad (7.2.6)$$

$$M\ddot{x} + \beta\dot{x} + cx = \sum_{j=1}^{k} m_j\varepsilon_j(\ddot{\varphi}_j\sin\varphi_j + \dot{\varphi}_j^2\cos\varphi_j) \qquad (7.2.7)$$

The main problem about the self-synchronization of vibro-exciters consists in establishing conditions at which the rotors of all the exciters rotate with the same absolute value of their average angular velocities, despite the absence of any direct links between them or the difference in the parameters characterizing the exciters and the torques acting on them. In other words, we are to ascertain the conditions of the existence and stability of the solutions of system (7.2.6), (7.2.7) of the form

$$\varphi_s = \sigma_s[\omega t + \alpha_s + \psi_s(\omega t)] \quad (s = 1, \ldots, k), \qquad x = x(\omega t), \qquad (7.2.8)$$

where ω is the absolute value of the average angular velocity of rotation, α_s are the constants (the initial phases of rotation), ψ_s and x are the periodic functions of time t with a period $T = 2\pi/\omega$, each of the values σ_s being equal either to 1 or to -1; in the first case the rotor of the s-th exciter rotates in the positive direction, in the second case — in the negative direction [c] The value of the synchronous frequency ω has not been known before, it is to be found in the course of solving the problem.

Motions of type (7.2.8) will be called by us *simple synchronous motions*; of interest may also be *multiple synchronous motions* when $< \dot{\varphi}_s > = \sigma_s p_s \omega/q_s$

---

[b]See footnote on p. 139

[c]The values σ_s are introduced for the universality of notation of the resultant relationships. They characterize the type of the synchronous motion under consideration. So, if $\sigma_1 = 1$, $\sigma_2 = 1$, $\sigma_3 = -1$ etc., that means that in the motion under investigation the first and the second rotors rotate in the direction which is taken to be positive, the third rotor — in the direction, taken to be negative, etc.

where p_s and q_s are positive integers. This more complicated case will be discussed later.

Apart from ascertaining the conditions of the existence and stability of synchronous motions it is of interest to establish at least approximately the law of motion of the rotors and the platform in the stable synchronous motions and to solve the inverse problem ("the *problem of synthesis*") which consists in such a choice of the parameters (and often of the structure) of the system which will provide the existence and stability of the *synchronous motion of a certain type*.

As has been shown by experience and by analytic estimations, the rotation of rotors in the synchronous motions of the real systems under investigation differs but little from the uniform motion. Therefore the functions $L_s(\dot{\varphi}_s)$ and $R_s(\dot{\varphi}_s)$, just like in 6.1.and 6.3, can be linearized near the value $\dot{\varphi}_s = \sigma_s\omega$, assuming that

$$L_s(\dot{\varphi}_s) - R_s(\dot{\varphi}_s) = L_s(\dot{\varphi}_s) - R_s^0(\,|(\varphi_s)|\,)\mathrm{sgn}\,(\dot{\varphi}_s)$$
$$= L_s(\sigma_s\omega) - \sigma_s R_s^0(\omega) - k_s(\dot{\varphi}_s - \sigma_s\omega) \qquad (7.2.9)$$

where

$$k_s = k_s^* + k_s^0, \quad k_s^* = -\left(\frac{dL_s}{d\dot{\varphi}_s}\right)_{\dot{\varphi}_s=\sigma_s\omega},$$
$$k_s^0 == \left(\frac{dR_s^0}{d|\dot{\varphi}_s|}\right)_{\dot{\varphi}_s=\sigma_s\omega} = \frac{dR^0(\omega)}{d\omega}, \qquad (7.2.10)$$

with k_s^* and k_s^0 being the *coefficients of electrical and mechanical damping* respectively; usually k_s^* and k_s^0 and consequently the resultant *coefficient of damping k_s* are positive, which will be always implied below.

Taking into consideration the proximity of the motions under consideration to those with the uniform rotation of rotors, and taking into account equality (7.2.9), we will present equations (7.2.6), (7.2.7) as

$$I_s\ddot{\varphi}_s + k_s(\dot{\varphi}_s - \sigma_s\omega) = \mu\Phi_s(\varphi_s,\ddot{x}) \quad (s = 1,\ldots,k), \qquad (7.2.11)$$

$$M\ddot{x} + \beta\dot{x} + cx = \sum_{j=1}^{k}(\ddot{\varphi}_j\sin\varphi_j + \dot{\varphi}_j^2\cos\varphi_j) \qquad (7.2.12)$$

where

$$\mu\Phi_s(\varphi_s,\ddot{x}) = L_s(\sigma_s\omega) - \sigma_s R_s^0(\omega) + m_s\varepsilon_s(\ddot{x}\sin\varphi_s + g\cos\varphi_s), \qquad (7.2.13)$$

$\mu > 0$ being a small parameter. Using the method of direct separation of motions, we will consider the solutions of system (7.2.11)(7.2.12) of the form

$$\varphi_s = \sigma_s[\omega t + \alpha_s(t) + \psi_s(t, \omega t)], \quad x = x(t, \omega t), \qquad (7.2.14)$$

where α_s are the slow components and ψ_s and x are the fast 2π-periodic in respect to ωt components, satisfying the conditions

$$< \psi_s(t, \omega t) >= 0, \qquad < x(t, \omega t) >= 0; \qquad (7.2.15)$$

we also suppose that $\dot{\alpha}_s \ll \omega$.

It is obvious that the synchronous motions (7.2.8) will be answered by the stationary values of the slow variables $\alpha_s = \text{const}$.

7.2.2   Main Equations of Vibrational Mechanics

Expression (7.2.13) according to (7.2.2), (7.2.4), (7.2.5), and (7.2.9) can be presented in the form

$$\mu \Phi_s = I_s \ddot{\varphi}_s + k_s(\dot{\varphi}_s - \sigma_s \omega) - E_{\varphi_s}(L) + Q_{\varphi_s} \qquad (7.2.16)$$

where

$$E_{\varphi_s} = \frac{d}{dt}\frac{\partial}{\partial \dot{\varphi}_s} - \frac{\partial}{\partial \varphi_s}$$

is the Euler operator, $L = T - \Pi$ is the Lagrange function of the system, and the generalized force $Q_{\varphi_s} = L_s(\dot{\varphi}_s) - R_s(\dot{\varphi}_s)$ is determined according to (7.2.9). Taking into consideration that equation (7.2.12) can be written as

$$E_x(L) = Q_x, \quad Q_x = -\beta \dot{x}, \qquad (7.2.17)$$

we conclude that system (7.2.11), (7.2.12) refers to the *systems with almost uniform rotations* which were examined by the method of direct separation of motions in 4.2.1. Therefore we can use here the results of that investigation.

The equations of slow motions (the main equations of vibrational mechanics) are presented in form (see equations (4.2.38) chapter 4):

$$I_s \ddot{\alpha}_s + k_s \dot{\alpha}_s = \mu \sigma_s < [\Phi_s] > \qquad (s = 1, \ldots, k) \qquad (7.2.18)$$

where

$$\mu \sigma_s < [\Phi_s] >= \sigma_s L_s(\sigma_s \omega) - R_s^0(\omega) + m_s \varepsilon_s < \ddot{x}^0 \sin \varphi_s^0 >$$

$$= \frac{\partial \Lambda}{\partial \alpha_s} + \sigma_s < [Q_{\varphi_s}] > + < [Q_x]\frac{\partial x^0}{\partial \alpha_s} > \qquad (7.2.19)$$

The square brackets, in which the functions are enclosed, indicate that those functions are calculated at $\varphi_s = \varphi_s^0 = \sigma_s(\omega t + \alpha_s)$ and at the value $x = x^0$, satisfying the equation

$$M\ddot{x}^0 + \beta\dot{x}^0 + cx^0 = \sum_{j=1}^{k} m_j \varepsilon_j \omega^2 \cos(\omega t + \alpha_j); \qquad (7.2.20)$$

$$\Lambda =< [T - \Pi] >, \qquad (7.2.21)$$

denoting like in 4.2. the average for the period value of the Lagrange function of the system, calculated at $\varphi = \varphi_s^0$ and $x = x^0$.

The asymptotically stable periodic solution of equation (7.2.20) looks as

$$x^0 = \sum_{j=1}^{k} \frac{m_j \varepsilon_j}{M \Delta} \cos(\omega t + \alpha_j + \gamma), \qquad (7.2.22)$$

where the symbols employed in (3.12) of chapter 6 are used. By means of averaging over the first of the formulas (7.2.19) in view of the equalities

$$\sigma_s = 1/\sigma_s, \quad \sin \sigma_s \alpha_s = \sigma_s \sin \alpha_s, \quad \cos \sigma_s \alpha_s = \cos \alpha_s,$$

$$< \sin(\omega t + \alpha_s) \cos(\omega t + \alpha_j) >= \frac{1}{2} \sin(\alpha_s - \alpha_j),$$

$$< \sin(\omega t + \alpha_s) \sin(\omega t + \alpha_j) >=< \cos(\omega t + \alpha_s) \cos(\omega t + \alpha_j) >$$

$$= \frac{1}{2} \cos(\alpha_s - \alpha_j) \qquad (7.2.23)$$

we obtain

$$\mu \sigma_s =< [\Phi_s] >= \sigma_s L_s(\sigma_s \omega) - R_s^0(\omega)$$

$$-\frac{m_s \varepsilon_s \omega^2}{2M \Delta} \sum_{j=1}^{k} m_j \varepsilon_j \sin(\alpha_s - \alpha_j - \gamma). \qquad (7.2.24)$$

As should be expected, the same formula can be obtained after more cumbersome calculations when the second expression (7.2.19) is used. Indeed, taking into consideration (7.2.21), (7.2.2), (7.2.4), (7.2.23) and the notation (3.12) of chapter 6 and the relationships

$$\sum_{s=1}^{k} \sum_{j=1}^{k} m_s \varepsilon_s m_j \varepsilon_j \sin(\alpha_s - \alpha_j) = 0, \qquad (7.2.25)$$

we find

$$\Lambda = \frac{1}{4}\frac{\omega^2 \cos\gamma}{M\Delta}\sum_{s=1}^{k}\sum_{j=1}^{k}m_s\varepsilon_s m_j\varepsilon_j \sin(\alpha_s - \alpha_j),$$

$$\frac{\partial\Lambda}{\partial\alpha_s} = -\frac{1}{2}\frac{\omega^2 \cos\gamma}{M\Delta}m_s\varepsilon_s \sum_{j=1}^{k}m_j\varepsilon_j \sin(\alpha_s - \alpha_j),$$

$$< [Q_x]\frac{\partial x^0}{\partial\alpha_s} >= \frac{m_s\varepsilon_s\omega^2 \sin\gamma}{2M\Delta}\sum_{j=1}^{k}m_j\varepsilon_j \cos(\alpha_s - \alpha_j), \qquad (7.2.26)$$

from whence the same expression (7.2.24) is obtained. If we denote the "Lagrangian of the supporting body" (the platform) by $L^{(I)}$, we can then easily arrive at the relationship

$$\frac{\partial\Lambda}{\partial\alpha_s} = -\frac{\partial\Lambda^{(I)}}{\partial\alpha_s}$$

$$(\Lambda^{(I)} =< [L^{(I)}] >=< [T^{(I)} - \Pi^{(I)}] >=< \frac{1}{2}M(\dot{x}^0)^2 - \frac{1}{2}c(x^0)^2 >).\,(7.2.27)$$

As a result the main equation of vibrational mechanics (7.2.18) for the problem under consideration acquires the form

$$I_s\ddot{\alpha}_s + k_s\dot{\alpha}_s = \sigma_s L_s(\sigma_s\omega) - R_s^0(\omega) + V_s(\alpha_1 - \alpha_k, \ldots, \alpha_{k-1} - \alpha_k, \omega) \quad (7.2.28)$$

where

$$V_s = -\frac{m_s\varepsilon_s\omega^2}{2M\Delta}\sum_{j=1}^{k}m_j\varepsilon_j \sin(\alpha_s - \alpha_j - \gamma) \equiv$$

$$\frac{\partial\Lambda}{\partial\alpha_s} - \beta < \dot{x}^0\frac{\partial x^0}{\partial\alpha_s} > \equiv -\frac{\partial\Lambda^{(I)}}{\partial\alpha_s} - \beta < \dot{x}^0\frac{\partial x^0}{\partial\alpha_s} > \qquad (7.2.29)$$

are the *vibrational torques*.

Formulas (7.2.22) and (7.2.29) can be simplified if, when considering the motion far from the resonance of the oscillatory part of the system (with $\lambda = p/\omega$ being fairly different from unity) , one neglects the dissipation of the energy under oscillations. Assuming that $n = \beta/2M\omega = 0$ and taking into account the same notation (6.3.12) of chapter 6, we obtain

$$x^0 = -\frac{\omega^2}{\omega^2 - p^2}\sum_{j=1}^{k}\frac{m_j\varepsilon_j}{M}\cos(\omega t + \alpha_j), \qquad (7.2.30)$$

$$V_s = \frac{1}{2}\frac{\omega^2}{\omega^2 - p^2}\frac{m_s\varepsilon_s}{M}\sum_{j=1}^{k}m_j\varepsilon_j\sin(\alpha_s - \alpha_j) \equiv \frac{\partial\Lambda}{\partial\alpha_s} \equiv -\frac{\partial\Lambda^{(I)}}{\partial\alpha_s},$$

$$\Lambda^{(I)} = -\Lambda = \frac{1}{4M}\frac{\omega^2}{\omega^2 - p^2}\sum_{s=1}^{k}\sum_{j=1}^{k}m_s\varepsilon_s m_j\varepsilon_j\cos(\alpha_s - \alpha_j). \qquad (7.2.31)$$

In other words, in this case the vibrational torques acquire a potential character. Introducing the potential function

$$D = \Lambda^{(I)}(\alpha_1 - \alpha_k, \dots, \alpha_{k-1} - \alpha_k, \omega) - \sum_{s=1}^{k}[\sigma_s L_s(\sigma_s\omega) - R_s^0(\omega)](\alpha_s - \alpha_k),$$

$$(7.2.32)$$

we will write down the main equation (7.2.28) in the form

$$I_s\ddot{\alpha}_s + k_s\dot{\alpha}_s = -\frac{\partial D}{\partial\alpha_s} \quad (s = 1, \dots, k), \qquad (7.2.33)$$

corresponding to the form of the equations for the potential on the average dynamic systems (see equations (4.1.3) of chapter 4).

7.2.3   Analysis of the Main Equations. Vibrational Torques, Partial Angular Velocities; Vibrational coupling between the rotors

Let us turn our attention to the analysis of the main equations. First of all it should be emphasized that system (7.2.28) is much more simple than the initial system (7.2.6), (7.2.7) not only because its order is less, but also because at the indicated condition it allows the notation in form (7.2.33).

Let us consider the expression for the vibrational torques, playing the main part in the theory under consideration. According to formula (7.2.31) the vibrational torque V_s, acting on the s-th rotor, presents a sum

$$V_s = \sum_{j=1}^{k}v_{sj}, \qquad (7.2.34)$$

whose terms

$$v_{sj} = -v_{js} = \frac{1}{2}\frac{m_s\varepsilon_s m_j\varepsilon_j\omega^4}{M(\omega^2 - p^2)}\sin(\alpha_s - \alpha_j) \qquad (7.2.35)$$

are "individual" vibrational torques; $v_{ss} = 0$ v_{sj} is the moment, acting upon the s-th rotor on the j-th one. Due to the property of reciprocity $v_{sj} = -v_{js}$,

$v_{ss} = 0$ which takes place there, the total sum of all the vibrational torques is identically equal to zero:

$$\sum_{s=1}^{k} V_s = \sum_{s=1}^{k} \sum_{j=1}^{k} v_{sj} = 0. \tag{7.2.36}$$

It should be noted that expressions (7.2.29), which involve the energy dissipation in the oscillatory part of the system, do not satisfy the relations (7.2.35), (7.2.36) which are in this case satisfied only by the "conservative parts" of the vibrational torques $V_s^{(K)} = \partial \Lambda / \partial \alpha_s = -\partial \Lambda^{(I)} / \partial \alpha_s$. As must be expected , the vibrational torques become zero if $M \to \infty$ that is when the platform becomes fixed, and also if there are no unbalanced rotors there $((m_s \varepsilon_s) = 0)$.

From expression (7.2.35) it also follows that the action of the individual vibrational torques v_{sj} that is of the "vibrational coupling" between the rotors is quite similar to that of the spring that might have been placed between the s-th and the j-th rotors, creating on the rotors a moment, proportional to the phase shift angle $|\alpha_s - \alpha_j|$. The turning rigidity c_{sj} corresponding to that invisible spring at $|\alpha_s - \alpha_j| = 0$ is in the absolute value equal to

$$c_{sj} = (v_{sj})_{\max} = \left| \frac{\partial v_{sj}}{\partial (\alpha_s - \alpha_j)} \right|_{\alpha_s - \alpha_j = 0} = \frac{1}{2} \frac{m_s \varepsilon_s m_j \varepsilon_j \omega^4}{M |\omega^2 - p^2|}, \tag{7.2.37}$$

that is to the largest value $(v_{sj})_{\max}$ of the individual vibrational torque v_{sj}. The stable mutual location of the rotors under the action of that spring alone corresponds either to the phase shifting angle $\alpha_s - \alpha_j = 0$ or to the angle $\alpha_s - \alpha_j = \pi$, depending on the sign of the difference $\omega - p$ (either the pre- or the post-resonance motion).

The observer $\mathbf{V}$, who perceives the platform as being fixed, explains the effect of self-synchronization by the action of those invisible springs which seem to connect the unbalanced rotors (see Fig. 7.1,b); this perception is often convenient in the qualitative analysis. It is essential that the moment, transformed from those "springs" to the rotors, that is the individual vibrational torques v_{sj}, first is of a non-linear character, and, secondly, is limited by the absolute value $(v_{sj})_{\max}$. As will be shown later, the feasibility of the effect of self-synchronization depends greatly upon those limiting values of vibrational torques.

It should be noted that expression (7.2.37) can be presented as (cf. formulas (6.1.21) and (6.3.15) chapter 6)

$$(v_{sj})_{\max} = F_s A_j, \tag{7.2.38}$$

where $F_s = m_s \varepsilon_s \omega^2$ is the amplitude of the driving force, developed by the s-th rotor at the fixed platform, and $A_j = \frac{1}{2} a_j$ is the effective amplitude of the oscillations of the axis of the s-th rotor, excited by the j-th rotor (($a_j = m_j \varepsilon_j \omega^2 / M |\omega^2 - p^2|$) — being the "usual" amplitude of those oscillations).

To make the analysis of equations (7.2.28) more clear, we will also use the idea of partial angular velocities of the rotation of exciters. By the *partial angular velocity* of the s-th exciter we will mean, like in 6.1 and 6.3, the angular velocity, provided its rotor is installed not on a vibrating body, but on a fixed base, and that the direction of rotation of the rotor of the s-th exciter in the synchronous motion under consideration is assumed to be positive.

The angular velocities of rotation of the rotors installed on a fixed base φ_{s0} are determined in the stationary regime from the equations

$$L_s(\dot{\varphi}_{s0}) = R_s^0(|\dot{\varphi}_{s0}|)\operatorname{sgn}\dot{\varphi}_{s0}, \qquad (7.2.39)$$

and the partial angular velocities, according to the given definition, will be of the value[d]

$$\omega_s = \dot{\varphi}_{s0}\sigma_s, \qquad (7.2.40)$$

that is they satisfy the equation

$$\sigma_s L_s(\sigma_s \omega_s) = R_s^0(\omega_s)\operatorname{sgn}\omega_s \qquad (7.2.41)$$

Hence it follows that the angular velocity ω_s is positive if the s-th rotor, being established on a fixed base, rotates in the same direction as in the synchronous motion under consideration. In case the indicated directions of the rotation are opposite, ω_s is negative; then the engine of the asynchronous type operates in the generating regime. The partial angular velocities of the vibro-exciters with their engines switched off (or in case they have no engines at all) are equal to zero.

Now let the partial velocities ω_s, being positive, differ not too much from each other. Then the expression $\sigma_s L_s(\sigma_s \omega_s) - R_s^0(\omega_s)$ can be linearized in the vicinity of $\omega_s = \omega$ and according to (7.2.10) can be written down as

$$\sigma_s L_s(\sigma_s \omega_s) - R_s^0(\omega_s) = \sigma_s L_s(\sigma_s \omega) - R_s^0(\omega) - k_s(\omega_s - \omega).$$

But due to (7.2.41) at $\omega_s > 0$ the left-hand side of that equality becomes zero, and therefore

$$\sigma_s L_s(\sigma_s \omega) - R_s^0(\omega) = k_s(\omega_s - \omega). \qquad (7.2.42)$$

---

[d]See the footnote on p. 182

On taking into account the latter equality, the main equations (7.2.28) can be presented in the form

$$I_s \ddot{\alpha}_s + k_s \dot{\alpha}_s = k_s(\omega_s - \omega) + V_s(\alpha_1 - \alpha_k, \ldots, \alpha_{k-1} - \alpha_k, \omega)$$
$$(\omega_s > 0, \qquad s = 1, \ldots, k). \tag{7.2.43}$$

7.2.4 Stationary Regimes of Synchronous Rotation and Their Stability. The Integral Sign of Stability (Extreme Property) of Synchronous Motions

At $\alpha_s = \alpha_s^* = \text{const}$, the main equations (7.2.28) lead to the following relations for determining the $k-1$-th phase shift $\alpha_1 - \alpha_k, \ldots, \alpha_{k-1} - \alpha_k$ and the frequency ω of rotation of the rotors in possible synchronous motions

$$\sigma_s L_s(\sigma_s \omega) - R_s^0(\omega) + V_s(\alpha_1 - \alpha_k, \ldots, \alpha_{k-1} - \alpha_k, \omega) = 0$$
$$(s = 1, \ldots, k), \tag{7.2.44}$$

and from equations (7.2.43) we accordingly obtain

$$k_s(\omega_s - \omega) + V_s(\alpha_1 - \alpha_k, \ldots, \alpha_{k-1} - \alpha_k, \omega) = 0 \quad (s = 1, \ldots, k). \tag{7.2.45}$$

The necessary condition for the vibro-exciters to get self-synchronized is that equations (7.2.44) and (7.2.45) should have the solutions, real with respect to $\alpha_1 - \alpha_k, \ldots, \alpha_{k-1} - \alpha_k$, and positive with respect to ω. As for ω, it is easy to see that this frequency can be determined from the indicated equations quite independently. Indeed, putting together these equations and taking into account equation (7.2.36), we obtain accordingly

$$\sum_{s=1}^{k} \sigma_s L_s(\sigma_s \omega) = \sum_{s=1}^{k} R_s^0(\omega), \tag{7.2.46}$$

$$\omega = \sum_{s=1}^{k} k_s \omega_s \bigg/ \sum_{s=1}^{k} k_s \tag{7.2.47}$$

Equality (7.2.46) (after being multiplied by ω) presents an equation of energy balance in the system: the total power that supplied to the rotors from the engines, is equal to the total power spent to overcome the resistance to the rotation of the rotors. The vibrational torques in the systems under consideration do not participate in the general energy balance, they only redistribute the energy between the vibro-exciters, adding it to the "slower" exciters and subtracting it from the "faster" ones. That results in the leveling of the rotation frequencies of the exciters — in their self-synchronization.

Indeed, one can see from equalities (7.2.45) that if the partial velocity of a certain vibro-exciter ω_s is lower than the synchronous velocity ω, then the vibrational torque accelerates its rotor $((V_s > 0))$, and if ω_s is greater than ω, then it retards it $((V_s < 0))$. The greater the difference $|\omega_s - \omega|$, the greater is $|V_s|$ and the corresponding power $|V_s|\omega$, delivered or taken away by the vibrational coupling. That provides a synchronous rotation of rotors in spite of the difference in their partial velocities. As was already marked, this distinction can be quite essential: even with the engine switched off (in that case $\omega_s = 0$) the vibro-exciter still does not drop out of synchronism. Moreover, the engines of certain exciters can operate in the generating regime (in that case $\omega_s < 0$, i.e. the corresponding rotor on a fixed base rotates in the direction opposite to that of its rotation in the synchronous regime under consideration). That effect, though in a different form, was discussed in 6.3.

The described regularities are closely connected with the fact that the power, transmitted in the regime of self-synchronization, can be sufficiently large, and it is due to that circumstance that the self-synchronized vibro-exciters can be used in the industrial vibrational machines.

Along with that, it follows from equations (7.2.45) that the greatest value of vibrational torque $|V_s|_{\max} = \tilde{V}_s$ determines that greatest distinction of the partial velocity from the synchronous velocity at which self-synchronization is still possible. As for the vibro-exciters with the equal and positive partial angular velocities, they are always self-synchronized in the systems of the type under consideration (we will see it later).

According to equality (7.2.47), the angular velocity of the synchronous rotation ω is equal to the weighted-mean value of the partial angular velocities, with the positive resultant coefficients of damping k_s playing the role of weight coefficients. Hence, (particularly), it follows that ω is not less than the lowest and not more than the highest of the partial velocities ω_s. When the partial velocities ω_s are equal, the synchronous velocity ω is equal to them.

Let us consider the stability of synchronous solutions. Let there be a certain real solution of equations (7.2.44)

$$\alpha_1 - \alpha_k = \alpha_1^* - \alpha_k^*, \ldots, \quad \alpha_{k-1} - \alpha_k = \alpha_{k-1}^* - \alpha_k^*, \ldots, \quad \omega = \omega^* > 0,$$

i.e. a stationary solution of the main equations (7.2.28). The equations in variations for those equations and for the indicated solution are the linear differential equations with constant coefficients. It is not difficult therefore to compose the corresponding characteristic equation of the $2k$-th degree, the investigation of stability being in fact reduced to the study of the roots of that equation. However, as was established in 4.2.1, taking into account the peculiarity of the problem, (the smallness of the right-hand sides of equations (7.2.28)), the thing is reduced to the consideration of the algebraic equation of the k-th degree

$$\left| \frac{\partial V_s}{\partial \alpha_j} - \delta_{sj} k \right| = 0 \quad (s, j = 1, \ldots, k)$$

where, as before, δ_{sj} is the Kronecker delta (this equation corresponding to equation (4.2.33) of chapter 4). One of the roots of that equation in the case under consideration is always equal to zero, which is due to the autonomy of the initial equations of the problem on self-synchronization. This circumstance follows specifically from the equality

$$\sum_{j=1}^{k} \frac{\partial V_s}{\partial \alpha_j} = 0 \quad (s, j = 1, \ldots, k),$$

which is obtained by differentiating the expression $V_s(\alpha_1 - \alpha_k, \ldots, \alpha_{k-1} - \alpha_k, \omega)$ with respect to α_k: on adding all the rows of the determiner in that algebraic equation to a certain definite line, we will obtain a line with the elements κ. Having separated the indicated root, not affecting the conclusion of stability, we arrive at the algebraic equation of the $(k-1)$-th degree

$$\left| \frac{\partial (V_s - V_k)}{\partial (\alpha_j - \alpha_k)} - \delta_{sj} \kappa \right| = 0 \quad (s, j = 1, \ldots, k-1). \tag{7.2.48}$$

In case all the roots of that equation have their real parts negative, then the motion under consideration is stable; in case there are roots with positive real parts, the motion is unstable; in case there are zero roots or purely imaginary roots, an additional investigation is required.

If the integral criterion of stability can be applied, i.e. the dissipation in the oscillatory part of the system can be neglected, so that equations (7.2.33) are valid, then the question about the existence and stability of the stationary regimes can be solved on the basis of this criterion: the stable synchronous motions answer the points of rough minimums of the function D with respect to the phase differences $\alpha_1 - \alpha_k, \ldots, \alpha_{k-1} - \alpha_k$. (Recall (see 4.2) that by "rough" we mean here strict minimums, discovered by analyzing terms of the 2-nd order in the decomposition of the function near the critical point.) In the absence of minimum, which can be also discovered on a "rough level", the corresponding motion is unstable, and the indefinite case requires an additional investigation. In the presence of the potential function, the conditions of stability can also be formulated as the conditions of the negativeness of the roots of equation (7.2.48); then this equation will look as

$$\left| \frac{\partial^2 D}{\partial(\alpha_s - \alpha_k)\partial(\alpha_j - \alpha_k)} + \delta_{sj}\kappa \right|_{\alpha_s - \alpha_k = \alpha_s^* - \alpha_k^*, \omega = \omega^*} = 0$$

$$(s, j = 1, \ldots, k - 1), \tag{7.2.49}$$

and due to the symmetry of its coefficients, all of its roots will be real.

The importance of the integral criterion of stability (the extreme property) of synchronous motions consists not only in simplifying the investigation of stability and in making it more clear, but also in the fact that due to it, it is possible to establish the presence of the *tendency to synchronization* for a wide class of dynamic objects if by that tendency we imply that the system has at least one stable, in one sense or other, synchronous motion [103, 114, 120]. When we mean stability with respect to the relative rotation angles of the body, the idea of proving it in this case and in other similar cases is quite simple. In the present case it consists in the following. If the partial velocities of the exciters are positive and not very different from each other, expression (7.2.32) for the potential function D according to (7.2.42) can be presented as

$$D = \Lambda^{(I)}(\alpha_1 - \alpha_k, \ldots, \alpha_{k-1} - \alpha_k, \omega) - \sum_{s=1}^{k-1} k_s(\omega_s - \omega)(\alpha_s - \alpha_k). \tag{7.2.50}$$

But $\Lambda^{(I)}$ is a continuous 2π-periodic function of the phase differences $\alpha_s - \alpha_k$ and then it surely has minimums, quite rough as a rule. But then if the partial velocities ω_s are not very different from each other, the second term in (7.2.50) is not too large, so the function D will also have rough minimums. So according to the integral criterion and to the above definition, there will be a tendency to synchronization.

It should be emphasized that in case of equal positive partial velocities of vibro-exciters, according to (7.2.50) the role of the potential function D will be played by the averaged Lagrangian of the oscillatory part of the system $\Lambda^{(I)}$:

$$D = \Lambda^{(I)}(\alpha_1 - \alpha_k, \ldots, \alpha_{k-1} - \alpha_k, \omega). \tag{7.2.51}$$

This formula serves as an "operating instrument" for the inventors and designers of vibrational machines with self-synchronized exciters.

7.2.5 A Case of Two Exciters. Once Again about the Phenomenon of Vibrational Maintenance of Rotation

Many important regularities of self-synchronization of vibro-exciters can be demonstrated by the simplest case of two exciters. Restricting ourselves to the consideration of the system with positive and not very different partial velocities in a non-resonance case, and in view of (7.2.31), we will write down equations (7.2.45) with $k = 2$ as

$$k_1(\omega_1 - \omega) = -V_{\max}(\omega)\,\text{sgn}\,(\omega - p)\sin\alpha,$$
$$k_2(\omega_2 - \omega) = V_{\max}(\omega)\,\text{sgn}\,(\omega - p)\sin\alpha \qquad (7.2.52)$$

where

$$\alpha = \alpha_1 - \alpha_2, \quad V_{\max}(\omega) = \frac{1}{2}\frac{\omega^4}{|\omega^2 - p^2|}\frac{m_1\varepsilon_1 m_2\varepsilon_2}{M} = F_1 A_2 = F_2 A_1,$$
$$F_1 = m_1\varepsilon_1\omega^2, \qquad F_2 = m_2\varepsilon_2\omega^2, \qquad (7.2.53)$$

with $V_{\max}(\omega)$ being the modulus (maximum value) of the vibrational torques V_1 and V_2, the value $A_1 = \frac{1}{2}a_1$ $(A_2 = \frac{1}{2}a_2)$ being the effective amplitude of oscillations of the axis of the first (second) rotor, generated by the rotation of the second (first) rotor (see formulas (7.2.38) and the explanations to them), and $F_s = m_s\varepsilon_s\omega^2$ being the amplitude of the exciting force, developed by the s-th rotor. From equations (7.2.52) we find

$$\omega = \frac{k_1\omega_1 + k_2\omega_2}{k_1 + k_2}, \qquad \sin\alpha = \frac{k_{1-2}(\omega_2 - \omega_1)}{V_{\max}(\omega)\,\text{sgn}\,(\omega - p)},$$
$$k_{1-2} = k_1 k_2/k_1 + k_2. \qquad (7.2.54)$$

Hence it follows that the condition of the existence of synchronous motion is expressed by the inequality

$$k_{1-2}|\omega_2 - \omega_1| < V_{\max}(\omega), \qquad (7.2.55)$$

which, in accordance with the general conclusion (see 7.2.4) is always satisfied, provided the partial velocities are close enough.

However, as has been marked, self-synchronization can also take place when those values are rather different, for instance at $\omega_1 = 0$ or $\omega_2 = 0$ that is when the engine of one of the exciters is switched off. We mean here the phenomenon of a *vibrational maintenance of rotation* of an unbalanced rotor, discussed, though from another angle, in 6.1.4. The corresponding condition can be easily obtained in this case by considering equations (7.2.44) at $k = 2$.

One can see that it is in complete agreement with the corresponding condition of 6.1.5. In particular, the conclusion proves to be true that quite considerable power can be transferred to the rotor with the switched off electric engine from the engine of another vibro-exciter. As was already noted, this circumstance is of great importance for applications.

Let, however, the partial velocities ω_1 and ω_2 be the same. Then the second equation (7.2.54) allows two essentially different solutions:

$$\alpha = (\alpha_1 - \alpha_2)_1 = 0 \qquad \alpha = (\alpha_1 - \alpha_2)_2 = \pi. \tag{7.2.56}$$

The rotation of rotors, answering the first solution, will be called by us the *synphase rotation*, and that which answers the second solution will be called by us the *antiphase rotation* . According to (7.2.51) in this case the potential function is $D = \Lambda^{(I)}$ and according to (7.2.31) this function has a minimum for the first solution at $\omega < p$; while at $\omega > p$ the minimum of that function corresponds to the second solution. Thus it follows from the integral criterion of stability that before the resonance the synphase synchronous rotation of rotors is stable, while after the resonance the antiphase rotation is stable (see Fig. 7.3,*b*).

It is remarkable that as one can see from (7.2.30), when the rotation is antiphase, the platform does not oscillate in the approximation under consideration. Here we come across one of the remarkable paradoxes of synchronization of the rotating bodies — the *paradox of the inoperative coupling* (also see 7.7).

At $\omega - 1 = \omega_2$ the stable phasing of the rotors will, according to (7.2.54), differ from the synphase or antiphase phasing.

The problem of global stability of solutions (7.2.56) is considered in article [322].

7.2.6   *About the Case of Almost Identical Exciters*

If the partial angular velocities of all vibro-exciters are positive and equal, then in view of expressions (7.2.31), equations (7.2.45) are recorded as

$$\sum_{j=1}^{k} m_j \varepsilon_j \sin(\alpha_s - \alpha_j) = 0 \quad (s = 1, \ldots, k). \tag{7.2.57}$$

These equations admit of a group of solutions

$$\alpha_s = q_s^* \pi + \alpha_0 \quad (s = 1, \ldots, k), \tag{7.2.58}$$

where q_s^* are the numbers, each of which can be equal to zero or to a unity, α_0 is an arbitrary constant. Such solutions will be called by us *solutions of the* *first group*, and the solutions in which m numbers of q_s^* are equal to unity will be called by us *solutions of type m*.

The investigation of the stability of such solutions by means of equation (7.2.48), in the assumption that the parameters $m_s \varepsilon_s$ and k_s are also the same for every exciter, leads to the following results [68]. In the pre-resonance region $(\omega < p)$ it is only the solution of type (0) that is stable, i.e. the synphase rotation of rotors. In the post-resonance region $(\omega > p)$ stable motions are possible only in case there is an even number of exciters $k = 2l$. And namely, solutions of type (l) can be stable; and if there are two exciters $(k = 2, l = 1)$, the motion is really stable if $\omega > p$ which agrees with the result, obtained in 7.2.5. In case $k > 2$, to solve the question about stability, it is necessary to make an additional investigation.

Returning to the general case of exciters with the equal positive partial velocities, we will note that equations (7.2.57) are satisfied provided the following equalities are fulfilled:

$$\sum_{s=1}^{k} m_s \varepsilon_s \cos \alpha_s = 0, \qquad \sum_{s=1}^{k} m_s \varepsilon_s \sin \alpha_s = 0. \qquad (7.2.59)$$

The corresponding solutions of equations (7.2.57) will be called by us *solutions of the second group*. A simple investigation shows [103] that in the case of two or three exciters, equations (7.2.59), generally speaking, allow only one solution. In case $k > 3$, the phase differences $\alpha_s - \alpha_k$ for the solution of the second group cannot be determined unambiguously from equations (7.2.59) (and consequently from equations (7.2.57) either); to find them it is necessary to consider the subsequent approximations. In this case the question of the existence and stability of synchronous motions also requires an additional investigation. Such a situation is in this case the consequence of a certain degeneracy of the system under consideration. This degeneracy disappears in view of the non-similarity of the partial velocities of the exciters. The investigation of the corresponding special case, for which it was necessary to overcome a number of analytical difficulties, was made by Malakhova [362].

7.3   The General Case. A Short Review of the Results of Investigations

7.3.1   *Equations of Motion and the Main Equations of Vibrational Mechanics in the General Case*

The scheme of system in the general case of the problem on synchronization of vibro-exciters is presented in Fig. 7.4. It differs from the simplest scheme shown in Fig. 7.3, first by the fact that the exciters may be not only the ordinary unbalanced rotors *l*, but also the *planetary exciters* — rollers 2 or other bodies of revolution, and also rings 3, running along the corresponding bodies (see also 6.2); such exciters are usually devoid of engines, they simulate the operating units of some machines; there may also be *piston exciters* 4 which present bodies that can reciprocate along the rectilinear guides. The general requirement to

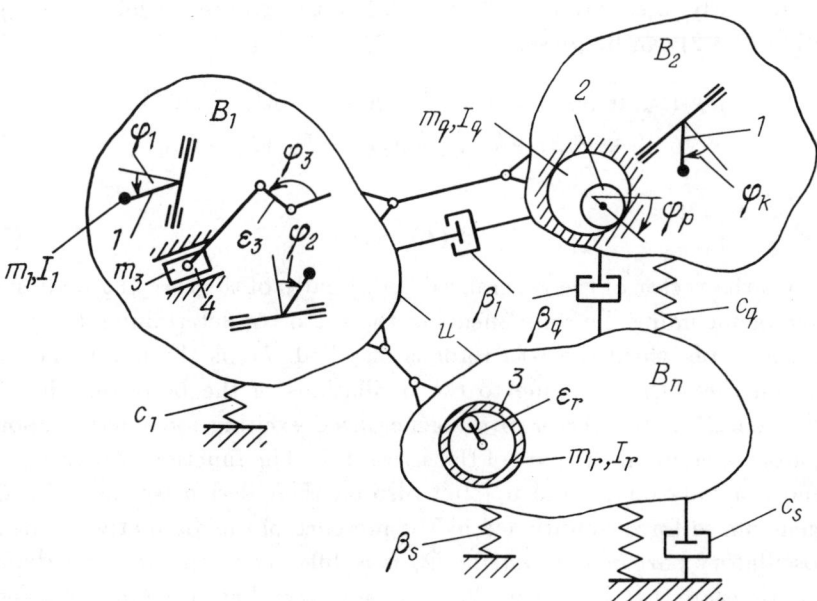

Figure 7.4. The general scheme of the system with self-synchronized vibro-exciters.

all the exciters is that their position in the motions under consideration should be determined by one angular coordinate φ_s which changes almost uniformly.

Secondly, the exciters are assumed to be placed not on one, but on several solid or deformable bodies $B_1, \ldots, B_n$ which may perform spatial oscillations and which are connected with each other, generally speaking, by non-linear elastic and damping elements; cases when in the process of motion the bodies $B_1, \ldots, B_n$ may collide are also of interest; those cases, however, require a special consideration. The *multiple synchronization* of vibro-exciters has a certain specific character and often also requires a special consideration. Therefore we will not consider those two cases here, we will only name below the publications in which they have been studied.

The equations of motion of the system described can be written in the following general form:

$$I_s \ddot{\varphi}_s + k_s(\dot{\varphi}_s - \omega) = \mu \Phi_s(\varphi_s, \mathbf{u}_{s1}) \qquad (s = 1, \ldots, k) \qquad (7.3.1)$$

$$D_u \mathbf{u} = \sum_{s=1}^{k} F_s(\varphi_s), \qquad (7.3.2)$$

where similarly to equations (7.2.11),(7.2.12), the expression $\mu \Phi_s(\varphi_s, \mathbf{u}_{s1})$ and equation (7.3.2) can be presented as

$$\mu \Phi_s(\varphi_s, \mathbf{u}_{s1}) = L_s(\sigma_s \omega) - \sigma_s R_s^0(\omega) + M_s(\varphi_s, \mathbf{u}_{s1})$$
$$\equiv I_s \ddot{\varphi}_s + k_s(\dot{\varphi}_s - \sigma_s \omega) - E_{\varphi_s}(L) + Q_{\varphi_s}, \qquad (7.3.3)$$

$$E_u(L) = Q_u. \qquad (7.3.4)$$

Here $\mathbf{u}$ is the vector of the generalized coordinates of the carrying bodies; $\mathbf{u}_s$ is the vector including the components of the vector $\mathbf{u}$, determining the position of the body on which the s-th rotor is installed; M_s is the moment of forces acting on the s-th rotor due to the oscillations of the body on which it is established; F_s is the vector of the generalized exciting force acting upon the oscillatory system on the part of the s-th rotor. The functions M_s and F_s may depend not only on φ_s and $\mathbf{u}_{s1}$ but also on their derivatives as well. Q_u is the generalized force conditioned by the presence of the dissipative elements in the oscillatory part of the system; D_u is a differential operator, all the other designations are the same as in 7.2. Note, however, that in the case of planetary exciters the moments $R_s(\dot{\varphi}_s)$ present only part of the moment of resistance to the rolling motion (see 6.2.); the other part which depends on the vector $\mathbf{u}_{s1}$ is supposed to be included into the expression M_s, so it also enters into the expression for the generalized force Q_{φ_s}.

Like in 7.2, the much more general system which is being considered here refers to systems with almost uniform rotations, studied in 4.2.1. So we will write now at once the equations of slow motions (the main equations of vibrational mechanics), corresponding to equations (7.3.1) – (7.3.4):

$$I_s \ddot{\alpha}_s + k_s \dot{\alpha}_s = \sigma_s L_s(\sigma_s \omega_s) - R_s^0(\omega) + V_s(\alpha_1 - \alpha_k, \ldots, \alpha_{k-1} - \alpha_k, \omega)$$
$$(s = 1, \ldots, k) \tag{7.3.5}$$

where

$$V_s =< M_s(\varphi_s^0, \mathbf{u}_{s1}^0) > \tag{7.3.6}$$

is the vibrational torque acting upon the s-th rotor, with $\varphi_s^0 = \sigma_s(\omega t + \alpha_s)$, and $\mathbf{u}_{s1}^0$ corresponds to the asymptotically stable 2π-periodic solution of the equation

$$D_u \mathbf{u}^0 = \sum_{s=1}^{k} F_s(\varphi_s^0). \tag{7.3.7}$$

If the operator D_u is non-linear, there may be several such solutions.

In case the planetary vibro-exciters are absent, the expression for the moment V_s can be also presented as

$$V_s = \frac{\partial \Lambda}{\partial \alpha_s} + < [Q_u] \cdot \frac{\partial \mathbf{u}^0}{\partial \alpha_s} > \tag{7.3.8}$$

where, as before, $\Lambda =< [L] >$, and the square brackets denote that the function, enclosed in them, is calculated at $\varphi_s = \varphi_s^0$ and $\mathbf{u} = \mathbf{u}^0$.

The analysis of equations (7.3.5) and expressions (7.3.6) and (7.3.8) allows to come to the same main results as the corresponding analysis, performed in 7.2 for the case considered there. In particular, the stationary regimes of the synchronous rotation of the rotors are determined as before from equations (7.2.44) or (7.2.45); the conditions of the existence of solutions of these equations, real with regard to the phase differences $\alpha_1 - \alpha_k, \ldots, \alpha_{k-1} - \alpha_k$ and positive with regard to ω, are the necessary conditions of a possibility of self-synchronization; the question of the stability of the stationary regime is solved, as before, depending on the signs of the real parts of the roots of equation (7.2.48).

7.3.2   *Potential Function and Integral Criterion of Stability*

If there is a potential function D, then the stable stationary synchronous regimes of the synchronous rotation of rotors correspond to the rough minima of that function.

In the case of the unbalanced vibro-exciters with the positive and not too different partial angular velocities and when the dissipation in the oscillatory part of the system can be neglected

$$D = -\Lambda(\alpha_1 - \alpha_k, \ldots, \alpha_{k-1} - \alpha_k, \omega) - \sum_{s=1}^{k-1} k_s(\omega_s - \omega)(\alpha_s - \alpha_k). \qquad (7.3.9)$$

Finally, if under those conditions the partial velocities are equal or the difference between them is very slight, and the oscillatory part of the system is linear (that is the operator D_u is linear), then, due to the relations

$$\frac{\partial \Lambda}{\partial \alpha_s} = -\frac{\partial \Lambda^{(I)}}{\partial \alpha_s}, \qquad (7.3.10)$$

which are analogous to relations (7.2.31), the averaged on the solutions $\varphi_s = \varphi_s^0$, $\mathbf{u} = \mathbf{u}^0$ Lagrangian of the oscillatory part of the system

$$D = \Lambda^{(I)}, \qquad V_s = -\partial \Lambda^{(I)}/\partial \alpha_s, \qquad (7.3.11)$$

can be taken to be the potential function.

More general, as well as the more specialized expressions for the potential function are given in [90, 103]. Here we will write out only the expression for the function D which refers to the case when under the same, mentioned above, conditions the rotors of the vibro-exciters are connected with each other by certain elastic and damping elements, between which there may also be certain concentrated masses. Following Nagaev [395], we will call such constraints between the rotors the *carried constraints* to distinguish them from the *carrying constraints* — solid bodies $B_1, \ldots, B_n$ on which the rotors are installed. Let $L^{(II)}$ be the Lagrangian of the system of carrying constraints, i.e. the Lagrangian of the system resulting from the initial system, with its carrying bodies being fixed. Then, as has been shown by Nagaev [395], instead of expressions (7.3.11), the following formulas will be valid (see also equality (4.2.42) chapter 4):

$$D = \Lambda^{(I)} - \Lambda^{(II)}, \qquad V_s = -\partial(\Lambda^{(I)} - \Lambda^{(II)})/\partial \alpha_s, \qquad \Lambda^{(II)} = < [L^{(II)}] > \qquad (7.3.12)$$

The example of employing an integral criterion of stability for investigating the self-synchronization of vibro-exciters in one system, frequently used in practice, is given in 7.5.

7.3.3   Short Review of the Results on the Theory of Synchronization of Mechanical Vibro-exciters

We will mention here briefly the investigations of the results which have not been stated or presented adequately enough in this chapter. Self-synchronization of the unbalanced and planetary vibro-exciters on a flatly-oscillating solid body was considered in the book of the author [103], and the self-synchronization of the unbalanced exciters on the spatially-oscillating softly vibro-isolated solid body — in the work by Lavrov [319], (also see [90]). Publications of the author [76], of Nagaev and the author [87], of Afanasyev, Makarov, Petchenev and the author [5] are devoted to the theory of the forced electric synchronization and also to the synchronization by means of elastic and other constraints between the rotors. The latter considers the idea of synchronization of rotors by means of a temporary switching off the "too fast exciters" from the circuit of asynchronous engines. It should be noted that the analytical investigation of such a system without using the method of direct separation of motions would have been most complicated and perhaps hardly possible at all.

Self-synchronization of the unbalanced vibro-exciters in systems with collisions of the bodies of the carrying oscillatory system is considered by Nagaev and Turkin [399] and Zaretsky [589, 590], and synchronization in the oscillatory system with the distributed parameters is discussed in the articles by Nagaev and Popova [396] and also by Sperling [506, 507]. The multiple synchronization of the unbalanced vibro-exciters was investigated by Ragulskis [449], Zaretsky [591] and Barzukov [47, 48], and that of the parametrically connected exciters — by Frolov and the author [109]. Self-synchronization of kinematic vibro-exciters was considered by Chunts [161] and later by Jakashov and Kryukov [172]. Self-synchronization of unbalanced vibro-exciters in the oscillatory systems with a considerable level of dissipation in the oscillatory part of the system was studied by Marchenko and the author [88, 90], by Barzukov and Vaisberg [49].

The use of the harmonic coefficients of influence in order to obtain expressions of the vibrational torques when solving problems on the synchronization of the mechanical vibro-exciters, connected with the linear oscillatory system, in different forms was suggested by Khodzhaev and Sperling [506, 507]. Using this method Sperling and his colleagues have considered in recent years a number of important problems on self-synchronization of vibro-exciters [373, 374, 375, 507, 508, 509, 510, 511, 512, 376]. One of the essential results is the solution of the problem on self-synchronization of both statically and dynamically unbalanced exciters [508, 509].

The so-called *quasi-conservative theory of the synchronization of unbalanced vibro-exciters* was developed by Nagaev, Khodzhaev, Guzev [230, 395, 403, 265].

Methods and the theory of the synthesis of vibrational machines with self-synchronized vibro-exciters were considered by Lavrov [319, 320, 321], by Nagaev and the author [87], by Guzev [230] and also by other researchers (see 7.6). Publications on establishing and developing the integral criterion of stability (the extreme property) in problems on synchronization of the rotating bodies were considered in 4.2.

A number of publications has been devoted to the investigation of a possibility of using self-synchronization of vibro-exciters in quite concrete systems of practical interest and also to the experimental investigations of the devices with self-synchronized vibro-exciters. A systematic statement of the theory and applications of the phenomenon of self-synchronization of vibro-exciters is given in the books of the author [90, 103] where, in particular, tables are given of the schemes of the devices and conditions are shown of the stability of different regimes of synchronous rotation of rotors of vibro-exciter. A more compact presentation of the theory with detailed tables is also given in the reference book [564, vol. 4].

7.4　Steadiness of Phasing the Vibro-exciters and the Adaptive Property of Vibrational Machines in case of Self-synchronization

7.4.1　On the Concept of the Steadiness of Phasing

The stability of a certain phasing of rotation of the rotors of vibro-exciters in the regime of self-synchronization is not sufficient for the practical use of self-synchronization. It is also often necessary that the phasing should not be too sensitive to various imperfections — such as a random scatter of the parameters of vibro-exciters (including their engines) with regard to their nominal values, caused by the inaccuracy of manufacture and mounting, or to the influence of the fluctuations of the technical load. If these deviations result in a considerable change in phasing, then the nature of oscillations of the operating unit of the machine also changes considerably, which in its turn may violate the technological process.

The property of a vibrational device with self-synchronized vibro-exciters to retain, within certain limits, the disbalance of the relative phases of the rotation of rotors of the exciters, affected by various factors, is called the *steadiness of phasing* of such a device.

7.4.2   Estimation of Steadiness. On the Relative Strength of Vibrational Coupling between the Unbalanced Rotors

Steadiness of phasing can be studied on the basis of the consideration of equations of type (7.2.44) or (7.2.45), determining the phases of the rotation of rotors under the stationary synchronous regimes. Knowing the possible deviations of the parameters of the system, one can find from those equations the deviations $\triangle \alpha_s$ of the relative phases of rotation of the rotors $\alpha_s - \alpha_k$ from their values $\alpha_s^* - \alpha_k^*$ in the stable synchronous motions in the absence of the deviations. Then in case the angles $\triangle \alpha_s$ and the parameter deviations are small, the equations can be linearized with regard to the deviations, as a result of which the finding of $\triangle \alpha_s$ will be reduced to solving the system of $k-1$ linear algebraic equations. After that it is not difficult to estimate the corresponding deviations in the law of the oscillations of the operating unit of the machine and their influence on the technological process.

The investigation of steadiness was performed for some simplest systems in view of the probability nature of certain deviations, and formulas, necessary for calculations, were obtained [90]; also see [597]. Here we will give a more crude but simpler method of estimating steadiness, proposed by Lavrov (see [103]). This method is based on the physical approach in the line of vibrational mechanics, consisting in the fact that the steadiness of phasing is determined by the struggle between of two factors. The stabilizing factor is the vibrational coupling between the rotors, the largest absolute value (modulus) of the vibrational torque $(V)_{\max}$ serving as the measure of that coupling, as was marked in 7.2.3. The destabilizing factor is presented by the above-mentioned, as a rule unregulated, imperfections of manufacture and also by the deviation of the technological loading from that of the nominal.

Those destabilizing factors in the first approximation can be considered proportional to the nominal moment L_0 of the engine, reduced to the rotor of the vibro-exciter (provided of course the engine has been selected correctly, that is it has a sufficient but not too large power). Then we come to the conclusion that a rough measure of the stability of phasing of self-synchronized vibro-exciters can be the value

$$k_\omega = \frac{(V)_{\max}}{L_0}.\tag{7.4.1}$$

For steadiness, the following conditions should be satisfied

$$k_\omega \geq k_\omega^* \tag{7.4.2}$$

where k_ω^* is the minimal admissible value of the coefficient k_ω which is called the *coefficient of vibrational coupling*. The value L_0 is determined from the catalogue data of engines. As for the absolute value of the vibrational torque $(V)_{\max}$, its expressions for some schemes can be found in [90, 103, 564, vol. 4] or can be estimated approximately according to formulas (1.21) and (3.15) of chapter 6 and according to formula (7.2.38) of this chapter. In accordance with those formulas we can assume that

$$(V)_{\max} = FA \qquad (7.4.3)$$

where $F = m\varepsilon\omega^2$ is the amplitude of the exciting force, developed by the rotor, installed on a fixed base, and A is the effective amplitude of oscillations of the axis of the rotor. Since in the majority of vibrational machines and devices the points of the carrying bodies perform oscillations along the elliptical trajectories, according to the above-mentioned formulas we have

$$A = \frac{1}{2}(a \pm b) \qquad (7.4.4)$$

where a is a major semi-axis and b is a minor semi-axis of the ellipse; the sign $(+)$ is taken when the direction of rotation of the rotor coincides with the direction of motion of its axis in the elliptical trajectory, and the sign $(-)$ is taken when those directions do not coincide. The value A should be calculated on the assumption that the oscillations of the carrying body are caused by all vibro-exciters except the one under consideration (see 7.2.3).

Of importance is the question of norming the values of steadiness, that is of fixing the maximal admissible values of the angles of phase disagreement $|\triangle\alpha|^*_{\max}$ or the minimal admissible values of the coefficient of vibrational coupling k^*_ω for different classes of vibrational devices. This question, whose consideration must be based not only on calculations, but also on the experimental data and on the practice of the use of the machines, still remains to be investigated. According to Lavrov 's recommendations (see [320]) it should be assumed for the vibrational machines, depending on their purpose, that $k^*_\omega = 0.5 - 4.0$, which roughly corresponds to the maximal angles of disagreement $|\triangle\alpha|^*_{\max}$ from 3 to 16°. At present, however, we have good grounds to consider these recommendations to be too cautious. So, for instance, cases are known of successful exploitation of vibrational screens with two self-synchronized vibro-exciters for which $|\triangle\alpha|_{\max} \approx 30^\circ$ and $k_\omega \approx 0.3$.

For a rough guess we can suggest a gradation of relative strength of the vibrational coupling between the vibro-exciters depending on the values of the coefficient k_ω (table 7.1). The second row of the table contains the interval of probability values of the occurrence of self-synchronization, calculated at the corresponding values of k_ω according to [114].

Table 7.1

Degrees of the relative strength of vibrational coupling between the
unbalanced rotors installed on a movable base

Degree of relative strength of vibrational coupling	Very weak coupling	Compara- tively weak coupling	Compara- tively strong coupling	Very strong coupling
Intervals of values of coefficient k_ω	0–0.01	0.001–0.1	0.1–0.2	> 0.2
Probability of the occurrence of self- synchronization	10%	10–50%	50–90%	> 90%

In case of a very strong coupling, one can be sure there will be
self-synchronization and all the attendant phenomena; in case of a comparatively strong coupling , one cannot be sure of it, but the possibility of such
phenomena must be taken into consideration. When the coupling is very weak,
those phenomena need not be taken into consideration, and when the coupling
is comparatively weak, they may be either regarded, or disregarded, depending
on the circumstances. It goes without saying that these recommendations are
but conventional.

7.4.3   *The Adaptive Property of Vibrational Machines with Self-synchronized Vibro-exciters*

Kosolapov [290] paid attention to the circumstance that the change in a stable
phasing of rotation of the rotors of self-synchronized vibro-exciters can play a
positive role in case of the change of a technological load on the machine. At
a certain position of the axes of rotation of the rotors this change takes place
in such a way that the machine "copies" with the increased load or with its
non-uniform distribution over the operating unit of the machine. In particular,
if two identical unbalanced co-axial vibro-exciters, installed on a softly vibro-
isolated planar oscillating solid body ^{e} B rotate in the opposite directions,
then their relative phasing is such that the resultant of the centrifugal forces,

---

^{e}Speaking about a *softly vibro-isolated solid body* we mean that the elastic elements,
connecting the body with a fixed base, are so "soft" that with the fixed positions of the
rotors, the free oscillation frequencies of the body are far less than the angular velocity of the
synchronous rotation of the rotors ω. In that case when investigating self-synchronization,
the stiffnesses of the elastic elements can be neglected.

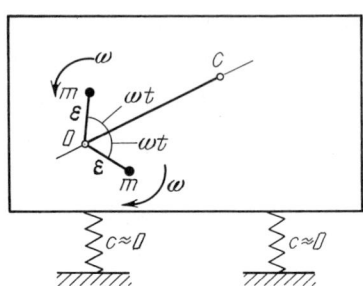

Figure 7.5. The adaptive property of vibrational machines with self-synchronized vibro-exciters.

developed by the rotors, always passes through the center of gravity of the body. As that takes place, the body performs rectilinear translational oscillations in the direction of the perpendicular dropped from the center of gravity of the body C upon the common axis of rotation of the rotors (Fig. 7.5). In other words, the above-mentioned resultant "watches" the position of the center of gravity of the body. That *"adaptive property of vibrational machines with self-synchronized vibro-exciters"* is to a certain extent retained provided the axes of rotation of the rotors are not too far apart from each other as compared to their mean distance to the center of gravity.

7.5   On Theoretic Investigation of the Devices with Self-synchronized Vibro-exciters. An Example of Using the Integral Criterion of Stability

Equations (7.2.44), (7.2.45), (7.2.48), (7.2.49), and also formulas (7.3.6) – (7.3.11) and (7.4.1) – (7.4.4) represent the main relations used when investigating and designing the machines and devices with self-synchronized unbalanced vibro-exciters.

Before one starts the investigation it is useful to make sure that the system had not been investigated before. Particularly, one should examine the tables given in [90, 103, 564, vol. 4] ; it is also necessary to bear in mind that some classes of systems were considered in the general form (see 7.3.3), and that a number of problems has been elaborated for making investigations by computer (including the derivation of formulas).

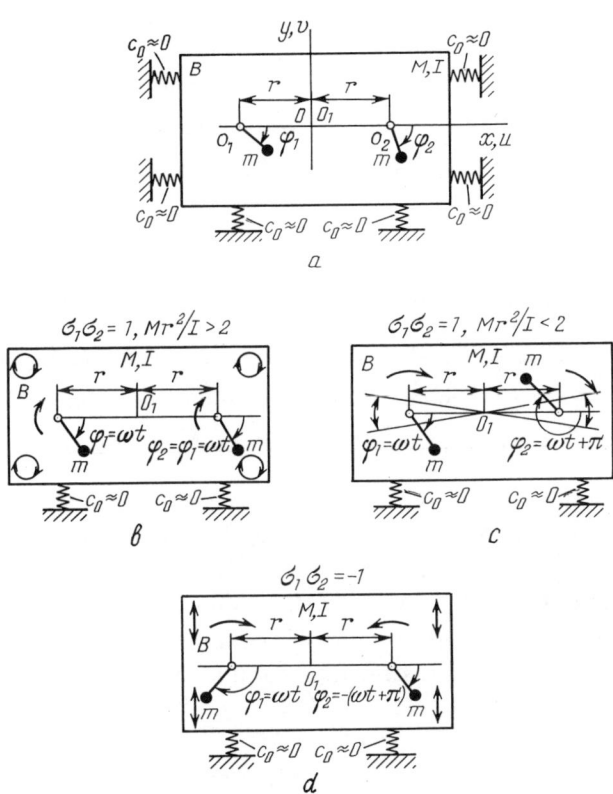

Figure 7.6. Self-synchronization of two identical unbalanced vibro-exciters, located on a softly-vibro-isolated flatly oscillating solid body.

As a simple, though important for practical purposes, example of investigating self-synchronization by using the integral criterion of stability, we will consider the system, shown in Fig. 7.6, a. Two unbalanced rotors, assumed to be identical, are placed symmetrically on a softly vibro-isolated solid body which can move parallel to the plane which is perpendicular to the axes of rotation of the rotors of vibro-exciters. The center of gravity of the carrying body B lies in the plane, passing through those axes, and is at the same distance r from all of them.

When the rotors rotate according to the law

$$\varphi_1 = \varphi_1^0 = \sigma_1(\omega t + \alpha_1), \quad \varphi_2 = \varphi_2^0 = \sigma_2(\omega t + \alpha_2), \tag{7.5.1}$$

the equations of oscillations of the body will have the form

$$M\ddot{x}^0 = F[\cos(\omega t + \alpha_1) + \cos(\omega t + \alpha_2)],$$
$$M\ddot{y}^0 = -F[\sigma_1 \sin(\omega t + \alpha_1) + \sigma_2 \sin(\omega t + \alpha_2)],$$
$$I\ddot{\varphi}^0 = Fr[\sigma_1 \sin(\omega t + \alpha_1) - \sigma_2 \sin(\omega t + \alpha_2)] \tag{7.5.2}$$

where M and I are respectively the mass and the moment of inertia of the body with the fixed rotors, and $F = m\varepsilon\omega^2$ is the amplitude of the exciting force, developed by each exciter. The stationary forced oscillations will be

$$x^0 = -F[\cos(\omega t + \alpha_1) + \cos(\omega t + \alpha_2)]/M\omega^2,$$
$$y^0 = F[\sigma_1 \sin(\omega t + \alpha_1) + \sigma_2 \sin(\omega t + \alpha_2)]/M\omega^2,$$
$$\varphi^0 = -Fr[\sigma_1 \sin(\omega t + \alpha_1) - \sigma_2 \sin(\omega t + \alpha_2)]/I\omega^2. \tag{7.5.3}$$

Since the partial velocities of the exciters ω_1 and ω_2 are assumed to be equal and positive, the role of the potential function D in this case, according to (7.2.51), is played by the average for the period value of the Lagrange function of the carrying body $\Lambda^{(I)}$. In view of the supposition that the elastic support is soft, it is equal to the mean value of the kinetic energy of that body. On performing the averaging, taking into consideration expressions (7.5.1) and (7.5.3), we obtain

$$D = \Lambda^{(I)} =< [T^{(I)}] >= \frac{1}{2} < M[(\dot{x}^0)^2 + (\dot{y}^0)^2] + I(\dot{\varphi}^0)^2 >$$
$$= \frac{F}{2M\omega^2}(1 + \sigma_1\sigma_2 - \sigma_1\sigma_2\frac{Mr^2}{I})\cos(\alpha_2 - \alpha_1) + C. \tag{7.5.4}$$

Here C is the value independent of the phases α_1 or α_2.

When the condition

$$1 + \sigma_1\sigma_2 - \sigma_1\sigma_2\frac{Mr^2}{I} < 0 \tag{7.5.5}$$

is satisfied, the function D has a rough minimum at $\alpha_2 - \alpha_1 = 0$ and when the opposite inequality is satisfied, at $\alpha_2 - \alpha_1 = \pi$. Therefore it follows from the integral criterion of stability that in the first case the synphase rotation of the rotors is stable and in the second case it is the antiphase rotation that is stable.

Let the rotors first rotate in the same directions, that is $\sigma_1\sigma_2 = 1$. Then if the condition

$$Mr^2/I > 2 \qquad (7.5.6)$$

is satisfied, stable is the synphase rotation, which as can be seen from (7.5.3) provides the circular translational oscillations of the carrying body with an amplitude $r_0 = 2m\varepsilon/M$ (Fig. 7.6,*b*). In case the opposite inequality $Mr^2/I < 2$ is satisfied, stable is the antiphase rotation, leading to the angular oscillations of the body with an angular amplitude $\varphi_0 = 2m\varepsilon r/I$ (Fig. 7.6,*c*).

When the rotors rotate in the opposite directions, when $\sigma_1\sigma_2 = -1$, condition (7.5.5) is never satisfied. Therefore it is the antiphase rotation that is always stable which according to (7.5.3) results in the rectilinear translational harmonic oscillations of the carrying body perpendicular to the plane in which the axes of rotation of the rotors lie. The amplitude in this case is $A_0 = 2m\varepsilon/M$ (Fig. 7.6,*d*).

According to (7.3.11) and (7.5.4), the expression for the absolute value of the vibrational torque will look as

$$V_{\max} = \left| \frac{\partial \Lambda^{(I)}}{\partial \alpha_1} \right|_{\max} = \left| \frac{\partial \Lambda^{(I)}}{\partial \alpha_2} \right|_{\max} = \frac{F^2}{2M\omega^2} \left| 1 + \sigma_1\sigma_2 - \sigma_1\sigma_2 \frac{Mr^2}{I} \right|. \qquad (7.5.7)$$

7.6   On the Synthesis of the Devices with Self-synchronized Vibro-exciters

The negative conclusion concerning the possibility of using the effect of self-synchronization at the pre-assigned structural scheme and parameters of the device does not mean that it is impossible to solve the problem by a certain modification of the scheme or by a suitable choice of the parameters. Nowadays a number of such purposeful methods are known that is methods of *synthesizing machines with self-synchronized vibro-exciters* [103, 230, 403]. These methods follow immediately from the regularities of the phenomenon of self-synchronization which will be discussed in the next section. They can be presented algorithmically so that they might be computerized. It should be noted however that the process of elaborating efficient and effective schemes of vibrational devices with self-synchronized vibro-exciters contains a number of heuristic elements, so it is quite just that copyrights and patents are awarded for the adequate solutions of such problems.

7.7   Regularities and Paradoxes of Self-synchronization of Unbalanced Vibro-exciters

In conclusion we will enumerate the main remarkable regularities and paradoxes, typical of self-synchronization of the unbalanced rotors and of interest for applications.

1. **Tendency of unbalanced vibro-exciters toward synchronization; vibrational maintenance of rotation.** The unbalanced vibro-exciters with close enough positive partial velocities ω_s, installed in a certain mechanical oscillatory system with a small dissipation, display a tendency toward synchronization. If the shafts of the unbalanced exciters, driven by the engines of an asynchronous type, being established on a fixed base, rotate with close angular velocities ω_s, then such exciters are sure to be self-synchronized, provided they are installed in the indicated system of oscillating bodies.

Though, under certain conditions , self-synchronization is possible even in case of a strong dissipation of the energy in the oscillatory system and also in case of sharp distinctions in ω_s, and even when certain $\omega_s = 0$ or $\omega_s < 0$. The first case answers the switched off engines of the corresponding exciters (*effect of vibrational maintenance of rotation*), and the second case answers the work of the indicated engines in the generating regime when these engines are involved in the rotation whose direction is opposite to that of their rotation on the fixed base.

2. **Effect of averaging the partial velocities .** The angular velocity of synchronous rotation of the vibro-exciters ω, under the conditions shown in i. 1, is not more than the largest and not less than the smallest of the partial angular velocities ω_s.

3. **Effect of the transmission of high powers.** The averaging of the angular velocities of rotation of the rotors which have different partial angular velocities ω_s, i.e. self-synchronization, can be interpreted as a manifestation of vibrational coupling between the rotors, i.e. the transfer of the rotating moments or powers via the oscillatory system from the "faster" exciters to the " slower" ones. For the practical applications of the phenomenon of self-synchronization and of the effect of vibrational maintenance of rotation, it is of primary importance that the indicated powers can be sufficiently large; they are of the order of the product of the amplitude of the exciting force $F = m\varepsilon\omega^2$, developed by the rotor on a fixed base, by the *effective amplitude of oscillations* of its axis A and by the angular velocity ω. In particular at $F = 10^6$ N, $A = 0.25 \cdot 10^{-2}$ m and $\omega = 314 \text{ s}^{-1}$, i.e. at the parameters, actual for the modern vibrational devices, the maximal possible power that can be transferred makes about 800 kWt.

4. Establishing certain relations between the phases of rotation of the rotors. In stable synchronous motions of the exciters, certain quite definite values of the relative rotation phases $\alpha_s - \alpha_k$ of the rotors are established. In some cases, especially in case of a large number of exciters, there may be several such stable (in the small) phasings. Every stable phasing corresponds to a certain law of motion of the oscillatory part of the system, that is of the carrying bodies on which rotors are installed.

5. Integral criterion of stability (extreme property) of synchronous motions. In case of self-synchronization of the unbalanced vibro-exciters with equal partial velocities far from the resonance and at a small energy dissipation in the linear oscillatory system, the phasings, stable in the small, correspond to the points of rough minimum with regard to the phase differences $\alpha_s - \alpha_k$ of the average for the period Lagrange function of the oscillatory part of the system, calculated in the adequate approximation. In many other cases the stable phasings correspond to the points of rough minimum of a somewhat more complicated in structure function of the differences $\alpha_s - \alpha_k$ — the *potential function D*.

6. Effect of a mutual balance of rotors installed on a softly vibro-isolated solid body(the *principle of minimum of the average kinetic energy*). Suppose the vibro-exciters are installed on a solid body which is connected with the fixed base by means of such soft elastic elements that the free oscillation frequencies of the body are considerably lower than the synchronous velocity ^{f} ω. Then the potential energy of the body can be neglected as compared to the kinetic energy, and if, besides, the exciters have the same positive partial angular frequences, then it follows from the integral criterion of stability that the stable synchronous motions will be answered by the minima of the average for the period values of the kinetic energy of the body.

In other words, those phasings of the rotors will be stable at which the unbalanced forces and the moments, generated by them, will be mutually compensated (canceled) in the sense that the averaged for the period kinetic energy takes the minimal value. In particular, if such phasing is possible at which there is a complete mutual compensation of the unbalanced forces and moments, then it is this phasing (answered by the zero value of the kinetic energy of the body) that is stable. The body in this case practically does not oscillate at all (Fig. 7.3,b and 7.7,a). This phasing will be called by us a *compensating phasing*.

It should be noted that this regularity can be considered as a peculiar generalization of the well-known *principle of Laval*, which consists in the self-

---

^{f} See the footnote on p. 205

balancing of the disc, fixed on an elastic shaft, in the post-critical region of rotation frequencies. Chapter 8 is devoted to this regularity and to its practical use.

7. **Paradox of inoperative coupling.** As follows from what was said in i. 6, when vibro-exciters with the same positive partial velocities are self-synchronized, there may exist and be stable such synchronous motions in which the carrying bodies (or body) in the synchronous motion remain immobile (Fig. 7.3,b and 7.7,a). It gives one an illusion that the oscillatory part of the system in such cases does not perform any functions at all and is not needed for self-synchronization. But in fact it is not so: at an accidental perturbation of the motion, for instance, when one of the phases α_s changes, there appear oscillations of the carrying bodies which do not stop until the perturbation is damped out.

The described effect is typical not only for the rotating objects, but for those oscillating as well. It was first described by Huygens who observed the self-synchronization of the pendulum clock [237].

8. **Dependence of the character of stable phasing of the exciters on the number of the degrees of freedom of the oscillatory system (the carrying bodies)** . The character of a stable phasing of vibro-exciters may change with the change of the number of the essential degrees of freedom of the carrying bodies. So, e.g. adding to the carrying body some more mass on a spring or a pendulum, etc., may change the rotation of the rotors — the stable synphase synchronous rotation may become unstable, while the unstable antiphase rotation may become stable and vice versa. (Fig. 7.7). The character of the stable phasing of vibro-exciters usually changes in that way when the angular velocity passes from one range between the frequencies of free oscillations of the carrying bodies or between some other separating frequencies to the adjacent range (see also chapter 8).

9. **Paradox of forcing.** To illustrate this paradox as well as the previous regularity, let us consider the following example. Assume that it is necessary to provide the rectilinear translational oscillations of a softly-vibro- isolated solid body according to the law close to the harmonic law (Fig. 7.7). For this purpose one can use two identical unbalanced vibro-exciters whose rotors rotate in the opposite directions on the face of it. To guarantee the oscillations of the body in the wanted direction, it seems expedient to place the body into the guides (Fig. 7.7,a). However, as investigation has shown (see 7.2.5 and Fig. 7.3,b) and as follows from what was said in 6, in this case the exciters will get self-synchronized steadily with such ratio of phases at which the exciting forces, developed by the exciters, become mutually balanced, and the body

remains practically motionless (this phasing is shown in Fig. 7.7,a by solid lines), while the wanted phasing (shown by dashed lines) is unstable.

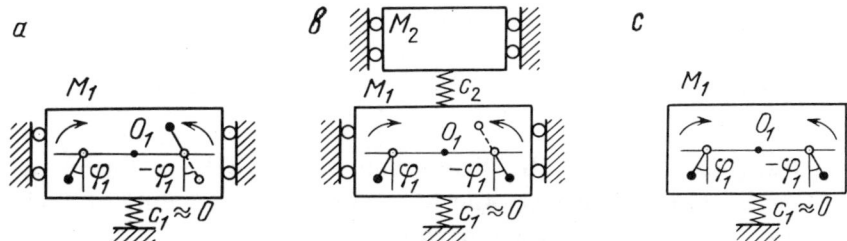

Figure 7.7.   The character dependence of stable phasing of rotors during self-synchronization of vibro-exciters on the number of degrees of freedom of the oscillatory system and the paradox of forcing.

Investigation shows that there are at least two ways to provide the stability of the wanted phasing: either to attach some additional body to the main body by means of a spring [9] (Fig. 7.7,b), or, no matter how paradoxical this may seem, merely remove the guides (Fig. 7.7,c). One can say that in both cases the desired result is obtained in the same way — by changing the number of the degrees of freedom of the oscillatory part of the system. Thus this paradox is closely connected with the regularity, described in 8.

10. **Dependence of the character of a stable phasing on the number of exciters.** The addition of one or several exciters may change essentially the character of their stable phasing while they are being self-synchronized [90, 103].

II. **Dependence of the character of stable phasing on the relative directions of rotation of the rotors of vibro-exciters.** The example considered in 7.5 gives an idea of this regularity (see also Fig. 7.6). The exception of that rule is the simplest system, considered in 7.2: in this case the vibrational torques do not depend on the numbers $\sigma_1, \ldots, \sigma_k$, characterizing the directions of rotation of the rotors (see equation (7.2.29)).

12. **Adaptive property of devices with self-synchronized vibro-exciters.** Under

---

[9]The stiffness of the spring, the masses of bodies and the frequency must in that case satisfy a certain inequality (see [90, 103]).

certain conditions the phasing of the rotation of rotors "in a favorable" way changes with the change of parameters of the oscillatory part of the system [290] (see 7.4.3). So for instance, under the conditions shown in Fig. 7.5 the resultant of the centrifugal forces, developed by the rotors, with the rotation being synchronous and stable, always takes the direction toward the center of gravity of the body on which the rotors are located.

All the regularities and paradoxes, discussed in this section, are widely used nowadays when creating new machines and devices and improving those that are in existence. Many important inventions are based on them [90, 103]. However, as was mentioned, the possibilities of using the peculiar regularities of self-synchronization are still a long way from being exhausted.

7.8   Supplement: On the Phenomena of Synchronization of the Oscillatory and Rotational Motions in Nature and in Technology. The modern state of the problem. General Definition of synchronization

The first observation and description of the phenomenon of self-synchronization was evidently made by Hristian Huygens. As far back as the beginning of the second half of the XY11-th century he discovered that a couple of pendulum clocks, ticking in a different way, got self-synchronized when they were attached to a light beam instead of a wall [237].

At the end of the X1X-th century Rayleigh noticed that two organ pipes, their apertures located close to each other, at tuning, would begin sound in unison, i.e. there would be mutual synchronization of oscillations. As that takes place, the pipes may practically "silence" each other. A similar phenomenon, was discovered by Rayleigh, in the case of two electric or mechanically connected tuning forks [520]. In the late XIX-th — early XX-th century, phenomena of synchronization in electric circuits and in some electro-mechanical systems were discovered. Mutual synchronization of the electric generators and the generators of electro-magnetic oscillations up to now has presented the main technical applications of synchronization; a great number of theoretical and experimental investigations are devoted to them.

As for the rotational motions of bodies, the synchronization of the rotational and orbital motions of the Moon has been known since time immemorial. The Moon is always turned to the Earth with the same side of its surface, which shows the equality of the average frequencies of rotation of the Moon round its axis and in its motion along the orbit round the Earth. Later many such (and more complicated) remarkable relations between the frequencies of rotational motions of bodies of the Solar System became known (see chapter 20); however, up to the recent time they were treated as isolated phenomena, having no common meaning. Apparently, it was only due to that fact that the self-synchronization of unbalanced rotors was discovered quite accidentally (see 7.1) as late as the end of the forties, that is about 300 years after Huygens had observed the synchronization of the oscillating objects (clocks) and many years after the discovery of synchronism ("resonances") in the motion of certain celestial bodies.

Subsequent investigations facilitated the understanding of the fact that self-synchronization of rotations is typical not only for the unbalanced rotors, located in a common oscillating system, but for many other classes of the rotating interacting bodies as well. It was established that for the appearance of the tendency to synchronization it was of principal importance that there should exist forces of interaction, depending, apart from anything else, on the angular coordinates of bodies. The synchronization of rotations, accompanied by the establishment of certain phase relations, can often appear even in case of very weak interactions. In particular it became clear that in the case of celestial bodies there is undoubtedly a certain general regularity — the tendency of the gravitationally interacting rotating bodies toward a mutual synchronization [56, 57, 58, 59, 90, 103, 392, 141, 421]. Besides, that tendency is defined by a mechanical principle, common for many classes of rotating bodies, — by the integral criterion of stability (extremal property) of synchronous motions (see chapter 4). In the field of mechanics and the theory of machines, for instance, an assumption was made and investigated about the possibility of self-synchronization of the cages of rolling bearings [31, 103, 600] and of turbine blades [179]. The dynamics of auto-balancing devices was considered from the positions of the theory of self-synchronization of the rotating bodies (see chapter 8).

Along with some obvious general elements, a number of essential distinctions were discovered: the distinctions of the synchronization of rotating bodies (particularly of the unbalanced rotors) from that of the oscillating objects (particularly of the pendulums) So, for the rotors there is no notion either of the frequencies of free oscillation or of the upper or lower positions, which are most essential for the pendulums, but there is a notion of the right-hand and left-hand rotation. From the twelve regularities, listed in 7.7, only the second, the fourth, the seventh, and partly the first are common for the rotors and pendulums, others are typical only for the rotors. Qualitative distinctions are also observed in the nature of phase spaces of the corresponding systems, as well as in the behavior of the trajectories, answering the synchronous motions in those spaces.

From the point of view of application, the effect of the transmission of greater power is of special importance (see 7.7): as estimations have shown, the power for the rotors was one or two orders higher than that for the pendulums. Perhaps it was due to that reason that the self-synchronization of mechanical objects of the type of pendulums, discovered more than three hundred years ago, has not yet found any serious application, while the self-synchronization of the rotors, discovered quite recently, is made wide use of.

From the principal point of view it is essential that the extreme property of stable motions, of the type characteristic of rotating bodies, has not been found for the oscillating objects. In this connection the existence of the tendency toward synchronization, proved for a wide class of rotating bodies, so far has not been established in a general form for the oscillating objects.

The discovery and investigation of the phenomenon of self-synchronization of the rotating bodies has stimulated the development of the general theory of synchronization of dynamic objects and helped to shape the view upon the tendency toward synchronization as a common property of material objects of different nature — the tendency toward self-organization. Publications of that line belong to Lurye, Nagaev, Neimark, Gurtovnik and also to the author. Those publications were partly mentioned in chapter 4. The investigation [23] is devoted to the mutual synchronization of chaotic motions.

Interesting investigations of the phenomena of synchronization appeared in biology and in physiology of animals [460, 495, 525, 162].

The results of most of those investigations on the theory of synchronization and on its experimental check and practical applications have been generalized in the books [90, 103, 114] and in the presentation [91]; a number of hypotheses about the role of the phenomena of synchronization in the micro-world, in the evolution of the Universe, in biology and in medicine, which have not yet been fully confirmed, are also discussed there.

Attention is seldom paid to the circumstance that the theory of synchronization of dynamic objects obviously belongs to modern conceptions of time self-organization in non-equilibrium systems [195], that is to thermodynamics and synergetics [447, 231]. Meanwhile, many principal moments, up to now discussed within the frames of those conceptions just speculatively, or as hypotheses, received a mathematical basis in the synchronization theory as far back as 1962–64.

In recent years the interest of the scientists of different trends towards problems of synchronization has greatly increased, which is proved by their numerous publications. Three main directions of the investigations can be specified:

the control of synchronization, the investigation of the regularities of the synchronization of chaotic motions, the study of the synchronization of certain systems, in particular the systems of radio-electronics. Some of these investigations as well as the references to other investigations can be found in, say, the Proceedings of the 1st International Conference "Control of Oscillations and Chaos" (St. Petersburg, Russia, August 1997) and also in the book [311]; see also [384].

In this connection the problem of elaborating a *general definition of synchronization* embracing as many well known applications as possible, is quite actual. Some suggestions are given in [6, 311].

It seems that while formulating the definition of synchronization it is expedient to proceed from the verbal encyclopedic understanding of the phenomenon of synchronization as the coordinated in time functioning of two or several processes or objects. The same term is used in the meaning of reducing the objects or processes to the coordinated functioning. Such a definition was suggested in the article [127] which also considered the case of controlled synchronization. At present the author can suggest a more simple and more general definition, following the above-mentioned principle:

Let there be a certain number k of objects or processes whose state in time is characterized in the sense that interests us (but necessarily is fully defined!) by the vectors $\mathbf{x}^{(s)}(t)$. (Let us call $\mathbf{x}^{(s)}(t)$ *selected motions*; t is either continuous or discrete time). Let there be given certain characteristics of these objects or processes

$$\mathbf{X}[t, \mathbf{x}^{(s)}(t)] : \mathbf{x}^{(s)}(t) \longrightarrow \mathbf{X}(t, s) \qquad (7.8.1)$$

These characteristics are supposed to be the same for all the objects or processes. They can be either scalar or vector values, constants or time-dependent. They can be frequencies of spectra of oscillations at the infinite or preassigned finite sliding intervals of time.

We will say that there is *synchronization of objects or processes with respect to the motions* $\mathbf{x}^{(s)}$ *and to the characteristics* $\mathbf{X}$ *if the equalities*

$$\mathbf{X}[t, \mathbf{x}^{(1)}(t + \tau_1)] = \ldots = \mathbf{X}[t, \mathbf{x}^{(k)}(t + \tau_k)] \qquad (7.8.2)$$

are satisfied, where $\tau_1, \ldots, \tau_k$ are certain constants ("the phase shifts").

By *full synchronization* we mean cases when relationship (7.8.2) are fulfilled approximately, and by *asymptotic synchronization* we mean cases when the equalities are fulfilled for $t \to \infty$.

The definitions of *self-synchronization* and of the *forced (controlled) synchronization* , given in the books [90, 113] as applied to the frequency synchronization refers to the general case as well.

Chapter 8

Generalized Principle of Auto-balancing

8.1 Laval's Principle and its Generalization, Ensuing from the Theory of Self-synchronization of Mechanical Vibro-exciters and from Some Other Investigations

As far back as 1884 the Swedish engineer Laval discovered that the unbalanced disk, fixed on an elastic shaft, in the post-critical region of the frequencies of rotation, i.e. at the frequencies ω, larger than the frequency of the free oscillations of the rotor p, gets self-centralized: its center of mass is located practically on the axis of rotation (Fig. 8.1,a). As a result, the unbalanced forces transferred to the bearings of the shaft, are reduced quite tangibly. This effect which was first theoretically explained by A. Fopple in 1895 [182] is successfully employed when designing various machines with fast-rotating rotors, for instance centrifuges. It manifests itself in many other systems as well, including more complicated rotor systems (Fig. 8.1,b and c) which also finds important applications (see, say, inventions [484, 485, 486], making it possible to essentially improve washing machines).

From what was said in 7.2.4. and in 7.3 about the integral criterion (extreme property) of the stability of synchronous motions of mechanical unbalanced vibro-exciters, we may obtain a number of important practical results which can be regarded as an essential generalization of the regularity, discovered by Laval. Recall that according to that criterion, valid under sufficiently general conditions, the stable synchronous motions of a certain number k rotors with the same positive partial angular velocities (see 7.2.3) possess a remarkable property of being answered by the rough minima of the average for the period values of the Lagrange function of the system $L^{(I)} = T^{(I)} - \Pi^{(I)}$, calculated for the oscillatory part of the system (that is for the system with the braked rotors) and in the assumption that the rotors rotate uniformly according to the law

$$\varphi_s = \sigma_s(\omega t + \alpha_s) \qquad (\sigma_s = \pm 1; \; s = 1, \ldots, k), \qquad (8.1.1)$$

while the oscillatory system (the carrying bodies) performs stationary oscillations under the action of the unbalanced forces, developed by the rotors in the indicated motion. From the condition of the minimum of the expression

$$D = \Lambda^{(I)} = < [L^{(I)}] > = < [T^{(I)} - \Pi^{(I)}] >, \qquad (8.1.2)$$

called the potential function, we determine the differences of the initial phases of the rotation of the rotors in the stable synchronous motions $\alpha_1 - \alpha_k, \ldots, \alpha_{k-1} - \alpha_k$.

If we take into consideration that in the post-critical region of the frequences $\omega > p_{max}$ (p_{max} being the largest frequency of free oscillations of the system, with the braked rotors; we consider here systems with a finite number of the degrees of freedom) where the potential energy can be neglected as compared to the kinetic energy, then we obtain the first important generalization of Laval's principle for the multi-rotor systems:

Principle of the minimum of kinetic energy in the theory of self-balancing. *In the post-critical region of frequencies the rotors with the same positive partial angular velocities are getting self-balanced in a sense that the average kinetic energy of the oscillatory part of the system has a minimum; if in that case such phasing of the rotors is available at which the carrying bodies are stationary, that is $T^{(I)} = 0$, then this phasing is certain to be stable. We will call it the compensating phasing.*

What has been said is illustrated by Figs. 7.3,*b*, 7.7,*a*, 8.1,*e* and 8.1,*i1*.

Another important regularity, ensuing from the integral criterion, consists in the fact that the whole frequency range of rotation $0 < \omega < \infty$ is split into intervals, separated by the free oscillation frequencies of the system $p_1 = p_{min}, \ldots, p_n = p_{max}$ and by some other frequencies $q_1, \ldots, q_r$ in which self-balance and increase of oscillations take place in turns. In that case, in the post-critical region $\omega > p_{max}$, as was noted, self-balancing is certain to take place, while in the region $\omega < p_{min}$ there is an increase of oscillations (in the latter case we may consider that $T^{(I)} \ll \Pi^{(I)}$, so that the stable motions are answered by the maximum of the potential energy). This regularity is illustrated in Fig. 8.1,*e* and 8.1,*i1*. Two identical unbalanced rotors, located in the middle of the simply supported beam (Fig. 8.1,*e*) rotate either synphasiously (the increase of oscillations) or antiphasiously (self-balancing), depending on the interval in which the rotation frequency ω lies. The role of the demarcating frequencies is played in that case by the odd free oscillation frequencies of the beam under consideration and also by some other values [90, 103, 507].

Above, to simplify our reasoning, we spoke about the rotors with the same partial angular velocities. For the rotors whose partial angular velocities are not the same, the formulated regularities are retained in form of a certain tendency. The same tendency toward the mutual compensation of unbalances in the post-critical region of the rotation frequencies was discovered for the rotor with suspended pendulums (Fig. 8.1,*g*), for the auto-balancers (Fig. 8.1,*f* see also 8.3) and for some other rotor systems, in particular for those peculiar rotors which are the rolling bearing cages, placed in the supports of a certain shaft (Fig. 8.1,*h*) [103, 600].

As was established, during self-synchronization in the post-critical region the cages display a tendency toward mutual compensation of the oscillations caused by them in the shaft rotating on the supports. The specific feature of this case lies in the fact that oscillations are caused here not by the unbalance of the cages, but by the inaccuracy of manufacturing the bearings, due to which the potential energy of the oscillations of the system depends on the angles of rotation of the cages [a] [600].

Regularities of a similar character were discovered by Ishlinsky and his pupils and colleagues in the behavior of another system which proved to be rather complicated - in that of a solid body, rotating on a string suspension (Fig. 8.1,*f*) (see [241, 242, 243]), and also [465] where bibliographic references are given).

Thus, several kinds of devices with rotating rotors, causing the oscillations of the system on which they were installed, display the same regularity of behavior which may be regarded as a certain *generalized principle of auto-balancing* and can be formulated as follows [114]:

The generalized principle of auto-balancing. *Single rotors or several rotors rotating synchronously, installed in the common oscillatory system and causing its oscillations due to the unbalance or to some other factors, display in the region of the frequencies of rotation higher than the largest free oscillation frequency of the system, a tendency to the weakening of oscillations, while in the region, lower than the lowest frequency — a tendency to the maximal strengthening of the oscillations of the system, and the intermediate range of rotation frequencies splits into intervals in which tendencies to weaking and to strengthening of oscillations take place in turns.*

---

[a]The idea of the possibility of using the effect of self-synchronization to reduce the oscillations of the rotors, caused by the imperfection of the rolling bearings, was suggested by Gyatsyavichus.

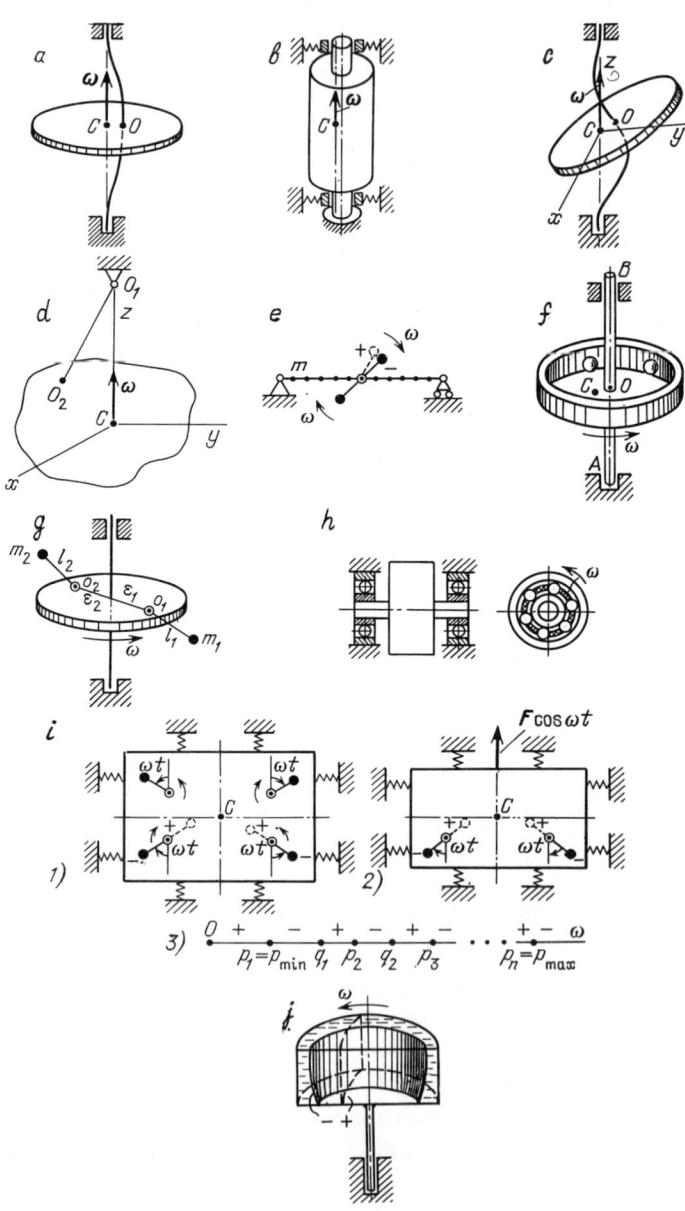

Figure 8.1. The generalized principle of auto-balancing.

Above we spoke about the behavior of rotors in the oscillatory systems with a finite number of degrees of freedom. Investigations show (see, e.g., [90, 103, 167]) that what has been said about the alternation of the ranges of weakening and strengthening the oscillations extends to the cases when the elastic shafts and a beam (Fig. 8.1,*f–g*) have a distributive mass (in these cases $p_{max} = \infty$). A similar behavior, at least when the rotation frequencies are not too high, is manifested by another system, not yet investigated well enough, that is the system with a rotor, containing some fluid (Fig. 8.1,*j*). The fluid which partly fills the vessel, rotating on an elastic shaft, can form, depending on the rotation frequency ω, either most balanced (-) or most unbalanced (+) configuration. This circumstance was used in particular in invention [96] (see also [164]).

Sperling and his colleagues made an important remark referring to the case when the carrying body presents a rotating rotor [511]. If in this case the polar moment of inertia of the rotor is larger than that of the equatorial, then due to the gyroscope effect eigenfrequency of the oscillations of the rotor λ, greater than the frequency of rotation ω. Therefore the relation $\omega > \lambda_{max}$ is not achieved in this case, and while the growth of ω is unlimited, the auto-balancing does not take place. It is not difficult, however, to conclude that the generalized principle of autobalancing, as it was formulated above, remains valid in this case too.

Another generalization refers to the case when the rotors are also connected by a certain system of the so called *carried constraints* (see 7.3). In this case according to equalities (7.3.12) chapter 7, we will have

$$D = \Lambda^{(I)} - \Lambda^{(II)} = < [L^{(I)}] > - < [L^{(II)}] > \qquad (8.1.3)$$

where $\Lambda^{(II)} = < [L^{(II)}] >$ is the Lagrangian of the system of carried constraints, averaged in the same way as the Lagrangian $L^{(I)}$. It should be noted that the expressions $\Lambda^{(I)}$ and $\Lambda^{(II)}$ are involved in the formula (8.1.3) with the opposite signs, as a result of which the carrying- and the carried constraints affect the stability of a certain synchronous motion in a somewhat opposite way. It goes without saying that in that case the given formulations must be modified accordingly.

It should be noted that within each type of systems considered above the thesis, formulated here, represents a fact, ascertained sufficiently strictly theoretically and checked up experimentally. However, there is no general proof of this statement which might embrace all the various rotor systems we have been speaking about. Therefore the indicated statement is referred by us to the category of principles.

Problems of the theory and application of auto-balancing devices aw well as of group foundations under unbalanced machines attract an unremitting interest of investigators [169, 178, 233, 128, 377, 378, 379, 489, 511, 376]. The dissertation [376] contains a review of these publications as well as a number of new interesting inventions.

8.2   Application to the Theory of Group Foundations for the Unbalanced Machines

In recent years the common (group) foundations under several unbalanced machines of the same type, rigidly bound with the foundation and actuated by engines of an asynchronous type, have been gaining more and more acceptance. The calculation of such foundations is based, as a rule, on the assumption that the phases of the unbalanced forces, developed by the machines, are of an accidental nature, and therefore the effects of unbalances of individual machines are to a certain extent mutually compensated [474]. Meanwhile cases are known when contrary to the expectations, the machines would work synchronously and synphaseously, which resulted in accidents. From the point of view of the above-stated regularities of self-synchronization of vibro-exciters such situations are quite understandable: under certain conditions the machines display a tendency to synchronous rotation, with not an accidental but quite a definite relationship of phases. As that takes place, in some cases the phases may prove to be close or coinciding (such a case must have taken place, causing the above-mentioned accidents), in other cases the phases are such that the unbalances are mutually compensated.

The above-mentioned results of the theory of self-synchronization make it possible to tackle the question about in which cases the vibrational coupling between the machines is weak, so that their self-synchronization is practically hardly probable at all, and therefore group foundations can be calculated according to the usual recommendations, and in which cases the vibrational coupling is, on the contrary, very strong and it is necessary to take into consideration the possibility of self-synchronization. In the latter case the foundation and the installation of the machine on it must be designed in such a way that there should be a mutual compensation, but not the amplification of the effect of unbalances [90, 97, 103, 104].

Tentative recommendations of that type, based on the principles, stated in 7.4 and 8.1, are given in table 8.1 which is a modification of table 7.1 (p. 196). This table is given in [97].

In this table k_ω is the coefficient introduced in 7.4, characterizing the relative strength of the vibrational constraint. As was marked, the case of a very weak vibrational coupling $k_\omega < 0.01$ is answered by less than 10% probability of self-synchronization, while the case of a very strong coupling $k_\omega > 0.2$ — by more than 90% (see table 7.1).

As before, in Table 8.1 ω is the synchronous angular velocity (frequency) which may be considered coinciding with the nominal velocity of rotation of the machine shafts; p_{min} and p_{max} are respectively the smallest and the largest frequencies of free oscillations of the foundation with the machines. The sign "−" shows that the calculation is recommended to be made without taking into account the vibrational coupling, the sign "+" shows that this coupling must be taken into account, and the signs "−+" show that the calculation may be made without taking into account the vibrational coupling, though it may give greatly overestimated values of oscillation amplitudes. The letters W and S in Table 8.1 mark the cases when at the occurrence of self-synchronization the effects of the action of individual unbalances are accordingly either weakened or strengthened. The strengthening or weakening is understood in the sense of the average for the period value of the kinetic energy of oscillations of the foundation.

Table 8.1

**Tentative recommendations on taking account of the phenomenon
of self-synchronization when designing group foundations for
unbalanced machines**

Relations between frequencies p_{min}, p_{max} and ω	Value regions of coefficient k_ω		
	$0 < k_\omega < 0.01$ (very weak vibrational coupling)	$0.01 < k_\omega < 0.2$ (either comparatively weak or comparatively strong vibrational coupling)	$k_\omega > 0.2$ (strong vibrational coupling)
$\omega > p_{max}$ (post-resonance region)	−	−+ W	+ W
$p_{min} < \omega < p_{max}$ (inter-resonance region)	−	+ W or S	+ W or S
$\omega < p_{min}$ (pre-resonance region)	−	+ S	+ S

When installing machines on a foundation so as to make a whole solid body, the weakening of oscillations in the sense it was indicated, will take place in case of a "soft" installation of the foundation on the base ($\omega > p_{max}$). Then in case it is possible to have such a phasing of the shafts in the synchronous motion that they will be perfectly balanced (the *compensating phasing*), the machines will display a tendency to perfect mutual balancing.

From what has been said it follows that when designing group foundations for the unbalanced machines, one should strive to provide such a situation at which the machines would be able, in principle, to get mutually balanced under the synchronous motion and that the mounting of the foundation onto the base would be soft.

Figures 8.2 shows certain schemes of mounting the unbalanced machines on a foundation as a whole solid body. In such cases perfect self-balancing

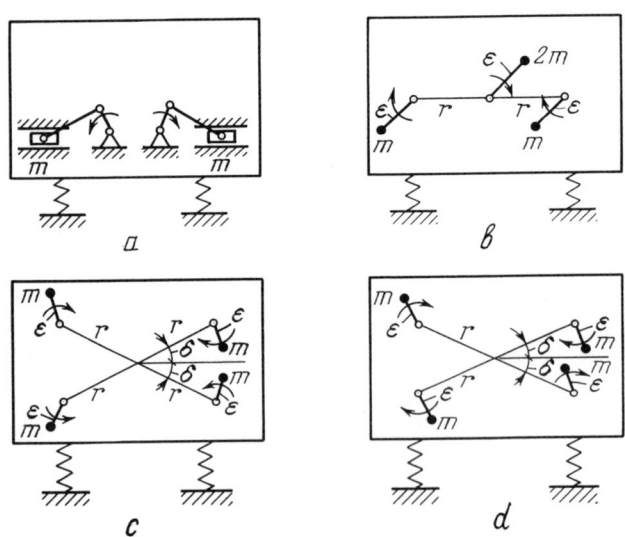

Figure 8.2. Examples of the schemes of installing several unbalanced machines on a single rigid foundation, at which a complete mutual balancing of machines is possible.

is possible, Fig. 8.2,*a* corresponds to the case when the unbalanced masses move rectilinearly, and Fig. 8.2,*b–d* — to the case when they rotate. All the pictures show the compensating phasings. According to what has been said, such phasings are certain to be stable when the mounting of the foundation on the base is "soft". It goes without saying that when the values of the static moments of the unbalanced masses $m_s \varepsilon_s$ are different, under the conditions of Fig. 8.2,a–d the compensation of unbalances will be, generally speaking, incomplete.

It should be noted that for the schemes with two machines the only stable synchronous regime in the post-critical region of frequencies is that with the compensating phasing. In the case of three or more machines, other, undesirable (that is " not compensating") stable synchronous motions may take place. Both the corresponding- and the compensating phasing will be answered by the minima of the function $< [T^{(I)}] >$. It should be also noted that there is a number of systems in which the compensating phasings answer the non-rough minima of the function $< [T^{(I)}] >$ or $< [L^{(I)}] >$. Both cases require a special consideration.

More detailed information about the calculation and design of the group foundations for the unbalanced machines in view of the phenomenon of self-synchronization can be found in [90, 103].

8.3   Supplement to the Theory of Auto-balance Devices

One of the possible constructive forms of the balancer for the automatic compensation of the unbalance of a rotating disk is presented schematically in Fig. 8.1,f. A disc is fixed on an elastic rotating shaft, the center of mass of the disc C not lying on the axis of the shaft AOB. The disc has a cylindrical or toroidal cavity, filled with oil. The axis of the cavity coincides with the tangent to the axis of the shaft at the point of fixing the disc O. Two small balls are placed into the cavity. Under certain conditions the balls take such positions in the rotating disc that they compensate the unbalance of the disc, the center of the cross-section of the shaft, in which the balancer is located, remaining on the axis of the bearings. As a result, the oscillations of the shaft and the dynamic loads on its supports are essentially reduced.

In other constructive versions of the auto-balancer, instead of the cavity in the disc and the balls, there are either rings, put on the shaft with a large gap, or levers, one of the ends of which is connected with a bush, freely revolving on the shaft, while the other end carries the unbalanced load. Those auto-balancers were invented by Thearle [537, 538] and they are employed in the machines of centrifuge types. They are especially effective in the cases when the unbalance of the rotor is liable to changes in the process of operation of the machine (also see [426, 428]).

The investigation of the dynamics of the auto-balancer, performed by different methods, leads to the conclusion which is in full agreement with the generalized principle of auto-balancing: the balls take in the disc a stable (compensating) position in the post-critical region of rotation frequencies of the shaft $\omega > p_{max}$. In case the shaft has distributed parameters, as was mentioned above, the frequency ranges where the unbalance is being compensated alternate with the ranges where oscillations are strengthened.

The dynamics of the auto-balancer was most fully investigated by Detinko [167]. Having studied by classic methods the case of a freely supported shaft with the distributed mass, a two-ball auto-balancer fixed in the middle of it, the author comes to the conclusion that the necessary condition of the normal work of an auto-balancer consists in the requirement that the angular velocity of rotation of the shaft ω should lie in the range between its odd critical velocities and the even critical velocities of the shaft with the intermediate support in the cross-section, supplied with the auto-balancer, that is it should satisfy the condition

$$2(k-1)^2 \pi^2 \sqrt{EI/Ml^4} < \omega < \Gamma_k^2 \sqrt{EI/M(\tfrac{1}{2}l)^4} \quad (k = 1, 2, \ldots) \qquad (8.3.1)$$

where EI is the bending stiffness of the shaft, l is its length, M is its linear mass, and the numbers Γ_k are the roots of the equation $\tanh \Gamma = \tan \Gamma$. The study of the case of the shaft with several auto-balancers leads to similar results. Detinko has also considered the shaft with the unbalanced mass distributed along its length.

The investigation of the work of an auto-balancer, based on the theory of self-synchronization of the rotating bodies, is given in [90, 103], its result coinciding with those of Detinko. In good agreement with them are also the results obtained by Sperling [507] who considered self-synchronization in the system, shown in Fig. 8.1,e, in the case of a beam with the distributed mass. And of course there is a good agreement between those results and the generalized principle of auto-balancing, formulated in 8.1.

The phenomenon of self-synchronization of rotors and the generalized principle of auto-balancing can be also used to solve the problem of compensating the given dynamic effect upon a solid body. A special case of that problem is shown in Fig. 8.1,$i2$: a softly vibro-isolated body is acted upon by a harmonic exciting force $F \sin \omega t$; it is necessary to provide the compensation of that force so that the body might remain practically immobile. As follows from the theory of synchronization of unbalanced vibro-exciters, the problem can be solved by means of placing on the body two similar symmetrically located unbalanced rotors, rotating in the opposite directions and developing the unbalanced forces $F/2$. According to the generalized principle of auto-balancing which refers to this system as well, the compensating phasing of the rotors, shown in Fig. 8.1,*i2*, is stable. The situation here is to a great extent similar to that taking place in the case of the auto-balancer.

Panovko drew the attention of the author to the fact that the described solution of the problem can substitute a much more intricate and expensive system of active vibro-isolation at which the compensating effect is achieved by means of special devices.

The investigation of the system, shown in Fig. 8.1,*i2* without using the results of the theory of self-synchronization, is made in [35].

Part III

Vibrational Mechanics of Processes (Vibrational Displacement and Shift)

Chapter 9

The Main Models and General Regularities of Processes of Vibrational Displacement from the Position of Vibrational Mechanics

9.1 On the Effect of Vibrational Displacement, its Theory and Applications

By the effect of *vibrational displacement* we mean the appearance of the "directed on the average" change (particularly - of motion) at the expense of the undirected on the average (oscillatory) effects. The following phenomena are based on this effect: vibrational transportation of single bodies and granular materials in the vibrating trays and vessels; the work of the devices called vibrational transformers of motion and vibro-engines; vibrational sinking of piles, sheet piles and shells; vibrational separation of particles of the granular material according to their density, size and some other parameters; the motion of vibrational coaches; the flight and swimming of living organisms. A harmful effect of vibrational displacement can be exemplified by the appearance under the action of vibration of the mobility of the nominally immobile parts of machines (particularly - the self-unscrewing of nuts)

The term "vibrational displacement" and its interpretation given above was suggested by Dzhanelidze and the author; this term was used as the title of our book [84], published in 1964, and gained acceptance in scientific and technical literature. Along with that, this term began to be used in vibro-measuring technology in quite another sense - in the group of terms "vibro-displacement", "vibro-velocity" and "vibro-acceleration" which refer to the kinematic characteristics of vibration. It is natural that below we will stick to the first interpretation.

Extensive literature is devoted to the theory of processes of vibrational displacement, especially to the vibrational transportation of single bodies and of granular materials. The review of the main results and the bibliographical data can be found in [84, 207, 315, 400, 443, 445, 513] and in the reference book [564, vol. 4]. Among the recent publications the following should be referred to [581, 177]; a number of references will be given below in the course of presenting the material.

The majority of problems on the theory of vibrational displacement are reduced to the investigation of the solutions of nonlinear differential equations with the periodic over the fast time $\tau = \omega t$ right sides, for which the velocities of the change of the generalized coordinates have the form

$$\dot{\mathbf{x}} = \dot{\mathbf{X}}(t) + \dot{\boldsymbol{\Psi}}(t, \omega t), \qquad (9.1.1)$$

where $\dot{\mathbf{X}}(t)$ is a slowly changing component, and $\dot{\boldsymbol{\Psi}}$ is a fast changing component, with

$$< \boldsymbol{\Psi}(t, \omega t) >= 0. \qquad (9.1.2)$$

The component $\dot{\mathbf{X}}(t)$ is called the *velocity of vibro-displacement*; in most cases its determination in the stable stationary motions (that is when $\dot{\mathbf{X}} = $ const) is of utmost interest for applications. In a wide class of systems an essential non-linearity of differential equations is caused by the presence in the system of dry friction forces and of unilateral constraints, though motions of such systems in certain regions of phase space can be linear as well.

The form of the solutions being sought (9.1.1) makes its quite natural and expedient to use concepts and methods of vibrational mechanics when solving problems of the theory of vibro-displacement. Meanwhile until the recent time this approach was practically not used at all; the only exception was the works by the author [95, 99, 114]. Therefore the task of this chapter is to illustrate the advantages of this universal approach to investigating the regularities of the processes of vibrational displacement and their numerous applications, mentioned at the beginning of this section.

A considerable place in the theory of vibrational displacement belongs to the problems on the motion of a material particle and of the simplest solids over a rough vibrating plane and to the problems on the motion of a body or of a particle under the action of vibration in the resistant medium. Being of a special interest for applications, they also play the role of the basic models for the problems on the theory of some technological processes, in particular, processes of vibro-sinking the piles and sheet piling, of vibrational separation of granular mixtures, and also of the motion of some vibrational coaches. Another group comprises problems on the processes of vibro-displacement in continuous and more complicated media — problems on slow flows, appearing in fluids, in gases and in granular materials under the action of vibration. In this chapter and in chapter 10 we will consider the models and applied problems of the first group; the models and problems of the second group will be assigned (to a certain extent arbitrarily) to part IV of this book, devoted to vibrorheology.

9.2   Simplest Model of the Process of Vibrational Displacement

9.2.1   *Motion of a Particle over a Rough Horizontal Plane under the Action of a Longitudinal Harmonic Force or the Longitudinal Vibration of the Plane*

Some important regularities of vibrational displacement can be made clear by solving a simplest problem — about the motion of a flat solid body (particle) whose mass is m over a rough horizontal plane under the action of a longitudinal harmonic exciting force $\Phi_0 \sin \omega t$ (Fig. 9.1,a) or a similar problem about the motion of a body over such plane, performing longitudinal harmonic oscillations. To be specific, we will first consider the first of these problems at a purely qualitative level,

It should be noted first of all that if the dry friction coefficients f_+ and f_-, while the body is sliding forward and backward along the plane, are identical and equal to f, then it is clear that at $\Phi_0 < m g f$ the body remains motionless, and at $\Phi_0 > m g f$ it will perform symmetrical oscillations with respect to a certain central position (g being the free fall acceleration). Though if we assume that $f_- > f_+$, then the symmetry will be broken. And as can be seen from the characteristic of dry friction force F, shown in Fig. 9.1,a, at $\Phi_0 > m g f_+$ the body will move in the positive direction, (in case $f_- < f_+$ at $\Phi_0 > m g f_-$ the body will move in the negative direction, that is just like it was before, in the direction where the resistant force is weaker.) [a]

Indeed, during a certain period of time of the first half-period $0 < \omega t < \pi$ when $\Phi_0 \sin \omega t > m g f_+$, the body will move to the right, and during the second half-period $\pi < \omega t < 2\pi$ when $\Phi_0 \sin \omega t < 0$ the body will stay where it is (if $\Phi_0 < m g f_-$), or it will move to the left to a distance shorter than that to the right, since the resistance to the motion to the left is stronger than when the body moves to the right. Thus during every period of the change of force,

---

[a]The case $f_+ \neq f_-$ may seem artificial, however, this is not so: it corresponds e.g. to the vibro-transportation of fish for which the friction coefficients are quite different when the sliding takes place in the direction of the scales and when it is opposite to them. The same refers to the sliding of a body over a fleecy surface or to the simplest model of the resistance of the ground in the problem of the vibrational sinking of piles and sheet pile (see 10.3).

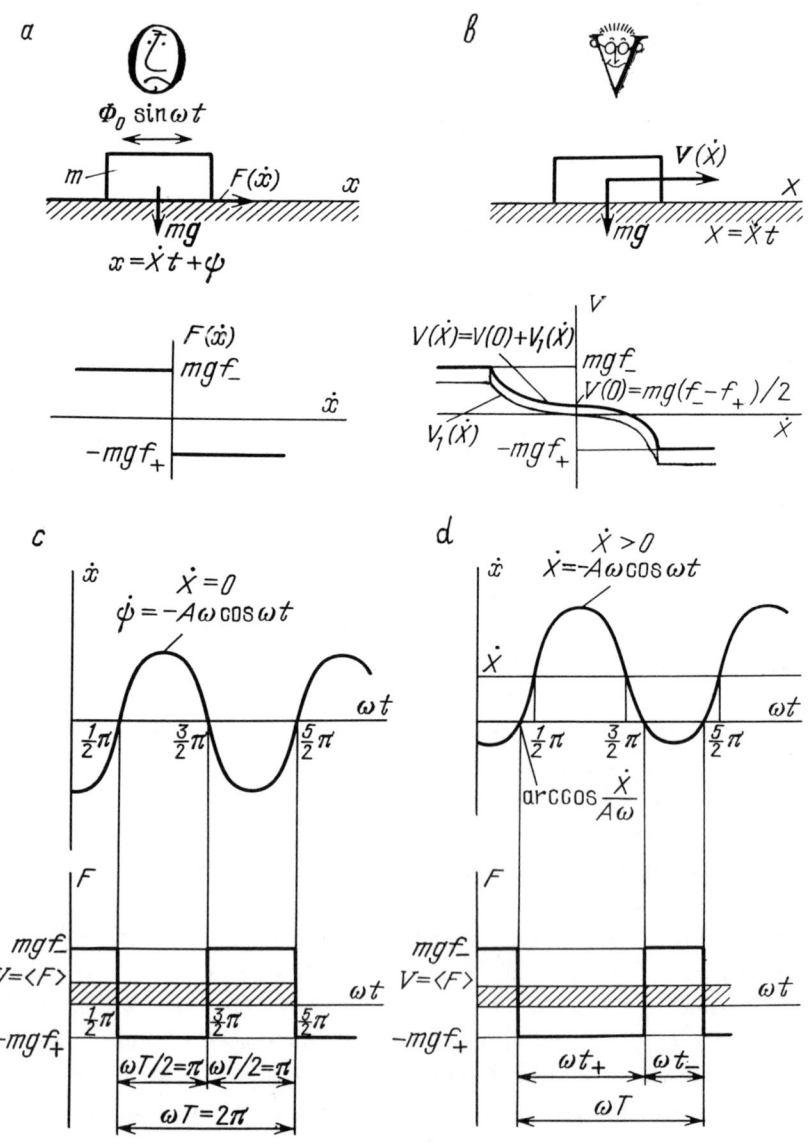

Figure 9.1. A flat body (a particle) on a rough horizontal plane at the asymmetrical law of dry friction under the action of a symmetrical (harmonic) vibration is displaced towards the lowest resistance.

a certain shift of the body to the right will take place. The observer **V**, who does not notice the fast force $\Phi_0 \sin \omega t$ or the force of dry friction which also changes " fast", will ascribe the motion of the body forward to the appearance of a certain force — the vibrational force **V**, acting in the positive direction (Fig. 9.1,*b*). This force, in particular, can make the body move even upwards along the plane, that is against the gravity force if the plane is tilted at a certain, not very large, angle.

Let us now consider the problem on the basis of the equation of motion of the body:

$$m\ddot{x} = F(\dot{x}) + \Phi_0 \sin \omega t, \qquad (9.2.1)$$

where m is the mass of the body and

$$F(\dot{x}) = \begin{cases} -m\,g\,f_+ & \text{at} \quad \dot{x} > 0 \\ m\,g\,f_- & \text{at} \quad \dot{x} < 0 \end{cases}$$

$$-m\,g\,f_+ < F(\dot{x}) < m\,g\,f_- \quad \text{at} \quad \dot{x} = 0 \qquad (9.2.2)$$

is the dry friction force, corresponding to the characteristic shown in Fig. 9.1,*a*. Equation (9.2.1) can be solved exactly by using either the method of fitting or the method of point mapping (see, say, [564, vol. 2 and 4]). However, a much more simple and clear approximate solution can be obtained by the method of direct separation of motions. The dry friction force $F(\dot{x})$ which can essentially change during the period, as well as the exciting force $\Phi_0 \sin \omega t$, will be referred to fast forces. Searching the solution of equation (9.2.1) of type (9.1.1), we will write down equations (3.2.55) and (3.2.56) of chapter 3 as

$$m\ddot{X} = V \qquad (9.2.3)$$

$$m\ddot{\psi} = F(\dot{X} + \dot{\psi}) - < F(\dot{X} + \dot{\psi}) > + \Phi_0 \sin \omega t \qquad (9.2.4)$$

where

$$V = < F(\dot{X} + \dot{\psi}) > \qquad (9.2.5)$$

is the vibrational force which, as can be seen, presents in this case the average for the period dry friction force acting upon the body.

We will solve this problem in a crude way, considering that the amplitude of the exciting force Φ_0 is much greater than the limiting values of the dry friction forces $F_+ = m\,g\,f_+$ and $F_- = m\,g\,f_-$, so that the latter can be neglected when solving the equations of fast motions (9.2.4). Then the periodic solution of this equation, satisfying condition (9.1.2), will be

$$\psi = -A \sin \omega t \qquad (9.2.6)$$

where

$$A = \Phi_0/m\omega^2. \qquad (9.2.7)$$

Now let us perform the averaging according to formula (9.2.5) in view of expressions (9.2.2) and (9.2.6). The process of this averaging is presented graphically in Fig. 9.1,c at $\dot{X} = 0$, and in Fig. 8.1,d — in the general case $\dot{X} \neq 0$. These figures illustrate very clearly the mechanism of appearance of the vibrational force. As a result we obtain

$$V = V(\dot{X}) = \begin{cases} -m\,g\,f_+ & \text{at} \quad \dot{X} \quad \geq \quad A\omega, \\ \dfrac{\omega}{2\pi} m\,g\,(f_-t_- - f_+t_+) & \text{at} \quad |\dot{X}| \quad \leq \quad A\omega, \\ m\,g\,f_- & \text{at} \quad \dot{X} \quad \leq \quad -A\omega \end{cases} \qquad (9.2.8)$$

where t_+ and t_- denote the intervals of time during which the body moves over the plane to the right ($\dot{x} = \dot{X} + \dot{\psi} > 0$) and to the left ($\dot{x} = \dot{X} + \dot{\psi} < 0$) respectively, with

$$\omega t_+ = 2\left(\pi - \arccos \frac{\dot{X}}{A\omega}\right), \qquad \omega t_- = 2\arccos \frac{\dot{X}}{A\omega}. \qquad (9.2.9)$$

In view of those expressions formula (9.2.8) for the vibrational force acquires the following form:

$$V(X) = \begin{cases} -m\,g\,f_+ & \text{at} \quad \dot{X} \quad \geq \quad A\omega, \\ \dfrac{m\,g}{\pi}[(f_+ + f_-)\arccos \dfrac{\dot{X}}{A\omega} - f_+\pi] & \text{at} \quad |\dot{X}| \quad \leq \quad A\omega, \\ m\,g\,f_- & \text{at} \quad \dot{X} \quad \leq \quad -A\omega, \end{cases} \qquad (9.2.10)$$

and the main equation of vibrational mechanics will be written as

$$m\ddot{X} = V(\dot{X}). \qquad (9.2.11)$$

The velocity of the stationary motion of the particle over the plane $\dot{X} = \dot{X}_*$ is the *velocity of vibro-displacement* which will be found from the equation $V(\dot{X}_*) = 0$, from which we obtain

$$\dot{X}_* = A\omega \cos \frac{\pi f_+}{f_+ + f_-} = \frac{\Phi_0}{m\omega} \cos \frac{\pi f_+}{f_+ + f_-}, \qquad (9.2.12)$$

and since $V'(\dot{X}_*) < 0$, the motion with this velocity is stable. At $f_+ = f_-$ according to formula (9.2.12) we obtain $\dot{X}_* = 0$ that is, as it must, the body "on the average" remains motionless (is in the state of quasi-equilibrium). It should be also noted that the value of the velocity of vibro-displacement $\dot{X} = 0$ is answered by the vibrational force

$$V(0) = \frac{1}{2}m\,g(f_- - f_+). \qquad (9.2.13)$$

All the obtained formulas remain also valid in the case when the body is acted upon by the exciting force $\Phi_0 \sin \omega t$ and the plane performs the preassigned harmonic oscillations with the amplitude A and the frequency ω.

It is not difficult to check a posteriori the applicability of the method of direct separation of motions; in this case, however, there is no necessity in it: the comparison of the expressions obtained here with those exact, on which we do not dwell here, displays the agreement of the results, provided the relations $\Phi_0/m\,g\,f_\pm$ are small, which had been assumed above.

9.2.2   *Analysis of the Solutions.   Effect of the Seeming Vibrational Transformation of Dry Friction into Viscous.   The Driving and the Vibro-transformed Vibrational Forces*

Let us turn to the analysis of the dependences we have obtained. First of all it should be noted that according to (9.2.8) the dependence $V(\dot{X})$ whose graph is presented in Fig. 9.1,*b*, is continuous — there has been a *vibrational smoothening of the discontinuous characteristic* of dry friction (Fig. 9.1,*a*) — the effect well known in the theory of automatic control (see, say, [279, 295, 442]). It should be noted further that the vibrational force $V(\dot{X})$ can be presented as a sum of two terms

$$V(\dot{X}) = V(0) + V_1(\dot{X}), \qquad (9.2.14)$$

where $V(0)$ is defined by equation (9.2.13). The force $V(0)$ can be called the *driving vibrational force* and the force $V_1(\dot{X})$ — the *vibro-transformed resistance force* [b], and since according to (9.2.14), $V_1(0) = 0$, the latter displays the nature of viscous friction [c] if in order to make a body move it is necessary to have a certain finite force. In case the motion can be caused by a force as small as small can be, we will speak about a *resistance force of the type of viscous friction*. Thus as a result of the action of vibration on the system that is being considered, there is not only a seeming (that is seen only by the observer **V**) transformation of dry friction into viscous, but also the generation of the driving

---

[b]We have used here the term, introduced by V. Andronov who thoroughly investigated a number of interesting mechanical systems with dry friction both in the absence and in the presence of vibration [22].

[c]Here and further on we say that *the force of resistance to the motion of the body is of the character of dry friction*

vibrational force. The latter circumstance is often forgotten, and only the effect of fluidization of the system with dry friction under vibration is spoken about. We will again emphasize that this fluidization occurs only with regard to slow motions; "in reality" (for the observer **O**), i.e., in fact the system remains that with dry friction (Fig. 9.1,*a,b*).

Thus, the consideration of this simplest system leads to the explanation and description not only of the effect of vibrational displacement, but also of other regularities which may be referred to vibrorheological effects. Chapter 10 is devoted to a more detailed consideration of such effects.

In conclusion we will mark that in case $\dot{X}/A\omega \ll 1$ when in expression (9.2.10) we can restrict ourselves to the first two members of the expansion of $\arccos \dot{X}/A\omega$ into a power series by $\dot{X}/A\omega$, the expression for the vibrotransformed vibrational force takes the form

$$V_1(\dot{X}) = -\frac{m\,g}{\pi}(f_+ + f_-)\frac{\dot{X}}{A\omega}, \qquad (9.2.15)$$

while the expression (9.2.13) for the driving vibrational force $V(0)$ remains the same.

9.3   Physical Mechanisms and Main Types of Asymmetry of the System, Causing Vibrational Displacement

The simplest model, considered in 9.2 shows the main condition of appearance of the effect of vibrational displacement — the asymmetry of the system: when $f_+ = f_-$, this effect is absent. However, the case considered here presents only one possible type of asymmetry, resulting in vibrational displacement. Other types of asymmetry and the physical mechanisms corresponding to them can be also characterized by the model of a flat body (material particle), moving relative to the vibrating rough plane; as was already marked, such a model is of basic importance for investigating the processes of vibrational displacement.

Fig. 9.2,*a* shows schematically six types of asymmetry and accordingly six ways of realizing the process of vibrational displacement (in this case — of vibrational transportation).

Asymmetry of type I will be called *force-asymmetry*. It can be caused either by the action of a constant force T (version 1), or by the slope of the plane relative to the horizon (version 2), which is in fact not different from version 1, or else by the fact that the absolute values of forces of resistance to the motion to the right F_+ and to the left F_- are different (version 3, a specific case of which was considered in 9.2.3 — see Fig. 9.1). The arrows on the axis x which

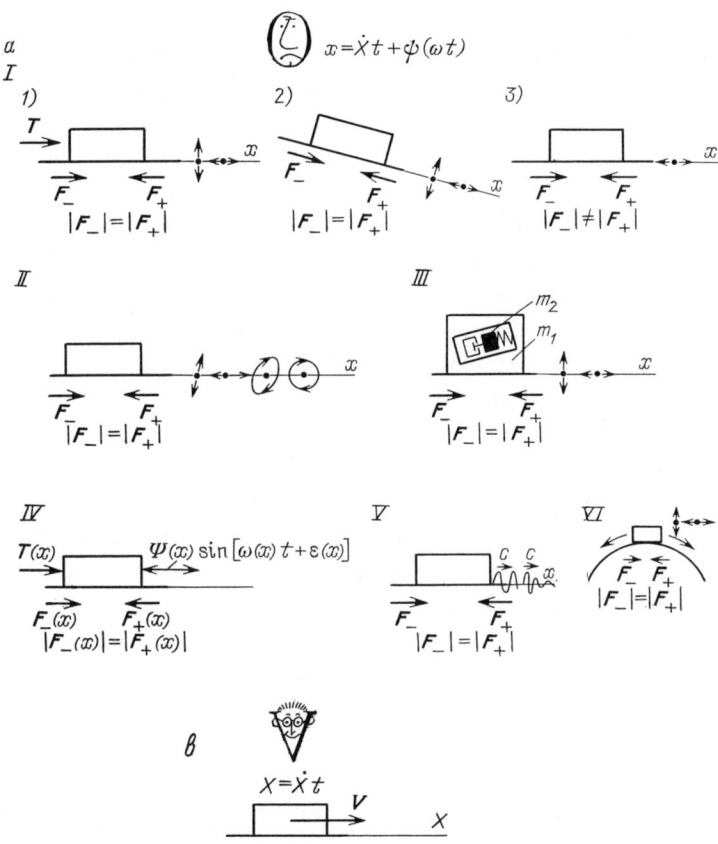

Figure 9.2. For the appearance of the effect of vibrational displacement it is necessary that there should be some or other asymmetry of the system.

have the same length in both directions show arbitrarily that the law of the vibration of the points of the surface is absolutely symmetrical, i.e. it presents e.g. purely harmonic oscillations [d]. Besides, in case of the asymmetry of type I, the trajectories of oscillations are symmetrical relative to the vibrating plane: they present segments of lines either perpendicular or parallel to the plane.

Asymmetry of type II will be called *kinematic asymmetry*. It is caused by

---

[d]The notions of symmetry and asymmetry of the law of oscillations in case of one-component vibration are specified in 9.4.1.

the asymmetry of one or another type of the trajectory of vibration of the points of the plane or of the law of motion along that trajectory. Fig. 9.2 shows the following examples: harmonic oscillations in the rectilinear trajectory, inclined to the plane at a certain angle, different from zero or from 90 degrees; longitudinal non-harmonic asymmetric oscillations — they are arbitrarily shown by arrows of different length in different directions; elliptical oscillations with the axes located asymmetrically relative to the vibrating plane; circular oscillations (in which case the asymmetry is formed by a definite direction of the run of the trajectory — clockwise or counterclockwise).

Asymmetry of type III can be called *structural or constructive*. The figure shows the case when this asymmetry appears due to the fact that the body has "the internal degree of freedom": a body of the mass m_2, connected with the body of the mass m_1 by elastic and damping elements, can move relative to the main body of mass m_1. The angle between the direction of a possible motion of the body and the plane is assumed to be different from zero and from 90° (that is what creates asymmetry). Systems similar to that presented in Fig. 9.2,*aIII* were briefly considered in 2.2.2 (Fig. 2.2,*c*).

Asymmetry of type IV will be called *gradient asymmetry*. In this case the vibrational displacement is caused by the essential dependence on the coordinate x of the parameters determining the motion of bodies, for instance on the amplitude Ψ, the frequency ω and on the phase ε of the vibrational effect of force, on the force of dry friction F, on the trajectory and the law of vibration of the points of the surface, on the slope of the surface towards horizon, on the force T, etc.

Asymmetry of type V can be called *wave asymmetry* — here the body moves in the direction of the propagation of either the longitudinal or transverse running regular wave or of single waves of an impulsive nature.

Finally, asymmetry of type VI can be called *initial asymmetry*: in this case the system as such is absolutely symmetrical, but the body can "prefer" the motion to the right or to the left under the action of the smallest initial shift or push. We will see that such, on the face of it highly improbable, type of asymmetry is met in the case of vibration of identical communicating vessels with the granular material (see 15.3).

It goes without saying that a combination of two or several types of asymmetry is possible.

Vibrational displacement (transportation) in all the enumerated cases can be interpreted in terms of vibrational mechanics: the observer **V** thinks that it takes place under the action of vibrational force **V** (Fig. 9.2,*b*).

We will explain in brief the appearance of the effect of vibrational dis-

placement of the body in different cases, presented in Fig. 9.2. For scheme
3 which refers to asymmetry of type I the mechanism of formation of the vibrational force is considered in detail in 9.2. As for scheme *1*, in the case
of the longitudinal oscillations it is reduced to scheme *3* if we assume that
$F_- = F + T, F_+ = F - T$ and scheme *2* presents a version of scheme *1* for which
$T = m g \sin \alpha$ where $m g$ is the weight of the body and α is the slope of the
plane towards the horizon. When the vibration of the plane is transverse in the
case of schemes *1* and *2*, the body during one half-period of oscillations either
breaks away from the plane, or the force pressing it to the plane decreases. In
both cases the force T can move the body along the axis x in spite of the fact
that in the absence of vibration the body remains motionless.

Asymmetry of type II is that which is most widely used in modern vibro-
technology. Here in case of the rectilinear harmonic oscillations at a certain
angle β to the plane, different from zero and from 90°, the mechanism of the
appearance of vibrational displacement consists in the following. During that
half-period when the force of inertia in the relative motion is directed to the
right-upwards, it either breaks the body away from the plane, or weakens the
pressure exerted on it, thus decreasing the normal reaction, and consequently
the force of friction. During the other half-period when the force of inertia is
directed to the left-downwards, this force, on the contrary, exerts an additional
pressure on the body, pressing it to the plane, increasing in that way the force
of dry friction. As a result, the body moves mainly to the right. Similarly, there
is a greater possibility for a body to move to the right in a certain half-period
of oscillations in the other three cases. Systems of this type of asymmetry are
considered in 9.4.

It is more difficult to understand the mechanism of the appearance of the
directed motion of the body m_1 at the asymmetry of type III. In this case the
body m_2 under the effect of vibration of the plane and of the motion of the
body m_1 performs oscillations with respect to the body m_1 and acts on it by
dint of the elastic and damping elements and also by dint of the guides. This
effect together with the immediate effect of the vibration of the plane upon the
body m_1 can make the body m_1 move either in the positive- or in the negative
direction, depending on the relation between the parameters, This effect will
be discussed in detail in chapter 13.

The mechanism of the appearance of the vibrational force at the asymmetry
of type IV in a special case $\omega = $ const, $\varepsilon = 0$ is discussed in detail in chapter
19. Asymmetry of type IV does not require any special explanations either.
As for the asymmetry of type V, in this case the running wave seems to be
pushing the body, or it "carrying" it in the direction of its propagation. In the

first case we may speak about the *asynchronous displacement of the body by a running wave*, in the second case — about its *synchronous displacement*.

Finally, at the asymmetry of type VI the role of vibration is reduced mainly to the liquidation or decrease of the dead zone, caused by dry friction(see 5.3, 9.4.1 and chapter 10).

9.4   More Complicated Main Models and Problems of the Theory of Vibrational Displacement

9.4.1   One-dimensional Motion of the Particle in the Resistant Medium under the Action of Vibration [95, 99, 198, 564, vol. 2 and 4]

As a model of the process of vibro-displacement, which is the generalization of the simplest model, studied in 9.2, we will consider the one-dimensional motion of a body (particle) of mass m_1 in the medium with the resistance, characterized by the force $F(\dot{x})$ depending on the relative velocity of the particle in the medium and which is not necessarily a force of dry friction. We will also assume that the particle is acted upon by a certain constant force T and by a fast changing force Φ which will be given as $\Phi = -m_2 \ddot{\xi}(\omega t)$ where $\xi(\omega t)$ is a certain 2π-periodic function of the fast time $\tau = \omega t$ with

$$< \dot{\xi} >= 0, \qquad (9.4.1)$$

and m_2 has a dimension of mass and in the general case $m_1 \neq m_2$; moreover, sometimes it will be more convenient to assume that m_2 can have negative values. However, in the case when, say, the motion of the particle over the plane is being considered (Fig. 9.3), $m_1 = m_2$.

The equation of the motion of the particle under the indicated assumptions has the form

$$m_1 \ddot{x} = T - m_2 \xi(\omega t) + F(\dot{x}). \qquad (9.4.2)$$

Many "one-dimensional" problems of the theory of vibrational displacement which are of practical importance are reduced to that equation. Among them are such as problems on vibrational separation of granular mixtures (10.2.5), on the movement of vibrational coaches (10.5), on vibrational driving in the piles (10.3), on vibrational pumps (10.6), on the motion of bodies in the oscillating fluid (17.2).

Searching solutions of equation (9.4.2) of type (9.1.1), we will pass from this equation to the corresponding equation of vibrational mechanics using the method of direct separation of motions. Assuming that the force T is slow,

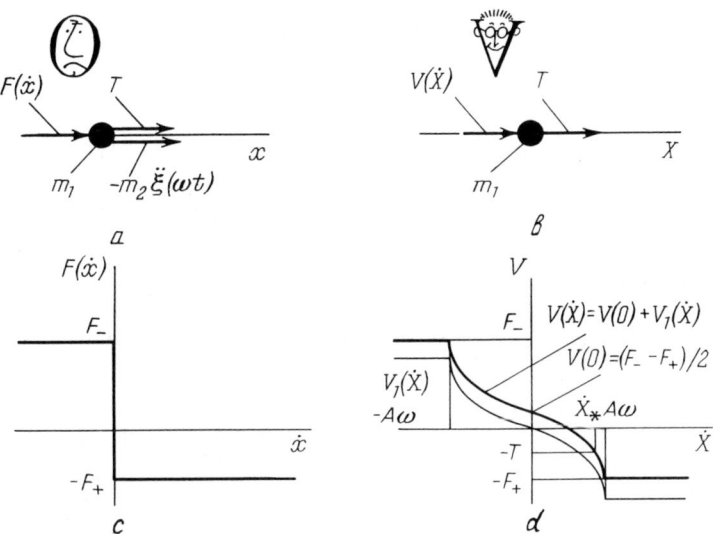

Figure 9.3. The motion of the body (particle) in the resistant medium under vibration; a) the scheme of the system; b) the picture as it is seen by the observer **V**; c) the characteristic of the resistant force of type of dry friction; d) the corresponding vibrational force in case of harmonic vibration.

and the forces $m_2\ddot{\xi}$ and F are fast, we will write down equations (3.2.55) and (3.2.56) of chapter 3 as

$$m_1\ddot{X} = T + V, \tag{9.4.3}$$

$$m_1\ddot{\psi} = -m_2\ddot{\xi} + F(\dot{X} + \dot{\psi}) - <F(\dot{X} + \dot{\psi})>, \tag{9.4.4}$$

where

$$V =< F(\dot{X} + \dot{\psi}) > . \tag{9.4.5}$$

Solving the equation of fast motions (9.4.4), we will assume that, like it is in 9.2, the force of resistance F is small as compared to the force $m_2\ddot{\xi}$ and it can be neglected in the first approximation. Then the periodic solution of equation (9.4.4), satisfying condition (9.1.2), will be

$$\dot{\psi} = -q\dot{\xi} \qquad (q = m_2/m_1). \tag{9.4.6}$$

In case of necessity (which, as a rule, never happens) this solution can be specified by means of the method of small parameter.

In view of (9.4.6) the expression (9.4.5) for the vibrational force can be presented in the form

$$V(\dot{X}) = < F(\dot{X} - q\dot{\xi}) >= V(0) + V_1(\dot{X}), \qquad (9.4.7)$$

where

$$V(0) = < F(-q\dot{\xi}) >, \quad V_1(\dot{X}) = < F(\dot{X} - q\dot{\xi}) - F(-q\dot{\xi}) >, \quad V_1(0) = 0, \quad (9.4.8)$$

are the driving- and the vibro-transformed forces respectively (see 9.2). As a result, the main equation of vibrational mechanics, obtained from the equation of slow motion (9.4.3), will acquire the form

$$m_1 \ddot{X} = T + V(\dot{X}), \qquad V(\dot{X}) = V(0) + V_1(\dot{X}). \qquad (9.4.9)$$

When studying all the applications, considered below, it is sufficient to assume that there is the following resistant force — velocity dependence:

$$F(\dot{x}) = -k_\pm |\dot{x}|^{n-1}\dot{x}, \qquad (9.4.10)$$

where the value of the coefficient $k_+ > 0$ corresponds to $\dot{x} > 0$, and the value $k_- > 0$ corresponds to $\dot{x} < 0$, $n \geq 0$, while the value $n \geq 1$ corresponds to the viscous friction and $n = 0$ — to the dry friction; n can be believed to have integer values.

Let us first consider a case of viscous friction ($n \geq 1$). Then according to (9.4.7) and (9.4.9)

$$V(\dot{X}) = - < k_\pm |\dot{X} - q\dot{\xi}|^{n-1}(\dot{X} - q\dot{\xi}) >,$$
$$V(0) = -q^n < k_\pm |\dot{\xi}|^{n-1}\dot{\xi} >, \quad V_1(\dot{X}) = V(\dot{X}) - V(0), \quad V_1(0) = 0. (9.4.11)$$

When the resistance is linear, when $n = 1$ and $k_+ = k_- = k$ so that $F(\dot{x}) = -k\dot{x}$, we have $V(0) = 0$ and $V_1(\dot{X}) = -k\dot{X}$ that is, as might be expected, the driving vibrational force does not appear and the slow motion does not depend at all on the character of vibration.

At $n = 3$ and $k_+ = k_- = k$ (the cubic resistance)

$$V(0) = -kq^3 < \dot{\xi}^3 >, \quad V_1(\dot{X}) = -k(X^3 + 3q^2\dot{X} < \dot{\xi}^2 >), \qquad (9.4.12)$$

that is the vibration results in the appearance of a driving vibrational force and of the additional linear resistance $3q^3k < \xi^2 >$. For instance at the biharmonic vibration

$$\xi = A\sin\omega t + B\sin(2\omega t + \delta) \qquad (9.4.13)$$

we have

$$< \dot\xi^2 > = \frac{1}{2}\omega^2(A^2 + 4B^2), \quad < \dot\xi^3 > = \frac{3}{2}A^2 B\omega^3 \cos \delta$$

and therefore

$$V(0) = -\frac{3}{2}kq^3 A^2 B \cos \delta,$$

$$V_1(\dot X) = -k[\dot X^3 + \frac{3}{2}q^2(A^2 + 4B^2)\omega^2 \dot X]. \tag{9.4.14}$$

Let us consider now the case of dry friction $n = 0$ when relation (9.4.10) can be presented as

$$F(\dot x) = \begin{cases} -F_+ & \text{at} \quad \dot x > 0 \\ F_- & \text{at} \quad \dot x < 0 \end{cases}$$

$$-F_+ < F(\dot x) < F_+ \quad \text{at} \quad \dot x = 0 \quad (F_\pm = k_\pm). \tag{9.4.15}$$

(In this case when the coefficients k_+ and k_- have the dimension and the meaning of forces, instead of $k_\pm$ we will use the symbol $F_\pm$). Suppose that in the absence of the vibrational effect ($\ddot\xi = 0$) the following condition is satisfied

$$- F_- < T < F_+, \tag{9.4.16}$$

as a result of which the particle remains fixed with respect to the medium. The value range of the force T, corresponding to condition (9.4.16) is sometimes called the *dead zone*.

Simple reasoning, similar to that given in 9.2, makes it possible to come to the conclusion that the necessary and sufficient condition of the appearance of vibrational displacement in the case of the resistance of dry friction which is being considered is the realization of one of the two inequalities:

$$\Phi_+ + T > F_+, \quad \Phi_- - T > F_-, \tag{9.4.17}$$

where

$$\Phi_+ = \quad \sup \Phi \quad = \sup(-m_2\ddot\xi) > 0,$$

$$\Phi_- = \quad -\inf \Phi \quad = \sup(m_2\ddot\xi) > 0 \tag{9.4.18}$$

are respectively the largest- and the smallest value of the function $\Phi = -m_2\ddot\xi$, taken with the opposite sign, characterizing the vibrational effect on the particle. When the first of the inequalities is satisfied, the particle will move forward ($\dot x > 0$), and when the second inequality is satisfied, the particle will move backward ($\dot{x} < 0$). In case both inequalities are satisfied simultaneously, the vibrational displacement of the particle will also take place, except a special case when the displacements of the particle for the period of the stationary motion will prove to be the same in both directions, i.e. $\dot{X}$ will be equal to zero.

Later on when either one or both conditions (9.4.17) are satisfied, we will say for the sake of brevity that there is a *sufficiently intensive* (for the appearance of vibrational displacement) *vibrational effect*.

In the process of the approximate solution of the equations of fast motions, we will generally assume (as we already assumed before) that the vibrational effect exceeds greatly the forces of resistance, that is we can restrict ourselves to the so called purely inertial approximation (see 3.2.8). As applied to the case of dry friction, that means that the inequality

$$\inf(\Phi_+, \ \Phi_-) \gg \sup(F_+, \ F_-). \tag{9.4.19}$$

is presumed to be fulfilled.

In this case we will speak briefly about a *highly- intensive vibrational effect* . It is easy to see that the fulfilment of condition (9.4.19) guarantees the validity of inequalities (9.4.17) provided the force T satisfies condition (9.4.16). In other words, the highly-intensive vibration is in this case sure to be sufficiently intensive for the appearance of vibrational displacement.

About the effect of low-intensive vibration on systems with dry friction see 13.1.

Let us now turn to finding the expression for the vibrational force in the case of resistance of the type of dry friction. The calculation by formula (9.4.7), similar to that made in 9.2 (see equation (9.2.8) and Fig. 9.1,c), leads to the following result:

$$V(\dot{X}) = V(0) + V_1(\dot{X})$$

$$= \begin{cases} -F_+ & \text{at} \quad \dot{X} > q \sup \dot{\xi}(t) > 0, \\ \dfrac{\omega}{2\pi}(F_- t_- - F_+ t_+) & \text{at} \quad q \inf \dot{\xi}(t) < \dot{X} < q \sup \dot{\xi}(t), \\ F_- & \text{at} \quad \dot{X} < q \inf \dot{\xi}(t) < 0, \end{cases} \tag{9.4.20}$$

where t_+ and t_- are the total duration of the intervals of time in every period during which the particle moves forward ($\dot{x} = \dot{X} - q\dot{\xi} > 0$) and backward ($\dot{x} = \dot{X} - q\dot{\xi} < 0$) respectively. Since t_+ does not decrease and t_- does not increase when $\dot{X}$ increases, $V(\dot{X})$ is a non-increasing function of $\dot{X}$, with the driving vibrational force $V(0)$ being, generally speaking, different from zero. The graph of the function $V(\dot{X})$, shown in Fig. 9.1,b for a specific case

$\xi = A \sin \omega t$, $q = 1$ and $F_\pm = m\,g\,f_\pm$ is also typical for the general case, considered here. Equation (9.4.20) in the indicated specific case becomes, as it must, (9.2.8).

Now let us turn to the analysis of the relations obtained. The velocity of the vibrational displacement in the stationary regimes is determined from the equation

$$T + V(0) + V_1(\dot{X}) = 0, \tag{9.4.21}$$

which is obtained from the main equation (9.4.9) at $\dot{X} = \text{const}$. The definite solution of $\dot{X} = \dot{X}_*$ of equation (9.4.21) is answered by a stable stationary motion if

$$V'(\dot{X}_*) < 0 \tag{9.4.22}$$

It is not difficult to make sure that this inequality is always fulfilled, since, as was assumed, $k_+ > 0$ and $k_- > 0$. Thus the appearance of the stationary regime of vibrational displacement is conditioned by the fact that equation (9.4.21) has a non-zero root $\dot{X}_*$.

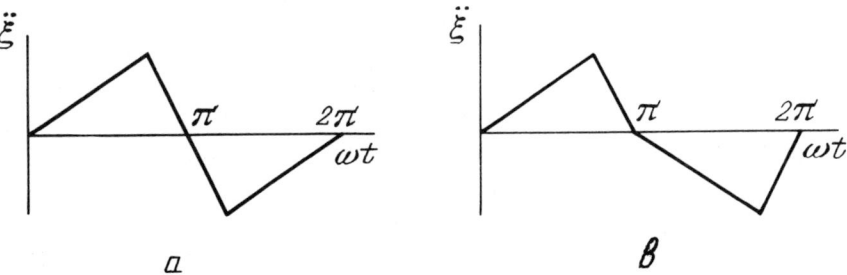

Figure 9.4. Examples of asymmetrical (a) and symmetrical (b) vibrational action.

If the friction is viscous, we are interested in the case when the constant force is absent ($T = 0$). Then due to the equality $V_1(0) = 0$ (see relation (9.4.8)) it turns out that for the possibility of vibrational displacement it is necessary that the driving vibrational force $V(0)$ should be different from zero. For that to take place, it is necessary, according to (9.4.11), that at $k_+ = k_-$ the law of the change of the vibrational effect $m_2\ddot{\xi}(\omega t)$ should be an "*asymmetrical-vibrational effect*" in the sense that the relation

$$\ddot{\xi}(\omega t + \pi) = -\ddot{\xi}(\omega t), \tag{9.4.23}$$

is not fulfilled, and consequently, the equality

$$\dot{\xi}(\omega t + \pi) = -\dot{\xi}(\omega t). \tag{9.4.24}$$

is not fulfilled either due to (9.4.1).

In other words, it is necessary that the curves $\ddot{\xi}(\omega t)$ and $\dot{\xi}(\omega t)$ during a certain half-period should not repeat (with the opposite sign) the run of these curves during the next half-period (Fig. 9.4). In particular, a simple harmonic effect is symmetrical. At $k_+ = k_-$, that is when the forces resistant to the motion of the particle in the positive and negative directions are not equal, vibrational displacement can also take place when the vibrational effect is symmetrical.

In case of dry friction, the appearance of vibrational displacement can be conditioned by three factors:

1. **By joint action of the constant force T and vibration.** This is the case when on the one hand the force T is comparatively not very large and in the absence of vibration does not lead to the appearance of vibrational displacement because the conditions $-F_- < T < F_+$ are satisfied, and on the other hand the presence of vibration at $T = 0$ does not lead to any vibrational displacement either, for instance due to the fulfillment of condition (9.4.23). It is possible to say about this case that the vibrational displacement is conditioned by vibrational smoothening of the characteristics of dry friction or by a seeming transformation of dry friction into viscous. In this case "the *dead zone*" seems to disappear and the force T together with $V(0)$ provides the course of the process. This case is of special importance for instance in enriching the mineral resources where vibration ,"diluting" the granular material or the structurized suspension, offers possibilities to manifest weak separating factors (see 10.2). A similar situation is characteristic of the process of vibro-sinking the piles or sheet piling(10.3) and also for a number of processes of vibrational transportation (10.1).

The described regularities were already paid attention to in 9.2. (see also Fig. 9.1); they can be referred to *vibro-rheological effects*. Part IV of this book is devoted to a more detailed consideration of those effects. These regularities are discussed there from another point of view - as a result of the seeming decrease of the dry friction coefficients under vibration (13.1).

2. **By the inequality of the resistant forces when the body moves in the positive and in the negative directions.** In this case even when the law of oscillations is symmetric and the force T is absent, that is when $t_+ = t_- = \pi/\omega$, according to equation (4.20) we obtain $V(0) = \frac{1}{2}(F_- - F_+) \neq 0$ which leads to the

vibrational displacement.

3. By the asymmetry of the law of vibrational effect. At $T = 0$, $F_+ = F_- = F$ and $\dot{X} = 0$ the intervals of time in this case are, generally speaking, unequal, and then according to (4.20), $V(0) = F(t_- - t_+) \neq 0$ that is the condition of the appearance of vibrational displacement is also satisfied.

The analysis which is made fully agrees with what was said in 9.2 regarding those types of asymmetry of the system which can condition the vibrational displacement. It is natural that vibrational displacement may also appear when there are several types of asymmetry at a time.

9.4.2   *The Problem in 9.4.1 in the Case of Harmonic Vibration*

A special case, corresponding to the vibrational effect of type $\xi = A \sin \omega t$ where $A > 0$ and $\omega > 0$, is often met in applications and deserves therefore a more detailed consideration. In that case the differential equation (9.4.2) takes the form

$$m_1 \ddot{x} = T + m_2 A \omega^2 \sin \omega t + F(\dot{x}). \tag{9.4.25}$$

We concretize the expression for the vibrational force $V(\dot{X})$ in the main equation (9.4.9) corresponding to this case. At $\xi = A \sin \omega t$ according to (9.4.11) we obtain

$$
V(0) = -\frac{(qA\omega)^n}{2\pi} \left[k + \int_{-\pi/2}^{\pi/2} \cos^n \tau \, d\tau + (-1)^{n-1} k_- \int_{\pi/2}^{3\pi/2} \cos^n \tau \, d\tau \right]
$$

$$
= \frac{(qA\omega)^n (k_- - k_+)}{\pi} \int_0^{\pi/2} \cos^n \tau \, d\tau
$$

$$
= \frac{1}{2^{n+1}} (qA\omega)^n (k_- - k_+) \frac{\Gamma(n+1)}{\Gamma^2\left(\frac{n}{2} + 1\right)} \tag{9.4.26}
$$

where $\Gamma(x)$ is the gamma-function.

Then it is not difficult to obtain the expressions $V(\dot{X})$ and $V_1(\dot{X})$. Let us dwell in detail on the case $n = 0$, corresponding to dry friction. As a result of calculation, similar to those made in 9.2 (see formula (9.3.9) and Fig. 9.1,d), we find

$$
\omega t_+ = 2\left(\pi - \arccos \frac{\dot{X}}{qA\omega}\right), \quad \omega t_- = 2 \arccos \frac{\dot{X}}{qA\omega}, \tag{9.4.27}
$$

and as a result the formula (4.20) acquires the form

$$V(\dot{X}) = V(0) + V_1(\dot{X})$$

$$= \begin{cases} -F_+ & \text{at} \quad \dot{X} \;>\; |q|A\omega, \\ \dfrac{1}{\pi}[(F_+ + F_-)\arccos \dfrac{\dot{X}}{qA\omega} - F_+\pi] & \text{at} \quad |\dot{X}| \;<\; |q|A\omega, \\ F_- & \text{at} \quad \dot{X} \;<\; -|q|A\omega, \end{cases} \qquad (9.4.28)$$

with

$$V(0) = \frac{1}{2}(F_- - F_+), \quad V_1(0) = 0. \qquad (9.4.29)$$

Equation (9.4.21) for determining the velocity of vibro-displacement $\dot{X} = \dot{X}_*$ at $|\dot{X}| < |q|A\omega$, according to (9.4.28), takes the form

$$T + \frac{1}{\pi}[(F_+ + F_-)\arccos \frac{\dot{X}}{qA\omega} - F_+\pi] = 0, \qquad (9.4.30)$$

from which we find

$$\dot{X}_* = qA\omega \cos \frac{\pi(F_+ - T)}{F_+ + F_-}. \qquad (9.4.31)$$

As it must, equations (9.4.27)–(9.4.31) change accordingly to (9.2.9), (9.2.10), (9.2.13) and (9.2.12) at $q = 1$ and $F_\pm = mg\,f_\pm$.

Equation (4.31) is valid when the values of T satisfy the condition $0 < \pi(F_+ - T)/(F_+ + F_-) < \pi$, that is when

$$-F_- < T < F_+ \qquad (9.4.32)$$

When this condition is not satisfied, then according to (9.4.9) and (9.4.28) there is an accelerated motion of the particle, the presence of vibration not affecting $\dot{X}$ at all. That case is of no special interest. What has been said follows also from considering Fig. 9.3,c: the point of intersection $\dot{X} = \dot{X}_*$ of the curve $V(\dot{X})$ and of the line $V = -T$, corresponding to the solution of equation (9.4.21), can exist provided the condition (9.4.32) is satisfied.

The results obtained also agree with the general conclusions of 9.4.1. In particular, condition (9.4.32) determines the so called *dead zone*: when it is satisfied and vibration is absent, the particle remains motionless.

About this problem also see the work by Gershman [198].

9.4.3   *The Motion of a Particle over the Tilted Rough Plane, Performing Harmonic Oscillations in Two Mutually Perpendicular Directions*

A considerable number of publications on the theory of vibrational displacement (see [84, 400, 564, vol. 4]) is devoted to the solution of the problem on the motion of a body (material particle) of the mass m over a rough vibrating plane, tilted to the horizon at a certain angle α and performing harmonic translational oscillations in two mutually perpendicular directions according to the law (Fig. 9.5,a)

$$\xi = A\sin(\omega t + \delta), \qquad \eta = B\sin\omega t. \tag{9.4.33}$$

Here ξ and η are the shifts of the points of the plane in the fixed axes of

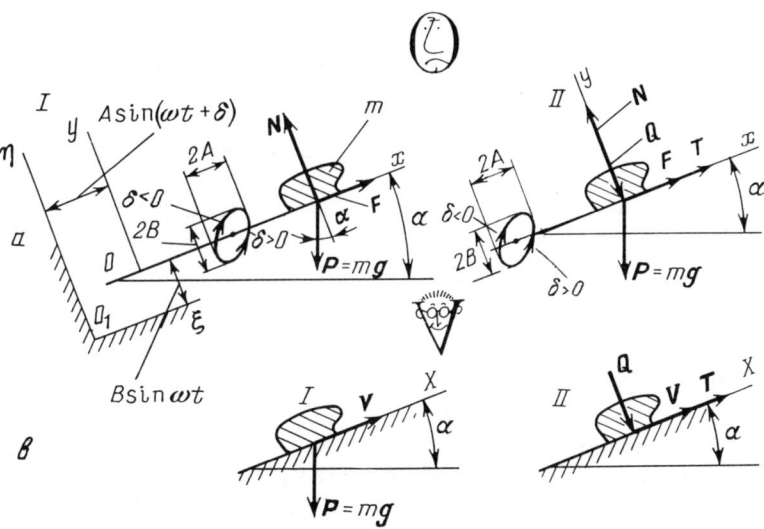

Figure 9.5. The motion of the particle over a tilted plane performing harmonic oscillations in two mutually perpendicular directions. a) I — the initial system; II — the system acted upon by the additional longitudinal and transverse constant forces; b) pictures seen by the observer **V**.

the coordinates $\xi O_1\eta$, with the axis $O_1\xi$ being parallel, and the axis $O_1\eta$ — perpendicular to the plane. A and B are the amplitudes of the longitudinal and transverse components of the vibration respectively, δ is the phase shift between the components (the values $\delta < 0$ being answered by the motion of the points along the trajectories clockwise, and $\delta > 0$ — counterclockwise); ω is the frequency of vibration; the trajectories of oscillations of the points of the plane being in this case ellipses.

This problem plays the main role in the modern theory of vibro-displacement: the study of many processes being based on its solution, such as vibrational transportation, vibro-separation, vibrational sinking of piles and many others which will be discussed in chapter 10.

In the mobile axes xOy rigidly bound with the vibrating plane, equations of motion of the particle have the form

$$m\ddot{x} = mA\omega^2 \sin(\omega t + \delta) - m g \sin \alpha + F(\dot{x}), \qquad (9.4.34)$$

$$m\ddot{y} = mB\omega^2 \sin \omega t - m g \cos \alpha + N, \qquad (9.4.35)$$

where $F(\dot{x})$ and N are respectively the force of dry friction and the normal reaction, g is the free fall acceleration.

At $y \equiv 0$, i.e. in the interval of time when the particle is on the plane, from equation (9.4.35) we obtain

$$N = m g \cos \alpha - mB\omega^2 \sin \omega t, \qquad (9.4.36)$$

and the dry friction force is determined by the relations

$$F(\dot{x}) = fN\mathrm{sgn}\dot{x} \quad \text{at} \quad \dot{x} \neq 0,$$
$$-f_1 N < F(\dot{x}) < f_1 N \quad \text{at} \quad \dot{x} = 0 \qquad (9.4.37)$$

where f and f_1 are the coefficients of sliding and of rest.

To these relations must be added the equalities, determining the connection between the longitudinal and transverse projections of the velocity of the particle before and after its collision with the plane. The following equalities are often used:

$$\dot{y}_+/\dot{y}_- = -R, \quad \dot{x}_+ = (1 - \lambda)\dot{x}_-, \qquad (9.4.38)$$

where $\dot{x}_-$ and $\dot{y}_-$ are the projections of the velocity of the particle before the impact, and $\dot{x}_+$ and $\dot{y}_+$ — after the impact; R is the restitution coefficient, and λ is the instantaneous friction coefficient (the adequate hypotheses of type (9.4.38) are described in more detail in [564, vol. 4]).

If the following equality is satisfied

$$w = B\omega^2/g \cos \alpha < 1, \qquad (9.4.39)$$

then according to (9.4.36) the normal reaction is $N > 0$ and the particle which happened to be on the plane remains there. The parameter w introduced here plays an essential role in the problems under consideration;it is called an *overload parameter*.

Thus the solution of this problem is reduced to integrating a rather complicated system of differential equations whose essential non-linearity is conditioned by the presence of the force of dry friction and of the unilateral constraint. Of utmost practical interest are the solutions of these equations, corresponding to the stationary regimes of vibro-displacement, i.e. solutions of the type

$$\dot{x} = \dot{X} + \dot{\psi}(\omega t), \quad \dot{y} = \dot{y}(\omega t), \tag{9.4.40}$$

where $\dot{X}$ is the constant (the velocity of vibro-displacement, in this case, — the *velocity of vibro-transportation*) and $\dot{\psi}$ and $\dot{y}$ are the periodic functions of ωt with a period $2\pi n \, (n = 1, 2, \ldots,)$, with

$$< \dot{\psi} >= 0, \quad < \dot{y} >= 0. \tag{9.4.41}$$

Here by a definite regime we mean the motion of type (9.4.40) which is characterized by a certain set and sequence of the stages of motion, that is intervals of time in which the motion is described by "smooth" (in this case — linear) differential equations (sliding forward, sliding backward, flight over the plane, relative rest). Finding the velocity $\dot{X}$ in the indicated stationary regimes is of special interest for applications.

At present this problem has been solved practically exhaustively by exact analytical methods for the regimes of motion of the particle without tossing, those regimes taking place provided condition (9.4.39) is satisfied (see for instance [84, 400, 564, vol. 4]). As for the regimes with tossing, they have not been studied exhaustively due to their exceptional diversity. It has been established that in certain space regions of the parameters, especially at the coefficient values R close to one, chaotic or rather complicated long periodic motions take place (see for instance publications by Gorbikov and Neimark [210, 211], Shchigel and Grinbaum [477]). The review of the results and the bibliography can be found in the cited books.

Every stationary regime of motion of the particle is answered by a certain region of existence and stability in space of the system parameters and a certain analytic expression for the velocity of vibro-transportation $\dot{X}$. For the motion with tossing, several stationary regimes can exist and be stable in the small in certain regions of space of the parameters.

Along with that, in spite of this complicating circumstance, the velocity of vibro-transportation $\dot{X}$ in the stationary regimes, being the integral characteristic of motion, displays, as a rule, a certain "stability" with regard to the type of the regime, continuously changing with the change of the parameters of the system, that "stability" having place even for most complicated regimes

[210, 211]. This circumstance made it possible to propose for the so called *regimes with sufficiently intensive tossing*, characterized by the parameter values w, satisfying the condition

$$w = \frac{B\omega^2}{g \cos \alpha} > 2\frac{1+R^2}{(1+R)^2}, \tag{9.4.42}$$

a   universal   approximate   formula   for   calculating   the   velocity   of vibro-transportation:

$$\dot{X}_* \approx A\omega \left[\frac{\pi p'(w,R)}{w} \frac{1-R}{1+R} \cos \delta - \sqrt{1 - \left[\frac{\pi p'(w,\dot{R})}{w} \right]^2 \left(\frac{1-R}{1+R} \right)^2} \sin \delta \right.$$
$$\left. - \frac{\pi p'(w,R)g}{\omega} \frac{2-\lambda}{\lambda} \sin \alpha \right] \tag{9.4.43}$$

where

$$\pi p'(w,R) = \frac{1}{2} \left[\frac{w(1+R)}{1-R} + \frac{\sqrt{w^2(1+R)^4 - 4(1+R^2)^2}}{(1-R^2)} \right]. \tag{9.4.44}$$

The graphs of this function are presented in Fig. 9.6.

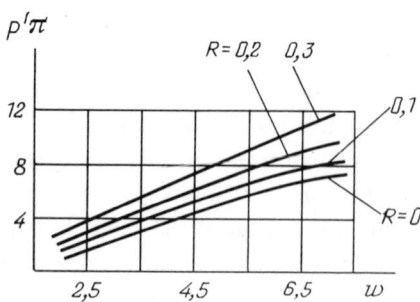

Figure 9.6. The dependence of the parameter p on w and R.

Formula (9.4.43) is changed for a simpler one when " tossing is still more intensive", when

$$w > 3.5\frac{1+R^2}{(1+R)^2}. \tag{9.4.45}$$

In this case

$$\dot{X}_* \approx A\omega \cos \delta - \frac{1+R}{1-R} \frac{2-\lambda}{\lambda} B\omega \tan \alpha. \tag{9.4.46}$$

For a number of applications, the generalization of the given relations is of a special interest when besides the forces already taken into consideration, the particle is acted upon by a constant longitudinal force T and a constant force Q, normal to the plane (Fig. 9.5,aII). These generalized relations are obtained from (9.4.43) and (9.4.46) by changing the expression $-mg \sin \alpha$ for $-mg \sin \alpha + T$, and the expression $mg \cos \alpha$ for $mg \cos \alpha + Q$:

$$\dot{X}_* = A\omega \left[\frac{\pi p'(w_1, R)}{w_1} \frac{1-R}{1+R} \cos \delta - \sqrt{1 - \left[\frac{\pi p'(w_1, R)}{w_1} \right]^2 \left(\frac{1-R}{1+R} \right)^2} \sin \delta \right]$$
$$+ \frac{\pi p'(w_1, R)g}{\omega} \frac{2-\lambda}{\lambda} \frac{T - mg \sin \alpha}{mg} \left(w_1 > 2\frac{1+R^2}{(1+R)^2} \right), \tag{9.4.47}$$

$$\dot{X}_* = A\omega \cos \delta + B\omega \frac{1+R}{1-R} \frac{2-\lambda}{\lambda} \frac{T - mg \sin \alpha}{Q + mg \cos \alpha}$$
$$\left(w_1 > 3.5\frac{1+R^2}{(1+R)^2} \right), \tag{9.4.48}$$

where the parameter

$$w_1 = \frac{mB\omega^2}{Q + mg \cos \alpha} \tag{9.4.49}$$

plays the same role as the overload parameter w in the initial problem.

Now let us show that having expressions (9.4.43), (9.4.46) – (9.4.48), it is not difficult to obtain the expression for the vibrational force **V** and compose the main equation of vibrational mechanics. This possibility was mentioned in 3.3 (the case when either the exact or approximate solution of the initial problem is known). For this purpose it should be noted that if we are interested in the solutions of the initial system of equations (9.4.34), (9.4.35) of type

$$\dot{x} = \dot{X}(t) + \dot{\psi}(t, \omega t), \quad \dot{y} = \dot{y}(t, \omega t) \tag{9.4.50}$$

where $\dot{X}(t)$ is the slow component and $\dot{\psi}$ and $\dot{y}$ are the fast components of motion which satisfy relations (9.4.41), then the main equation of vibrational mechanics for the generalized problem must have the form

$$m\ddot{X} = -mg \sin \alpha + T + V \tag{9.4.51}$$

(the gravitational force and the forces Q and T being the slow forces). Therefore the velocity of vibro-transportation in the stationary regimes $\dot{X} = \dot{X}_* = $ const is determined from the equation

$$m g \sin \alpha - T = V. \tag{9.4.52}$$

But this equation must necessarily have a solution $\dot{X} = \dot{X}_* = $ const coinciding (at least approximately) with expression (9.4.47) or in the corresponding case — with (9.4.48). Hence it follows that in order to obtain the formula for the vibrational force V in the indicated expressions, one should change $m g \sin \alpha - T$ for V and solve the equalities obtained with regard to V. As a result we will have

$$V \approx m \nu (\kappa A \omega - \dot{X}), \tag{9.4.53}$$

where

$$\nu = \frac{\omega}{\pi p'(w_1, R)} \frac{\lambda}{2 - \lambda},$$

$$\kappa = \frac{\pi p'(w_1, R)}{w_1} \frac{1 - R}{1 - R} \cos \delta - \sqrt{1 - \left[\frac{\pi p'(w_1, R)}{w_1} \right]^2 \left(\frac{1 - R}{1 + R} \right)^2} \sin \delta$$

$$\text{at} \quad w_1 > 2(1 + R^2)/(1 + R)^2; \tag{9.4.54}$$

$$\nu = \left(\frac{Q}{m} + g \cos \alpha \right) \frac{\lambda}{2 - \lambda} \frac{1 - R}{1 + R} \frac{1}{B\omega}, \quad \kappa = \cos \delta$$

$$\text{at} \quad w_1 > 3.5(1 + R^2)/(1 + R)^2. \tag{9.4.55}$$

It should be noted that the condition of stability of the stationary regime which in this case too is expressed by inequality (9.4.42), is certain to be satisfied, since according to (9.4.53) we have $V'(\dot{X}) = -m \nu < 0$.

Expression (9.4.53) will be used in chapter 15 in the process of an approximate examination of a number of more complicated problems.

9.4.4   On Vibrational Transportation of a Body (Particle) Upwards along the Tilted Plane. The Limiting Angle of Rise

It is not difficult to notice that in both cases, considered above, the vibrational transportation of a body upward along a tilted rough vibrating plane is quite possible. This effect is of great practical importance.

So in the case of the system, considered in 9.4.1, the velocity of the vibrational displacement upward over the plane at the angle α in the stationary

regime will be determined from equation (9.4.21) if we assume that
$T = -mg \sin \alpha$:

$$- mg \sin \alpha + V(0) + V_1(\dot{X}) = 0. \tag{9.4.56}$$

The marginal angle of the rise of the body $\alpha = \alpha_*$, that is the angle at
which $\dot{X} = 0$ will then be (recall that $V_1(0) = 0$)

$$\sin \alpha_* = V(0)/mg. \tag{9.4.57}$$

Using formula (9.4.20) and considering that in this case it is necessary to assume
that $F_\pm = mg f_\pm \cos \alpha$ where f_+ and f_- are the corresponding coefficients of
friction, we obtain

$$\tan \alpha_* = \frac{\omega}{2\pi}(f_- t_- - f_+ t_+). \tag{9.4.58}$$

In the case of symmetrical, say harmonic, vibration $t_+ = t_- = \pi/\omega$ and
then

$$\tan \alpha_* = \frac{1}{2}(f_- - f_+). \tag{9.4.59}$$

In the case of system 9.4.2 in order to determine the stationary value of
the velocity $\dot{X}$ at the rise upward along the plane, according to the scheme in
Fig. 9.5,aII, one should assume that $T = 0$. Then from (9.4.52) one obtains
the equation

$$mg \sin \alpha = V(\dot{X}), \tag{9.4.60}$$

and for the *limiting angle of the rise of a body upwards along the tilted plane*
$\alpha = \alpha_*$ — we get the relationship

$$mg \sin \alpha_* = V(0). \tag{9.4.61}$$

In the case of the regimes with intensive tossing, according to (9.4.53) we
will have

$$\dot{X} = \kappa A \omega - \frac{g \sin \alpha}{\nu}, \quad \sin \alpha_* = \frac{\nu \kappa A \omega}{g}. \tag{9.4.62}$$

Results of a more detailed investigation of the question about the limiting
angle of the rise of the body upwards over the plane can be found in [84,
400, 564, vol. 4]. In practice it is possible to transport single bodies and
granular materials upward along the planes tilted to the horizon up to 20° and
more, though the velocity of vibro-transportation falls down rather quickly
with the growth of the angle of inclination. In 10.1 other schemes of devices
will be considered enabling the transportation at much larger angles and even
vertically upwards.

9.4.5   About Other Models and Problems

The models of the processes of vibro-displacement, considered above, and the results of their investigation are most often used when studying the applications of the theory. Data about other models and problems can be found in [84, 400] and in the reference book [564, vol. 4]. A number of such models will be considered in chapter 10 conformably to different applied problems.

The models, described and mentioned above, belong to the cases when a discreet idealization of the system is possible. Some models of the processes of vibro-displacement in fluids, gases and in granular media, as well as the corresponding applications are considered in the book in Part IV.

Chapter 10

Effects of Vibrational Displacement in Technique, Technology and in Nature

10.1 Vibrational Transportation

One of the processes of vibrational displacement, most widely used in practice, is *vibrational transportation* — the directed motion of bodies inside the tubes, trays or vessels under the action of vibration. In different industries they use vibrational conveyers, feeders and dosing apparatus; in some machines, such as vibrational screens, dryers, concentrators, separators — transportation is combined with technological operations, making their important integral part.

An essential peculiarity of vibrational transportation facilities consists in the fact that the transportation of loads them is realized not due to their joint motion together with the load-carrying unit — tube, tray, vessel — but due to the vibration of the latter. This circumstance predetermines a number of important technological and operational advantages of the vibrational mode of transportation. The rectilinear translation of harmonic oscillations which are relatively easily realized are used most often. Their trajectories are inclined to the axis of the tube or the tray at an acute angle, which corresponds to the first case shown in Fig. 9.2, *aII*.

The principal schemes, descriptions of the design, and photos of different transport- and transport-technological vibrational machines can be found in [84, 90, 103, 145, 560, 207, 208, 209, 214, 215, 588, 443, 445, 513, 547, 227, 229] and in the reference-books [563, 232, 564, vol. 4]; the fundamentals of the theory and the methods of calculation are also to be found there. Therefore the aim of this section is not to give a systematic presentation of the question, but to emphasize the advantages of using the conception of vibrational mechanics when explaining the theory and describing the processes of vibrotransportation.

These advantages are especially appreciable in complicated cases; however, in more simple cases as well, when there are exact solutions of problems, these advantages are quite obvious and very convenient in using the results.

Generalizing what was said in 9.4, we can write the main equation of vibrational mechanics for one-dimensional processes of vibrational transportation of individual solid bodies (particles) in the following form

$$m\ddot{X} = -m\,g\,\sin\alpha + T + V(\dot{X}) \tag{10.1.1}$$

where m is the mass of the body, α is the angle of the slope of the surface to the horizon, g is the free fall acceleration, T is a certain constant or slow longitudinal force, and $V(\dot{X})$ is the vibrational force which in case of regimes with tossing greatly depends on the *overload parameter*

$$w_1 = \frac{m|\ddot{\eta}|_{\max}}{m\,g\,\cos\alpha + Q}. \tag{10.1.2}$$

Here Q is a constant or slowly changing transverse force, pressing the body to the plane, and $|\ddot{\eta}|_{\max}$ is the maximal value of the transverse component of the acceleration of the plane. The forces T and Q are introduced for generality: they can represent, say, the action of the air during the vibropneumatic transportation, the action of the centrifugal forces of inertia, the action of electromagnetic forces etc. Equation (10.1.1) corresponds to the picture seen by the observer **V**, shown in Fig. 9.5,*b*.

The velocity of vibrotransportation under the stationary regimes $\dot{X} = |dot X_*$ is obtained from the equation

$$m\,g\,\sin\alpha = T + V(\dot{X}). \tag{10.1.3}$$

The condition of stability of such regimes is the inequality $V'(\dot{X}_*) < 0$

For the regimes with an intensive tossing in the case of a two-component harmonic vibration of the plane and with forces T and Q being constant, the vibrational force is defined by expression (4.53) of chapter 9. The *limiting angle of lifting the body along the plane* in the absence of the forces T and Q can be calculated by formulas (4.59) and (4.62) of the same chapter.

As was marked in 3.3 and shown by an example in 9.42, the expressions for V can be obtained in other cases as well, when either the exact or approximate expressions for the velocity of vibrotransportation are known.

A simple method of the experimental determination of the vibrational force when transporting a body along a rough plane, performing arbitrary two-component periodic oscillations (Fig. 10.1), follows from equality (10.1.3). Between support *1* and load *2*, moving with a preset constant velocity $\dot{X}$, a

sufficiently soft spring *3* should be placed. When in the course of time the
average velocity of the load will get the stationary value $\dot{X}$, the average stress
in the spring T according to equality (10.1.1) at $\alpha = 0$ will be equal to $V(\dot{X})$.
In particular, at the stationary support this stress will be equal to $V(0)$. The
value $V(0)$ can also be found by experimental determination of the marginal
angle of the rise of the body along the plane: to do it, it is enough to use
expression (4.61) of chapter 9.

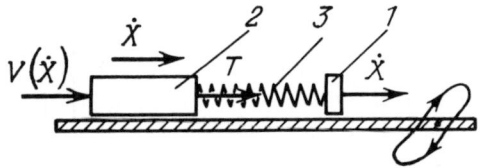

Figure 10.1. To the experimental determination of the vibrational force in the case
of the transportation of bodies along a vibrating plane.

To design vibrational devices with translational oscillations of the operating
unit transporting the granular medium, the layer of which is not too thick (of
the order of about 20–30fold average size of the particles), at the overload
coefficient $g < w < 10g$ and the oscillation frequencies of 200 oscillations per
minute $< n = 30\omega/\pi < 3000$ oscillations per minute the above equations
for the case of a single solid body (material point) fit quite well. Sometimes
they are to be somewhat corrected by means of taking into account additional
factors, in particular — the air resistance [84, 564, vol. 4]. Continual models
of slow motions of the granular medium under vibration for some other, more
complicated cases, are proposed in chapter 15, which also considers a number
of adequate problems of the theory of vibrotransportation.

Here we will dwell on the question of vibrational transportation of bodies
either vertically upward or at a big angle to the horizon, and we will illustrate
the efficiency of this approach by investigating one of the devices proposed for
this purpose.

The usual solution of the problem consists in using spiral trays with a
vertical axis, the average angle of the slope of the bottom of the tray to the
horizon not exceeding the limiting values for the vibrating plane. Angular
and axial harmonic oscillations are imparted to the tray. Such conveyers have

received wide recognition [446]; to drive them, self-synchronized vibro-exciters can be used very successfully [90, 103] (also see chapter 7).

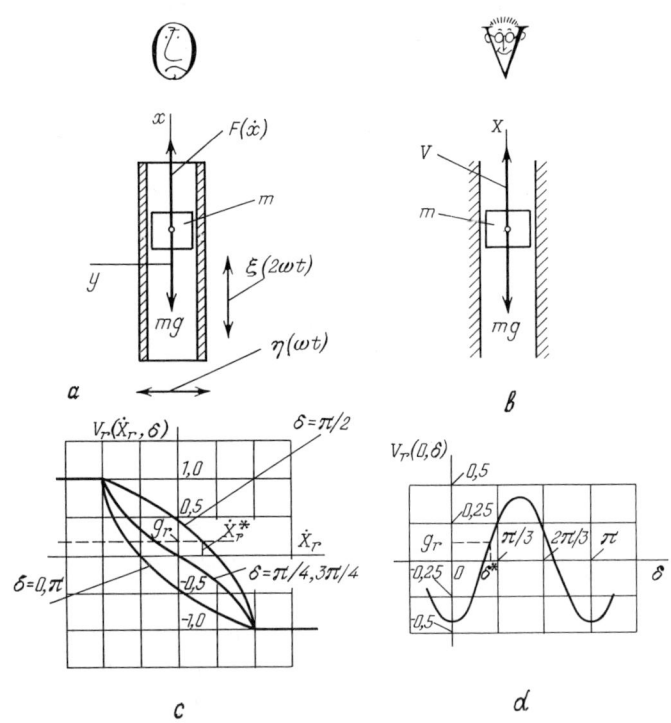

Figure 10.2. Device for the vibrational transportation of a body vertically upwards [142, 564]. a)Scheme of the system; b)the picture as it is seen by the observer **V** ; c) graphs of the dimensionless vibrational force $V_r(X_r, \delta)$ in case $\xi = A_x \sin 2(\omega t - \delta)$, $\eta = A_y \sin \omega t$; d) the $V_r(0, \delta)$ dependence in that case.

An unconventional device for vertical vibrotransportation was proposed by Brumberg [142, 564, vol. 4]. It is a tube (Fig. 10.2,a) to which transverse and longitudinal oscillations are imparted, the frequency of the longitudinal oscillations being twice as large as that of the transverse oscillations ω. The idea of the device lies in the following. At the proper phasing of oscillations the longitudinal force of inertia acting on the transported body m in the relative motion is directed upward just in those intervals of time when the action of the transverse force of inertia, pressing the body to the walls of the tube, is the weakest. In the intervals of time when the longitudinal force of inertia is directed downward, the load is pressed to the walls of the tube most strongly.

As a result, a vibrational force appears which is directed upward, overcoming the weight of the body and providing its rise up the tube.

The investigation of the operation of the described device made by Brumberg [142] was based on obtaining an exact solution of the differential equation of the motion of the load. This solution is rather complicated. It is much simpler to obtain an approximate solution by composing the main equation of vibrational mechanics by method of direct separation of motions if we make a certain additional supposition which seems quite natural.

Assuming that there is no gap between the body m and the walls of the tube, we will write down the differential equation of the relative motion of the body along the tube as

$$m\ddot{x} = -m\,g - f\,m|\,\ddot{\eta}(\omega t)\,|\mathrm{sgn}\dot{x} - m\ddot{\xi}(2\omega t) \qquad (10.1.4)$$

where x is the coordinate of the body, counted upward along the axis of the tube; m is the mass of the body; g is the free fall acceleration; f is the sliding friction coefficient; $\xi(2\omega t)$ and $\eta = \eta(\omega t)$ are the periodic functions of their arguments with a 2π-period, characterizing the law of the transverse and longitudinal oscillations of the tube respectively.

Now we are interested in the solutions of equation (10.1.4) of the type

$$\dot{x}(t) = \dot{X}(t) + \dot{\psi}(t, 2\omega t) \qquad (10.1.5)$$

where $\dot{X}(t)$ is the slow component of the velocity (which is the velocity of vibrotransportation), and $\psi(t, 2\omega t)$ is the fast component, 2π-periodical with respect to the argument $2\omega t$, satisfying the condition

$$< \dot{\psi}(t, 2\omega t) >= 0. \qquad (10.1.6)$$

Equations (2.55) and (2.56) of chapter 3 will then be written as

$$m\ddot{X} = -m\,g + V \qquad (10.1.7)$$

$$m\ddot{\psi} = f\,m[|\,\ddot{\eta}(\omega t)\,|\mathrm{sgn}(\dot{X} + \dot{\psi}) - <|\,\ddot{\eta}(\omega t)\,|\mathrm{sgn}(\dot{X} + \dot{\psi}) >] - m\,\ddot{\xi}(2\omega t) \quad (10.1.8)$$

where

$$V = -f\,m < |\,\ddot{\eta}(\omega t)\,|\,\mathrm{sgn}(\dot{X} + \dot{\psi}) > \qquad (10.1.9)$$

Now let us assume that the friction force $f\,m|\ddot{\eta}|$ is small as compared to the longitudinal component of the force of inertia $m\ddot{\xi}$. Then the 2π-periodic with respect to $2\omega t$ solution of the equation of fast motions (10.1.8), satisfying condition (10.1.6), will be

$$\dot{\psi} = -\dot{\xi} \qquad (10.1.10)$$

and expression (10.1.9) for the vibrational force will take the form

$$V = V(\dot{X}) = -f\,m\, <\,|\,\ddot{\eta}(\omega t)\,|\mathrm{sgn}[\dot{X} - \dot{\xi}(2\omega t)]\,>\,. \qquad (10.1.11)$$

In the case of the harmonic oscillations of the tube according to the law

$$\xi = A_x \sin 2(\omega t - \delta), \quad \eta = A_y \sin \omega t \qquad (10.1.12)$$

introducing the dimensionless vibrational force V_r and the dimensionless velocity $\dot{X}_r$ in accordance with the formulas

$$V_r = V/\left(\frac{2}{\pi}f\,m\,A_y\omega^2\right), \quad \dot{X}_r = \dot{X}/2A_x\omega \qquad (10.1.13)$$

after averaging the expression (10.1.11) we obtain $(0 \le \delta \le \frac{1}{2}\pi)$:

$$V_r = V_r(\dot{X}_r, \delta) \begin{cases} 1 & \text{at} \quad \dot{X}_r \le -1 \\ -\cos\delta\sqrt{2(1+\dot{X}_r)} + 1 & \text{at} \quad -1 \le \dot{X} \le \cos 2\delta \\ \sin\delta\sqrt{2(1-\dot{X}_r)} - 1 & \text{at} \quad \cos 2\delta \le \dot{X}_r \le 1 \end{cases} \quad (10.1.14)$$

From this equation it is easy to get expressions for any δ if we take into consideration the relationship

$$V_r(-\dot{X}_r, \delta \pm \frac{1}{2}\pi) = -V_r(\dot{X}_r, \delta)$$

following from (10.1.11) and (10.1.12).

The dependences $V_r(\dot{X}_r, \delta)$ and $V_r(0, \delta)$ are presented in Fig. 10.2,c,d. As one can see, the largest positive values V_r are obtained at $\delta = \frac{1}{2}\pi$, which, taking into account (10.1.12), corresponds to the given above qualitative explanation of the work of the device; at $0 < \delta < 1/4\pi$ and $3/4\pi < \delta < \pi$ the vibrational force is negative and so the transportation of the body upward is impossible.

The main equation of vibrational mechanics (10.1.7) when taking into consideration (10.1.13) will be written as

$$q\ddot{X}_r = -g_r + V_r(\dot{X}_r, \delta) \qquad (10.1.15)$$

where

$$q = A_x/\frac{f}{\pi}A_y\omega, \quad q_r = g/\frac{2f}{\pi}A_y\omega^2 \qquad (10.1.16)$$

The values of the velocity in the stationary regime of motion will be determined from the equation

$$V_r(\dot{X}_r, \delta) = g_r. \tag{10.1.17}$$

This regime is stable since according to (10.1.14) $V_r'(\dot{X}_r) < 0$.

The solution of equation (10.1.17) will be determined as the abscissa of the point of intersection of the curves in Fig. 10.2,c and the straight line, corresponding to equality (10.1.17). One can see in the figure that at $\delta = \frac{1}{2}\pi$ the velocity is positive if $g_r < 0.4$, that is the transverse vibration is intensive enough.

The examining of Fig. 10.2,c and d also shows that changing the parameters δ and $A_y\omega^2$, one can regulate the speeds of lifting the body X within rather wide limits, in particular it is possible to have $\dot{X}$ become zero, that is have the body "hover" in the tube. The values of the parameter $\delta = \delta_*$, corresponding to that situation, can be easily determined by the known value g_r by means of the graph of Fig. 10.2,d.

In the book by the Kobrinskys [274] conditions are obtained under which in such a device the ball moves up the tube, the diameter of the ball being smaller than the internal diameter of the tube. Another device is describes in [490].

10.2   Vibrational Separation of the Components of Granular Mixtures

10.2.1   *Factors, Determining the Efficiency of Using Vibration in the Processes of Separating the Components of Granular Mixtures*

Technological processes whose aim is to separate the particles of the granular material according to their size (classification), density, shape, magnetic-, electric- and other parameters (they are called *separation parameters*) play a big role in mechanical ore dressing and in benefication, in the production of building materials, in chemical industry, in powder technology, in the processing of grain and grain products, in food industry and in some other industries.

Beginning with the most ancient methods — the classification by sieves — a great role in forming the separation processes is played by vibration, which is conditioned by at least four factors.

1. In the sense, shown in 9.2.2, vibration transforms forces of the type of dry friction, typical for the interaction of the particles of granular material, into those of the type of viscous friction. As a result, conditions are created to display the differences (contrasts) in the parameters of separation: often these

differences are comparatively small and under static conditions they do not manifest themselves.

2. As a result of the action of vibration, along with the transformation of friction, the particles of the granular mixture are acted upon by the driving vibrational forces, which under the adequate conditions also lead to the growth of intensity of the separation process and its "resolving ability", that is to the possibility to separate particles whose separation parameters differ very little. Vibrational forces also provide the transportation of the initial mixture along the working surfaces, say, sieves, and the products of separation — to the corresponding receivers.

3. Vibration is an important element of the process of screening, since providing the intensive motion of particles with respect to the holes of the sieve, it increases (under certain conditions of course) the probability for the particles of small fraction to pass through the holes.

4. Particles of the granular material with different properties move along the vibrating surfaces in different trajectories, which provides the basis for separating them without using the sieves.

When creating and improving the devices for the vibrational separation of the granular mixtures, it is necessary to take into consideration that the increase of intensity of vibration may result not only in the increase of the factors, facilitating the separation, but also in the factors, facilitating the stirring. However, when separating particles by means of sieves, which are a kind of absorbing screens for the particles of the passable fraction, the intensification of stirring may prove to be useful since it increases the probability for the passable particles to get to the sieves.

The theory and practice of the classification of the granular material by means of vibrating sieves is the subject matter of a great number of investigations and of some monographs (see, say, [554, 214, 215, 588, 553], and also reference books [232, 564, vol. 4]). Here we will consider two other ways of vibrational separation of particles of granular mixtures — in a layer of the material and on the vibrating surfaces. The use of the approaches of vibrational mechanics when investigating and describing those methods seems especially expedient.

10.2.2   *On Separation of Particles in a Layer of Granular Material under the Action of Vibration (Segregation, Stratification, Self-classification)*

The separation of particles under the action of vibration is used both as an independent process and as a component of a more complicated separation process, for instance in the vibrational screens which were mentioned above, and also in the concentration tables, in the jigging machines and in other devices.

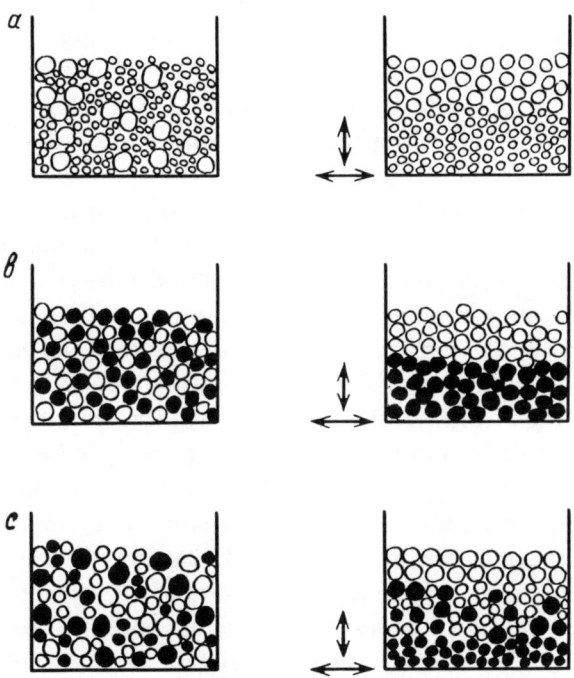

Figure 10.3. Under the action of horizontal symmetrical or vertical vibration of the vessel with the granular mixture a separation of particles takes place, according to their size and density (segregation and self-sorting).

Certain regularities of the separation in a layer which are also called *segregation, stratification, and self-classification*, are shown schematically in Fig. 10.3. In the absence of vibration the mixture of heterogeneous particles of the granular material which is in the vessel in the gravity field or in some other stationary force field may, due to the action of forces of the type of dry friction, have an infinite number of the continuously distributed equilibrium positions: the mixture being arranged in the same way or almost in the same way as when it was poured into the vessel. In case, say, a mixture of big and small particles of the same density is subjected to vibration (not too intensive, lest the chaotic component of the process, that is the mixing, should be predominant), then as a result of the effect of vibration, the big particles will take the position above the small particles (Fig. 10.3,a). In the case of the mixture of particles of the same size but with different densities, the light particles will take the positions above the heavy ones (Fig. 10.3,b); and finally in the case of a mixture of big and small particles of different densities the lower position will be taken by small heavy particles, they will be followed by small light ones, then the big heavy particles (or a mixture of small light and big heavy particles); the big light particles will appear to be in the upper position (Fig. 10.3,c) (we mean here the case when due to vibration there are no slow streams of the medium — see, say, 15.1).

Thus in all the cases we have considered the granular mixture under vibration tends to a certain quasi-equilibrium state. This occurs, as was marked before, due to the transformation of the forces of dry friction; along with that the equilibrium state can be greatly affected by the driving vibrational forces which appear during the vibration. As a result, the equilibrium position of the mixture in the potential field of forces may not correspond to the minimum of the potential energy, which would have happened if there had been only forces of viscous friction alone, and as is sometimes erroneously believed [368]. This quite important circumstance can be partly seen in Fig. 10.3,c.

The simplest model, illustrating the behavior of this system, is a body, lying on the concave rough surface (Fig. 10.4,a). In the gravity field this body may be in the state of stable equilibrium at any point of the surface where the angle of the tilt of the tangent to the horizon α does not exceed in absolute value the angle of friction $\rho = \arctan f$ (f being the friction coefficient of sliding).

In case of vibration, the picture changes abruptly. In the case of horizontal or vertical symmetric, for instance harmonic vibration of the surface (see 9.4) the positions of the stable quasi-equilibrium ("equilibrium on the average") are the points of the minimums on the surface, i.e. those which answer the minimums of the potential energy in the gravity field (Fig. 10.4,b). This situation answers the simplest case. In more complicated cases when there is one or other asymmetry of the system (Fig. 9.2), the quasi-equilibrium positions do not correspond to the points of the minimums of the potential energy (Fig. 10.4,c). We can also say that in such cases when the system with dry friction is acted upon by vibration, an effect of vibrational drift takes place (see chapter 11).

The picture shown in Fig. 10.3 is idealized: usually in practice we have to deal not with a small, but with a rather large number of different classes of

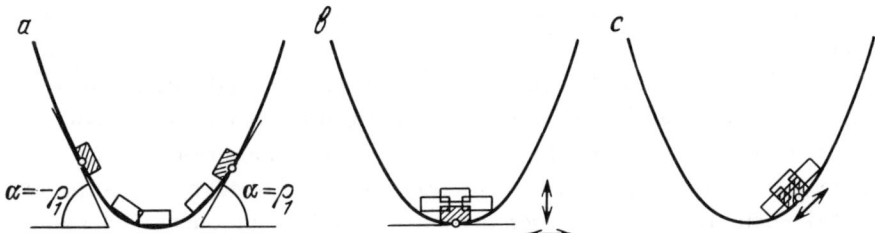

Figure 10.4. Simplest model of the behavior of the system with dry friction under the action of vibration — a body on the concave rough surface.

particles, which differ not only in size and density, but also in shape and other parameters. Besides, the behavior of the system is affected by a number of random factors. So the borders of the location of particles in their final states (Fig. 10.3) are more or less vague. It should be noted that the gaseous or fluid medium, where the mixture to be separated is in, greatly affects the character of the processes under discussion.

10.2.3   *General Statement of the Problem on the Separation of Particles of the Granular Mixture in a Vibrating Vessel. Brief Characteristic of the State of the Problem*

The general statement of the problem on separating the particles (fractions, classes) of the granular material in a vibrating layer follows from the qualitative description, given in 10.2.2. There is a vessel with a mixture of particles which differ in size, density and other parameters, whose totality is denoted by vector $\mathbf{a}$. We have got preset the law of the oscillations of the points of the surface of the vessel, the field of the external forces and the average for the period of oscillations density of the distribution $f(x, y, z, \mathbf{a}, \mathbf{t}_0)$ of the particles by their parameters at each point of the vessel at a certain initial moment of the "slow" time t_0. It is necessary to find the distribution density $f(x, y, z, \mathbf{a}, \mathbf{t}_0)$ at any moment of the time t; in particular the final distribution

$$f^*(x, y, z, \mathbf{a}) = \lim_{t \to \infty} f(x, y, z, \mathbf{a}, t)$$

A simpler statement of the problem is also possible when the mixture is characterized by a certain set of classes of particles $b_1, \ldots, b_n$. The initial concentrations $c_1(x, y, z, t_0), \ldots, c_n(x, y, z, t_0)$ of particles of each class are given and it is necessary to find their values at an arbitrary moment of time $t > t_0$ and in particular the limiting values $c_1^*, \ldots, c_n^*$ at $t \to \infty$.

In most cases the problem stated in that way cannot be solved without taking into account both the slow and fast components of the field of velocities of the medium in the vessel. One also has to take into consideration the presence of the gaseous or liquid phase in the gaps between the particles, and sometimes of both phases simultaneously, i.e. to study the motion of the three-phase system "solid-fluid-bubbles of gas". In other words, then there appear problems about the penetration of vibration into one- or multi-component media and also about the slow streams in those media.

At present only a few special cases of the stated problem have been considered, for instance those, corresponding, say, to the situation when the concentrations of the particles $c_1, \ldots, c_{n-1}$ which interest the investigator are so small in comparison with that of the particles of a certain class c_n ("medium", "bed") that the interaction of particles of the first $n - 1$ classes can be neglected (or taken into account only approximately) and the motion of the isolated particles of those classes can be considered in the medium of the particles of the "bed". Besides the deterministic approach (see below and also [85, 132, 214, 215, 225, 294, 416, 171, 368]), the process can be regarded as random: in the latter case the mathematical model is provided by the equations of the type of equations of diffusion or of the Focker-Plank-Kolmogorov equations. The main contribution in the development of this direction belongs to Nepomnyashchy [411]. The disadvantage of the first approach lies in the fact that it does not consider the random factors, the drawback of the second approach is the difficulty in establishing the functional dependence of the parameters, involved in the differential equation of the accidental process, on the parameters of vibration. The attempt to combine those two approaches, taking into account the interaction of certain classes of the particles that are to be separated was made by Hainman and the author [85]. A popular consideration is given in [236].

The part of the problem, concerning the motion of multiphase media has been studied comparatively well. Investigations and interesting results achieved in that field have been generalized in the monograph by Nigmatullin [412]. The penetration of vibration into different media has been studied far less (see chapter 16).

From the very statement of the problem under consideration follows the

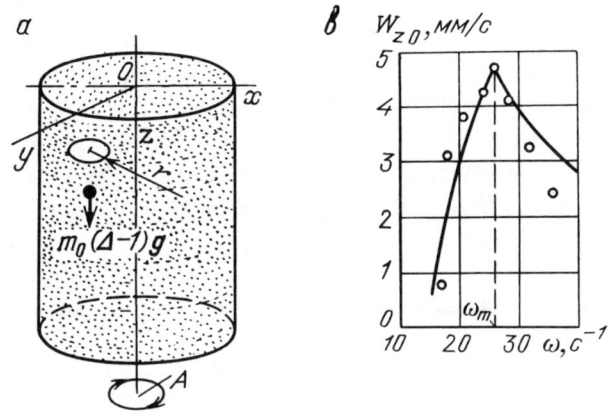

Figure 10.5. Pseudoresonance effect .

expediency of using when solving it the approaches of vibrational mechanics and, in particular, the theory of vibrational displacement: most of all we are interested in the slow changes in the system; for their consideration, however, it is necessary to know the fast processes.

In this section we use this approach to consider two problems about the motion of a particle in the homogeneous vibrating medium with the resistance of type of dry friction, and also a problem on the vibrational separation of the multi-component mixture in view of both determinate and casual factors; in the latter case the results of the above-cited work by Hainman and the author are stated [85]. Investigations, devoted to the penetration of vibration into different media are briefly considered in chapter 16.

10.2.4   *Motion of a Heavy Particle in the Medium with a Resistance of Dry Friction Type, Performing Circular Horizontal Oscillations. Pseudoresonance Effect*

Let us consider the motion of a heavy particle, placed into the medium which performs horizontal circular translational oscillations with the frequency ω and the radius of the trajectory r and exerts resistance to the particle similar to that of dry friction (Fig. 10.5). The corresponding problem was considered by Gortinsky, Ptushkina and the author [82]. Here we give the main results of the solution and discuss them from the position of vibrational mechanics.

We will designate the force of resistance to the relative shift of the particle in any horizontal direction by F_h, and in the vertical direction by F_v. The mass of the particle with the associated mass of the medium will be designated by m_1 and the mass of the medium in the volume equal to the volume of the particle by m_0, the ratio of the average densities of the particle to the medium will be designed by $\delta = \rho/\rho_0$. Let $\dot{x}, \dot{y}, \dot{z}$ be the projections on the axis of the rectangular coordinate system xyz of the relative velocity of the particle in the medium. Then the differential equations of motion of the particle relative to the medium can be written in the form

$$m_1\ddot{x} = m_0(\Delta - 1)r\omega^2 \cos\omega t - F_h \frac{\dot{x}}{\sqrt{\dot{x}^2 + \dot{y}^2 + \dot{z}^2}}$$

$$m_1\ddot{y} = m_0(\Delta - 1)r\omega^2 \sin\omega t - F_h \frac{\dot{y}}{\sqrt{\dot{x}^2 + \dot{y}^2 + \dot{z}^2}}$$

$$m_1\ddot{z} = m_0(\Delta - 1)g - F_v \frac{\dot{z}}{\sqrt{\dot{x}^2 + \dot{y}^2 + \dot{z}^2}}$$

$$\sqrt{\dot{x}^2 + \dot{y}^2 + \dot{z}^2} \neq 0. \tag{10.2.1}$$

Let the following relations be satisfied

$$\frac{F_v}{m_0 g} > |\Delta - 1| > \left[\left(\frac{m_0 r\omega^2}{F_h}\right)^2 + \left(\frac{m_0 g}{F_v}\right)^2\right]^{-1/2}. \tag{10.2.2}$$

They express the condition that on the one hand the weight of the particle in the medium does not exceed the force of resistance to its motion in the vertical direction, on the other hand that the force of inertia in the relative motion is sufficient to overcome the resistant force of the type of dry friction. Then, as is easy to see, equations (10.2.1) allow the exact solution

$$\dot{x} = R\omega\cos(\omega t + \beta) \quad \dot{y} = R\omega\sin(\omega t + \beta) \quad \dot{z} = \dot{Z}_0 \tag{10.2.3}$$

corresponding to the motion of the particle with respect to the medium on a spiral path. Then expressions are obtained for the radius of the spiral R and for the velocity of the vertical submersion of the particle $\dot{Z}_0$

$$R = r\sqrt{\left[\frac{m_0}{m_1}(\Delta - 1)\right]^2 - \left[\frac{F_h}{m_1 r\omega^2}\right]^2 (1 - \delta^2)} \tag{10.2.4}$$

$$(\delta = m_0(\Delta - 1)g/F_v)$$

$$\dot{Z}_0 = \frac{\delta}{\sqrt{1 - \delta^2}} R\omega, \qquad (10.2.5)$$

while the expression for the phase shift β is further on unessential. It is possible to prove without much difficulty that the obtained motion is stable [82].

An important result follows from formula (10.2.5): particles, more dense than the medium ($\Delta > 1$) are submerged, and particles less dense ($\Delta < 1$) float up. That is the *effect of the seeming transformation of dry friction into viscous friction*. If we compare equation (10.2.5) with the well known expression for the free fall velocity of a spherical particle in the viscous fluid at the small numbers of Reynolds

$$\dot{Z}^* = \frac{m_0(\Delta - 1)g}{3\pi\mu^* d} \qquad (10.2.6)$$

(d is the diameter of the particle, and μ^* is the coefficient of the fluid), we find the effective (seeming) viscosity coefficient of the medium:

$$\mu^* = \frac{F_v\sqrt{1 - \delta^2}}{3\pi d R\omega} = \frac{F_v}{3\pi d r\omega} \left\{ \left[\frac{m_0(\Delta - 1)}{m_1\sqrt{1 - \delta^2}} \right]^2 - \left[\frac{F_h}{m_1 r\omega^2} \right]^2 \right\}^{-1/2} \qquad (10.2.7)$$

The system under consideration is remarkable by the fact that the stationary motions are separated in it in a natural way: the motion of the particle in the horizontal plane is fast, and its vertical motion is slow. Besides, equations of fast motions (the first two equations (10.2.1)) at $\dot{Z} = $ const allow an exact periodic solution of the type

$$\dot{x}_* = R_1\omega\cos(\omega t + \beta_1) \quad \dot{y}_* = R_1\omega\sin(\omega t + \beta_1)$$

Substituting this solution into the third equation (10.2.1) we come to the equation of the slow vertical motion of the particle (the main equation of vibrational mechanics):

$$m_1\ddot{Z} = m_0(\Delta - 1)g + V(\dot{Z}) \qquad (10.2.8)$$

where

$$V(\dot{Z}) = -F_v\dot{Z}(R_1^2\omega^2 + \dot{Z}^2)^{-1/2} \qquad (10.2.9)$$

is the vibrational force for whose obtaining in this case no operation of averaging is wanted. As one can see this force is of the character of a nonlinear viscous resistance [$V(0) = 0$]. In the purely inertial approximation (see 3.2.8), when we can assume that $F_h \ll m_0(\Delta - 1)r\omega^2$ we obtain

$\beta_1 \approx -\pi/2$, $R_1 \approx r(\Delta - 1)m_0/m_1$ from the first two equations (10.2.1), and according to (10.2.9) we get

$$V(\dot{Z}) = -F_v \dot{Z} \left\{ \left[\frac{m_0}{m_1}(\Delta - 1)r\omega \right]^2 + \dot{Z}^2 \right\}^{-1/2} \tag{10.2.10}$$

To complete the investigation we have to express the radius of oscillations of the points of the medium r by the parameters of oscillations of the walls of the vessel that contains it. Let the vessel perform circular translational oscillations in the horizontal plane with a frequency ω and an amplitude A. We will make the simplest assumption that the motion of the medium can be considered as the motion over a rough bottom of the vessel of an absolutely solid plane body. This assumption is acceptable if the thickness of the layer is small as compared to the width of the vessel and not too large as compared to the sizes of the particles of the medium. The solution of the corresponding problem was obtained by Zhukovsky [599] who found that in the stationary regime the body performs stable circular translational oscillations with the frequency ω and with the radius of the trajectory, defined by the relationships

$$r = \begin{cases} A & \text{at} \quad A\omega^2 < fg \\ fg/\omega^2 & \text{at} \quad A\omega^2 > fg, \end{cases} \tag{10.2.11}$$

where f is the friction coefficient of sliding. According to (10.2.11), with the increase of the frequency ω up to a certain value $\omega_m = \sqrt{fg/A}$, the radius of the trajectory remains equal to the amplitude of oscillations of the vessel, and then decreases rather abruptly. As was marked before, due to sluggishness, the body "has no time" to follow the oscillations of the vessel. In view of expression (10.2.11), formula (10.2.5) for the average velocity of the sinking or floating of the particle will acquire the form

$$\frac{\dot{Z}_0}{\dfrac{m_0(\Delta - 1)g}{F_v}} = \begin{cases} A\omega \sqrt{\left[\dfrac{m_0(\Delta - 1)}{m_1\sqrt{1 - \delta^2}} \right]^2 - \left(\dfrac{F_h}{m_1 A\omega^2} \right)^2} & \text{at} \quad A\omega^2 < fg \\[4mm] \dfrac{fg}{\omega} \sqrt{\left[\dfrac{m_0(\Delta - 1)}{m_1\sqrt{1 - \delta^2}} \right]^2 - \left(\dfrac{F_h}{m_1 A\omega^2} \right)^2} & \text{at} \quad A\omega^2 > fg \end{cases} \tag{10.2.12}$$

The graph of the dependence $\dot{Z}_0 = \dot{Z}_0(\omega)$ plotted for the values $\rho = 2.65\text{g/cm}^3, \rho_0 = 1.35\text{g/cm}^3, m_0 = 0.1\text{g}, m_1 = 0.43\text{g}, F_v = 1.67 \cdot 10^{-2}\text{N}$, $F_h = 0.39 \cdot 10^{-2}\text{N}, f = 1$ and $A = 1.5\text{cm}$ is presented in Fig. 10.5,b; the points corresponding to the experimental data are obtained under the same conditions. The obtained dependence looks like a resonance curve, which in fact it is not: as is clear from what has been said, the peak character of the dependence is connected here with a peculiar game of forces of the type of dry friction and those of inertia. So this effect can be called the *pseudo-resonance effect* .

The physical explanation of this effect, following from relationship (10.2.12), consists in the following. Let the amplitude of oscillations of the vessel A be fixed. Then when the oscillation frequencies ω are small, the amplitude of the acceleration $A\omega^2$ is also small, and the mixture moves practically together with the vessel. As that takes place, the acceleration of the medium $A\omega^2$ increases with the growth of ω, as a result of which the intensity of the relative motion of particles which differ, say, in density increases, which in its turn leads to the increase of the tempo of separation. With any further increase of the frequency ω and the acceleration of oscillations of the vessel $A\omega^2$, the medium inside it "has no time" to follow the oscillations of the walls of the vessel and when $A\omega^2$ are large enough remains practically motionless in space; as a result the process of separation stops. The maximum of the speed of the process is near the value $\omega = \omega_m$, corresponding to the greatest acceleration $A\omega^2$, at which the medium is still moving together with the vessel.

The pseudo-resonance effect is used particularly in the machines for refinement of the grain from the mineral particles of a similar size. If such impurities happen to get into the grain before grinding, the bread, baked from such flour, gives a most unpleasant crackle on the teeth.

10.2.5   *Problem of 10.2.4 in the Case of Vertical Harmonic Oscillations of the Medium. The Effect of Floating up of a Heavy Particle in the Medium of Light Small Particles*

The equation of motion of the particle in the medium, performing vertical harmonic oscillations with the frequency ω and the amplitude A will be written as

$$m_1 \ddot{z} = m_0(\Delta - 1)(g + A\omega^2 \sin \omega t) + F(\dot{z}) \qquad (10.2.13)$$

where

$$F(\dot{z}) = \begin{cases} -F_+ & \text{at} \quad \dot{z} > 0 \\ F_- & \text{at} \quad \dot{z} < 0 \end{cases}$$

$$-F_+ < F(\dot{z}) < F_- \quad \text{at} < \dot{z} = 0 \quad (F_+ \geq F_- > 0) \qquad (10.2.14)$$

is the force of resistance to the relative motion of the particle in the medium. That force, generally speaking, is assumed to be greater when the particle moves downward, i.e. in the direction of the bottom of the vessel. This assumption corresponds to a number of experimental data; the difference $F_+ - F_-$ increases as the bottom of the vessel is approached, which, however, in equation (10.2.13) will be considered only parametrically, since the change of the coordinate z, and consequently F_+ and F_- for the oscillation period is insignificant. The symbols m_1, m_0, and Δ have the same meaning as in 10.2.4

On the basis of the exact solution of equation (10.2.13) this problem was considered in the work by Hainman and the author [84, 85]. This problem also corresponds to that, studied in 9.4.2 by method of direct separation of motions. In that case equation (10.2.13) is answered by the following values of the parameters in equation (9.4.25) of chapter 9:

$$m_2 = m_0(\Delta - 1), \; T = m_0(\Delta - 1)g, \; m_2/m_1 = q = m_0(\Delta - 1)/m_1. \quad (10.2.15)$$

Therefore according to (9.4.28) and (9.4.29) the main equation of vibrational mechanics (9.4.9) will look as

$$m_1 \ddot{Z} = m_0(\Delta - 1)g - \frac{1}{2}(F_+ - F_-) + V_1(\dot{Z}) \quad (10.2.16)$$

with

$$V(\dot{Z}) = V(0) + V_1(\dot{Z})$$

$$= \begin{cases} -F_+ & \text{at} \quad \dot{Z} \; > \; |q|A\omega \\ \frac{1}{\pi}[(F_+ + F_-)\arccos\dfrac{\dot{Z}}{qA\omega} + F_+\pi] & \text{at} \quad |\dot{Z}| \; < \; |q|A\omega \\ F_- & \text{at} \quad \dot{Z} \; < \; -|q|A\omega \end{cases}$$

$$V(0) = -\frac{1}{2}(F_+ - F_-), \quad V_1(0) = 0. \quad (10.2.17)$$

Equation (10.2.16) can be interpreted as the equation of motion of a heavy particle in some viscous fluid; the role of the weight of the particle in that fluid being played by the value

$$P^* = m_0(\Delta - 1)g - \frac{1}{2}(F_+ - F_-). \quad (10.2.18)$$

In other words, in the slow motion a certain positive value $\frac{1}{2}(F_+ - F_-)$, corresponding to the driving vibrational force $V(0)$, is subtracted from the "true" weight of the particle in the medium $m_0(\Delta - 1)g$. This force is directed toward the smaller resistance to the motion of the particle i.e. upward (also see 9.2). If $P^* > 0$, the particle sinks, and if $P^* < 0$, it floats up.

The obtained result makes it possible to explain the remarkable effect — *floating up of a big heavy particle in a layer of small light particles.* Indeed, despite the fact that for such a particle $\Delta > 1$ the value P^* may prove to be negative at the expense of the second summand in expression (10.2.18) — the additional buoyancy force $\frac{1}{2}(F_+ - F_-)$ appearing under vibration due to a greater resistance to the motion of the particle downward than to that upward. A particle may also float up in the medium of smaller particles of the same material since a big particle is more dense than the medium ($\Delta > 1$). What has been said is illustrated in Fig. 10.6, a, b where pictures seen by the observers **O** and **V** are compared. It goes without saying that what has been said is true if the oscillation intensity $A\omega^2$ is sufficient to provide, at the given difference of the densities of the particle and the medium $\rho_0 |\Delta - 1|$, a relative motion of the particle in the medium (at $\Delta = 1$ according to (10.2.13) such motion does not take place at any $A\omega^2$).

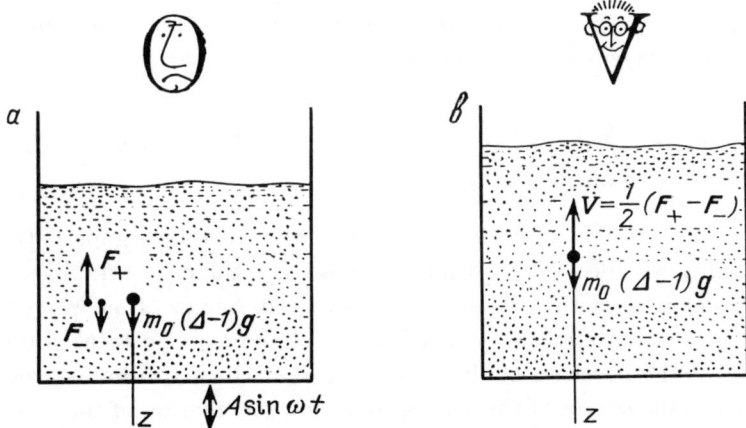

Figure 10.6. Under certain conditions a heavy particle (say, a steel ball) floats up in a less dense medium (say, in the sand), placed into a vertically-vibrating vessel.

In accordance with relations (10.2.15), expression (9.4.31) of chapter 9 for the average velocity of the stationary motion of the particle in the medium in the case under consideration will take the form

$$\dot{Z}_* = \frac{m_0(\Delta - 1)}{m_1} A\omega \cos \frac{\pi[F_+ - m_0(\Delta - 1)g]}{F_+ + F_-}$$

This formula is valid at

$$-F_- < m_0(\Delta - 1)g < F_+$$

in the opposite case an accelerated motion of the particle takes place. It should be noted that according to this equation at $\Delta = 1$ the velocity is $\dot{Z}_* < 0$, i.e. the particle floats up if

$$\pi[F_+ - m_0(\Delta - 1)g]/(F_+ + F_-) > \pi/2$$

that is if

$$P^* = m_0(\Delta - 1)g - \frac{1}{2}(F_+ - F_-) < 0$$

this corresponds to the conclusion made before by means of a direct consideration of equation (10.2.16)

An interesting experimental investigation of the effect of floating up of a heavy ball in a layer of sand was performed by Levin [327].

10.2.6   Kinetics of Vibrational Separation of Multi-Component Mixture (the Continual Description)

Let us consider the n-component granular mixture whose gaps between the particles may be filled up with fluid or gas. Let there be established in the medium, under the action of vibration, excited in one or other way ("the fast process"), a certain stationary distribution of the average total volume concentration of the particles of the medium $c < 1$ which will be assumed to be unchanged in the course of the process of a slow segregation of the components of the medium. This concentration is considered high enough so that particles are located rather densely and "compete for the volume they occupy": the change of positions of the particles taking a certain volume, proceeds as an exchange for other particles with the same total volume. This assumption naturally imposes a certain restriction on the size distribution of the mixtures under consideration. Vibration is assumed to be rather intensive so that there is an effect of fluidization of the mixture.

Under the indicated assumption Hainman and the author [85] obtained the following nonlinear partial differential equations, describing the one-dimensional slow process of changing the volume concentrations $c_1, \ldots, c_n$ of the components of the granular mixture, i.e. the corresponding equations of vibrational mechanics:

$$\frac{\partial c_i}{\partial t} = \frac{\partial}{\partial t}\left[\frac{\partial c_i}{\partial z}\sum_{j=1}^{n}a_{ij}c_j\right] - \frac{\partial}{\partial z}\left[c_i\left(\sum_{j=1}^{n}b_{ij}c_j + \sum_{j=1}^{n}a_{ij}\frac{\partial c_j}{\partial z}\right)\right]$$
$$(i = 1,\ldots,n). \tag{10.2.19}$$

Here z is the space coordinate, and a_{ij} and b_{ij} are the functions of z, whose determination is based on the study of the "fast" process or they can be determined experimentally. They depend on the properties and characteristics of the particles and on the parameters of vibration in the vicinity of the given point. Then the following relation takes place:

$$a_{ij} = a_{ji}, \qquad b_{ij} = -b_{ji} \tag{10.2.20}$$

The function a_{ij} characterizes the intensity of the diffusive motion, and the function b_{ij} characterizes the velocity of the ordered motions of particles with a_{ij} in its meaning being proportional to the sum, and b_{ij} — to the difference of the probabilities of the exchange of particles of the i-th component for the equal in volume number of the particles of the j-th component for a certain characteristic period of time τ near the point with the coordinate z (for more detail see [85]).

The sum

$$D_i = \sum_{j=1}^{n}a_{ij}c_j \tag{10.2.21}$$

in equations (10.2.21) plays the role of the diffusion coefficient, answering the particles of the i-th component, and the sum

$$\dot{Z}_i = \sum_{j=1}^{n}b_{ij}c_j + \sum_{j=1}^{n}a_{ij}\frac{\partial c_j}{\partial z} \tag{10.2.22}$$

plays the role of the velocity of the motion of particles of the i-th component under the corresponding conditions. With the use of designations (10.2.21) and (10.2.22) equations (10.2.19) can be presented as

$$\frac{\partial c_i}{\partial t} = \frac{\partial}{\partial z}\left(D_i\frac{\partial c_i}{\partial z}\right) - \frac{\partial}{\partial z}(c_i\dot{Z}_i) \quad (i = 1,\ldots,n) \tag{10.2.23}$$

Adding up the right and left parts of equations (10.2.19) and (10.2.23) by virtue of equalities (10.2.20), we obtain

$$\partial(c_1 + \ldots + c_n)/\partial t = \partial c/\partial t = 0 \tag{10.2.24}$$

which agrees with the assumption about the stationary distribution of the concentration of the solid phase $c = c(z)$. Thus system (10.2.19) contains $n - 1$ independent equation for determining the unknown concentrations. The closing relation of this system is the equality

$$c_1 + \ldots + c_n = c(z) < 1 \qquad (10.2.25)$$

in which the distribution of the total phase concentration $c(z)$ must be determined from the solution of a separate problem about the action of vibration on the system under consideration. The volume concentration of the medium filling the gaps between the particles $c_{n+1}(z)$ is connected with $c(z)$ by an obvious equality $c(z) + c_{n+1}(z) = 1$.

Nonlinear equations (10.2.19) and (10.2.23) differ greatly from the Focker-Plank-Kolmogorov equations, used in the above-cited work by Nepomnyashchy [411] and in the publications by his followers. It is not difficult to show that only in the assumptions about an essential predominance of the particles of one component (let it be the n-th component, we will call it the *bed-component of the mixture*) i.e. at

$$c_n \gg c_i \quad (i = 1, \ldots, n - 1); \qquad c_n \approx c(z) \qquad (10.2.26)$$

systems (10.2.19) and (10.2.23) decompose into $n - 1$ disconnected equations of the type of Focker-Plank-Kolmogorov's for each of the separated components. Indeed, in the indicated assumption the particles of the components are displaced at the expense of exchanging places with the particles of the bed. Therefore in equations (10.2.19) and (10.2.23) the terms, containing the products of the concentrations of c_i and c_j and the derivative of c_j with respect to the spatial coordinate $(i, j = 1, \ldots, n - 1)$. Then the first $n - 1$ indicated equations will acquire the form

$$\frac{\partial c_i}{\partial t} = \frac{\partial}{\partial z}\left(D_i \frac{\partial c_i}{\partial z}\right) - \frac{\partial}{\partial z}(c_i \dot{Z}_i)$$

$$D_i = a_{in} c_n \quad \dot{Z}_i = b_{in} c_n + a_{in}\frac{\partial c_n}{\partial z} \qquad (i = 1, \ldots, n - 1) \qquad (10.2.27)$$

answering the Focker-Plank-Kolmogorov equations. As for the latter equation, it becomes an identity due to supposition (10.2.26) and equalities (10.2.20).

The equations given here are generalized for the case of a spatial problem when apart from that there are also slow streams of the totality of all fractions as a single medium. These streams are characterized by a certain velocity V_e, which may depend on the spatial coordinates; we will call the velocity V_e the *velocity of the transportation motion of the medium.* Under the assumption that the functions of the coordinates a_{ij} are still of the scalar character, and b_{ij} are replaced by the vector $\mathbf{b}_{ij}$, the corresponding differential equations have the form

$$\frac{\partial c_i}{\partial t} = -\mathrm{div}(\mathbf{V}_e + \mathbf{V}_{ri})c_i + \mathrm{div}(D_i \mathbf{grad}\, c_i) \tag{10.2.28}$$

where

$$\mathbf{V}_{ri} = \sum_{j=1}^{n}(\mathbf{b}_{ij}c_j + a_{ij}\mathbf{grad}\, c_j) \tag{10.2.29}$$

is according to its meaning the velocity of particles of the i-th component with respect to the medium, and the diffusion coefficients D_i are still defined by the expressions (10.2.21). As it should be, adding up equations (10.2.28) we come to the equation of continuity for the medium as a whole (we still believe that $\partial c/\partial t = 0$):

$$\mathrm{div}(\mathbf{V}_e c) = 0. \tag{10.2.30}$$

Let us consider two most simple examples of the use of the indicated equations.

Example 10.2.1.*Stationary distribution of particles of a two-component mixture in a closed vibrating vessel.* When solving this problem, we will make the simplest assumption: we will consider a one-dimensional case in the absence of any slow flows of the medium on the whole and will consider that the coefficients a_{12} and $b_{1,2}$ do not depend on the spatial coordinate z. We also assume that the total volume concentration $c = c_1 + c_2$ does not depend on that coordinate either. In this case, using the relative concentrations c_1/c and c_2/c and retaining for them the same designations, we will have $c_1 + c_2 = 1$.

In the case under consideration equations (10.2.19) have the form

$$\frac{d}{dz}\left(\frac{dc_1}{dz}a_{12}c_2\right) - \frac{d}{dz}[c_1(a_{12}\frac{dc_2}{dz} + b_{12}c_2)] = 0$$

$$\frac{d}{dz}\left(\frac{dc_2}{dz}a_{21}c_1\right) - \frac{d}{dz}[c_2(a_{21}\frac{dc_1}{dz} + b_{21}c_1)] = 0 \tag{10.2.31}$$

In view of the relations $a_{12} = a_{21}$, $b_{12} = -b_{21}$, and $c_1 = 1 - c_2$ this system is reduced to the equation

$$\frac{d}{dz}[a_{12}\frac{dc_1}{dz} - c_1 b_{12}(1 - c_1)] = 0 \tag{10.2.32}$$

On integrating the last equation and in view of the fact that the particle flows through the lower and upper bases of the vessel are equal to zero, we obtain

$$c_1 = \left\{1 + \frac{1 - \exp[b_{12}(h - h_1)/a_{12}]}{\exp(-b_{12}h_1/a_{12}) - 1}\exp\left(-\frac{b_{12}}{a_{12}}z\right)\right\}^{-1} \tag{10.2.33}$$

where

$$h_1 = \int_0^h c_1\, dz \qquad (10.2.34)$$

designates the height of the layer corresponding to the total volume of particles of the first component and h is the total height of the layer of particles of both components.

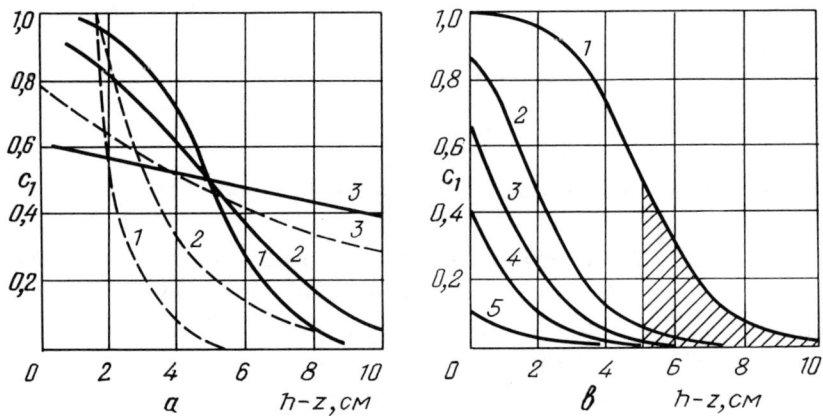

Figure 10.7. Stationary distribution of "heavy" particles of the two-component granular mixture in a closed vibrating vessel .

Fig. 10.7,a shows the calculated in accordance with (10.2.33) dependences of the concentration of particles of the first component c_1 on the height $h - z$ over the level of the bottom of the vessel, with the total thickness of the layer $h = 10 cm$ and the values of the relation $b_{12}/a_{12} = 10.5$ and $1 cm^{-1}$ (the respective solid lines *1, 2, 3*). Thus the curves answer the assumption that $b_{12} > 0$. According to the above-mentioned meaning of the coefficient b_{12}, this may mean that, for example, the first component consists of the particles of approximately the same size as the second component, but they are heavier (the axis z being assumed to be directed vertically downward). The total content of particles of both components was assumed to be the same $h_1 = 0.5h$. For comparison the same figure gives the results of the calculations of the concentration without taking into account "the competition for the volume taken" (the respective interrupted lines). As one can see, the results in the conditions of a satisfactory

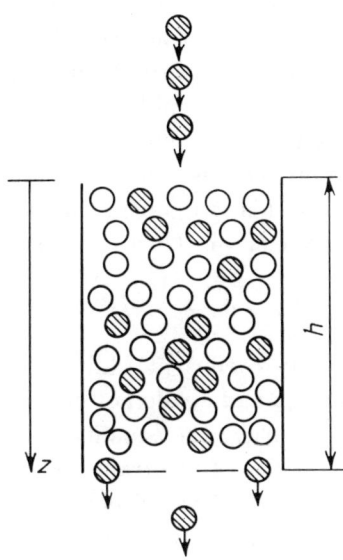

Figure 10.8. Scheme of a reactor .

separation (curves *1* and *2*) appear to be quite different and approach each other only under the conditions of a relatively weak separation.

Fig. 10.7,*b* shows the calculated by the same formula dependences of the distribution of the concentration of heavy particles according to the height, their concentration in the mixture being 50, 20, 10, 5 and 1% (the respective curves are *1–5*). Sometimes the separability of particles is proposed to be characterized by the value of the area, limited by the ordinate $c_1 = 0.5$, by the distribution curve and by the x-axis (the shaded area in Fig. 10.7,*b*; see e.g. the work by Vinogradov [568]). It is easy to show that in the conditions of the example under consideration, this area is equal to $b_{12} \ln 2/a_{12}$, i.e. it characterizes the relationship of the coefficients, defining the velocities of both the ordered and the chaotic motions.

Example 10.2.2.*Passing of homogeneous particles through a vibrating layer of granular medium.* As a second simplest example we will consider a one-dimensional case of the passing of homogeneous particles through a layer of other particles. Let the operational volume of the reactor (Fig. 10.8) be filled with the particles of the first component, their quantity answering the thickness of the layer h_1 (see equation (10.2.34)). Particles of this component do not get into the operational volume and do not spill through the sieve which is formed at the bottom of the vessel. Particles of the second component (they are shaded in the figure) get to the upper part of the volume, pass through the layer of particles of the first component and spill through the sieve. The stream of particles of the second component when there is no spilling is such that it raises the level of particles in the operational volume by q cm/s. We assume that the stream of particles of this component through the sieve is proportional to their concentration near the sieve. Then in the stationary regime we have

$$c_2|_{z=h} = q/\beta \qquad (10.2.35)$$

where β is the proportionality coefficient. Like in the first example, we will assume that the coefficients a_{12} and b_{12} do not depend on the coordinate z and the concentrations c_1 and c_2 satisfy the equality $c_1 + c_2 = 1$.

It is obvious that the velocity of the transportation motion V_e in the stationary regime that interests us is directed down the axis z and is equal to q. As that takes place, the equation for the concentration of the particles of the first component according to (10.2.28) can be written as

$$\frac{d}{dz}\left[a_{12}\frac{dc_1}{dz} - c_1[q + b_{12}(1 - c_1)]\right] = 0 \qquad (10.2.36)$$

The boundary conditions consist in the absence of any stream of particles of that component across the sections $z = 0$ and $z = h$; besides, according to (10.2.35)

$$c_1|_{z=h} = 1 - c_2|_{z=h} = 1 - q/\beta$$

Integrating equation (10.2.36) in view of the indicated conditions and designations (10.2.34), we obtain

$$c_1 = \left(1 + \frac{q}{b_{12}}\right)\left\{1 + \left[1 - \left[\frac{1 + q/b_{12}}{1 - q/\beta}\left(\exp\left(-\frac{b_{12}}{a_{12}}\right) - 1\right) + 1\right]\right.\right.$$
$$\left.\left. \times \exp\left(-\frac{b_{12}h_1}{q + b_{12}}\right)\right]\left[\exp\left(-\frac{b_{12}h_1}{a_{12}}\right) - 1\right]^{-1}\exp\frac{(q_1 + b_{12})z}{a_{12}}\right\}^{-1} \qquad (10.2.37)$$

The analysis of this expression shows that the restriction to the value of the stream q, imposed by the condition of the existence of the stationary regime, is connected not only with the limited sieve capacity (the trivial condition being $q \leq \beta$) but also with the capacity of the layer of particles of the first component.

This restriction appears if the particles of the indicated component "sink" ($b_{12} > 0$); as that takes place, the maximal value of the stream q depends essentially on the values of the coefficients, characterizing both the ordered (b_{12}) and the chaotic (a_{12}) motions of particles and also on the thickness of the layer h.

It is impossible to solve such problems by considering the linear models because the process is determined by the maximal possible concentrations of the particles; in such cases the linear model cannot be used. In particular, the successive application of the linear model to the system under consideration leads to the conclusion that the capacity of this system is unlimited at any values of the coefficients a_{12}, b_{12}, β and h.

10.2.7   Separation of Particles on the Vibrating Surfaces

The method of dry separation on the vibrating surfaces [16, 17, 438, 439] is based on the use of the circumstance that the particles of the granular material, possessing different friction coefficients with respect to the surface, different shapes, different elastic properties and under certain conditions — different densities and different sizes, move on the rough vibrating surface in different trajectories. Two versions of such surfaces (decks) — concave and plane — are indicated in Fig. 10.9. The translational vibration, as a rule rectilinear

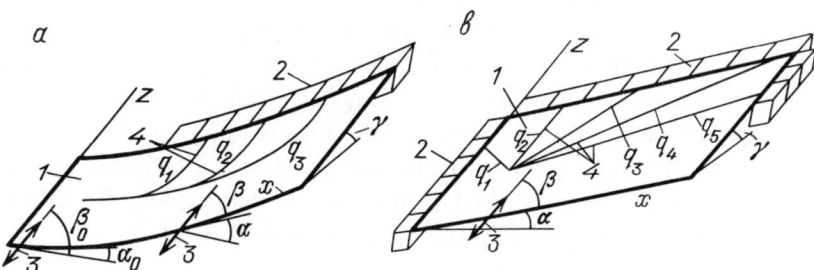

Figure 10.9. Separation of particles of the granular material on vibrating surfaces. .

and harmonic, is in some way or other assigned to the frame on which the decks are fixed. The designs of the corresponding machines — the *vibrational separators* have special provisions to regulate the longitudinal and transversal angles of the tilt of the frame to the horizon α_0 and γ; the latter, as a rule, not exceeding 10°. The material being separated is transferred to the zone near the lower edge of the deck.

In case of the concave deck (Fig. 10.9,a), since the angle of the transversal tilt γ is small, the particles of the material move mostly along the deck. As they move forward-upward, the angle of the tangent to the surface of the deck α increases, therefore the longitudinal motion gets slower and the transversal motion becomes predominant, due to which the particle begins to slip down into one of the receiving meshes on the edge of the deck. An approximate view of the trajectories of the particles with different parameters is shown in Fig. 10.9,a. In the case of a flat deck all the points on its surface are equivalent. As a result, the vector of the average velocity of the stationary motion of the particles at every point of the deck has the same value and direction and so the trajectories of particles are direct lines, different for the particles with different parameters (Fig. 10.9,b).

The investigation shows (see below) that if we consider the particle as a material point without considering the air resistance to its motion, then the form of its trajectory on the deck does not depend on the sizes or the mass of the particle and is defined only by the coefficients of friction against the surface, by the restitution coefficient R and by the coefficient of instant friction at the impact λ, i.e. in the long run by the properties of the material of the particle. If the particle is considered to be a solid body, then one discovers a dependence of the character and trajectory of its motion on its shape — the motion of the sphere-shaped particles is different from that of the flat particles: the first usually roll down the plane, while the latter move approximately in the way the material point does, especially in the regimes without tossing. If the air resistance is taken into consideration (which is quite necessary for the particles whose size is sufficiently small, see [84]), then the trajectories begin to depend quite essentially on the mass and sizes of the particle. The type of that dependence can be regulated in the desired direction provided the deck is made porous and the air is pressed through it; as that takes place, big heavy particles can be made to move up the plane with the small light ones crawling down [114, 215, 276, 277, 356].

Vibrational separators are successfully applied to separate diamond grains according to their shape, to separate the spherical particles which are used when creating cermet filters, to classify the fine grain or fine powders, to separate fine mica from the pegmatite ores. The vibrational separators which supply air through the deck are applied when processing the grain [215, 356]. The advantages of this method of separation are the following: it does not need using sieves or water, it is highly sensitive to some parameters of the particles, which cannot be distinguished by any other method, this method provides a possibility of easy regulation and of readjusting the process. The drawbacks of this method are its comparatively low productivity, caused by the superficial character of the process, and the necessity to stabilize the frequency and amplitude of vibration, the moisture content of the material, etc.; though when dividing into a small number of fractions, the last drawback is less essential. To overcome the former drawback, separators with a greater number of decks are used [16, 564, vol. 4].

Vibrational separators work at the amplitudes of acceleration $6g - 10g$; to separate larger particles, the larger amplitudes and smaller frequencies of vibration prove to be more efficient, to separate smaller particles — vice versa. The separation of particles larger than 0.3–0.5 mm is realized better on a surface with a great coefficient of friction, for instance on that, covered with polishing sandpaper or with a layer of rubber. Fine powders are better separated on smooth metallic surfaces.

The initiator of creating vibrational separators is Pliss, the author of a number of important inventions and investigations in this field. The theory of dry vibro-separation is the subject-matter of his publications [?, 17, 67, 84, 203, 588, 276, 277, 438, 439] and of a number of other investigations (also see [564, vol. 4]). We will briefly dwell here on the approach to this theory, based on the consideration of the trajectories of the slow motions of particles on vibrating surfaces, i.e. the approach, characteristic of vibrational mechanics.

We will assume that from the solution of the corresponding problem on vibrotransportation (see 10.1) we know the velocity of the stationary process of vibrotransportation of a particle over the plane with a longitudinal and transversal tilt towards the horizon, as it takes place in the case of a flat deck (Fig. 10.9,b). The projections of this velocity onto the longitudinal axis X and the transversal axis Z will be designated by us by $\dot{X}$ and $\dot{Z}$ respectively. In the case of a translational vibration of the flat deck these projections are constant for all the points of the deck. In the case of a concave deck the indicated projections can be considered parametrically dependent on the coordinates X and Z of the point of the deck since the radius of curvature of the deck is considerably large as compared to the sizes of the trajectory of vibration of the points of the deck and to the displacement of the particle over the deck for the period of its oscillations. In other words, the motion of the particle near any point of the concave deck can be regarded as the motion over the corresponding tangent plane.

Then the differential equation of the trajectories of slow motions of the particle over the deck, i.e. the corresponding equation of vibrational mechanics will look as

$$\frac{dZ}{dX} = \frac{\dot{Z}}{\dot{X}} \tag{10.2.38}$$

As an example let us consider the case when the deck performs rectilinear translational harmonic oscillations with the frequency ω and the amplitude A, whose trajectories make a certain angle β with the axis X (Fig. 10.9). Let the air resistance be not taken into consideration, and let the overload parameter (also see formulas. (9.4.42) and (9.4.46) of chapter 9)

$$w = A_1 \omega^2 \sin \beta / g \cos \alpha \tag{10.2.39}$$

satisfy the relationship

$$w > 3.5(1 + R^2)/(1 + R)^2 \tag{10.2.40}$$

where α is the angle of the inclination of the surface towards the horizon, g is the free fall acceleration, and R is the restitution coefficient at the impact of the particle against the plane. Then motion takes place with an intensive tossing of the particle, in which $\dot{X}$ and $\dot{Z}$ are defined by the approximate formulas [84, 564, vol.4]

$$\dot{X} = \frac{\pi p'(w, R)g}{\omega} \left(\frac{1 - R}{1 + R} \cot \beta \cos \alpha - \frac{2 - \lambda}{\lambda} \sin \alpha \right)$$
$$\dot{Z} = \frac{\pi p'(w, R)g}{\omega} \left(\frac{1 - R}{1 + R} - \frac{2 - \lambda}{\lambda} \sin \gamma \cos \alpha \right) \tag{10.2.41}$$

analogous to formulas (4.43) chapter 9 where γ is the angle of the transverse inclination of the deck, λ is the instant friction coefficient at the impact, and $p'(w, R)$ is the function of the parameters w and R which is in this case unessential.

In view of the expressions (10.2.41), equation (10.2.38) takes the form

$$\frac{dZ}{dX} = \frac{(1 - q) \sin \gamma}{q \cot \beta - \tan \alpha} \tag{10.2.42}$$

where

$$q == \frac{\lambda(1 - R)}{(2 - \lambda)(1 + R)} \tag{10.2.43}$$

is the value which plays the role of the generalized parameter according to which the separation of particles takes place in the regime under consideration (*vibroseparation parameter*). In order to find by means of equation (10.2.42) the trajectories of the particles at different values of q it is now sufficient to preset the angles α and β. In the case of a flat deck the angles α and β are constant and according to (10.2.42) the trajectories of the particles are direct lines (Fig. 10.9,*b*). In the case of a concave deck we have $\alpha = \alpha(X)$ and $\beta = \beta(X)$. Knowing these dependences and being given the initial conditions $X = X_0$ and $Z = Z_0$, it is not difficult by means of equation (10.2.42) to construct the trajectories of the particle for different values of the parameter q.

Fig. 10.9,*a* shows schematically a series of such trajectories, with $q_1 < q_2 < \ldots < q_n$. As one can see, when the particle approaches a section with a certain definite inclination α, depending on q, the trajectory of the particle bends abruptly in the direction of the inclination of the deck. As was already mentioned, this is connected with the decrease of the longitudinal projection of the velocity $\dot{X}$ when approaching the limiting angle α_* at which the particle can move upwards along the tilted plane (see 9.4.4).

When enriching the mineral resources, the method of separating particles in the thin layers of the fluid on the vibrating surfaces is widely used in the food industry and in some other industries. It is used in the concentrators, in sluice boxes, in some centrifugal devices. It is characteristic of this method to combine the separation of particles in the vibrating volume and on a rough vibrating surface. One can read about the theory and practice of the use of this method in the monograph by Kisevalter [268] and also in the book by Goldina and Karamsin [203], in which the results of the fundamental investigations in this field, made by Goldin, are set forth very thoroughly. Among the recent essential investigations, the work by Krasnov [294] should be mentioned.

10.2.8   On Separating the Particles in the Vibrational and Wave Fields, Created in the Rarefied Suspensions

The possibility of separating particles in the fast oscillating force fields was briefly mentioned in 5.3; also see chapter 19. The theoretic and experimental investigations, made by Ganiev, Ukrainsky and their followers exposed new possibilities of separating heterogeneous particles which form rarefied suspensions in fluids and in the gaseous media by imparting the oscillatory or wave motions to them. A detailed account of investigations made in this field and the adequate reference can be found in [190, 191, 192].

10.3   Vibrational Sinking and Intrusion, Vibrational Cutting

An important field of using vibrational technology is the sinking of piles, sheet piling and shells and also geological drilling; in many cases the application of vibration makes it possible to abruptly reduce the expenditure of time and means on conducting these operations.

By *vibrational sinking* one implies the penetration (vertically downward as a rule) of a solid body into the resisting medium under the action of the constant force and of forces with alternating signs. By *vibrational intrusion* we will understand the penetration of a solid body into the resisting medium with the given average velocity.

The dynamic scheme of one of the simplest vibro-sinkers of piles is shown in Fig. 10.10. The element *1* which is being sunk (for brevity sake we will call it a *pile*) is connected rigidly with exciter *2*, generating the harmonic exciting force $\Phi_0 \sin \omega t$. By means of soft springs the vibro-exciter is connected with

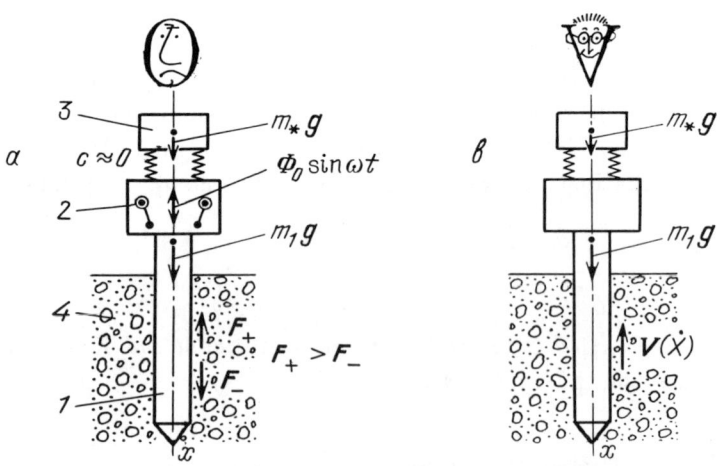

Figure 10.10. Vibrational sinker [473, 550] .

additional load *3*, exerting upon the system a purely static effect by its weight $m_* g$ (the weight of the exciter and that of the pile being designated by $m_1 g$).

To explain the main regularities of the work of the sinker we will make the

simplest supposition relating to the resistance to the sinking of the pile into the
ground 4: we will consider that when the pile moves downward, the resistance
is equal to $-F_+$ and when it moves upward it is equal to F_-, with $F_+ > F_-$
since F_- is conditioned by forces of resistance alone, distributed over the lateral
surface, while F_+ considers also the forces of resistance, acting on the bit-end
of the pile.

The differential equation of motion of the pile under the assumption made
has the form

$$m_1 \ddot{x} = (m_1 + m_*)g + \Phi_0 \sin \omega t + F(\dot{x}) \qquad (10.3.1)$$

where

$$F(\dot{x}) = \begin{cases} -F_+ & \text{at} \quad \dot{x} > 0 \\ F_- & \text{at} \quad \dot{x} < 0 \end{cases}$$

$$-F_+ < F(\dot{x}) < F_- \quad \text{at} \quad \dot{x} = 0. \qquad (10.3.2)$$

If we assume that

$$T = (m_1 + m_*)g, \quad m_2 A \omega^2 = \Phi_0 \quad \frac{m_2}{m_1} = \frac{\Phi_0}{m_1 A \omega^2} = q, \qquad (10.3.3)$$

then equation (10.3.1) coincides with equation (4.25), considered in 9.4.2. Therefore we can write down the corresponding equation of vibrational mechanics
(see equation (4.9) and formulas (4.28) and (4.29) of chapter 9):

$$m_1 \ddot{X} = (m_1 + m_*)g + V(0) + V_1(\dot{X}) \qquad (10.3.4)$$

where

$$V(\dot{X}) = V(0) + V_1(\dot{X})$$

$$= \begin{cases} -F_+ & \text{at} \quad \dot{X} \geq \Phi_0/m_1\omega \\ \dfrac{1}{\pi}[(F_+ + F_-)\arccos \dfrac{\dot{X} m_1 \omega}{\Phi_0} - F_+\pi] & \text{at} \quad |\dot{X}| \leq \Phi_0/m_1\omega \\ F_- & \text{at} \quad \dot{X} \leq -\Phi_0/m_1\omega \end{cases}$$

$$V(0) = -\frac{1}{2}(F_- - F_+) < 0 \quad V_1(0) = 0 \qquad (10.3.5)$$

Equation (10.3.4) and expressions (10.3.5) are valid at the highly-intensive
vibration (see p. 248). We must also pay attention to the coincidence of equations (10.3.4) and (10.2.13) — the latter describing the corresponding motion of
the particle. They coincide with an accuracy to the values of the coefficients.
This coincidence is quite natural because under the assumption made, both
processes are similar.

When the following conditions are satisfied

$$(m_1 + m_*)g < F_+, \tag{10.3.6}$$

the pile will not sink in the absence of vibration, and in case of vibration, the velocity of the stationary process of vibro-sinking $\dot{X} = \dot{X}_* = \text{const}$ is determined by the expression

$$\dot{X}_* = \frac{\Phi_0}{m_1 \omega} \cos \frac{\pi[F_+ - (m_1 + m_*)g]}{F_+ + F_-} \tag{10.3.7}$$

which is analogous to expression (10.2.19) for the corresponding velocity of the particle. The pile will be sunk if $\dot{X}_* > 0$. Hence from (10.3.7) we obtain the condition of vibro-sinking

$$(m_1 + m_*)g > \frac{1}{2}(F_+ + F_-) = -V(0) \tag{10.3.8}$$

which can also be easily got from equation (10.3.4) directly.

Note that in this case the process of vibrational displacement is realized in accordance with the scheme of Fig. 9.2,*a*,*I*,*1* (force-asymmetry).

It should be also noted that the vibrational force $V(\dot{X})$ actually depends on the depth of the sinking, that is on the coordinate x since the forces F_+ and F_- are, generally speaking, increasing with the depth of sinking. However, since the displacement of the pile during the period is small as compared to x, the indicated dependence can be taken into consideration in the final formulas (10.3.5) – (10.3.8) only parametrically.

Formulas (3.6) – (3.8) show the positive value of the additional load $m_* g$. It also follows from those formulas that the vibrational effect makes it possible to reduce the static force, necessary to sink the pile, from the value F_+ to that of $1/2(F_+ - F_-)$. That makes one of the main advantages of the vibrational method of sinking. To overcome the force of resistance to the sinking F_+ just by the weight of the system or by any other static pressing-in would have been most difficult and hardly expedient at all. Another advantage results from the fact that, as was discovered experimentally, the forces of resistance F_+ and F_- during the sinking into the water-saturated ground may decrease quite considerably due to the vibrational effect. The physical mechanism of that phenomenon has not yet been studied well enough. The most obvious proof of that reduction of the resistant forces is the consumption of energy. It is easy to see that should the forces F_+ and F_- remain unchanged, the consumption of energy under the vibrational sinking would be at least not lower than that under the static pressing-in since under vibrational sinking the pile during the cycle moves as a rule not only downward, but upward as well.

We have given here one of the simplest though quite expedient schemes of a vibro-sinker and have roughly considered only the simplest model of the process (the *purely plastic model of vibrational sinking*). Many other schemes have also been proposed, studied theoretically and realized quite successfully, for instance those with a simultaneous use of both longitudinal and rotational vibrational effects, those combining vibration and impact, etc. The idea of vibrational intrusion is used not only in the sinkers, but also in the machines, used for laying the underground communications — in the so called piercing aggregates and vibrational moles, and also in the machines for caltivating, loosening and cutting the ground.

The vibrational and percussive-vibrational methods of sinking were developed in Russia mostly due to the publications by Barkan, Savinov, Tseitlin and their colleagues and followers. The first theoretic investigations were made by Neimark [408] and by the author [70, 84]. The more detailed data about this trend of using vibration and the references can be found in [46, 84, 454, 473, 550, 564, vol. 4]. Among the recent publications, investigations [566, 519, 567] should be mentioned.

Here we will dwell only upon the use of the relationships we have obtained in conformity with the problem on vibrational intrusion (Fig. 10.11). Under the former assumptions the main equation (10.3.4) for the scheme, shown in this picture, will have the form

$$m_1 \dot{X} = T + V(\dot{X}). \tag{10.3.9}$$

Hence it follows that the force T necessary for providing the intrusion of the body into the medium with a certain constant velocity $\dot{X}_*$ is expressed by the formula

$$T = -V(\dot{X}_*) \tag{10.3.10}$$

where the vibrational force is determined, as before, in accordance with (10.3.5). Like in the case of vibrational sinking, $V(\dot{X})$ in fact does not remain constant in the process of intrusion, increasing, as a rule, with the increase of x, however in this case too, this dependence can be taken account of in the formula for $V(\dot{X})$ only parametrically.

The graph of the function $T = -V(\dot{X})$ is given schematically in Fig. 10.11,*b*. The graph shows that to provide the process of intrusion under a highly-intensive vibrational effect $[\Phi_0 \gg \sup(F_+, F_-)]$ it is necessary to have a pressing-in force

$$T > T(0) = \frac{1}{2}(F_+ - F_-) \tag{10.3.11}$$

while at the static pressing-in this force must satisfy the condition $T > F_+$. From examining the graph it also follows that the intrusion with the average velocity $\dot{X}_* > \Phi_0/m_1\omega$ is inexpedient since in this case it must be again $T \geq F_+$.

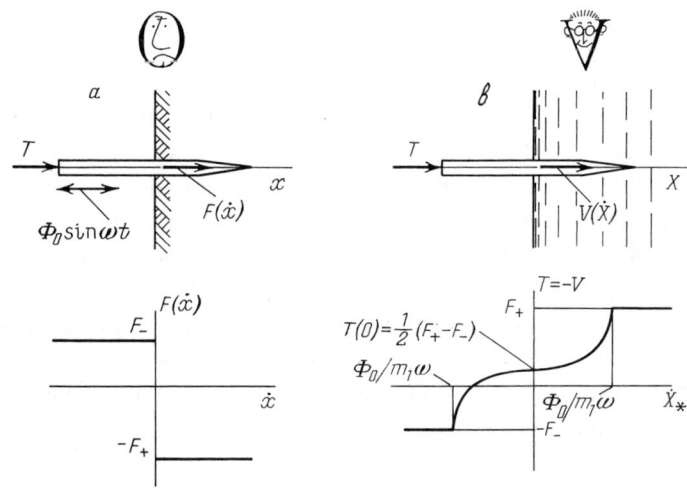

Figure 10.11. Vibrational intrusion into the medeum whose resistance is of dry friction type.

The stated results with small modifications and additions are transferred to the simplest model of the process of *vibrational cutting*; recently this process has become practised on a large scale (see, say, monographs by Podurayev [440] and Kumabe [302]), while its theory has not yet been developed well enough.

The scheme of the process is shown in Fig. 10.12. In the process of ordinary cutting (Fig. 10.12,a) the force of cutting $T = F_+$ is applied to the cutter; when intensive oscillations of high frequency are imparted to the cutter (Fig. 10.12,b), this force is somewhat greater than the value $\frac{1}{2}(F_+ - F_-)$. The oscillation frequency in different mechanisms makes from 50 Hz to tens of Hz. The oscillation frequency in different mechanisms makes from 50 Hz to tens of

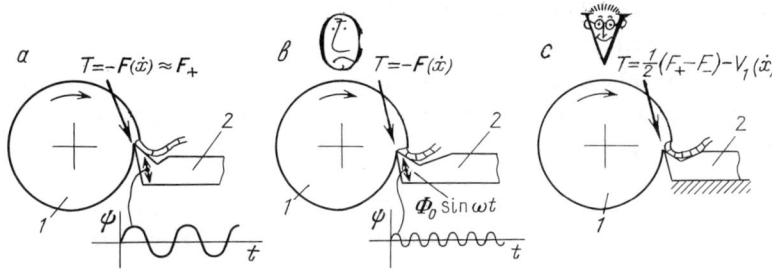

Figure 10.12. Ordinary and vibrational cutting:I — the part that is being processed; 2 — the cutting instrument.

kHz, the mechanical, electromagnetic, electrohydraulic, hydromechanical, and also electro- and magnetostrictional devices being used as generators. For the efficiency of the vibrational effect it is necessary that, like above, the speed of the cutting $\dot{X}$ should not exceed that of the vibration of the cutter $A\omega$ (in the above-considered model — the value $\Phi_0/m_1\omega$):

$$\dot{X} < A\omega = \Phi_0/m_1\omega \qquad (10.3.12)$$

Besides the decrease of the necessary force of cutting, another important motive of using the vibrational cutting is the reduction of the level of the summary oscillations of the cutter, appearing in the process of cutting and thus improving the quality of the surface that is being processed. The thing is that in case of ordinary cutting, due to the finite stiffness of the system "machine – device – detail" and to the dropping "real" nonlinear characteristic of the force $F(\dot{x})$, resistant to the cutting, in the system there appear self-oscillations [a]. It suggests an idea that the reduction of the total level of oscillations, observed in practice during the vibrational cutting, is explained

---

[a]Up to now, using the approximate methods of solving the equation of fast motions, we have not in fact taken into consideration the difference between the dry friction forces in motion F_+ and F_- and the corresponding friction forces of rest F_{1+} and F_{1-}; while it is by this difference the indicated "dropping" character of the dependence $F(\dot{x})$ is but approximately taken into account. This makes it possible to study self-oscillations in the system (the classical example may be provided by self-oscillations of the pendulum of Froud [21, 166, 358]). Interesting and very essential regularities of the processes of vibrational displacement which allow to reveal and take into account a less idealized characteristic of the force $F(\dot{x})$ are investigated in the monograph of Gudushauri and Panovko [226]. The

by a well-known phenomenon of *asynchronous suppression of self-oscillations* [21, 308]: for comparatively small (as compared to those being excited) oscillations the dry friction force $F(\dot{x})$ is transformed into the force $V(\dot{X})$, the component of which $V_1(\dot{X})$ presents the additional viscous resistance which suppresses self-oscillations (see 13.2.2).

Though in some cases the self-oscillations, appearing in the process of cutting, play a positive role, and they are therefore specially provided and amplified (see, say, [573, 574, 575]).

10.4   Vibrational Transformation of Motion; Vibro-engines

In technical engineering and in instrument making industry, of greatest practical importance are such devices in which either the rotation or the translational motion of their parts or nodes is obtained by means of exciting vibration either in those units, or in the units which are in contact with them. Such devices are called *vibrational transformers of motion and vibro-engines*.

First of all let us mark that the transportation of bodies over vibrating surfaces, considered in chapter 9 and in section 1 of this chapter, is nothing else but the transformation of the oscillatory motion into translational. The body (Fig. 10.13,*a*) plays the part of the rotor *1*, and the vibrating plane — that of the stator *2*.

Fig. 10.13,*b* presents the schemes of two vibrational transformers of the oscillatory motion into rotational, meant for driving certain rotor machines [10, 216]. In both transformers cylinder *1* (the rotor) is enveloped by an elastic belt *2*, for instance by a wedge belt. In the first device one end of the belt is fixed to traverse *4* directly, and the other end — via an elastic element *5*, maintaining, when working an almost constant tensile force of the belt. Let harmonic angular oscillations be given to the traverse; suppose the cylinder in this case remains motionless. Then it is clear that as the traverse is deflected to the right and the belt is slipping over the cylinder, there will be developed a greater moment of dry friction forces than when the traverse is deflected to the left. As a result, there appears a driving vibrational torque $V(0)$ which in the conditions of the picture will tend to rotate the cylinder counter-clockwise.

In another device, shown in Fig. 10.13,*b*, the traverse is absent, elastic element *5* connects the right end of the belt with the fixed base while the left end is given longitudinal harmonic oscillations. In this case, the cylinder being

---

difference between the forces of the friction of motion and those of rest is taken by us into consideration in chapter 13 in this book.

fixed, there must also appear a vibrational torque which will tend to rotate the cylinder counter-clockwise.

When the described devices are in operation, the motion of the cylinder presents an imposition of the "fast" rotational oscillations upon the "slow" uniform rotation with a certain frequency Ω. It is this motion that the operating devices of some technological vibrational machines must be ensured. Another unexpected and efficient application consists in the following. The balls of the rotor bearings of heavy electric machines, due to the shaking when transported by railway, cause dents in the raceways of the bearings. Should the shaft be supplied with a device, similar to one of those described above, the shaking will cause a slow rotation of the shaft and will prevent the formation of dents.

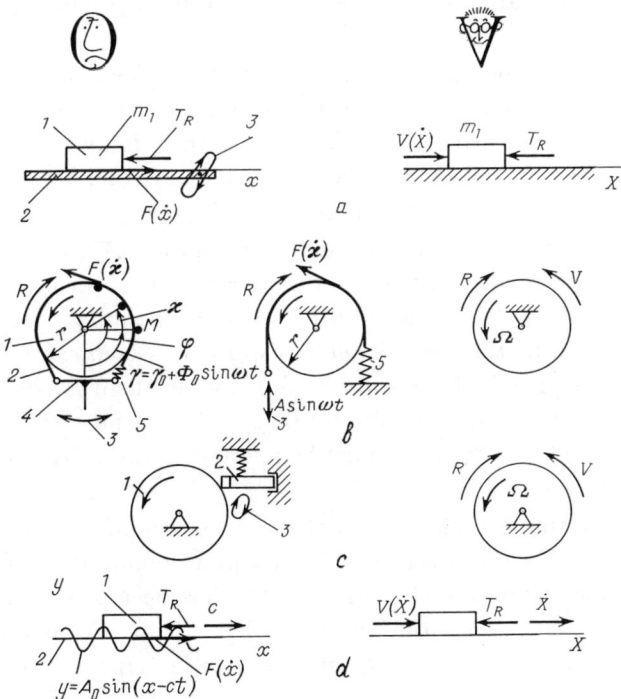

Figure 10.13. Vibrational transformers of motion and vibroengines.

A number of such vibrational transformers of motion has been proposed, investigated and realized by Gorgtinsky and his pupils [10, 216].

Figs. 10.13,c and d show the principal schemes of vibro-engines of two types. In the first case the rotation of rotor *1* is excited by shock-vibrational effects of the piezoelectric ceramic rod *2*, the tip of which, contiguous to the rotor, performs oscillations along the trajectories *3*, close to elliptical. In the second case the translational motion of body *1* (the rotor) is induced by the running wave, created in unit *2* (the stator) which it is in contact with. A similar principle is also used in some vibro-engines of rotational motion. In Fig. 10.13, on the right, pictures are shown, as before, seen by the observer **V**. To obtain the effect of vibrational displacement, practically all the types of asymmetry are used in the devices, considered in 9.3.

Vibro-engines were taken up in precise-instrument engineering because they facilitate accomplishing the positioning and the assigned discreet displacements with the resolving ability of about thousandth fractions of μ at the velocity up to 200 mm/s. There are structures ensuring the reverse and the multidimensional motions.

Different schemes, designs and the theory of vibro-engines were created in Russia and in Lithuania mostly due to the works of Lavrinenko, Ragulskis, Bansevichus, Vasilyev, Kurilo, Ragulskene and their followers. This trend of the use of vibration is reflected in the books [43, 44, 140, 559, 306, 318, 430] which also contain numerous references.

It should be noted that vibrational transformation of motion in fact takes place at the vibrational maintenance of the rotation of an unbalanced rotor as well. It is considered in detail in 6.1. The corresponding devices, however, should be referred to the *vibrational transformers (engines) of synchronous type*, since the frequency of the rotation of rotors in them is bound by integer relationships with the frequency of rotation ω. There is no such bond in all the devices enumerated here, so they must be referred to the *vibrational engines of asynchronous type*.

We will show that simple models of the processes of vibrational displacement are suitable for the description of the main regularities of the work of the vibrational transformers of motion. What has been said does not require any special explanations for the scheme in Fig. 10.13: the main equation of vibrational mechanics for such a transformer can be written in the form

$$m\ddot{X} = V(\dot{X}) - T_R \qquad (10.4.1)$$

where $V(\dot{X})$ is the vibrational force, defined in accordance with the formulas of the type given in chapter 9 and in 10.1, and $T_R \geq 0$ is the useful load which is believed to be constant. The average velocity of the body (anchor) under the stationary regime $\dot{X} = \dot{X}_*$ is found from the equation

$$V(\dot{X}) = T_R, \tag{10.4.2}$$

such regimes being stable, since, at least in all the cases we have studied $V'(\dot{X}_*) < 0$. From equation (10.4.2) we can obtain the "static characteristic" of the transformer; this characteristic allows to establish the limiting ability of the transformer (the engine).

As an example we will consider the case of purely longitudinal harmonic oscillations of the plane (the stator) according to the law $\dot{X}_* = \dot{X}_*(T_R)$, the characteristic of the dry friction force $F(\dot{x})$ being asymmetrical. This case was studied in 9.4, the graph of the vibrational force $V(\dot{X})$ shown in Fig. 9.3,c corresponding to it, while the abscissa of the point of intersection of the curve $V(\dot{X})$ with the line $V = T = -T_R$ corresponds to the solution of equation (10.4.2). The equation of the characteristic of the vibro-engine is obtained from formula (4.31) of chapter 9 at $T = -T_R$, $q = 1$ and has the form

$$\dot{X}_* = A\omega \cos \frac{\pi(F_+ + T_R)}{F_+ + F_-}. \tag{10.4.3}$$

According to relationship (4.28) of chapter 9 this formula is valid at

$$T_R < F_-. \tag{10.4.4}$$

From formula (10.4.3), and from Fig. 9.3,c, it follows that the positive values of the velocity $\dot{X}_*$, i.e. the normal work of the transformer, correspond to the values of the load T_R lying within the range

$$0 < T_R < V(0) = \frac{1}{2}(F_- - F_+). \tag{10.4.5}$$

As that takes place, $\dot{X}_*$ cannot exceed the velocity of vibration $A\omega$:

$$\dot{X}_* < A\omega. \tag{10.4.6}$$

We have already come across this general regularity of the processes of vibrational displacement (see, for instance, 10.3). One can see from the formula (10.4.5) that in order to increase the potentiality of the vibro-engine of this type it is expedient to increase the asymmetry of the characteristic of friction, i.e. the difference $F_- - F_+$. That is usually done in practice; there are even designs with the self-braking of the reverse motion of the anchor, which corresponds to $F_- \to \infty$. The latter case, however, requires a special consideration, because vibration in this case cannot be considered to be highly intensive in the sense, mentioned in 9.4 (see p. 248)

The applicability of the investigated models to the schemes with the transformed rotational motion, shown in Fig. 10.13,*b* and *c*, is at first glance less evident. However, to make sure of it, it is sufficient to note that the role of the vibrating plane, that is the stator, in the devices shown in Fig. 10.13,*b* is in this case played by the belt, while in the device in Fig. 10.13,*c* it is played by the tip of the rod. Let us consider, say, the devices with the belt. The equations of the motions of the rotors of those devices have the form

$$I\ddot{\varphi} = F(\dot{\varphi} - \dot{\gamma})r - R \tag{10.4.7}$$

where φ is the angle of rotation of the rotor,

$$\gamma = \gamma_0 - \Phi_0 \sin \omega t \tag{10.4.8}$$

is the angle, defining the position of a certain fixed point of the belt ($\gamma_0 = $ const, Φ_0 is the amplitude, ω is the oscillation frequency of the indicated angle), I is the moment of inertia of the rotor, r is its radius, R is the useful load torque and the dry friction force is defined by usual relationships, similar to formulas (9.4.15) of chapter 9:

$$F(\dot{\varphi} - \dot{\gamma}) = \begin{cases} -F_+ & \text{at} & (\dot{\varphi} - \dot{\gamma}) > 0 \\ F_- & \text{at} & (\dot{\varphi} - \dot{\gamma}) < 0 \end{cases}$$
$$-F_+ < F(\dot{\varphi} - \dot{\gamma}) < F_- \quad \text{at} \quad (\dot{\varphi} - \dot{\gamma}) = 0. \tag{10.4.9}$$

Here, as was marked, $F_- > F_+$. If in the equation of motion (10.4.7) we pass over from the angle φ to the angle of rotation of the rotor with respect to the belt

$$\kappa = \varphi - \gamma = \varphi - \gamma_0 + \Phi_0 \sin \omega t.$$

then this equation will take the form

$$I\ddot{\kappa} = I\Phi_0 \omega^2 \sin \omega t + F(\dot{\kappa}) - R. \tag{10.4.10}$$

The equation (10.4.10) with an accuracy to the designations coincides with the differential equation (4.25) of chapter 9, describing the motion of a body (rotor) in the case of the scheme of Fig. 10.13,*a*. Relationships (10.4.9) will coincide with formulas (9.2.2) of chapter 9 also with an accuracy to designations.

Thus, the investigation of the devices under consideration is reduced to the study of the device, shown in Fig. 10.13,*a* — the case, corresponding to the longitudinal harmonic oscillations of the plane. In particular, instead of formula (10.4.3) for the velocity of the stationary rotation of the rotor Ω the following equation will be obtained

$$\Omega_* = \Phi_0 \omega \cos \frac{\pi(F_+ r + R)}{(F_+ + F_-)r} \qquad (10.4.11)$$

and instead of formulas (10.4.5) and (10.4.6) we will have the relationships

$$0 < R < V(0) = \frac{1}{2}(F_- - F_+)r \quad \Omega_* < \Phi_0 \omega. \qquad (10.4.12)$$

If the rotor and the tip, shown in Fig. 10.13,c are considered to be absolutely solid, and if the rotor is considered to be installed in soft supports (see the footnote on page 205), and the oscillations of the tip are considered to be preassigned, then the investigation of this device in the system of the coordinates, connected with the tip, will also be reduced to the consideration of the equations of motion of the body of revolution over the vibrating plane (see, for instance, [84]). Studying such a model of vibro-engine, which is as simple as simple can be, we can describe a number of important regularities of its work, in particular, we can obtain approximate equations of the characteristic $\Omega_* = \Omega_*(R)$. Meanwhile, the investigators of such engines prefer to use much more complicated models with many degrees of freedom, which consider the rheological properties of the bodies of the rotor and of the tip; that practically excludes a possibility of analytic investigation. The consideration of such complicated models has certain grounds due to high-frequency oscillations (of the order of tens of kHz and higher) used in those vibro-engines. But we would like to emphasize that the possibilities, offered by the simplest models, are far from being exhausted. The approach to the theory of vibro-engines and vibro-transformers from the position of vibrational mechanics also seems fruitful, which we have tried to show in this section.

We did not dwell here upon the theory of vibro-engines, which use the running waves of deformation in the adjoining units of the rotor or stator (Fig. 10.13,d); this theory demands a special consideration [43, 44, 140, 430].

10.5   Vibrational Movement, Vibrational Coaches

10.5.1   Definitions. Preliminary Remarks

By *vibrational movement* we mean the movement of a body in a certain medium or in a force field which occurs due to the periodic motions of bodies, connected with the moving body. Devices or living organisms, moving like that, will be arbitrarily called by us *vibrational coaches* [114].

It is natural that the interaction of the device or the organizm with the environment or the presence of the external field is absolutely necessary for changing the average velocity of motion of the center of inertia of such a coach, for otherwise a well known thesis of mechanics that such a change is impossible at the expense of the internal forces alone would be violated. It was already mentioned that sometimes researchers forget about the presence of the external forces or fields, and then it causes misunderstanding, and even doubts about the validity of the laws of mechanics (in the preface and in 1.7 see about the observer $\underset{\sim}{W}$).

Vibrational movement is one of the manifestations of the effect of *vibrational displacement* — obtaining the "directed on the average" motions at the expense of "non-directed" effects. The energy, necessary for the movement, can be received either from the internal- or from the external source.

In this section we first give the description of various coaches and of the physical mechanism of their movement on a purely qualitative level from the position of the observer $\mathbf{V}$. Then we consider the theory of these coaches on the basis of the conception of vibrational mechanics.

10.5.2   *Vibrational Movement over a Rough Surface. Self-propelled Vibro-compactors of the Ground, Travel on Skateboards*

Figure 10.14 shows two schemes of coaches for the movement over a rough surface; these schemes correspond, in particular, to the self-propelled vibrational contractor of the ground [54, 84, 563, 475, 564, vol. 4]. In the scheme in Fig. 10.14,*a* the vibro-exciter (a device with periodically moving masses) installed in the body of the coach, generates a harmonic exciting force $\Phi_0 \sin \omega t$ whose direction makes a certain acute angle β with the surface. The mechanism of the appearance of the vibrational force corresponds in this case to that presented in Fig. 9.2,*aII* and described in 9.3 (kinematic asymmetry).

The scheme in Fig. 10.14,*b* is more elaborate, but more complicated. There is an additional mass m_2 there, connected with the mass m_1 by means of an elastic element with a stiffness c_{12}, and an "additional" static load m_3, established on springs whose stiffness is c_{23}. This additional load is of help in case the coach is used as a compactor. The stiffness c_{12} can in this case be chosen from the condition of the proximity to resonance, which makes it possible not only to amplify the vibrational effect on the ground, but also, in case of necessity, to raise the velocity of the movement of the compactor. The stiffness c_{23} is chosen sufficiently small, so that it might provide a practically static action of the additional load. The vibro-exciter (not shown in the figure) is linked with the mass m_2. The vibrational force is in this case formed simultaneously according to the schemes in Fig. 9.2,aII and $aIII$ (kinematic and structural asymmetry). Fig. 10,14,b shows a picture seen by the observer **V** in both cases.

The differential equations of the movement of the coach according to the scheme in Fig. 10.14,a, as one can easily see, are obtained from equations (9.4.34) and (9.4.35) of chapter 9, if we assume that

$$\alpha = 0 \quad \delta = 0 \quad A = \Phi \cos \beta / m\omega^2, \quad B = \Phi_0 \sin \beta / m\omega^2 \qquad (10.5.1)$$

In view of these relationships, everything said in 9.4.3 about the motion of

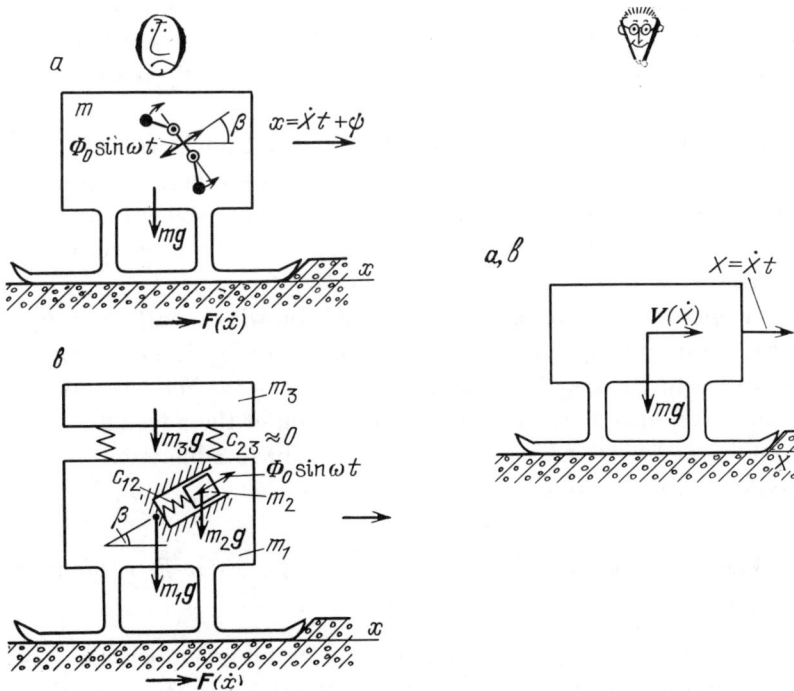

Figure 10.14. Vibrational coaches for the movement over a rough surface.

the particle over a vibrating plane is directly referred to the case of the coach under consideration. In particular, the main equation of vibrational mechanics will have the form

$$m\ddot{X} = V(\dot{X}) \tag{10.5.2}$$

and the velocity of the stationary motion of the coach $\dot{X} = \dot{X}_*$ will be established from the equation

$$V(\dot{X}) = 0 \tag{10.5.3}$$

The ways of obtaining the expression for the vibrational force $V(\dot{X})$ were also discussed in 9.4.3. Expressions for $V(\dot{X})$, answering the motion with an intensive tossing of the particle (as applied to the given case — with an intensive jumping of the compactor), were also given there.

As for the second scheme, shown in Fig. 10.14,*b*, when studying it, it is possible to make use of the solutions of the corresponding problem about vibrational transportation (see publications by Nagaev and Yakimova [400, 583] and also Molasyan [391]). As was already mentioned, the resonance effects are manifested here very clearly. They are reflected on the expression for the vibrational force $V(\dot{X})$ in equation (10.5.2) which naturally refers to this case as well.

A simple and elegant theory of the travel on a skateboard was suggested by Smolnikov and Ispolov [245]. The vibrational coach here is represented by a man together with a skateboard. The periodic movements of the man's legs generate a force acting from the road and directed mostly forward, which leads to the appearance of the corresponding vibrating force.

Nagaev and Tamm have shown that a coach on wheels with a vibro-exciter, established on it, can also move over a rough surface [529]. The theory of such a coach, called in Russian "vibrokhod" i.e. *"vibro-traveller"*, is more complicated, but it can also be reduced to the definition of the adequate vibrational force, conditioned by the kinematic (under certain modification — constructive) asymmetry of the system.

10.5.3   *Vibrational Movement in Fluid and in Gas. Vibro-flier, the movement of living organisms*

Fig. 10.15 shows several schemes of movement in fluid or in gas.

The scheme in Fig. 10.15,*a* corresponds to the so called *vibro-flier*. The main element of such a "coach" is a body whose form is such that the resistance of the medium to its motion along the axis upward is considerably less than that to its motion downward (in this sense the body is asymmetrical; it is shown in the figure in form of an umbrella). As a result of the axial oscillations of a body, excited in one way or another, there appears a lifting vibrational force $V(\dot{X})$ which may under certain conditions exceed the weight of the system mg, so that the apparatus will be lifted upwards. The principle of the appearance of the vibrational force in this case is similar to that shown in Fig. 9.2,*aI,3* (the force-asymmetry). The theory of vibro-flier has been considered by Nagaev and Tamm and they suggested the name of the adequate coach [401, 529].

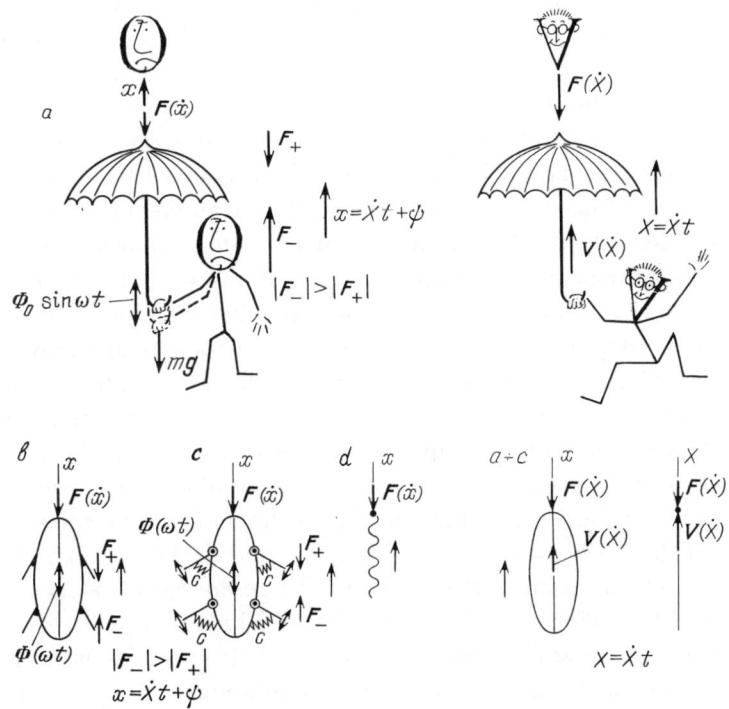

Figure 10.15. Vibrational coaches for the movement in a fluid or gas and simplest models of motion of living organisms.

According to the scheme in Fig. 10.15,*b* the coach is also characterized by the fact that the resistance to the motion forward is different from that to the motion backward, due to the presence of fins, connected with it. The vibro-exciter, not shown in the picture, is supposed to generate a periodic exciting force (not necessarily symmetrical). Here the mechanism of the appearance of the vibrational force is similar to that shown in 9.2,*aI,1,3* and 9.2,*aIII*. The device, shown by this scheme can be called *vibrational ship*. Note that if the resistance force $F(\dot{x})$ is nonlinear, then the vibrational force can be generated even without the fins (i.e. at $|F_+|$) at the expense of the asymmetry of the law of the change of the exciting force $\Phi(\omega t)$. So, by abrupt motions of his body in one direction and by smooth motions in another, the man may make the boat he is in start moving in the direction he wants.

A more general scheme *2* in Fig. 10.15,*b* differs from scheme *1* by the fact that the fins are fixed to the body by hinges and by elastic elements of a certain stiffness C which can be chosen from the condition of the resonance of the fin with an exciting force frequency ω [99]. Exciters, working synchronously and sinphaseously, can also be established directly on the fins. It is natural that the coach, made according to that scheme, will be able to develop a much greater speed. It should be noted that this scheme is answered by the mechanism of the formation of thrust (and of the lifting force) in aircraft with the waving wings and in the flight of insects and birds and in the swimming of some organisms.

It's quite remarkable that devices, similar to those shown in the schemes of Fig. 10.15,*b,c*, can move without any consumption of the internal energy and without any internal source of excitation. Indeed, suppose there are no vibro-exciters in these devices, i.e. $\Phi(\omega t) = 0$, but their bodies systematically suffer occasional pushes from outside, equally probable in all directions. Due to the sloping fins, it is "easier" for such devices to move forward than backward. As a result, a vibrational force appears in this case as well, which makes the device move slowly forward. This effect can be amplified if the stiffness of the elastic elements in the device (Fig. 10.15,*c*) is chosen from the condition of the proximity of their free oscillation frequency in fluid to the dominant frequency of the turbulent pulsation in the reservoir. There are data that the cetaceans can regulate the stiffness of their fins and tail, increasing or decreasing the inflow of blood to certain groups of muscles [429]. It is not excluded that these animals get part of the energy , spent on the movement, from the "renewable source of energy" — the turbulent flow [b].

The scheme in Fig. 10.15,*d* corresponds to the mechanism of swimming of the flagellums. These organisms move in fluid, giving to their extended bodies or to their parts the transverse oscillations in form of a running wave. The mechanism of formation of the vibrational force is in this case similar to that shown in Fig. 9.2,*aV* (wave-asymmetry).

It should be noted that the study of the regularities of the movement of the living organisms — a most interesting and intricate problem — attracts the

---

[b]Another mechanism of receiving by animals the energy from the flow — the wave mechanism — is considered in [579] (see also [293])

attention of well-known researchers (see, for instance, [293, 278, 317, 338, 352, 437, 459, 528, 329, 330, 461, 531, 582, 579]).

Extensive literature is devoted to the theoretical and experimental investigation of the movement of the oscillating bodies in streams of fluid and gas as well, and also to the theory of the corresponding aircraft and sailing apparatuses (see, say, [437, 504, 448, 479]). Here we pursued a very modest aim — to point to the organic connection of these problems with the theory of vibrational displacement and to the expedience of considering them from the position of vibrational mechanics.

Let us dwell here on the theory of coaches, schematically shown in Fig. 10.15, a and b. We will assume that the medium resistance is quadratic and not the same when the coach moves forward or backward and that the exciting force $\Phi(\omega t) = 0$ is harmonic. Then the equation of motion of the coach can be written as

$$m_1 \ddot{x} = -G + \Phi_0 \sin \omega t + F(\dot{x}) \qquad (10.5.4)$$

where m_1 is the mass of the body of the coach together with associated mass, g is the free fall acceleration, Φ_0 is the amplitude of the exciting force, ω is the exciting force frequency, and

$$F(\dot{x}) = -k_{\pm}|\dot{x}|x = \begin{cases} -k_+ \dot{x}^2 & \text{at} \quad \dot{x} > 0 \\ k_- \dot{x}^2 & \text{at} \quad \dot{x} < 0. \end{cases}$$

Equation (10.5.4) is a special case of equation (9.4.2) of chapter 9, corresponding to

$$T = -m_1 g, \quad -m_2 \ddot{\xi}(\omega t) = \Phi_0 \sin \omega t \quad n = 2. \qquad (10.5.5)$$

Therefore the main equation of vibrational mechanics — equation (9.4.9) of chapter 9 — will in this case have the form

$$m_1 \ddot{X} = -G + V(0) + V_1(\dot{X}) \qquad (10.5.6)$$

where in accordance with formulas (10.5.5) and (10.4.11) of the same chapter

$$V(\dot{X}) = - < k_{\pm}|\dot{X} - q\dot{\xi}|(\dot{X} - q\dot{\xi}) > \qquad (10.5.7)$$

$$V(0) = < k_{\pm}|q\dot{\xi}|q\dot{\xi} >, \quad V_1(\dot{X}) = V(\dot{X}) - V(0)$$

with

$$q\dot{\xi} = \frac{\Phi_0}{m_1 \omega} \cos \omega t \quad (q = m_2/m_1). \qquad (10.5.8)$$

Introducing the designation

$$A = \Phi_0/m_1\omega^2, \tag{10.5.9}$$

we will write down expression (10.5.7) as

$$V(\dot{X}) = - < k_{\pm}| \dot{X} - A\omega \cos\omega t \,|(\dot{X} - A\omega \cos\omega t) >$$
$$V(0) = (A\omega)^2 < k_{\pm}| \cos\omega t \,| \cos\omega t > . \tag{10.5.10}$$

Recall that these approximate expressions are valid provided $\Phi_0 \gg F(\dot{x})$.

Having made the calculations in accordance with formulas (10.5.10), we obtain ($\tau_1 = \arccos \dot{X}/A\omega$):

$$V(\dot{X}) = V(0) + V_1(\dot{X})$$

$$= \frac{1}{2\pi}[-k_+ \int_{\tau_1}^{2\pi-\tau_1} (\dot{X} - A\omega \cos\tau)^2 d\tau + k_- \int_{2\pi-\tau_1}^{2\pi+\tau_1} (\dot{X} - A\omega \cos\tau)^2 d\tau]$$

$$= \frac{1}{2\pi}\{-k_+[(2\dot{X}^2 + A^2\omega^2)\cdot(\pi - \arccos \dot{X}/A\omega) + 3A\omega\dot{X}\sqrt{1 - \dot{X}^2/A^2\omega^2}]$$

$$+k_-[(2\dot{X}^2 + A^2\omega^2)\arccos \dot{X}/A\omega - 3A\omega\dot{X}\sqrt{1 - \dot{X}^2/A^2\omega^2}]\}$$

$$= \frac{1}{2\pi}\{(k_+ + k_-)[(2\dot{X}^2 + A^2\omega^2)\arccos \dot{X}/A\omega$$

$$-3A\omega\dot{X}\sqrt{1 - \dot{X}^2/A^2\omega^2}] - \pi k_+(2\dot{X}^2 + A^2\omega^2)\} \quad \text{at } |\dot{X}| < A\omega$$

$$V(\dot{X}) = -k_{\pm}\left(\dot{X}^2 + \frac{A^2\omega^2}{2}\right)\text{sgn}\dot{X} \quad \text{at } |\dot{X}| > A\omega$$

$$V(0) = \frac{1}{4}(A\omega)^2(k_- - k_+) = \frac{1}{4}(\Phi_0/m_1\omega)^2(k_- - k_+) \tag{10.5.11}$$

From equation (10.5.6) and expressions (10.5.11) it follows that to lift the coach upward the following condition should be provided

$$\frac{1}{4}(\Phi_0/m_1\omega)^2(k_- - k_+) > G \tag{10.5.12}$$

i.e. the resistance to the motion of the body downward must exceed quite sufficiently the resistance to its motion upward. In case the law of vibrational effect is asymmetric $-m_2\ddot{\xi}(\omega t)$, this condition, as was mentioned, is unnecessary — the lifting of the body is also possible when $k_+ = k_-$. A detailed consideration of this case, based on the exact solution of the corresponding equation, is given in the above-cited work of Nagaev and Tamm [401].

Figure 10.16 shows the most exotic coach of those considered above, called by its authors — Beletsky and Ghiverts — the *gravi-flier* [56]. This coach is meant for the travels in the cosmic space. It is a cosmic ship, moving along an initially elliptic orbit round the Earth. There is a dumb-bell on the ship, formed by two equal masses $m/2$, whose axis is, say, perpendicular to the plane of the orbit.

Suppose that the length of the dumb-bell $2l$ changes according to the harmonic law with a certain amplitude A and a period $T = 2\pi/\omega$ coinciding with the period of revolution of the ship along the orbit. Let the masses of the dumb-bell be moving apart at the half-coil of the orbit which is closest to the Earth (i.e. in the vicinity of perigee), and during the other half-coil (i.e. in the vicinity of apogee) they are coming closer. It is easy to guess that due to the heterogeneity of the gravitational field (the mass, located closer to the Earth, is attracted to it stronger), to provide such motion of the masses of the dumb-bell it is necessary to perform some work against the gravitational forces, i.e. to consume some energy. This energy, obtained at the expense of the internal (with respect to the ship) source of energy, will be transferred to the energy of orbital motion of the ship, i.e. the ship will move away from the Earth and may go even beyond the Earth's gravitational field, despite the fact that it has not got a traditional reactive engine!

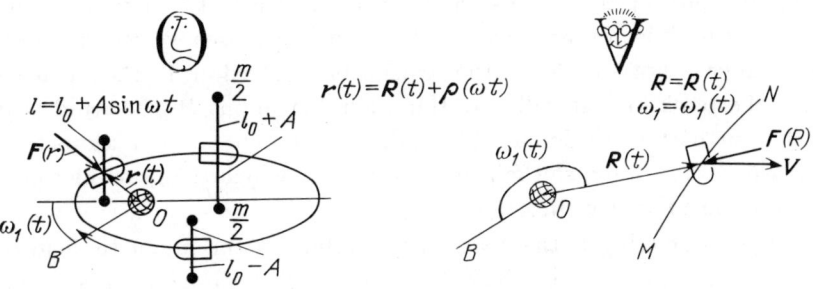

Figure 10.16. Vibrational apparatus for the movement in the gravitational field (gravicraft). To the right — the picture as it is seen by the observer **V**.

The fact that we include the gravi-flier into the category of vibrational coaches and that in this case we differentiate the fast- and slow motions requires a special explanation. Indeed, the period of the revolution of the Earth's satellite, and consequently the period of oscillations of the masses of the dumb-bell are of the order of several hours and more. Fast forces are not seen at first glance. But it is necessary to bear in mind that in our definitions of vibration, of fast and slow motions, of vibrational displacement and of vibrational coaches we have never mentioned the absolute values of frequencies or of the periods of oscillations, we meant only the relative values (see the footnote on p. 38 and 3.2.4). From this point of view it is the motions of the ship along its orbit and the change in the distance between the masses of the dumb-bell that are fast, while the evolution of the orbit, i.e. the change of the mean distance R from the ship to the Earth and " the longitude of perigee" — the angle $\omega_1(t)$, shown in Fig. 10.16, that are slow. The observer $\mathbf{V}$ will see in this case the ship moving slowly along a certain curve MN (see the right part of the figure), which differs greatly from the actual almost elliptical orbit of the ship (compare it with the trajectory in the left part of the picture). Fast are those parts of the gravitational force which are connected with the positions of the masses of the dumb-bell.

The observer $\mathbf{V}$ will think that the described slow motion of the vibro-flier is the result of the appearance of the vibrational force V presenting in this case a thrust force. (Note that should the phasing of motion of the masses of the dumb-bell be opposite to that indicated above, there will appear not the thrust, but the braking of the ship.)

It is not very difficult to obtain the expression for the thrust V. Calculations show that thrust appears not only in case of resonance, i.e. when the period of the change in the length of the dumb-bell and that of the revolution of the ship in the orbit coincide, but also at any other relation of the periods. Unfortunately, this thrust is not large even in the described resonance regime, therefore the gravi-flier is not simple for the realization: according to the calculations, given in the book [56], it will take the gravi-flier about 10 thousand years to leave the Earth's gravitational field, its dumb-bell being even a kilometer long. But that does not diminish the principal interest which the gravi-flier presents from the point of view of mechanics.

Not less interesting is the coach for traveling in the magnetic field of the planet — *magneto-flier*, proposed by Urman. The thrust is created in this case by periodic turns of the magnet, placed in the coach. Like in the case of a gravi-flier, there is every ground to refer the magneto-flier to the vibrational coaches, so the author is grateful to Professor Urman for the replenishment of the collection.

The idea of a gravi-flier is closely connected with a remarkable delusion that the gravi-flier presupposes a change in the velocity of motion of the mechanical system at the expense of the action of internal forces, which is absolutely erroneous and impossible according to the laws of mechanics. It goes without saying that it is not so: there are external forces acting here - forces of the gravitational attraction of the masses of the dumb-bell to the Earth. Without those forces the evolution of the orbit of the gravi-flier would have been impossible. This delusion is characteristic of the observer **W**, whom we spoke about in the preface and in 1.7.

10.6   The Vibro-jet Effect, Vibrational Pumps; Some More about Vibrational Coaches

10.6.1   On Vibro-jet Effect, Vibrational Pumps

By *vibro-jet effect* we mean the appearance of the "directed on the average" flow of the fluid due to the "non-directed on the average" (vibrational) actions. We mean here one of the manifestations of the effect of vibrational displacement. Devices, designed to create a flow of fluid by using a vibro-jet-effect, are called *vibrational pumps*.

The simplest and most commonly used in applications method of realizing the vibro-jet effect consists in imparting oscillations to a solid or deformable body, placed in gas or fluid. These oscillations cause in the general case a certain directed on the average slow flow of the medium near this body. In the previous section we mentioned that such oscillations also lead, as a rule, to the appearance of a vibrational lifting force or thrust (Fig. 10.15). This combination of effects is not accidental: according to the theses of hydromechanics, the appearance of a directed flow and of an averaged force, acting upon the body, usually accompany each other. It is not accidental, therefore that there is a likeness of the schemes of some vibrational coaches, shown in Fig. 10.15, and the schemes of vibrational pumps, given in Fig. 10.17.

Figure 10.17,*a* shows a case of exciting a flow by an asymmetric body (in the same sense as in the previous section) in form of an umbrella, under symmetric, purely harmonic oscillations. Along with that, the directed flow may appear in case of a symmetric body as well, due to the asymmetry in the law of oscillations of the body or the force, acting on it (Fig. 10.17,*b*). However, as that takes place, just like in case of a vibrational coach, the resistance to the motion of

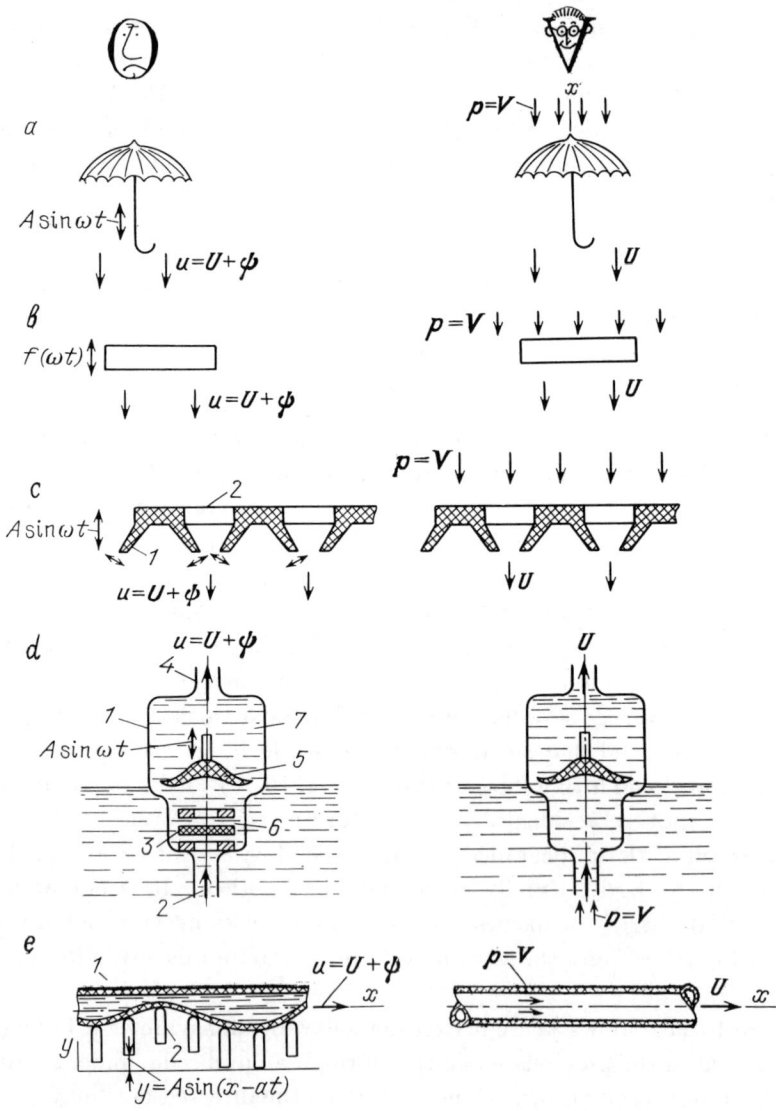

Figure 10.17. Vibrational pumps. A body vibrating in the fluid or gas may serve as a vibrating pump.

the body in the medium must be of a nonlinear character.

Figure 10.17,c shows a widely spread method of creating a directed flow of fluid — by means of imparting oscillations (symmetrical, as a rule,) to the perforated plate I with the narrowing openings or slotted canals 2. Due to an essential asymmetry of the system (the resistance to the motion of the fluid when it moves in the direction of the narrowing of the openings or slots is less than when it moves in the direction of the widening) there appears a pulsating flow of the fluid, directed on the average toward the narrowing. The observer V will connect the appearance of the constant part of that flow with the appearance of a certain additional difference in pressure $(\Delta p)_V$ which being combined algebraically with the existing difference $(\Delta p)_0$ will either amplify the flow or weaken it or it may even result in stopping the flow altogether (effect of *vibrational shutting the openings in the plate*) or to the appearance of a reverse flow. We may believe that it is this effect that underlies the phenomenon, described in literature, when the petrol tanks stop supplying fuel to the aircraft, which sometimes resulted in accidents.

The vibro-jet effect can be amplified if the narrowing canal is performed in form of deformable tabs whose stiffness is chosen from the condition of the proximity to the resonance with the frequency of the excited oscillations. The tabs play the role of certain valves which at the longitudinal oscillations of the plate let a greater quantity of if fluid pass toward the narrowing than toward the widening of the canal. As a result, a directed on the average flow is formed which may prove to be more intensive than when the walls of the canal are nondeformable.

This method is used in the vibrational flotation machine [9] and in the machine for washing off particles of the granular material or the metallic chip. It is necessary to provide in these devices a pulsating stream of fluid in which on the one hand are provided intensive oscillations of particles with respect to the fluid, on the other hand — a slow circulating motion of the medium being processed. The vibrating perforated plates of the type of those shown in Fig. 10.15,c are also used in the devices for mixing the heterogeneous fluids or suspensions, schematically shown in Fig. 10.18 [564, vol. 4].

Figure 10.17,d shows a scheme of the so called *volume-inertial pump*. It consists of a body 1, which has a receiving opening 2 with a reverse valve 3 and a releasing opening 4 to which an elastic hose is fixed. Inside the body there is an elastic piston of a special shape 5 to which vibrations are imparted along the axis of the body from a hermetic (usually electromagnetic) vibro-exciter. The fluid in which the body is immersed enters the receiving chamber 6 through the reverse valve 3. When the piston moves downward, the reverse valve closes and the fluid is squeezed through the gap between the piston and the body into the chamber over piston 7. When the piston moves upward, it presses itself to the body along the circumference. Then the pressure in the receiving chamber drops, valve *3* opens and the fluid gets into the pump. At the same time the elastic piston pushes away a certain volume of fluid from the chamber above it through the release opening from the chamber.

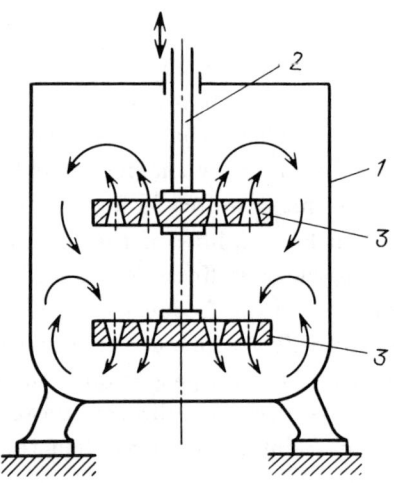

Figure 10.18. Scheme of the apparatus for intermixing heterogeneous fluids or suspensions, based on using the vibro-jet effect. *1* —the frame of the apparatus; *2* — the rod; *3*— the plates with conic holes. Curvilinear arrows indicate the scheme of the streams.

Pumps, similar to that described here, are produced in Russia and in some other countries serially. Their great advantages are simplicity and reliability of design, their fitness (if manufactured specially) for the supply of the aggressive fluids and pulps with abrasive particles. A considerable contribution into the elaboration of the theory and methods of calculation and design of vibrational pumps was made by Usakovsky [551, 564, vol. 4].

There is a special group of pumps and feeders for the strictly measured supply of fluids and powders in very small quantities. Such devices are very important, say, in medicine and in the powder metallurgy. For this purpose in some devices oscillations of the type of running wave are imparted to the tubes

with the medium that is being supplied. As an example Fig. 10.17,*e* shows
a peristaltic pump for the supply of the fluid. In this pump the elastic tube
1 is pinched by rods *2* extending in series. A number of precision vibrational
devices for the measured supply of fluids and granular materials was elaborated
and introduced by Ragulskis and his colleagues (see, say, [44, 143]). It is
easy to see that the methods of generating the directed flows in the devices,
shown in Fig. 10.17 agree very well with the principles of creating vibrational
displacement, shown in Fig. 9.2,*aI–IV*.

In the right part of Fig. 10.17, like it was before, pictures are shown, as they
are seen by the observer **V**. From his point of view the directed flow of the
fluid appears under the action of a certain additional pressure **V**, presenting
in this case the vibrational force.

10.6.2   *Lishansky's Vibrational Pump; The Simplest vibroflighter: An Approximate Theory*

We will consider a device, shown schematically in Fig. 10.20. The absolutely
solid disc of radius R placed in the liquid, performs harmonic oscillations perpendicularly to its plane with the amplitude A and frequency ω. In its neutral
position the disc is a distance h_0 apart from the fixed flat wall. Due to vibration
in the space between the wall and the disc there appears a region of high pressure. Therefore if, say, a pipe of the radius $a << R$ is mounted into the wall,
the other end of the pipe being placed into the zone of a lower pressure, then
some liquid will flow inside the pipe, i. e the device will act as a pump. The
idea of such a pump was proposed by Lishansky [336, 337], and the hydrodynamic theory of such a pump was developed by Korotkin [289]. Besides, in case
there is no pipe, the disc will acted upon by a certain constant force V which
will compensate the weight of the disc. So the disc will "hover" (i.e. hang up)
at some distance from the wall, i.e. the device will act as a vibrational coach
named "vibroflier". We will show that the above effects can be explained in a
rather simple way and roughly described on the basis of the foregoing general
approach. We will assume the liquid to be ideal and incompressible, and the
flow to be symmetrical to the axes, with a radial velocity u_r, independent of
the coordinate z. Though these suppositions might be abandoned. With those
assumptions the equation for the velocity of the liquid u_r and the continuity
equation will be written as follows:

$$\frac{\partial u_r}{\partial t} + u_r \frac{\partial u_r}{\partial r} = -\frac{1}{\rho}\frac{\partial p}{\partial r}, \qquad \frac{\partial u_r}{\partial r} + \frac{u_r}{r} = 0 \qquad (10.6.1)$$

where p denotes the pressure, ρ denotes the density of liquid. We will assume that

$$u_r = U + u', \qquad p = P + p' \tag{10.6.2}$$

where U and P, u' and p' are the respective slow and fast components of velocity and pressure with $< u' >= 0$ and $< p' >= 0$. Then for the velocity U (supposing the slow motion is stationary) we will obtain the following differential equation

$$U \frac{dU}{dr} + < u' \frac{du'}{dr} >= -\frac{1}{\rho} \frac{dP}{dr} \tag{10.6.3}$$

We have not written here the second equation since within the framework of this approximate investigation the fast component u' can be defined from the consideration of continuity (mind that that component is as a rule defined approximately). Indeed, having written down the condition of constancy of the volume of liquid inside the cylinder of a certain medium radius r (neglecting the presence of holes in the wall, believing that $r \gg a$), we have $(h_0 + a \sin \omega t)(r + \eta)^2 = h_0 r^2$ where η is the radial shift of particles of the liquid under vibration.

Hence, supposing that $A/h_0 \ll 1$ and $\eta/r \ll 1$, we obtain $\eta = Ar \sin \omega t / 2h_0$, $u' = d\eta/dt = Ar\omega \cos \omega t / 2h_0$ and

$$< u'du'/dr >= 1/8 \cdot rA^2\omega^2/h_0^2 \tag{10.6.4}$$

Upon integrating the equation (10.6.3) we obtain

$$\rho U^2/2 = -(P + P_v - P_*) \tag{10.6.5}$$

where $P_v = \rho/16 \cdot (rA\omega/h_0)^2$ denotes vibrational pressure, and P_* denotes the integrating constant. Thus, the vibrating disc is acted upon by a hoisting vibrational force

$$V = \int_0^R 2\pi r P_v dr = \frac{\pi}{32}\rho(A\omega R^2/h_0)^2 = \frac{\pi}{32}\gamma \cdot A\omega^2/g \cdot AR^4/h_0^2 \tag{10.6.6}$$

where $\gamma = \rho g$ is the specific gravity of the liquid. The disc hangs up in the liquid, provided

$$V/G = 1/32 \cdot \gamma/\gamma_d \cdot A\omega^2/g \cdot AR^2/\delta h_0^2 > 1 \tag{10.6.7}$$

where G is the weight of the disc, δ is the thickness and γ_d is the specific gravity of the material the disc is made of.

The height of the disc hanging up in the liquid h_0^* is determined from the prerequisite $V = G$:

$$h_0^* = R\sqrt{1/32 \cdot \gamma/\gamma_d \cdot A\omega^2/g \cdot A/\delta} \qquad (10.6.8)$$

So, e.g. with $R = 20$ cm, $\gamma/\gamma_d = 2.7$, $\quad A\omega^2/g = 4$, $\quad A/\delta = 0.5$, we obtain $h_0^* \approx 3$ cm.

From the expression (10.6.4) one can see that the effect under consideration is caused by the presence of the nonlinear convective term in the first equation (10.6.1). To determine the integrating constant P_* we can make use of the condition that at a certain distance from the edge of the disc, i. e. when $r = kR$ where $k > 1$, the pressure in the liquid is leveled off and becomes equal to P_∞, the velocity V being equal to zero. The publication [289] recommends for rough calculations to assume that $k = 1.4$. As a result we obtain $P_* = P_\infty + \rho \cdot (kAR\omega/4h_0)^2$ and, according to (10.6.5)

$$P - P_\infty = \rho(A\omega/4h_0)(k^2 R^2 - r^2) - \rho U^2/2. \qquad (10.6.9)$$

Using ordinary formulas of hydraulics, expression (10.6.9) makes it possible to determine the liquid discharge Q provided the device is used as a vibrational pump.

The given consideration is a convincing confirmation of the circumstance that the solution of the equation of fast motions can be found very crudely without a considerable loss accuracy in the description of slow motions (see 3.3.2).

10.6.3   *Generating the Flow of Fluid by Means of a Vibrating Plate with Holes of Changing Cross-sections*

As an example of using the approach of vibrational mechanics to simulating and investigating the vibro-jet effect and vibrational pumps one can consider the theory of devices, shown in Fig. 10.17,c and Fig. 10.18, assuming that the walls of the openings or canals in the vibrating plates are undeformable [115]. The case of the deformable walls of the canal, formed by rigid lamellar flaps, fixed on the resilient hinges, has been studied in [9].

For the sake of convenience let us convert the problem — we will consider the motion of a free perforated plate in a fixed cylindrical vessel (Fig. 10.19), i.e. in fact the adequate problem about a vibrational coach. Let the velocity of the plate with respect to the vessel $\dot{x}$ be connected with the discharge of the fluid through the plate Q by the relationship

$$Q = -S\dot{x} \qquad (10.6.10)$$

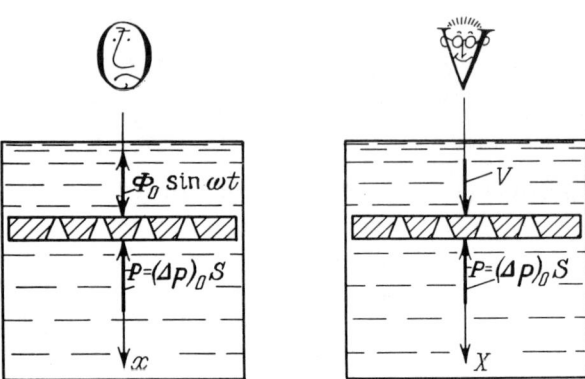

Figure 10.19. Generating the flow of fluid by means of a vibrating plate with holes of changing cross-setion.

where S is the area of the plate. It is believed that the plate is acted upon by a harmonic exciting force $\Phi_0 \sin \omega t$ and by a certain constant force P which is answered by a static pressure differential $(\Delta p)_0$ along the thickness of the plate, so that

$$P = -(\Delta p)_0 S. \qquad (10.6.11)$$

The resistance to the motion of the plate will be considered quadratically dependent on the velocity of the motion of the plate with respect to the fluid $\dot{x}$. And when the plate moves in the positive direction of the axis x, the coefficient of the resistance k_+ is less than when it moves in the negative direction k_-. This corresponds to the direction of the narrowing of the openings or slots in the plate, shown in Fig. 10.18, a. In other words, the force of resistance will be defined by the formula

$$F(\dot{x}) = -k_\pm \dot{x}^2 \operatorname{sgn} \dot{x} = -k_\pm \dot{x} |\dot{x}| \quad (k_- > k_+) \qquad (10.6.12)$$

Note that this formula is in agreement with the well known formula of hydraulics for the discharge Q of the fluid through the opening

$$Q = \mu s_0 \sqrt{2\Delta p / \rho} \qquad (10.6.13)$$

where s_0 is the area of the opening in its narrow part, ρ is the density of the fluid, Δp is the pressure difference, $\mu < 1$ is the dimensionless coefficient of the discharge of the fluid. When the fluid flows in the direction of the narrowing of the opening, this coefficient is greater than when it flows in the direction of the widening, so $\mu_- > \mu_+$. Formula (10.6.13) changes to (10.6.12) if we assume that

$$Q = -S\dot{x}, \quad \Delta p = -F/S \operatorname{sgn} \dot{x}, \quad \mu = \mu_\pm$$

and if we designate

$$k_- = \frac{\rho S^3}{2\mu_+^2 s_0^2}, \quad k_+ = \frac{\rho S^3}{2\mu_-^2 s_0^2}, \quad (\mu_+ < \mu_-, \, k_+ < k_-). \tag{10.6.14}$$

It is natural that the coefficients μ_+ and μ_- and also k_+ and k_- may differ essentially from their static values and may depend on both the frequency and the amplitude of vibration of the plate.

Under these assumptions the equation will have the form

$$m_1 \ddot{x} = P + \Phi_0 \sin \omega t + F(\dot{x}) \tag{10.6.15}$$

where m_1 is the mass of the plate and of the parts rigidly fixed to it, and the forces P and F are defined by the expressions (10.6.11) and (10.6.12). This equation at $P = -m_1 g$ coincides with the equation of motion of the vibrational coach (10.5.4), considered in the previous section; therefore we can take advantage of the solution of the problem obtained there.

The main equation of vibrational mechanics will have the form

$$m_1 \ddot{X} = P + V(\dot{X}) \tag{10.6.16}$$

where the vibrational force $V(\dot{X})$ is defined by the expression (10.5.11) and as before it is believed that $\Phi_0 \gg F(\dot{x})$ and we have $A = \Phi_0/m_1\omega^2$. In case $\dot{X} \ll A\omega$ the equation is considerably simplified

$$V(\dot{X}) = \frac{1}{4}(A\omega)^2(k_- - k_+) - \frac{3}{2\pi}A\omega(k_+ + k_-)\dot{X} \tag{10.6.17}$$

As a result, according to (10.6.10), (10.6.11), and (10.6.16) we obtain the following formula for the stationary average discharge of fluid through the plate:

$$Q_* = -S < \dot{x} >= -S\dot{X} = \frac{\pi}{3}\left[\frac{2(\Delta p)_0 S^2}{A\omega(k_- + k_+)} - \frac{1}{2}A\omega S\left(\frac{k_- - k_+}{k_- + k_+}\right)\right] \tag{10.6.18}$$

At $(\Delta p)_0 = 0$, i.e. in case of the absence of the static difference pressure,

$$Q_* = Q_V = -\frac{\pi}{6}A\omega S\left(\frac{k_- - k_+}{k_- + k_+}\right). \tag{10.6.19}$$

The sign "minus" before this expression shows that the flow Q_V is directed toward the narrowing of the openings in the plate (recall that $k_- > k_+$).

From formula (10.6.18) we also obtain the following simple expression for the velocity $A\omega = (A\omega)_* = (\Phi_0/m_1\omega)_*$, corresponding to the effect of a *vibrational shutting the openings in the plate* :

$$(A\omega)_* = (\Phi_0/m_1\omega)_* = 2\sqrt{(\Delta p)_0 S/(k_- - k_+)} \qquad (10.6.20)$$

This formula may be used for the experimental definition of the difference of the resistance coefficients $k_- = k_+$.

10.6.4   Five Principle of Generating Vibrational Flows and Forces

As we saw above, the same principle can be used to create both a vibrational pump and a vibrational coach, e. g. a vibroflier.

Figure 10.20 shows five types of installations and accordingly five *physical principles which can be used to create vibrational coaches and pumps*. The vectors **V** in Fig.10.20 indicate the directions of vibrational lifting forces, and the vectors Q — the directions of flows.

The installation shown in Fig. 10.20,*1* was considered in section 10.6.2. The principle used here is named by us the *principle of boundary-effect.*In Fig. 10.20.2 we can see three versions of the installations based on the principle named by us the *principle of umbrella*. The effect is achieved here due to the difference in the resistance of the liquid to the motion of bodies when they are moving in the opposite directions (in this case upward and downward.) In Fig. 10.20.3 we see three versions of installations whose principle of action can be named the *principle of valve*. In the installation shown in Fig. 10.20.3,*a* the valve is closed when the body moves upward and it is open when the body moves downward. The installations shown in Fig. 10.20.3,*b,c* differ from those in Fig. 10.20.2,*b,c* by the fact that the plates (petals) are connected with the oscillating body not rigidly but by means of elastic hinges. As a result, the plates perform vibrations facilitating the motion of the liquid downwards and opposing its motion upwards. This effect can be amplified if we select the rigidity of elastic elements which is to be based on the conditions of resounance. It is remarkable that installations similar to those shown in Fig. 10.20.2,*b* and in Fig. 10.20.3,*b* can move "by themselves", i. e. without any vibroexciter installed on them, at the expense of casual pushes caused by the vorticity of the liquid. Figure 10.20.4 shows the principle which can be named the *principle of asymmetry of excitation*. Unlike all other installations in this case it is assumed that a body which is considered to be symmetrical is acted upon by

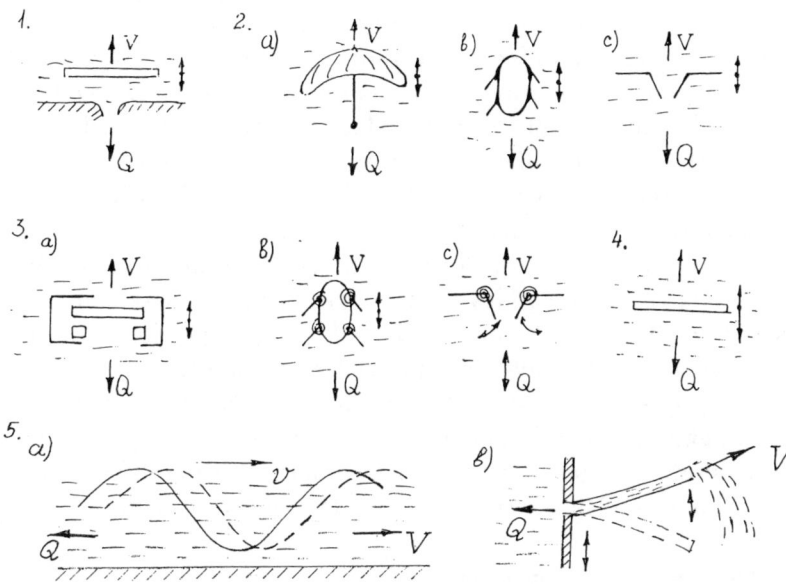

Figure 10.20. Principles of creating vibrational pumps and coaches: *1* — principle of boundary-effect, *2* — principle of umbrella (three versions), *3* — principle of valve (three versions), *4* — principle of asymmetry of excitation, *5* — wave principle (two versions).

an asymmetric exciting force. As a result, the body is repelled from the liquid much more strongly when it is moving upward than when it moves downward. In Fig. 10.20 the symmetric oscillations are arbitrarily designated by arrows of equal length in both directions while the asymmetric oscillations are designated by the arrows of different lengths. Though it should be noted that the asymmetry of systems in one form or another is used in all installations shown in Fig. 10.20. That circumstance reflects the principle characteristic of all processes of vibrational displacement (see 9.3). Figure 10.20.5 shows two versions of installations based on the principle named the *wave principle*. In the first case the flow is generated by the running wave, and in the second case — by the standing wave.

It should be noted that just like in the installations shown in Fig. 10.20.3,*b* and *c* too several principles can be shown simultaneously.

The elaboration of the theory of the devices described above is far from being completed. The theory of installation shown in Fig. 10.20.1 was considered in the work [289] as well as in section 10.6.2. The installations of type shown in Fig. 10.20.2,c were considered in [115] and in 10.6.3. The work [504] contains the experimental confirmation of the umbrella-principle. The book [551] is devoted to installations of valve-type (Fig. 10.20.3,a). The work [401] is devoted to the theory of installations with asymmetric excitation (Fig. 10.20.4). The authors of that work proposed the term "vibroflier". The theory of the valve installation shown in Fig. 10.20.3,c is considered in the paper [9], which establishes a considerable growth of water discharge and accordingly of the driving force in the region which is in the vicinity of the resonance. Installations shown in Fig. 10.20.3,c are considered in the paper [9]. The theory of installation based on the wave principle in case of the running wave is considered in book [44], and is based on the same principle in the case of the standing wave — in publications [422, 249, 455]. In the last case the action of vibration increases considerably in vicinity of the resonance with elastic oscillations of the tube. The work [249] contains an extended review of the problem.

10.6.5   *On Vibrational Convection*

The problem of vibrational convection is related to the problems considered in this section. In particular, publications [593, 199, 200, 594] are devoted to this problem. To solve adequate most interesting problems the authors have used the approach which is also based on the works by Kapitsa [257, 258]. This approach in this class of problems is mathematically based in [325].

10.6.6   *On the Main Equations of Vibrational Hydromechanics and Hydraulic*

As one can see from this chapter, the main statement of chapter 3 about the structure of the equations of vibrational mechanics also refer to the equations describing the slow motion of the fluids under vibration. And namely, the terms corresponding to the vibrational pressures, stresses and volume forces are added to those equations (see, e. g. equations (10.6.5, ...) and also the corresponding equations in [593, 199, 200, 594]

Chapter 11

Vibrational Shift (Drift)

11.1   On the Notion of Vibrational Shift (Drift)

By *vibrational shift, or drift* , we mean a change in the positions of a stable equilibrium of a mechanical system, caused by vibration. The equilibrium positions under vibration imply in the general case the positions of equilibrium for the system, seen by the observer **V** who "does not notice" the fast motions in the vicinity of those positions. Such positions of equilibrium are called *quasi-equilibrium positions* (see page 92)

From the point of view of the observer **V** the effects of vibrational shift are explained by the appearance of vibrational forces or moments.

11.2   Effect of Vibrational Shift in Applications; Special Features of the Effect in Systems with Dry Friction

We have already come across the phenomenon of vibrational shift in the previous chapters and will come across them below. Here we will enumerate such phenomena which are of essential importance for applications.

In 5.1.5 it was mentioned that the pendulum with a rectilinearly vibrating axis of suspension when tends to get located along the direction of vibration as if it were attracted to those positions by invisible springs. As a result, the pendulum with a vibrating axis of suspension will, generally speaking, trace not the vertical, but some other direction (Fig. 11.1,*a*). The pointer of the device under vibration will experience a similar shift (drift) (Fig. 11.1,*b*). The magnetic compass needle, acted upon not only by the magnetic field of the earth, but also by a weak alternating magnetic field, caused by the proximity of the electric devices of the alternating current, can instead of the North show the direction to the South [358].

The mechanism of drift in similar systems was considered in the general form in 5.3. In chapter 19 it is discussed from a different position — in connection with the behavior of a material particle in a fast oscillating stationary field. As will be shown, the particle is attracted to the points of the minimum of the amplitude of a standing wave $|\Psi(x)|$ (Fig. 19.1). As a result, while in the

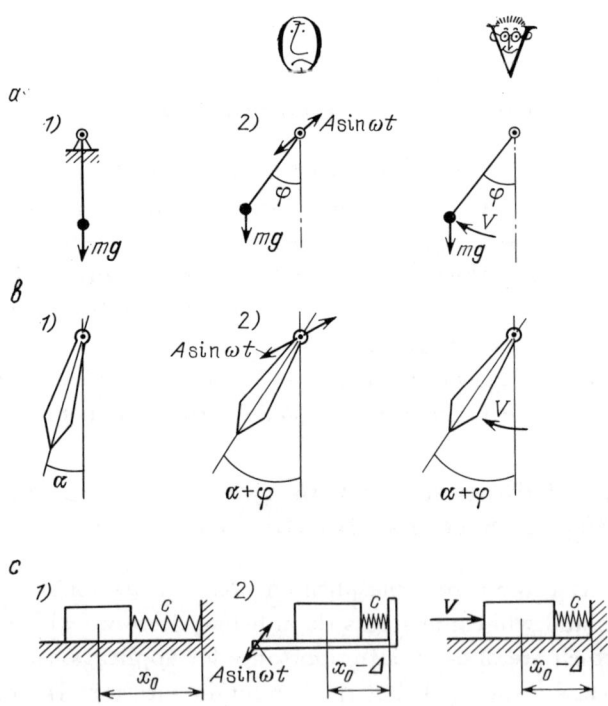

Figure 11.1. Vibrational shift (drift).

absence of the oscillations of the field the particle had certain positions of stable equilibrium, in case of their presence these positions will be shifted in a certain way in the direction of the indicated points of the minimum of the function $|\Psi(x)|$. The pendulum with a rectilinearly vibrating axis of suspension can be considered to be a special case of that system. In 5.3. this situation was also discussed from the position of the conception of the potential on the average dynamic systems — as a consequence of a possibility of the appearance in such systems under vibration of new "potential pits", or of the shift or even disappearance of the former "potential pits" of the system.

It is characteristic of the discussed above and of some other effects of vibrational shift that there is a positional force in the system, which is clear from the definition given in 11.1 and from the examples considered above. In this connection it should be noted that there is a link between the effects of vibrational displacement and shift. Suppose the main equation of vibrational mechanics for the system has the form

$$m\ddot{X} = T + V(0) + V_1(\dot{X}), \tag{11.2.1}$$

where T is a certain constant force, and $V(0) + V_1(\dot{X}) = V(\dot{X})$ is a vibrational force, with $V(0) \neq 0$ and $V_1(0) = 0$. This equation is typical for the process of vibrational displacement, whose velocity $\dot{X} = \dot{X}_*$ is determined from the equation

$$T + V(0) + V_1(\dot{X}) = 0 \tag{11.2.2}$$

(see, say, equations (9.4.9) and (9.4.51) of chapter 9). Such a system has no positions of equilibrium. In case there is a positional force $F(x)$ changing smoothly enough with the change of x, the relationship (11.2.1) is replaced by the equation

$$m\ddot{X} = T + V(0) + V_1(\dot{X}) + F(X), \tag{11.2.3}$$

which will be answered by the positions of equilibrium $X = X_*$ (i.e. positions of quasi-equilibrium for the initial system), determined from the equality

$$T + V(0) + F(X) = 0, \tag{11.2.4}$$

while the positions of equilibrium of the system in the absence of vibration are defined by the equation

$$T + F(X) = 0. \tag{11.2.5}$$

In case of a smooth positional force, satisfying, say, conditions $F(0) = 0$, $F'(x) \neq 0$, equations (11.2.4) and (11.2.5) have one solution each, those solutions being different from each other, i.e. there is an effect of vibrational shift (drift) in the system.

Effects of vibrational drift in gyroscopic systems require a special consideration. They were already mentioned in 5.3.

Effects of vibrational shift with dry friction are characterized by essential peculiarity. It consists in the fact that in the absence of vibration in such systems, unlike the systems with viscous friction having isolated positions of equilibrium, there is an infinite number of the positions of equilibrium, continuously filling a certain section. In such cases we speak of the *continuum of equilibrium positions*; the corresponding section is called a *dead zone*. If vibration is intensive enough, this zone disappears — as if dry friction were transformed into viscous, and one or several positions of quasi-equilibrium appear. These positions, however, are shifted, as a rule, with respect to the equilibrium positions which the system might have in the absence of dry friction, and they can be located beyond the dead zone of the system with dry friction. The cause of the vibrational shift here is the appearance of the driving vibrational force $V(0)$; this force is conditioned by the presence of one or another kind of asymmetry of the system. What has been said is illustrated by Fig. 11.1,*c* and also by Fig. 10.4; the latter is discussed in 9.2.2 in connection with the problems of the theory of vibrational separation of particles of the granular mixtures. A bright example of vibrational shift in systems with dry friction is the abnormal behavior of the granular materials in the communicating vessels, discussed in 15.3 (see Fig. 15.6).

It should be noted that sometimes vibration is recommended to be used to liquidate the dead zone in the devices. However it is clear from what has been said that such recommendations must be treated with caution, because one or another type of asymmetry of the system may cause the appearance of a driving vibrational force $V(0)$ leading to an error comparable to that, or even exceeding that, caused by the presence of the dead zone. Measures must be taken that there should be no such asymmetry. This trend of the use of vibration is considered in the book by Kanapenas [256].

Part IV

Vibrorheology

Chapter 12

On Rheology and Vibrorheology

12.1   Rheology as a Section of Mechanics

By rheology we imply a section of mechanics which studies the deformation and fluidity of substance. Classical models of rheology are the elastic body of Hooke, the viscous fluid of Newton, the plastic body of Saint-Venant. The linear resistance and the dry friction when a body moves in a medium or with respect to other bodies can be considered to be the simplest degenerated cases or as the discreet analogs of the latter two models (i.e. the viscous fluid and the plastic body) — the cases much spoken about in this book.

Of major importance in rheology are the so called *rheological equations*, establishing the connection between the kinematic and force parameters, characterizing the state of the systems under consideration. In the above-mentioned classical models these are respectively the equations, expressing Hooke's generalized law, Newton's law and the law of ideal plasticity of Saint-Venant and in their discreet analogs — Hooke's law in its simplest form, the law of resistance, proportional to the velocity of the body, and Coulomb-Amonton's law of dry friction.

Constants involved in the rheological equations (the modulus of elasticity, the viscosity coefficient, and the dry friction coefficients) are sometimes called *rheological constants* or *rheological characteristics* .

One distinguishes *macrorheology* which deals with heterogeneous or quasi-heterogeneous materials, lacking structure, and *microrheology* which considers the rheological behavior of two- or polyphase systems, depending on their rheological characteristics and the properties of their components.

12.2   Definition of Vibrorheology. Macro- and Microvibrorheology

As was shown above by a number of examples, under the action of vibration on the nonlinear mechanical systems, not only the rheological characteristics of a body but also the character of a rheological equation can change essentially with respect the slow actions. Particularly, as was already mentioned, a system with dry friction can behave as that with viscous friction.

By *vibrorheology* (or more precisely by *macrovibrorheology*) we will imply a section of mechanics which studies changes, caused by vibration in the rheological properties of bodies with respect to slow forces, and also the relevant slow motions of bodies. In other words, macrovibrorheology can be defined as the rheology for the observer V, who was mentioned repeatedly in this book.

This definition is given in publications [95, 99, 114], though the term "vibrorheology" was apparently first suggested by Rebinder. It should be noted that this term is sometimes used in another sense (see e.g. [208, 420]). We must emphasize that this definition refers just to macrovibrorheology.

The study of the specific features of the behavior of polyphase systems under vibration can be referred to *microvibrorheology*.

Vibrorheology is correlated with vibrational mechanics the way rheology is correlated with mechanics, and vibrorheology is correlated with rheology the way vibrational mechanics is correlated with mechanics. This circumstance is reflected schematically in Fig. 1.1. Though the vibrorheological effects (as well as the rheological ones) are not necessarily of a purely mechanical character: they can be associated with thermal phenomena, with chemical transformations etc. Therefore rheology is often considered to be a branch of physics.

It goes without saying that vibrorheological effects play an important role in technology. They make the principal basis of many technical applications of vibration.

Two essential moments should be emphasized once more: 1) in most cases it is expedient to speak of macrovibrorheological effects as of the seeming ones, which take place only for the observer **V** ; 2) as a result of the action of vibration on the nonlinear mechanical systems, in the general case there is not only a change in the rheological characteristics or properties of the body with respect to slow actions, but there also appear either driving or shifting forces or torques.

12.3   Vibrorheological Equations, Vibrorheological Properties and Effective Vibrorheological Characteristics

Like it was in the previous sections of the book, the main idea when investigating vibrorheological effects is the transfer from the initial differential equations of the motion of the system to more simple differential equations, describing the slow motions — the main equations of vibrational mechanics. When considering problems of vibrorheology, we will call these equations *vibrorheological equations*.

In some cases vibration does not cause any major qualitative changes of the rheological properties of the system with respect to slow actions, and the rheological characteristics of the body suffer only quantitative changes. So if systems with dry friction are acted upon by vibration of comparatively small intensity, there is no seeming change in the type of friction as yet and only the main characteristics — the dry friction coefficients — are changing. The type of viscous friction does not change either, it remains viscous.[a] In such cases we will speak about the *effective rheological characteristics* of bodies under vibration.

Note that when the action of vibration is comparatively long, the rheological (consequently — the vibrorheological too) characteristics and even properties of bodies can change very slowly (much more slowly than the slow coordinates of the system; see, e.g. [283]). Apparently the same effect can take place when driving-in the piles and the sheet piling (see page 294).

---

[a]The main qualitative distinction of the dry friction from viscous friction is given in the footnote on page 239.

Chapter 13

Effective Rheological Characteristics under the Action of Vibration

13.1 Effective Coefficients of Dry Friction under the Action of Vibration or Shock; Some Applications

13.1.1 *Effective Frictional Coefficients of Rest. The simplest Model — an Absolutely Solid Body under a Harmonic Action*

The effect of the seeming reduction of the friction coefficient of rest is the simplest manifestation of vibrorheological regularities, allowing the elementary consideration [74, 84].

Let an absolutely solid body be pressed by a force $\mathbf{N}$ to a rough surface and let it be acted upon by a longitudinal harmonic force $\mathbf{S}$ directed along the plane (Fig. 13.1,a). Let the body be also acted upon by a force $\Phi = \Phi_0 \sin \omega t$; then for the body to begin moving along the plane it is necessary that there should be not the force $S = S_0 = f_1 N$ like it is in the absence of the force Φ, but only the force $S = f_1 N - \Phi_0$ (f_1 being the friction coefficient of rest). Therefore it will seem to the observer $\mathbf{V}$ "who does not see" the fast force Φ (Fig. 13.1,c) that the friction coefficient of rest has reduced with respect to the slow force $\mathbf{S}$, and has become

$$f_1^{(=)} = \frac{S^{(=)}}{N} = f_1 \left(1 - \frac{\Phi_0}{N}\right). \tag{13.1.1}$$

Similarly when the force Φ acts perpendicularly to the plane,

$$f_1^{(\perp)} = f_1 \left(1 - \frac{\Phi_0}{N}\right). \tag{13.1.2}$$

If the force Φ is parallel to the plane and is perpendicular to the force $\mathbf{S}$, i.e. to the plane of the Figure, then

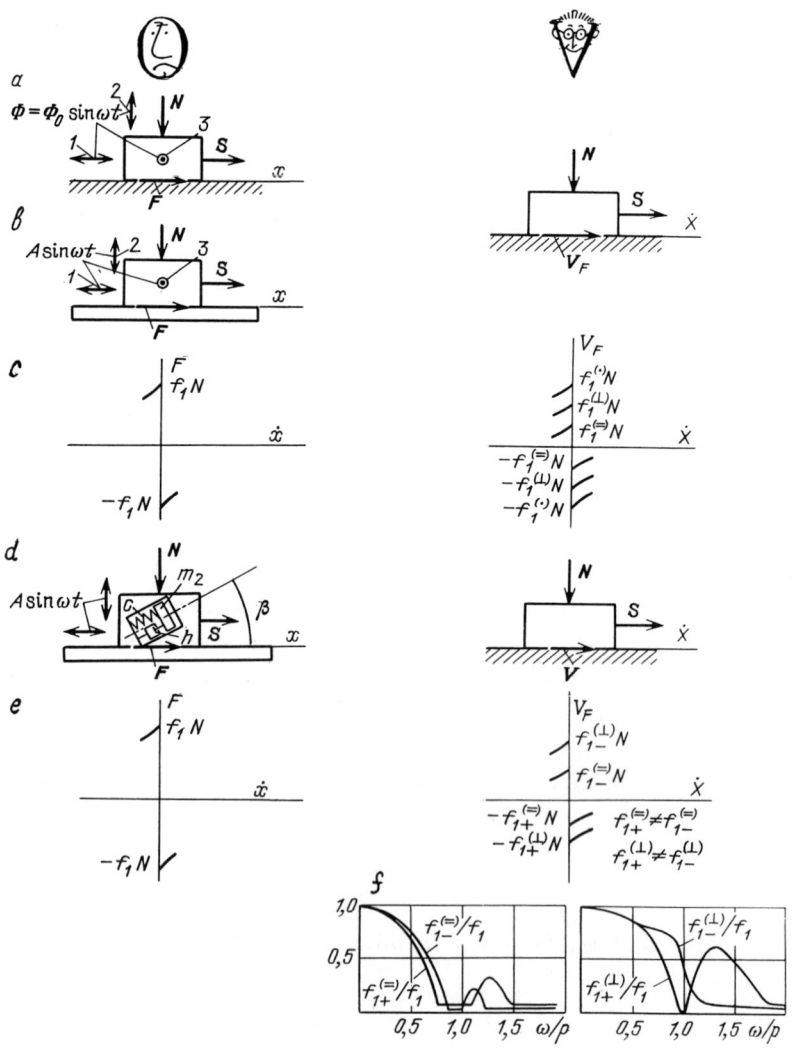

Figure 13.1. Effective frictional coefficients of rest under vibration.

$$f_1^{(\cdot)} = f_1 \sqrt{1 - (\Phi_0/f_1 N)^2}. \qquad (13.1.3)$$

Let us introduce the parameter

$$w = \Phi_0/N, \qquad (13.1.4)$$

characterizing the relative intensity of vibration and called the *parameter of overload*, or, simply, *overload*. Then formulas (13.1.1)–(13.1.3) will be written as

$$f_1^{(=)} = f_1 \left(1 - \frac{w}{f_1}\right), \quad f_1^{(\perp)} = f_1(1 - w), \quad f_1^{(\cdot)} = f_1 \sqrt{1 - (w/f_1)^2}. \quad (13.1.5)$$

Formulas (13.1.5) remain valid in case the force Φ is absent, but the plane performs harmonic oscillations in the adequate directions (Fig. 13.1,*b*); then it is only necessary to calculate the parameter of the overload according to the formula

$$w = mA\omega^2/N, \qquad (13.1.6)$$

i.e. to assume that in (13.1.4) $\Phi_0 = mA\omega^2$ where m is the mass of the body, A is the amplitude and ω is, as before, the frequency of vibration. Finally, if the normal force N is the weight of the body mg, then

$$w = A\omega^2/g, \qquad (13.1.7)$$

Formulas (13.1.1)–(13.1.3) and (13.1.5) make sense only while the values $f_1^{(=)}$, $f_1^{(\perp)}$ and $f_1^{(\cdot)}$ called the *effective coefficients of friction under vibration* are positive; at the larger values of the overload parameter w there is a seeming change in the character of friction (see 14.2). In that case the effective coefficients of friction must be considered to be equal to zero.

13.1.2   A More Complicated Model is a Solid Body with an Internal Degree of Freedom

A more complicated and more interesting result is obtained if the body on a rough vibrating surface possesses an "internal degree of freedom", i.e. it is linked with a certain body with a mass m_1 (Fig. 13.1,*d*) by means of an elastic element with a rigidity c and a damping element with the coefficient of viscous resistance h; let that additional body be able to move with respect to the main body along a certain fixed direction, making the angle β with its foundation. The investigation, made by Molosyan and the author [89], similar to that given above, shows that in the case under consideration due to the *structural asymmetry* of the system (see 9.3. and Fig. 9.2,*aIII*) which takes place provided $\beta \neq 0$, $\pi/2$ and π, the conditions of the beginning of sliding of the main body to the right with the increase of the force S do not coincide with the conditions of the beginning of its sliding to the left. Therefore it is necessary to distinguish here the effective coefficients when there is sliding to the right (f_{1+}) or to the left (f_{1-}) (Fig. 13.1,*e*).

Fig. 13.1,*f* shows the graphs of the dependence of the relative effective coefficients of friction on the ratio of the oscillation frequency of the plane ω to the free oscillation frequency of the internal body $p = \sqrt{c/m}$, with the main body fixed. From the graph it follows that the character of the effects under investigation is that of resonance: the distinction of the coefficients $f_{1+}^{(=)}$ from $f_{1-}^{(=)}$ and $f_{1+}^{\perp}$ from $f_{1-}^{\perp}$ is clearly manifested within the frequency range $0.6p < \omega < 1.7p$. At $p/\omega \to \infty$, i.e. in the case when the internal body is rigidly linked with the main body as it must be, $f_{1+}^{(=)} = f_{1-}^{(=)}$ and $f_{1+}^{\perp} = f_{1-}^{\perp}$.

From Fig. 13.1,*f* it also follows that depending on the value ω the presence of the internal degree of freedom may lead either to the decrease or to the increase of the effective coefficients of friction. Then if within certain ranges the frequency changes are $f_{1+}^{(=)} > f_{1-}^{(=)}$ (or $f_{1+}^{\perp} > f_{1-}^{\perp}$), then in other ranges the opposite inequalities may be satisfied. This proves that by changing the oscillation frequency ω at the fixed values of the other parameters we may have the direction of motion of the body on the plane be changed: if $f_{1+}^{(=)} = 0$ and $f_{1-}^{(=)} > 0$, then in the absence of the force S the system moves over the plane to the right, and if $f_{1-}^{(=)} = 0$ and $f_{1+}^{(=)} > 0$, then it moves to the left. This conclusion has been proved experimentally. Finally, it follows from the graphs that when ω are larger than a certain value, the effective coefficients of friction become zero, which shows the "change in the character of friction" — the *seeming transformation of the dry friction into viscous* (see 14.2).

13.1.3   Several Other Models; Shock Effect; Experiment of Tolstoy

Besides the systems, described above, some other systems have been considered which simulate the reduction of the friction coefficient of rest under the action of vibration and impact. We will briefly describe several of such models.

Fig. 13.2,*a* shows two absolutely solid bodies m_1 and m_2 pressed to each other by elasic elements c with a force which at the state of rest of the bodies is equal to N_0. The bodies are acted upon by the shifting forces S; the yoke in which the system is located performs the preassigned harmonic oscillations with an amplitude A and the frequency ω.

Such a system is the simplest model which considers the elastic and inertial properties of the contact zone of the contiguous bodies. The investigation made

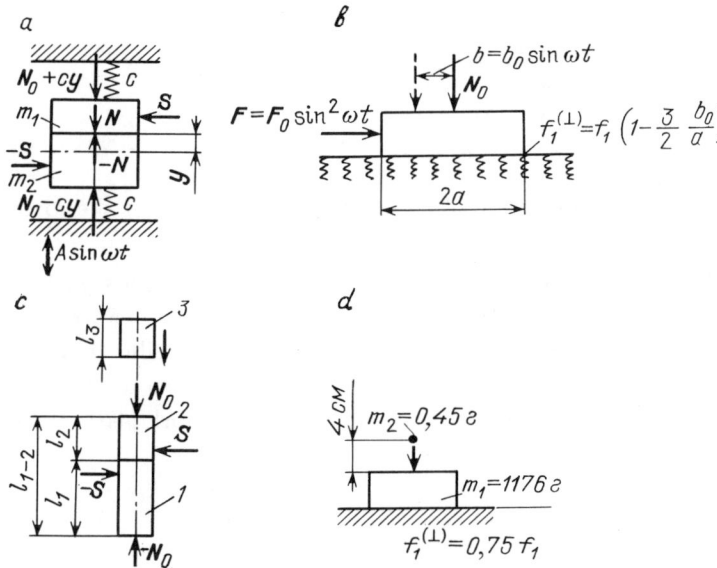

Figure 13.2. Effective frictional coefficients of rest under vibration and shock effects.

by Ryspekov [470] and the author results in the following expression for the effective coefficient of friction:

$$f_1^{(\perp)} = f_1 \left(1 - \frac{N_\sim}{N_0} \right). \tag{13.1.8}$$

Here $N_\sim = m_* wr\eta$ is the amplitude of the variable component of the normal pressure between the bodies,

$$m_* = \frac{m_1 m_2}{m_1 + m_2}, \quad w = A\omega^2, \quad r = \frac{|p_1^2 - p_2^2|}{p^2}, \quad \eta = \frac{p^2}{p^2 - \omega^2},$$

$$p^2 = \frac{c_1 + c_2}{m_1 + m_2} \quad (p \neq \omega), \quad p_1^2 = \frac{c_1}{m_1}, \quad p_2^2 = \frac{c_2}{m_2}. \tag{13.1.9}$$

As one can see, the coefficient $f_1^{(\perp)}$ can be substantially reduced and can become zero even under a relatively weak vibrational action if the frequency ω is approaching the eigen frequency p.

Fig. 13.2, *b* shows a model, shaped as a rigid stamp, lying on an elastic base under the action of a variable longitudinal force $S = S_0 \sin^2 \omega t$ and a constant normal force N_0 with a variable point of application. Florina [187] obtained the following expression for the effective coefficient of friction:

$$
f_1^{(\perp)} = \begin{cases} f_1 \left(1 - \dfrac{3}{2} \dfrac{b_0}{a} \right) & \text{at} \quad \dfrac{3}{2} \dfrac{b_0}{a} < 1 \\ 0 & \text{at} \quad \dfrac{3}{2} \dfrac{b_0}{a} > 1. \end{cases}
\tag{13.1.10}
$$

Here b_0 is the shift amplitude of the point of application of the force N_0, and a is half the size of the body.

Fig. 13.2, *c* shows a system with a shock: two homogeneous elastic rods with a distributed mass, l_1 and l_2 long, are pressed to each other with a force N_0; the shifting forces S are applied to the rods. Rod *2* collided with a certain velocity v with homogeneous rod *3*, l_3 long. The expression for the coefficient $f_1^{(\perp)}$ is in this case obtained in the above-mentioned work [470]; it looks as

$$
f_1^{(\perp)} = f_1 \left(1 - \frac{\sigma_{\sim}}{\sigma_0} \right).
\tag{13.1.11}
$$

Here $\sigma_{\sim} = N_{\sim}/F$ is the amplitude of the variable, and $\sigma_0 = N_0/F$ is the constant component of the normal pressures in the boundary cross-section between rods *1* and *2* (F being the cross-section area of the rods), with

$$
\sigma_{\sim} = \frac{v}{2c} E, \quad c = \sqrt{E/\rho},
\tag{13.1.12}
$$

where c is the velocity of the propagation of sound in the rods, E is the modulus of elasticity, and ρ is the density of the material of the rods. Formula (13.1.11) reflects that in certain periods of time after collisions there appear dynamic tensile stresses in the cross-section under consideration.

The conclusion, following from formula (13.1.11), which seems unexpected on the face of it and which is of a considerable practical importance, lies in the fact that a relatively weak impact can make the effective friction coefficient of rest $f_1^{(\perp)}$ become zero, for a certain, may be even for a very short period of time. This conclusion, however, agrees very well with the result of a quite remarkable experiment, made by Tolstoy [548, 549]: when a ball whose mass was as small as $m_1 = 0.45$ g fell on a "slider" whose mass was $m_2 = 1176$ g from a height of 4 cm, the coefficient $f_1^{(\perp)}$ was decreased by 25%! (See Fig. 13.2, *e*).

13.1.4   Vibrational Conception of the Sliding Friction; Vibrational Control of Dry Friction

In 13.1.1. we spoke about the friction coefficient of rest f_1. As for the sliding friction coefficient f, we know that it is always somewhat smaller than f_1. However, until recently there has not been given any explanation of this important fact. At present there is every ground to say that the "physical" sliding friction coefficient does not differ from the friction coefficient of rest, and the measurable sliding friction coefficient f is nothing else but the effective coefficient of friction under vibration' $f_1^{(\perp)}$ according to the definition given above.

The possibility of such interpretation apperas due to the idea of a close connection of sliding friction and vibration, developed by Tolstoy [114, 201, 224, 548, 549]. According to that idea, when rough deformable bodies are sliding (see the model in Fig. 13.3,a), random wide-band perturbations are sure to be generated. As that takes place, perturbations whose frequencies approach

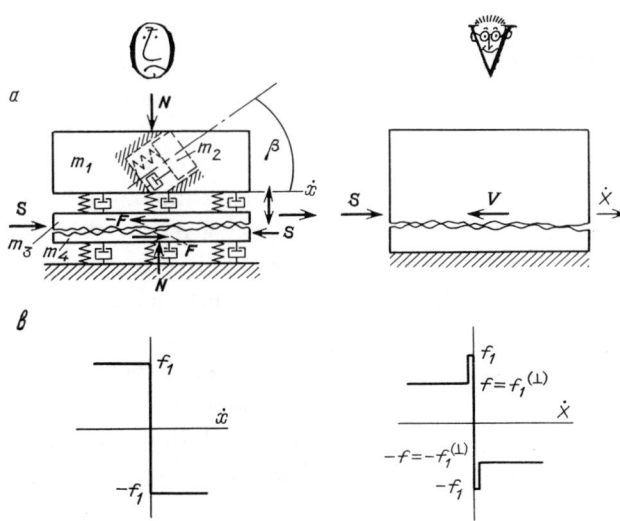

Figure 13.3. Vibrational conception of sliding friction.

those of eigen frequencies, conditioned by the elasticity of the contact, lead, due to the resonance amplification, to the essential, normal with respect to the contiguous surfaces, relative oscillations of bodies (in fact to self-excited -oscillations).

In the process of those oscillations, which were actually observed experimentally, the contact between the bodies may be broken, just as it may happen in the model, shown in Fig. 13.1,d. As a result, there is a seeming reduction of the friction coefficient f_1 to the effective value $f_1^{(\perp)}$ which is usually interpreted as the sliding friction coefficient f (Fig. 18.3,b).

It is remarkable that in the case under consideration the point of view of the observer $\mathbf{V}$ who does not see the vibration is shared by all specialists in mechanics. It should be noted that this effect may occur not only due to the break of contact between the contiguous surfaces in the process of oscillations, but also due to the nonlinear character of forces of elasticity in the contact (springs in Fig. 13.3,a; also see [201]).

The design and analysis of wave models, explaining the nonlinear character of the dependence of dry friction force on the velocity of sliding is the subject matter of the investigation [418].

If we assume that the model under consideration has the "internal degree of freedom" (a certain mass on a spring, shown in Fig. 13.3,a by hatching) with an asymmetry of structure (the angle β is different from 0 or from $\pi/2$), then it is possible to predict different values of the sliding friction coefficient during the motion in different directions, just like in the case of the model in Fig. 13.1,d. A proper experiment might become another indirect confirmation of the vibrational conception of the sliding friction. As for the effective (seeming) reduction of the dry friction force under a specially excited vibration, this effect was proved and is successfully used in technology (the so called *vibrosupports*, see e.g. [44, 256]). In a sense we may even say that due to vibration the sliding friction coefficient may become negative, i.e. the force of friction may be transformed into a driving force — we mean vibro-engines (vibrational transformers of motion), considered in 10.4.

In accordance with what has been said we cannot help mentioning the way of controlling the process of vibro-displacement of a body, lying on a rough vibrating surface, by means of controlling the sliding friction coefficients and friction coefficients of rest. This method was suggested by Fedaravichus [180]; his theoretical investigation is given in the article by Rotovas, Fedaravichus, and the author [122]. The method consists in the fact that the plane, vibrating with the main frequency ω is given an additional transverse vibration of a much higher frequency Ω and of smaller amplitude, the latter changing during the period $T = 2\pi/\omega$ according to a certain given law. In particular, at certain intervals of the period the additional exciter may be switched off altogether, and then the indicated amplitude is equal to zero. As a result, the friction coefficients f_1 and f become the controlled functions of time with the indicated period. Choosing the law of the control, it is possible to provide the greatest speed of vibro-transportation in the selected easily changeable direction at the given character of the main vibration (with the frequency ω). In this case the law of the main vibration can be absolutely symmetrical, in particular, harmonic; the force asymmetry is provided here by the law of the change of the coefficients f_1 and f in time (see 9.2 and 9.3). The solution of other problems of optimal control is also possible. The idea, presented here, seems to be most promising for various applications of the effects of vibrational displacement.

It should be noted that in this case we deal with a *three-level hierarchy of rates of the course of processes*: "very fast" refers to the additional microvibration of the frequency Ω which controls the friction coefficients f_1 and f; "fast" refers to the main macrovibration of the frequency ω, and finally, "slow" refers to the process of vibrational displacement with the velocity $\dot{X}(t)$. The asymptotic methods of the theory of differential equations, describing such hierarchical processes, have been elaborated by Petchenev [435] with the use of the idea of averaging. (See also the book by Nayfeh [405])

13.1.5   Application to the Theory and to Projecting Designs, Operating in Conditions of Impacts and Vibration

Machines often have detachable joints of parts whose relative immobility is provided on the basis of forces of dry friction. Among them there are various thread joints, joints, provided by interference fits and many others. Mechanics-operators have noticed that if the machine works in the conditions of impacts and vibration, then the "nominally immobile" parts in reality possess a certain micromobility — with time the nuts get unscrewed, the contacting surfaces of the parts of machines get ground in and even worn out which results in undesirable consequences and sometimes in accidents.

What has been said in this section, as well as in chapters 9 and 10, makes it possible to explain these facts and in some cases to make estimate calculations. But the main qualitative conclusions consist in the following:

1) In the designs, working in the conditions of impacts and vibration one should be most careful when counting upon the action of the forces of dry friction, even in case there is a specially made interference fit;

2) In such designs it is desirable to avoid any asymmetry of structure of the system, except the cases when such asymmetry guarantees facilitating the so called self-tightening joint.

It is surprising that the mistakes connected with the effects under consideration are sometimes made when designing machines, even by experienced designers, not withstanding the fact that they have enough facilities at their disposal to prevent or at least essentially diminish the influence of those unwanted phenomena.

Meanwhile the reduction of the effective coefficient of dry friction under vibration and under impacts is successfully used in special devices for screwing the nuts — in the so called *power nut-drivers* (see, say, [564, vol. 4]).

13.1.6 Possible Applications to Seismology and to the Theory of Explosive Effects

The aim of the present item is to draw the attention of specialists who are involved in investigating earth-quakes and consequences of explosions to the results of the investigations of models, shown in Figs. 13.1 and 13.2. At least two conclusions follow from those results, which may be of interest for such specialists:

1) Even under the action of seemingly slow impacts, and in case of resonance phenomena — in case of a weak vibration as well — the resistance to the processes of shift may in fact disappear: the indicated effects are able to initiate serious seismic events. In other words, the idea that one must be very cautious when dealing with explosions in seismic regions, or vice versa that one must organize special "discharging" explosions fixed for a definite time, has certain grounds.

2) Objects or structures with internal degrees of freedom (see, say, Fig. 13.1,*d*) which stand freely on the ground or on another foundation, may, under the action of a rather weak vibration of the foundation acquire an unwanted mobility, especially if there is asymmetry of structure of the system and resonance situations [89].

It should be noted that ideas and hypotheses of that kind were suggested by Babeshko and Frolov [33].

At the international conference which held in Baku in June 1991 it was suggested by Nikolayev and Kerimov that it is quite possible that there is a connection between the earth-quakes which occurred in the Caucasus and in Central Asia and the underground nuclear explosions and heavy bombing during the hostilities in the Persian Bay [254].

13.2   Effective Friction when the System with a Positional-Viscous Resistance is acted upon by Vibration

13.2.1   *Vibrational Transformation of the Characteristic of the Resistance of the Oscillatory System with One Degree of Freedom*

As was mentioned, in systems with nonlinear viscous friction under vibration there is no qualitative change of the seeming character of friction — at slow motions the friction still remains viscous, however, characteristics of resistance may in that case change quite essentially. We have already come across such vibrational transformation of the characteristics of viscous friction when we considered the motion of a particle in the medium with nonlinear-viscous resistance under the action of vibration (see 9.3 and 10.5). Now we will consider the case of an oscillatory system with one degree of freedom, described by the equation of the type

$$\ddot{x} + \mu f(x)\dot{x} + p^2 x = A\omega^2 \cos\omega t \tag{13.2.1}$$

where

$$f(x) = a_0 + a_1 x + a_2 x^2 + a_3 x^3 + a_4 x^4, \tag{13.2.2}$$

μ, p^2, A and ω are the positive constants, with μ being a small parameter; we also assume that the vibration frequency ω is much greater than the frequency of free oscillations of the system p in the absence of the force of resistance; the constants $a_0, \ldots, a_4$ can have either positive or negative values; some of them may even be zeros. The force of resistance which in equation (13.2.1) is answered by the term $\mu f(x)\dot{x}$ will be arbitrarily called by us *positionally-viscous resistance*; the polynomial $f(x)$ is in this case "cut off" by us at the fourth degree of x since that is enough to reveal the main regularities, essential for applications.

We are interested in the solutions of equation (13.2.1) of the type

$$x = X(t) + \psi(t, \omega t) \tag{13.2.3}$$

where X is the slowly changing and $\psi(t, \omega t)$ — the fast changing functions of time, with ψ periodic with respect to ωt with a period of 2π and it satisfies the condition

$$< \psi(t, \omega t) >= 0. \tag{13.2.4}$$

Using the method of direct separation of motions, we will write equations (3.2.55) and (3.2.56) of chapter 3 as

$$\ddot{X} + p^2 X = \mu < (\dot{X} + \dot{\psi})f(X + \psi) >, \tag{13.2.5}$$

$$\ddot{\psi} + p^2\psi = A\omega^2\cos\omega t + \mu[(\dot{X}+\dot{\psi})f(X+\psi) - \, <(\dot{X}+\dot{\psi})f(X+\psi)>]. \quad (13.2.6)$$

With X and $\dot{X}$ frozen, and μ small enough, the equation of fast motions (13.2.6) has a single 2π-periodic with respect to ωt solution

$$\psi = \frac{A\omega^2}{p^2-\omega^2}\cos\omega t + O(\mu), \quad (13.2.7)$$

satisfying condition (13.2.4). Therefore with an accuracy to the terms of the order of μ and in view of the obvious equality

$$< \psi^n\dot{\psi} >= 0 \quad (13.2.8)$$

we will obtain

$$< (\dot{X}+\dot{\psi})f(X+\psi) >= \dot{X}f(X)$$
$$+ \left[\frac{1}{2}B^2a_2 + \frac{3}{8}B^4a_4 + \frac{3}{2}XB^2a_3 + 3X^2B^2a_4\right]\dot{X} \quad (13.2.9)$$

where

$$B = A\omega^2/(p^2-\omega^2) = A/(\lambda^2-1), \quad \lambda = p/\omega. \quad (13.2.10)$$

As a result, the vibrorheological equation (13.2.5) (the main equation of vibrational mechanics) will be presented in the form

$$\ddot{X} + \mu f_1(X)\dot{X} + p^2X = 0. \quad (13.2.11)$$

Here $f_1(X)\dot{X}$ is the function which characterizes the positional- viscous resistance to the slow motions of the system, i.e. it is the *effective characteristic of friction*. As a matter of convenience for comparison with the initial characteristic $f(x)\dot{x}$ we will write out these characteristics one under the other

$$f(x)\dot{x} = (a_0 \quad + \quad a_1x \quad + \quad a_2x^2 \quad + \quad a_3x^3 \quad + \quad a_4x^4)\dot{x},$$
$$\quad (13.2.12)$$
$$f_1(X)\dot{X} = (A_0 \quad + \quad A_1X \quad + \quad A_2X^2 \quad + \quad A_3X^3 \quad + \quad A_4X^4)\dot{X}.$$

Hence we see the character of transformation, suffered by the initial characteristic of resistance: the coefficient of the linear part of the resistance a_0 has been changed for the expression $A_0 = [(a_0 + \frac{1}{2}B^2a_2 + \frac{3}{8}B^4a_4)$, the coefficient a_1 at $x\dot{x}$ has been changed for the term $A_1 = a_1 + \frac{3}{2}B^2a_3$, $A_2 = a_2 + 3B^2a_4$, $A_3 = a_3$ $A_4 = a_4$ etc. These changes explain the important regularities of slow motions of the system which are considered below.

Equation (13.2.11) is answered by the only position of equilibrium $X = 0$; this position will be either stable or unstable depending on the type of the function $f_1(X)\dot{X}$. If all coefficients $a_0, \ldots, a_4$ are positive, then from (13.2.12) it follows that the effective damping in the system may become much greater than the initial, and if they are negative, then, accordingly, the "negative resistance" grows and it leads to the instability of the equilibrium position $X = 0$. Most interesting, however, are cases when the indicated coefficients have different signs. In such cases the positive damping near the point $x = 0$ in the initial system (13.2.1) ($a_0 > 0$) may be answered by the negative damping near $X = 0$ in the system, described by equation (13.2.10) ($a_0 + \frac{1}{2}B^2 a_2 + \frac{3}{8}B^4 a_4 < 0$), and vice verse. Two such cases are considered below.

It should be noted that the a posteriori check of the validity of assumptions of the method of direct separation of motions seems unnecessary here, since equalities (13.2.3), (13.2.7) and equations (13.2.11) can be considered as just a form of presenting the solution of the initial equation according to the degrees of the small parameter μ.

It should be also remarked that when solving this problem we did not use directly the expression for the vibrational force

$$V =< (\dot{X} + \dot{\psi})f(X + \psi) - \dot{X}f(\dot{X}) >$$

since there was no necessity in it.

13.2.2   Special Cases — the Asynchronous Suppression and Excitation of Self-oscillations; Some Applications

When solving the classical problem on the *asynchronous suppression of self-oscillations*, the following differential equation is usually considered:

$$\ddot{x} - \mu(1 + \beta x - \alpha x^2)\dot{x} + p^2 x = A\omega^2 \cos\omega t \quad (\mu > 0,\ \alpha > 0,\ \beta > 0), \quad (13.2.13)$$

which in equation (13.2.1) is answered by the values of the coefficients

$$a_0 = -1, \quad a_1 = -\beta, \quad a_2 = \alpha, \quad a_3 = a_4 = 0. \quad (13.2.14)$$

In this case the vibro-rheological equation (13.2.12) takes the form

$$\ddot{X} - \mu(1 - \frac{A^2}{A_*^2} + \beta X - \alpha X^2)\dot{X} + p^2 X = 0 \quad (13.2.15)$$

where

$$A_* = \sqrt{2/\alpha}\,|1 - \lambda^2|, \quad \lambda = \frac{p}{\omega}. \quad (13.2.16)$$

A rather simple investigation shows (see e.g. [308, 363]) that due to the negative damping at small x and in case there is the term $\alpha x^2 \dot{x}$, the initial equation (13.2.13) in the absence of vibrational action ($A = 0$) describes the stable self-oscillations with a frequency approaching p. The position of equilibrium $x = 0$ is in this case unstable. Equation (13.2.15), provided the following condition is satisfied

$$A > A_* \qquad (13.2.17)$$

and provided X are small enough, is answered by positive damping, due to which the position of equilibrium ($X = 0$) becomes asymptotically stable, and self-oscillations are suppressed. The value A_* plays the role of the *critical amplitude* of vibrational action at which asynchronous suppression of self-oscillations can commence.

Thus, the well known result has been obtained by us as a consequence of changing the effective rheological characteristic of the system — the appearance due to vibration of additional positive damping.

Usually, when studying the *asynchronous excitation of self-oscillations* one considers the following differential equation [308]

$$\ddot{x} + \mu(1 - \alpha x^2 + \gamma x^4)\dot{x} + p^2 x = A\omega^2 \cos\omega t \quad (\mu > 0, \, \alpha > 0, \, \gamma > 0), \quad (13.2.18)$$

which is answered in equation (13.2.1) by the values of the coefficients

$$a_0 = 1, \quad a_1 = 0, \quad a_2 = -\alpha, \quad a_3 = 0, \quad a_4 = \gamma. \qquad (13.2.19)$$

Then the vibrorheological equation is reduced to the form

$$\ddot{X} + \mu(1 - \frac{1}{2}\alpha B^2 + \frac{3}{8}B^4\gamma + 3X^2 B^2\gamma - \alpha X^2 + \gamma X^4)\dot{X} + p^2 x = 0 \quad (13.2.20)$$

where B as before is defined by equality (13.2.10).

When A=0, the point $x = 0$ is for equation (13.2.18) a point of stable equilibrium — self-oscillations are not excited. While for the vibrorheological equation the condition of instability of the equilibrium point $X = 0$ and at the same time the condition of exciting self-oscillations is the inequality

$$B^4 - \frac{4}{3}\frac{\alpha}{\gamma}B^2 + \frac{8}{3\gamma} < 0, \qquad (13.2.21)$$

which can be satisfied if

$$\alpha^2 > 6\gamma. \qquad (13.2.22)$$

When this condition is satisfied, inequality (13.2.21) will be valid provided the amplitude A lies within the range

$$(1 - \lambda^2)^2 B_1^2 < A^2 < (1 - \lambda^2)^2 B_2^2 \qquad (13.2.23)$$

where B_1^2 and B_2^2 are the roots of the equation

$$B^4 - \frac{4}{3}\frac{\alpha}{\gamma}B^2 + \frac{8}{3\gamma} = 0, \qquad (13.2.24)$$

i.e.

$$B_{1,2}^2 = \frac{2}{3}\frac{\alpha}{\gamma}\left(1 \pm \sqrt{1 - 6\gamma/\alpha^2}\right). \qquad (13.2.25)$$

So in the case under consideration the realization of inequalities (13.2.22) and (13.2.23) leads to the excitation of self-oscillations under the action of vibration, while in the absence of vibration, self-oscillations are not excited.

The results obtained coincide with the well known results, obtained in another way (see e.g. monographs [308, 363]); though in the book [363] there is an arithmetic mistake in the record of the solutions of the quadratic equation (13.2.24).

Touching upon the applications of the effect of vibrational suppression of self-oscillations, we will mark that it is used when suppressing the running strata in the gas-discharge devices by means of imposing a field of super-high frequency on the discharge. The relevant theory is considered in the monograph by Landa [309] which also contains the bibliographic data.

In the work of Machabeli and the author [105] a hypothesis is put forward that the *mechanism of automaticity and redundancy in the system of exciting the cardiac rhythms* can be explained by the phenomenon of the asynchronous damping of self-oscillations. Normally the "guide of the rhythm" is a sinusoid node exciting the contraction of the muscles of the heart with a frequency of about 60–80 contractions per minute. These contractions suppress self-oscillations which may excite the reserve guides of the rhythm — the atrioventricular node and the Purkyne fibers. In case of failure of the sinusoid node, the atrioventricular node acts as the guide of the rhythm with a frequency of about 40–60 contractions per minute, and in case the latter fails as well, then it is the Purkyne fibers (the frequency of contractions 30–40 per minute). In connection with the above-mentioned hypothesis it should be noted that simulating the work of the heart by nonlinear oscillators, particularly by the so called *relaxational generators*, goes as far back as the classic investigations by Van der Pol and Van der Mark [558].

The supposition that the reduction of the level of vibration of the instrument during vibrational cutting is explained by the phenomenon of asynchronous suppression of self-oscillations was advanced in 10.3.

13.3   Equation of Reinolds' as a Vibrorheological Equation. Effective Viscosity of Fluid in Turbulent Motion; Effect of the External Vibrational Action

It is not difficult to show [95, 99] that Reynolds' classical equation can be interpreted as a vibrorheological equation. As is known, the velocity $\mathbf{u}$ and the pressure p in the turbulent motion of the fluid in many case can be presented as

$$\mathbf{u} = \mathbf{U} + \mathbf{u}', \quad p = P + p' \tag{13.3.1}$$

where $\mathbf{U}$ and P are the slow- and $\mathbf{u}'$ and p' are the fast pulsation components. Let us consider the equation of isothermal motion of viscous incompressible fluid in the absence of volume forces and the equation of continuity

$$\rho\frac{\partial \mathbf{u}}{\partial t} + \rho(\mathbf{u}\cdot\nabla)\mathbf{u} = -\nabla p + \mu\nabla^2\mathbf{u}, \quad \nabla\cdot\mathbf{u} = 0 \tag{13.3.2}$$

where ρ is the density, μ is the coefficient of the viscosity of the fluid and $\nabla = \frac{\partial}{\partial x}\mathbf{i} + \frac{\partial}{\partial y}\mathbf{j} + \frac{\partial}{\partial z}\mathbf{k}$ is the Hamilton operator. The equation of motion in this case does not contain any fast forces and therefore equations (3.2.55) and (3.2.56) of chapter 3 are presented in the form

$$\rho\frac{\partial \mathbf{U}}{\partial t} + \rho(\mathbf{U}\cdot\nabla)\mathbf{U} = -\nabla P + \mu\nabla^2\mathbf{U} + \mathbf{v}, \quad \nabla\cdot\mathbf{U} = 0; \tag{13.3.3}$$

$$\rho\frac{\partial \mathbf{u}'}{\partial t} + \rho[(\mathbf{u}'\cdot\nabla)\mathbf{u}' + (\mathbf{u}'\cdot\nabla)\mathbf{U} + (\mathbf{U}\cdot\nabla)\mathbf{u}')]$$
$$= -\nabla p + \mu\nabla^2\mathbf{u}' - \mathbf{v}, \quad \nabla\cdot\mathbf{u}' = 0,$$
$$\frac{1}{T}\int_0^T \mathbf{u}'dt = 0, \quad \frac{1}{T}\int_0^T p'dt = 0 \tag{13.3.4}$$

where

$$\mathbf{v} = \mathbf{v}^{(i)} = -\frac{1}{T}\int_0^T (\mathbf{u}'\cdot\nabla)\mathbf{u}'d\tau \tag{13.3.5}$$

is a purely induced vibrational force and T is the so called *period of averaging*.

The equation of slow motions of the fluid (13.3.3) which has been obtained, according to the terminology of 3.3. is a vibrorheological equation, exactly coinciding with the Reynolds equation (see e.g. [340]), the vibrational force $\mathbf{v}$ corresponding to the *turbulent stresses*. Here, however, the Reynolds equation belongs to the system of integro-differential equations (13.3.3), (13.3.4) which

like the initial equations (13.3.2) is closed. It is necessary to bear in mind
that the solution of equations (13.3.4) for the pulsatory components $\mathbf{u}'$ and p'
even at the "frozen" $\mathbf{U}$ and P and their derivatives presents certain difficulties,
though much lesser than when solving directly the initial equations (13.3.2).
Besides, as was marked more than once, the approximate solution of equations
(13.3.4) is also permissible here. Taking into consideration what has been said
the author has but a faint hope that in this case the approach to the problem
from the position of vibrational mechanics will in this case be useful. The aim
of this section was just to illustrate the connection of the indicated approach
with the well known classical results.

As it must be, equations (13.3.4) allow for the pulsatory components to
have a trivial solution $\mathbf{u}' = 0$, $p' = 0$ which in case of its stability answers
the laminar motion. Therefore we spoke about seeking the solutions of these
equations corresponding to the self-oscillations of a chaotic character.

By analogy with equation (13.3.2) the turbulent stresses in the applied
theory of turbulence are sometimes presented as

$$\mathbf{v} = \mu^* \nabla^2 \mathbf{U}, \qquad (13.3.6)$$

introducing the *coefficient of turbulent viscosity* μ^*; the latter, according to
our terminology being nothing else but the *effective viscosity coefficient* in the
turbulent flow of the fluid; it is answered by the viscosity, perceived by the
observer $\mathbf{V}$ (see Fig. 1.2,c); it greatly exceeds the usual viscosity of the fluid.
In this connection academician Novozhilov used to say that in the turbulent
motion the water behaves as if it were transformed into pitch.

Quite interesting is the possibility of changing (end even reducing !) the
turbulent viscosity of the fluid under the ultrasonic or vibrational action upon
the turbulent jet [569]. Here one can see the analogy with what was said in
13.2.2, though the mechanism of the phenomenon seems to be different, at any
rate more complicated than that considered by us. We think, however, that
the vibrorheological approach will be of use when studying this phenomenon.
Data about numerous experimental investigations made in this field are given
in a detailed review by Vlasov and Ginevsky [570].

It should be also noted that the transfer to the vibrorheological equations
is successfully used in hydromechanics in considering the effect of vibration
on convective flows and on their stability (see in particular monograph [200]
and publications [38, 263]). There, unlike the classical problem, in many cases
which are of practical interest it proves possible to construct a satisfactory approximate solution of the equations of fast motions and to obtain an expression
for the vibrational forces.

13.4   On Other Cases of Using Concepts of Effective Viscosity under Vibration

It is sometimes convenient to introduce the coefficients of effective viscosity in case systems with dry friction are acted upon by intensive vibration (see 10.2.4, and also chapter 14). It is also expedient to use the concept of effective viscous resistance when studying the vibration of vessels with suspension and also when solving problems of synthesis of dynamic materials (see chapter 17 and 18.2).

Chapter 14

Vibrorheological Transformation of Nonlinear Mechanical Systems with Discontinuous Characteristics into Systems with Viscous Friction

14.1 Vibrorheology of Systems with Dry Friction

When vibration becomes intensive enough, the effective coefficients of dry friction, calculated according to formulas of 13.1, become zero, which shows the qualitative change (for the observer **V** !) of the character of friction in the system — the dry friction is transformed into viscous (about the principal distinction of dry friction from viscous — see the footnote on page 239).

The *effect of the vibrational transformation of dry friction into viscous* was considered in detail in the previous chapters. So now we will just recall the main theses and will give the main references. First of all it should be emphasized again that we speak here about the *seeming effect* which refers only to slow forces and which disappears with the end of vibration. This effect is, as a rule, accompanied by the appearance of a driving vibrational force, and as far as the slow actions are concerned, there is a smoothening of the discontinuous characteristic of dry friction. These effects were illustrated in 9.2.2 on a very simple model (see Fig. 9.1). As was marked, they are known in the theory of automatic control [279, 295, 442].

These effects play an important role in the process of vibrational displacement and shift (chapters 9–11), and also in the process of separating the particles of granular mixtures (10.2). In particular, they explain the *phenomenon of segregation (self-sorting)* of the particles of granular mixtures under vibration, including the so called *pseudo-resonance effect* (see 10.2.4). The same effects play a negative role, causing the self-unscrewing of nuts and other undesirable processes in the machines where the conjugation of parts is counted on the action of dry friction forces (13.1.5).

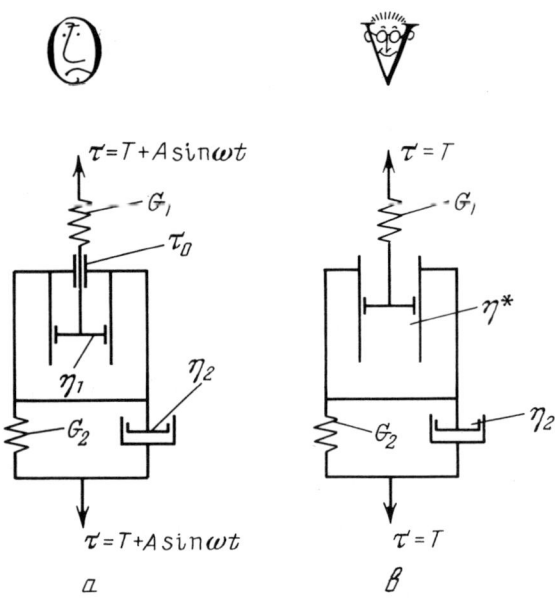

Figure 14.1. The ordinary (a) and vibrorheological (b) models of concrete mixture under shear deformations (G_1 and G_2 are the elastic elements, η_1 and η_2 are the elements with viscous friction, τ_0 is the element with dry friction and η^* is the element with effective viscous friction).

Different cases of the transformation of dry friction which are of practical interest and which are sometimes called *fluidization* have been considered by Andronov [22]. In publications by Albert and Savinov [13], and Kondakov [287] (see also [475]) this effect was investigated for the media of the type of concrete mixtures. Figure 14.1,a shows a usual concrete model, and Fig. 14.1,b — a vibrorheological model of concrete mixture under shear deformations. As one can see in the picture, the element τ_0 with dry friction and the element η_1 with viscous friction in the usual rheological model are answered by the element η^* in the vibrorheological model. Publications [13, 287] contain relationships and graphs, establishing the connection between the characteristics of the usual and vibrorheological models and the parameters of vibration.

When considering the effects of vibrational transformation of dry friction, it is often convenient to introduce the effective coefficients of viscous friction [82, 84, 221, 442, 475]; it is natural that such coefficients depend essentially on the parameters of vibration.

14.2   Vibrorheology of Systems with Periodic Collisions

14.2.1   On Vibrorheological Simulation of Vibroshock Interactions by Forces of Viscous Friction

We have already come across a case when while studying the slow motions, it proved possible to simulate the vibro-shock interactions of the elements of the system by means of viscous resistance forces and a certain constant force. We mean here the process of a vibrational transportation of body under intensive tossing, considered in 9.4.3: the vibrational force in this case appeared to be approximately presentable by expression (9.2.53) of that chapter. Here we will study another very simple system, also interesting for a number of applications.

Let us consider the motion of a body of mass m in the space between two flat surfaces *1* and *2* under the action of a certain force Q which may also include the force of dry friction against the surface *1* (Fig. 14.2,*a*); this surface is assumed to be fixed. Surface *2* performs harmonic oscillations of the frequency ω whose swing of the transverse component A is such that once during each period of oscillations $T = 2\pi/\omega$ when the surfaces approach each other, the body is squeezed between them and stops instantly. When surface *2* is removed from surface *1*, the body is released and again moves along surface *1* under the action of the force Q. We will consider this force to be constant or changing but very little during the oscillation period T. All the other parameters will be also considered by us to be either constant or changing very slowly.

We will designate by $T_1 \leq T = 2\pi/\omega$ the period of time of the free motion of the body under the action of the force Q. Then the displacement of the body at such motion is

$$x = \frac{Q}{m} \frac{t^2}{2} \qquad (0 \leq t \leq T_1),\qquad\qquad (14.2.1)$$

and the average for the period velocity (the velocity of slow motion) is

$$\dot{X} = <\dot{x}> = \frac{1}{T} \int\limits_0^{T_1} \dot{x}\,dt = \frac{1}{2}\frac{Q}{m}\frac{T_1^2}{T}. \qquad\qquad (14.2.2)$$

It is obvious that the same value of the velocity of the stationary motion must be obtained from the corresponding main equation of vibrational mechanics:

$$m\ddot{X} = Q + V. \qquad (14.2.3)$$

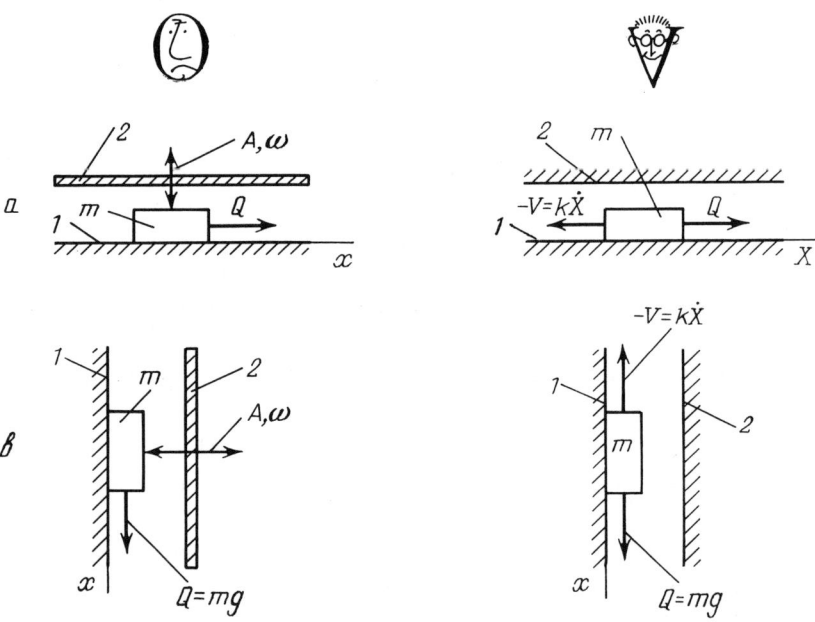

Figure 14.2. Modeling vibro-shock interactions by viscous friction forces when studying slow motions. *a*) The general scheme of the system; *b*) the motion of the body by gravity. On the right there are the pictures seen by the observer **V**.

Comparing this equation at $\dot{X} = \text{const}$ with relationship (14.2.2), we find the following expression for the vibrational force, obtained by Krasnov and the author:

$$V = -k\dot{X} \qquad \left(k = 2m\frac{T}{T_1^2} \right). \qquad (14.2.4)$$

Thus we come to the conclusion about a possibility of simulating the considered vibro-shock system when studying its slow motions by means of a simplest system with viscous friction (see the right-hand part of Fig. 14.2). It should be noted that formula (14.2.4) remains valid in case planes *1* and *2* are vertical, and $Q = m\,g$, i.e. it presents the gravity force (Fig. 14.2,b). Then according to formula (14.2.2) we obtain

$$X = \frac{1}{2}g\frac{T_1^2}{T}. \tag{14.2.5}$$

We must note that all the relationships which refer to the above-considered example are exact: they follow from the exact solution (14.2.1). We should also note that the motion of the body, corresponding to this solution, practically immediately becomes stationary — after the moment the body is first squeezed between the surfaces.

If we recall the derivation of the formula for the viscous resistance force, based on the molecular-kinetic representation, the conclusion about the possibility of simulating the vibro-shock interactions when considering the slow motions of "micromechanical" systems will not seem paradoxical. Nevertheless, as far as we know, such a conclusion has not yet been formulated. We can only refer to certain publications where the corresponding representation was successfully used when solving some concrete problems. Here we mean in particular the work [99] in which formula (9.2.63) of chapter 9 was obtained; the publications by Marchenko and the author [88, 90] and Barzukov and Vaisberg [49, 554] where the vibro-shock interactions were simulated by forces of resistance when investigating the self-synchronization of vibro-exciters on the basis of equations, characteristic of vibrational mechanics i.e. the equations of slow forces.

Fidlin has shown the applicability of the method of averaging to the study of vibro-shock systems of the type under consideration. As a result of the use of that method, slow motions are described by differential equations with smooth right sides. The investigation made by him [186] can be considered as a mathematical ground of the vibrorheological model for a wide class of vibro-shock systems (see also the work by Perestyuk [432]).

14.2.2   Applications to the Calculation of the Capacity of Crushers

Apart from the above-listed applications, vibrorheological modeling the vibro-shock interactions by forces of viscous friction makes it possible to come up to the solution of some important practical questions of the theory of cone- and jaw crushers for solid materials.

The process of crushing the material in such crushers is performed in the crushing chamber — the space narrowing downwards between the two working surfaces (linings). Figure 14.3,a shows the cross-section of such surfaces — the transverse cross-section for the jaw crusher and the one along the axis for the cone crusher. For the sake of simplicity the figure shows the case of the linings with the rectilinear cross-sections *1* and *2*; in reality these sections are somewhat distorted, however, in those cases as well these linings can often be considered to be "locally rectilinear" because the displacement of the material during one oscillation period is small as compared to the length and the radius of the curvature of the working part of the linings.

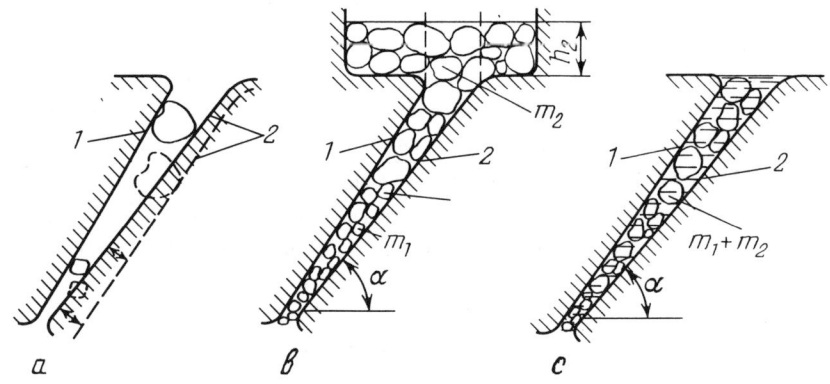

Figure 14.3. On the estimation of the capacity of the cone- and jaw crushers.

One of the linings (*2*) performs oscillations, excited kinematically or (in new designs) — by means of a mechanical unbalanced vibro-exciter [90, 103, 592, 456, 232, 457]. The other surface (*1*) in most designs can be considered to be practically fixed (though the generalization for the case of immobility of both linings does not cause any special difficulty). When the linings approach each other, a considerably big piece of material is squeezed between them, stops and is crushed, and when the plates drift apart, the material is released and falls down in the direction of the narrowing of the chamber, then it is squeezed and crushed again and so on.. With a certain approximation we may assume that when the surfaces come most closely to each other, i.e. once per the oscillation period, all the material to be crushed is stopped, i.e. we have here a situation similar to one considered in 14.2.1 (See Fig. 14.2). Thus the main dissipative factor — systematic stops of the material in every period of motion is taken account of.

Along with that, smaller fractions, formed in the course of crushing, in fact keep on moving without stopping in the space among big pieces of the material, which provides a certain growth of the capacity as compared to the estimation based on the indicated assumption.

Under the formulated assumption, as well as under some other simplifying assumptions, the problem about the displacement and crushing of the materials in the crushing chambers is considered in [50, 98, 100, 107, 262] on the basis of the theory of vibrational displacement (chapter 9). Programs have been made to calculate the capacity of the crusher, the deformations, strength and intensity of the wear of the plate in different cross-sections of the crushing chamber. Along with that when designing and operating the crushers, problems arise concerning an approximate estimation of the influence of some additional factors on the capacity of the machine Q. Such estimations can be easily performed in view of what has been said in 14.2.1.

One of the problems is the estimation of the increase of capacity of the crusher if above the crushing chamber there is a "cap" of the material (Fig. 14.3,*b*). Experience shows that such an increase in fact does take place and is quite considerable — up to 50% and more, as compared to the capacity without the cap. However, up to now there had been no theoretical explanation of this fact.

Considering the slow motion of the material in the crushing chamber as the motion of the viscous fluid along the respective tube and neglecting the driving vibrational force and the dry friction force when the material moves between the squeezes [a], we can conclude that the capacity of the crushers (the discharge of the "fluid") is proportional to the longitudinal projection of the weight of the material $m_1 g$ inside the crushing chamber together with the weight $m_2 g$ of the column of the material over the chamber:

$$Q = k'(m_1 \sin \alpha + m_2) g \qquad (14.2.6)$$

(k' is the proportionality factor). It goes without saying that this formula is valid only in case the height of the column of the material does not exceed the depth of the penetration of vibration inside the column, i.e. the depth at which fluidization takes place (see chapter 16); in the opposite case in formula (14.2.6) the mass m_2 denotes the mass of the column whose height is equal to the corresponding depth of the penetration.

---

[a]In the lower, determining zones of the crushing chambers of modern crushers, most of the time between the squeezes the material falls down freely in the space between the linings, and only during a very short period of that time slides along the lining.

Let Q_0 be the capacity of the crusher in the absence of the "cap", i.e. at $m_2 = 0$. Then, according to (14.2.6)

$$Q_0 = k'm_1 g \sin \alpha \qquad (14.2.7)$$

$$Q = Q_0 \frac{m_1 \sin \alpha + m_2}{m_1 \sin \alpha} = Q_0 \left(1 + \frac{m_2}{m_1 \sin \alpha} \right)$$
$$= k'm_1 g_* \sin \alpha = Q_0 \frac{g_*}{g} \qquad (14.2.8)$$

where

$$g_* = \left(1 + \frac{m_2}{m_1 \sin \alpha} \right) g. \qquad (14.2.9)$$

Formulas (14.2.7) – (14.2.9) in some other form were obtained by Denisov; it follows from them, in particular, that to estimate the effect of the column of the material on the capacity of the crusher in the context of the described vibrorheological model it is possible to use the above-mentioned programs, substituting according to formula (14.2.9) the value g_* for the earth gravity.

Such calculations were made by Katsman and Titova for the crusher KMDT-2200 of the plant "Uralmash". They gave the following results. In the absence of the "cap" the capacity of the crusher (the capacity of the narrowest section of the crushing chamber located at the top of the so called parallel zone) made 115 m³/hr; a similar computation made with the "cap", corresponding to the value $g_* = 15$ m/sec², brought to the value $Q = 159$ m³/hr, while according to the approximate formula (14.2.8) it is $Q' = 115 \cdot 15/9.81 = 176$ m³/hr. A comparatively good agreement of the latter two results shows the adequacy of the vibrorheological model underlying the basis of the approximate formula (14.2.8) to the more exact "micromodel" underlying the basis of the programs for the computer. These results also agree with the experimental data.

Another problem refers to the estimation of the capacity of crushers at the so called wet crushing (Fig. 14.3,c). Theoretic considerations and the preliminary experimental data lead to the conclusion that the water supply to the crushing chamber may provide a number of essential technological advantages, in particular the increase of capacity of the crusher. When solving the problem we will assume, as before, that the capacity of the crusher is proportional to the weight of the material and to the water inside the crushing chamber, and we will roughly assume that the average velocity of the motion of the water does not differ from that of the material. Then

$$Q_w = k'(m_1 + m_w) g \sin \alpha. \qquad (14.2.10)$$

Here, like above, m_1 is the mass of the material in the crushing chamber, and m_w is the mass of the water in it. Denoting the capacity of the crusher at the dry process by $Q_0 = k' m_1 g \sin \alpha$, we will have, similarly to (14.2.8)

$$Q_w = Q_0 \frac{m_1 + m_w}{m_1} = Q_0 \left(1 + \frac{m_w}{m_1}\right) = k' m_1 g_w \sin \alpha = Q_0 \frac{g_w}{g} \qquad (14.2.11)$$

where

$$g_w = \left(1 + \frac{m_w}{m_1}\right) g. \qquad (14.2.12)$$

If we assume that the material fills 0.45 of the volume of the chamber V_0, then the maximum volume which water can take will be $0.55 V_0$ and $m_1 = 0.45 \rho_w V_0$, $m_w = 0.55 \rho_w V_0$ where ρ and ρ_w are respectively the density of the material and of the water. Then we come to the following estimation of the maximum capacity of the crusher Q_w at the wet crushing:

$$Q_w = Q_0 \left(1 + \frac{0.55}{0.45} \frac{\rho_w}{\rho}\right). \qquad (14.2.13)$$

Assuming here that $\rho / \rho_w = 2.5$, we will have

$$Q_w = Q_0 \left(1 + \frac{1.22}{2.5}\right) \approx 1.5 \, Q_0. \qquad (14.2.14)$$

Thus, on the basis of the presented approximate approach, a sesquilateral increase of the capacity of crusher can be expected at the transfer to the wet crushing. This conclusion is in good agreement with the results of the preliminary experiments.

Chapter 15

Vibrorheology of Granular Materials

15.1 Vibrorheological Models of a Layer of Granular Medium

The effective mathematical description of the behavior of a granular medium under vibration is of a considerable interest for applications. A number of interesting publications has been devoted to this problem (see for instance [11, 27, 84, 207, 208, 174, 296, 297, 316, 453, 469, 500, 513, 181, 154, 32, 253, 298, 299, 564, vol. 4]), but it is still far from being solved satisfactorily. This chapter considers on the basis of a vibrorheological approach certain models describing the slow streams of granular material, generated by vibration.

First we will dwell on describing some major regularities of the behavior of the granular medium in vibrating vessels. A certain distribution of vibration in the granular medium (including both the determined on the average component as well as the random component) is established in such vessels sufficiently fast — after several oscillation periods. This is a "fast" process on the background of which some slow processes take place: there appear certain streams and processes of segregation (self- sorting). A number of such processes was considered in 10.1–10.2. It is natural that for the investigation of these slow processes, seen by the observer $\mathbf{V}$, and in fact most interesting, it is of utmost importance to study the fast process, i.e. the stationary vibrational field. We mean here the problem of the penetration of vibration into the granular medium, considered in chapter 16. Now we will dwell on certain qualitative regularities and will make some remarks. Let A be the amplitude and ω be the frequency of the harmonic vibration of the vessel with granular material. At the accelerations $A\omega^2 < g$ for the vertical- and $A\omega^2 < f_1 g$ for the horizontal oscillations (f_1 being the friction coefficient of rest, and g — the free fall acceleration) the material moves mainly together with the vessel. At the accelerations $A\omega^2 \approx g$ the particles of the material acquire a certain mutual mobility — a process of fluidization takes place, which leads first to the condensation and then, with the further growth of $a\omega^2$, to the loosening and mixing. The processes of separation (segregation, self-sorting) take place at the stages of fluidization and loosening.

As the experiment has shown, the air resistance (i.e. its filtration through the layer of particles) is an important factor, affecting the behavior of the granular material in the vibrating vessels when the vertical vibration $A\omega^2 > g$ is intensive, i.e. in motions with tossing [469, 32, 298, 299]. So, for example, while a single big particle or a relatively thin layer of big particles of the material at $A\omega^2 > 3.3g$ can be in the state of breaking away from the bottom of the vessel for the intervals of time longer than the oscillation period $T = 2\pi/\omega$, for a relatively thick layer these intervals as a rule do not exceed the oscillation period. The pressure differences of the air at the height of a layer of the material can exceed the atmospheric pressure by one order of magnitude. In case of small particles the air resistance has a still greater significance [84, 564, vol. 4].

What has been said here gives an idea of how difficult it is to simulate the behavior of the granular material under vibration.

In 9.4 and 14.2 it was shown that when studying the slow motions of the bodies, interacting either by means of dry friction forces or by collisions the indicated interactions can be simulated by means of viscous friction forces in view of the driving vibrational force. This leads to the vibrorheological approach to the modeling of the behavior of the granular material in the vibrating trays and vessels [114]. This approach is discussed below.

As was mentioned in 10.1, when the thickness of the layer h does not exceed a certain value h_* and when the parameters of vibration are within certain limits, the motion of the granular material over a vibrating flat surface can be considered as the motion of a solid body (particle). In this case any body acted upon besides other slow forces by the vibrational force V will be a vibrorheological model. We will call this model "model **A**" (Fig. 15.1, a).

The vibrational force V can be found as a result of solving the corresponding problem on vibrotransportation or else by using the experimental data. So, for instance, in the case of the regimes with a sufficiently intensive tossing of the material over the flat surface, performing translatory oscillations along the elliptical trajectories, the vibrational force is defined by an approximate formula (4.53) of chapter 9. In view of the further consideration, it should be noted that on passing in this formula from the force V to the tangent stresses $\tau_v = V/F$ where F is the area of the surface of the body, contiguous to the vibrating surface, we can present the indicated formula in the following way:

$$\tau_v = \tau_v(v_\tau) = \tau_{v0}[1 - (v_\tau/v_{\tau 0}].\tag{15.1.1}$$

Here the constants τ_{v0} and $v_{\tau 0}$ have a simple physical meaning: $\tau_{v0} = \tau_v(0)$ i.e. it is equal to the stress at the zero slow tangent component of the velocity v_τ, and $v_{\tau 0}$ is the value of the velocity v_τ at which $\tau_v = 0$, i.e. $\tau_v(v_{\tau 0}) = 0$.

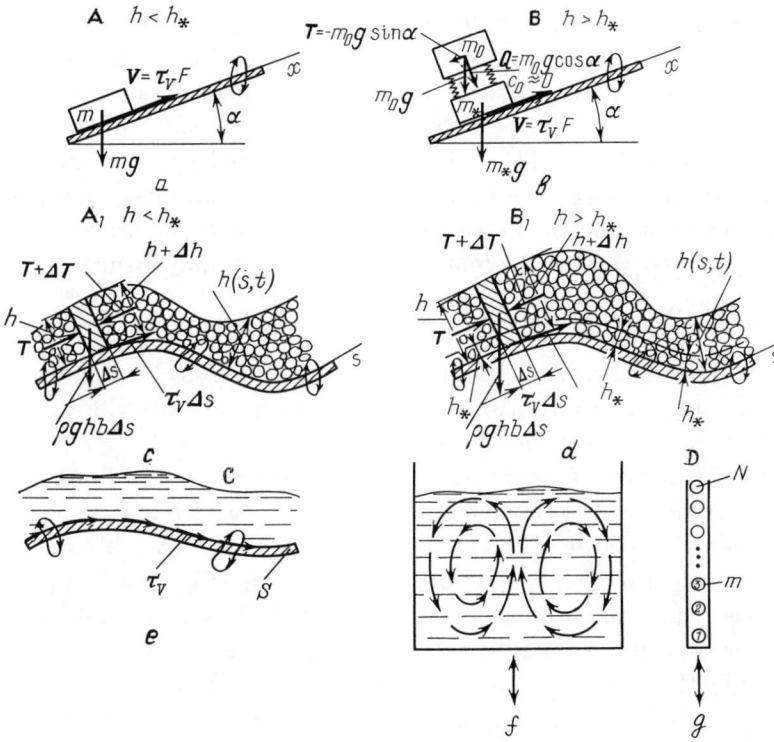

Figure 15.1. Vibrorheological models of granular material layer moving along the vibrating surface.

Then, according to formula (9.4.53) of chapter 9

$$\tau_{v0} = \frac{m\nu\kappa A\omega}{F}, \quad v_{\tau 0} = \kappa A\omega,\tag{15.1.2}$$

where the coefficients κ and ν are defined by the relationships (9.4.54) and (9.4.55) of chapter 9 at $Q = 0$; all the other designations are given in 9.4.

Let us emphasize that formula (15.1.1) reflects the circumstance, mentioned before more than once, that under vibration it is not only the fluidization of the granular material that takes place, but there is also the appearance of a driving vibrational force. If it were only the fluidization, then generally speaking, no vibrational transportation of the material over a rough surface would take place. Meanwhile such a transportation is actually observed and it underlies a number of devices and technologies.

In the case the thickness of the layer of the granular material h exceeds the above-mentioned value h_*, a more complicated model can be used (Model **B**), shown in Fig. 15.1,b. A body of a mass m_*, contiguous to the vibrating flat surface, is connected in this case with a body of mass m_0 by means of some elastic elements of a very small stiffness c_0; then $m_* + m_0 = m$ where m is the mass of the whole layer. Thus, the body m_* is acted upon by an additional constant force $Q = m_0\,g\,\cos\alpha$. In that case the same formulas are valid for the vibrational force V and the tangent stress τ_v as those in the previous case, but at $Q = m_0\,g\,\cos\alpha \neq 0$. In the first approximation we may assume that

$$m_* = \rho F h_*, \quad m_0 = \rho F(h - h_*) \quad (m = m_* + m_0 = \rho F h) \qquad (15.1.3)$$

where ρ is the bulk density of the granular medium, h_*, as above, is the limiting value of the layer at which the motion of the layer over the vibrating surface can still be regarded as the motion of a solid body (particle), $h \geq h_*$ is the total thickness of the layer, F is the area of the flat surface, taken by the granular material; for the sake of the definiteness we will consider that the values ρ, h and h_* refer to the state of the medium in the absence of vibration.

The described model, in spite of its utmost simplicity, makes it possible to explain the reduction of the average velocity of the layer, and in many cases of the specific capacity as the thickness of the layer h increases. This reduction is explained by the decrease — as the height h grows — of the overload parameter w_1 to which the force Q is proportional (see formula (9.4.49) of chapter 9 and also formula (15.1.14), given below). The model also explains the increase of the "phase of detachment of the material" $\delta_0 = \arcsin 1/w_1$ as h grows (see e.g. [158, 564, vol. 4]). In certain cases this model should be specified by introducing a damping element between the masses m_* and m_0.

It is natural that when studying the above simplest models, the motion (including the slow motion which is of interest to us) can be found by means of a direct use of the solution of the problem on the vibrotransportation of a body (particle). The importance of the approach under consideration is determined, however, by the possibility of its application for the approximate solution in more complicated cases.

One of such cases is the motion of a layer of granular material over a not uniformly vibrating nonplanar surface, when the thickness of the layer, generally speaking, is different at different points of that surface. The corresponding problem is of interest for a number of applications (see 15.2). Studying the one-dimensional motion of the medium in the tray of a rectangular cross-section, we assume that the points of the tray perform periodic oscillations whose parameters may depend on the arc coordinate s of the point of the surface (Fig. 15.1, c and d).

Let us first consider the case when the thickness h of the layer is everywhere smaller than the value h_* introduced by us above (Fig. 15.1, c). Having made the equation of the motion of the element of the layer Δs, we arrive at the following vibrorheological equation, describing the small ("creeping") motion of the layer of the medium (we will call the corresponding model — model $\mathbf{A_1}$):

$$\rho h \frac{\partial v_\tau}{\partial t} = -\rho g h \sin \alpha - \frac{\partial T}{\partial s} + \tau_v \qquad (15.1.4)$$

where $v_\tau = v_\tau(s,t)$ is a slow component of the velocity of the medium, and $T = T(s,t;h)$ is the interacting force between the elements of the medium referred to a unit of the "width" of the layer. The force T will be assumed to be similar to the hydrostatic force, acting upon the lateral surface of the element and it will be defined by the formula

$$T = \frac{1}{2} \frac{\rho g h^2}{\cos \alpha} \qquad (15.1.5)$$

approximately valid at considerably small values of the slope α and the derivatives $\partial h/\partial s$ and $h \partial \alpha/\partial s$ as compared to a unit.[a] It is also natural to assume that the change of the slope α and of the parameters of vibration of the surface during the interval of time of the order of the oscillation period $T = 2\pi/\omega$ is very small and therefore when calculating the vibrational stress τ_v the indicated values can be considered to be "frozen". As before, we will refer the bulk density of the medium ρ and the height of the layer h to the state of the medium in the absence of vibration. The vibrational stress τ_v depends on the velocity $v_{\tau 0}$, on the thickness of the layer h, and on the coordinate s.

It is necessary to add to equations (15.1.4) the relationship

$$\frac{\partial h}{\partial t} = -\frac{\partial(v_\tau h)}{\partial s} \qquad (15.1.6)$$

following from the condition of the conservation of mass.

---

[a]It is not difficult to obtain the expression for T under more general assumptions as well.

In the case of oscillations of the vibrating points along the elliptical trajectories for the regimes with an intensive tossing when

$$w_1 = B\omega^2/g \cos\alpha > 3.5(1 + R^2)/(1 + R)^2, \qquad (15.1.7)$$

according to formulas (9.4.53) and (9.4.55) of chapter 9 and according to (15.1.1) and (15.1.2) we have the following expression for the vibrational stress (taking into consideration that at $h < h_*$ we must have $Q = 0$):

$$\tau_v = \rho g h q_1 \cos\alpha \left(1 - \frac{v_\tau}{v_{\tau 0}}\right) \qquad (15.1.8)$$

where

$$q_1 = \frac{\lambda}{2 - \lambda}\frac{1 - R}{1 + R}\frac{A}{B}\cos\delta, \quad v_{\tau 0} = A\omega\cos\delta. \qquad (15.1.9)$$

In view of the expressions (15.1.5) and (15.1.8) the system of differential equations (15.1.4), (15.1.6) can be presented in the following way:

$$\frac{h}{g}\frac{\partial v_\tau}{\partial t} = -h\sin\alpha - \frac{1}{2}\frac{\partial}{\partial s}\left(\frac{h^2}{\cos\alpha}\right) + hq_1\cos\alpha\left(1 - \frac{v_\tau}{v_{\tau 0}}\right)$$

$$\frac{\partial h}{\partial t} = -\frac{\partial(v_\tau h)}{\partial s} \qquad (15.1.10)$$

It should be noted that the values α, q_1 and v_{τ_0} are, generally speaking, functions of the arc coordinate s. The initial distribution of the thickness of the layer $h(s, 0)$ and of the velocity $v_\tau(s, 0)$ can be given as initial conditions.

The stationary flow of a layer of granular material under the action of vibration is of major interest for many applications. In this case, assuming that in the equations (15.1.10) $\partial v_\tau/\partial t = 0$ and $\partial h/\partial t = 0$, we obtain for the height of the layer h the following ordinary differential equation:

$$\frac{dh}{ds} = -\frac{1}{2}\sin 2\alpha - \frac{1}{2}h\cos\alpha\frac{d}{ds}\left(\frac{1}{\cos\alpha}\right) + q_1\cos^2\alpha\left(1 - \frac{c}{bhv_{\tau 0}}\right) \qquad (15.1.11)$$

where $c = v_\tau hb = \text{const}$ denotes the volume discharge of the medium through the cross-section of the tray (b being the width of the tray). Equation (15.1.11) presents a special case of the Abel equation of the 2nd kind, speaking generally, it cannot be solved in quadratures.

Now let us consider a case when $h > h_*$ throughout the layer. Like in model **B**, we will assume that the inertia is concentrated in the lower part of the layer h_* which is in direct contact with the vibrating surface, while the part lying above acts upon the lower part by means of forces, similar to the hydrostatic pressure; the corresponding model (model **B₁**, is presented in Fig. 15.1,*d*).

Due to the fact that the thickness of the "inertial part" of the layer is fixed in the case under consideration, instead of the equation of motion (15.1.4) we will have the relationship

$$\rho h_* \frac{\partial v_\tau}{\partial t} = -\rho g h \sin\alpha - \frac{\partial T}{\partial s} + \tau_v \qquad (15.1.12)$$

where the approximate expression for the force T is, as before, defined by formula (15.1.5).

Turning to the expression for the vibrational stress τ_v, we will note that in this case in formulas (9.4.49), (9.4.53) and (9.4.55) of chapter 9 and in (15.1.2) we must assume that

$$m = m_* = \rho h_* F, \quad Q = \rho g(h - h_*)F \cos\alpha,$$

$$w_1 = \frac{h_*}{h}\frac{B\omega^2}{g\cos\alpha}, \qquad \frac{Q}{m} = \frac{Q}{m_*} = \frac{g(h - h_*)}{h_*}\cos\alpha \qquad (15.1.13)$$

and for the regimes with the intensive tossing, at which

$$w_1 > 3.5(1 + R^2)/(1 + R)^2 \qquad (15.1.14)$$

formula (15.1.8) as before appears to be valid for τ_v. Substituting expressions (15.1.5), (15.1.8) into equation (15.1.12) and adding equation (15.1.6), we come to the system

$$\frac{1}{g}\frac{\partial v_\tau}{\partial t} = -\frac{h}{h_*}\sin\alpha - \frac{1}{2h_*}\frac{\partial}{\partial s}\left(\frac{h^2}{\cos\alpha}\right) + \frac{h}{h_*}q_1\cos\alpha\left(1 - \frac{v_\tau}{v_{\tau 0}}\right),$$

$$\frac{\partial h}{\partial t} = -\frac{\partial(v_\tau h)}{\partial s} \qquad (15.1.15)$$

describing the motion of the layer in the case under consideration.

For the stationary motion when $\partial v_\tau/\partial t = 0$, $\partial h/\partial t = 0$ and $v_\tau h b = \text{const}$ the first equation of the system (15.1.15) becomes an ordinary linear differential equation, coinciding with the corresponding equation (15.1.11) for the model $\mathbf{A}_1$.

It should be expected that the models $\mathbf{B}$ and $\mathbf{B}_1$ will prove to be most suitable for studying the stationary motions of a layer of the granular material and also in the cases when the angle α and the parameters of vibration change with the coordinate s rather slowly. The discreet analogs of the models $\mathbf{A}_1$ and $\mathbf{B}_1$ were considered in [95, 99, 101].

It should be noted that the models $\mathbf{A}_1$ and $\mathbf{B}_1$ are easily generalized for the two-dimensional case when the granular medium moves over a non-cylindrical

vibrating surface. Equations (15.1.10) and (15.1.12) are changed in this case
for the corresponding vector relationships

$$\rho h \frac{\partial \mathbf{v}_\tau}{\partial t} = -\rho g h - \mathbf{grad}\, T + \underline{\tau}_v \qquad (15.1.16)$$

$$\rho h_* \frac{\partial \mathbf{v}_\tau}{\partial t} = -\rho g h - \mathbf{grad}\, T + \underline{\tau}_v \qquad (15.1.17)$$

$$\frac{\partial h}{\partial t} = -\mathrm{div}\,(\mathbf{v}_\tau h) \qquad (15.1.18)$$

Here the force T is defined by formula (15.1.5) and to obtain expressions
for the vibrational stress τ_v it is necessary to solve a two-dimensional problem
on vibrotransportation (see, e.g. [84, 564, vol. 4]).

Besides the cases when $h < h_*$ and $h > h_*$ considered above, situations may
be of interest when on some sections along the tray it is the first inequality that
is satisfied, while on the other sections along the tray — it is the second. Then it
is necessary to have a conjugation of solutions of the corresponding differential
equations.

The above models do not let us study the intra-layer motion in comparatively thick layers of the medium, which is quite essential for a number of
applications. Such processes can be investigated on the basis of the most universal of the models, considered above (the model **B**; see Fig. 15.1,e). In
accordance with this model [114] the slow motion of the granular material under a sufficiently intensive vibration of its particles is considered to be a motion
of viscous (not necessarily Newtonian) fluid, whose rheological characteristics
as well as its density depend on the character of vibration. In those regions
where the vibration is not intensive enough, and fluidization does not appear,
the granular material may be considered as a solid body; under a more precise
analysis it is possible to consider a transitional zone, in which there is no fluidization of the medium, but there is a reduction of the effective coefficients of
dry friction. As for the conditions at the boundary S, instead of the usual for
the viscous fluid conditions of adherence $\underline{v}|_s = 0$, just like it was in the models,
considered above, at the sections where the medium is contiguous to the walls,
the tangent stress is assumed to be preassigned as $\underline{\tau}|_s = \underline{\tau}_v(\underline{v}_\tau)$.

As was mentioned, the expression for $\underline{\tau}_v(\underline{v}_\tau)$ can be found either analytically, or on the basis of rather simple experiments; in some cases the expression
of type (15.1.1), obtained for the regimes of a certain character, can be considered as an approximation, suitable for the study of other regimes as well.
Though, as is clear from what was said, the value $m_*/\rho F = h_*$ (the thickness
of the layer interacting with the vibrating walls) in fact can be considered to

be a certain empirically determined coefficient. It should be also noted that in some cases the dependence of $\tau_v(v_\tau)$ on the value m_* is relatively weak, and sometimes can be neglected altogether (see e.g. formula (9.2.16), following from relationship (9.2.15) of chapter 9).

As was repeatedly mentioned, the researchers have been paying attention to the analogy in the behavior of fluid and granular material under vibration. This analogy was grounded and proved theoretically by Raskin [453] for the sufficiently rarefied medium of "almost elastic" particles. The insufficiency of this analogy was already mentioned, in particular the necessity of taking into account the vibrational forces when studying the slow motions. This circumstance is of primary importance when formulating the boundary conditions and it makes the main idea of the above-made proposal (the model **B**). In some cases it may prove necessary to take into account not only the surface-, but also the volume vibrational forces in the fluid, simulating the granular material under vibration. It is necessary in particular when studying the "slow" processes of vibrational separation (segregation) of the components of the granular mixtures (also see 10.2).

Taking into account the volume vibrational forces made it possible for Fidline to describe and explain the slow "convective" flows appearing in the symmetrical and symmetrically vibrating vessel [185] (Fig. 15.1,f). Fidline simulates the granular material as a Newtonian compressible fluid with the density and other parameters depending on the characteristic of vibration in the given point of the medium. The field of vibration in the medium is supposed to be described by just one scalar parameter — *"vibro-temperature"* which according to the theory of *vibroconductance*([63, 424]; also see 16.6) is governed by the heat equation in case there is heat discharge. In such assumptions the role of volume vibrational forces is played by forces of the Archimedes type which appear here via the dependence of density on vibro-temperature.

To estimate the parameters of the model Fidline considered a chain N of identical particles, moving in the gravity field above a vibrating plane and not absolutely-elastically colliding with each other (Fig. 15.1,g). The motion of such a chain has been studied numerically by computer which made it possible to obtain the indicated estimations.

Numerical experiments with a chain of colliding particles served as a basis for the physical model of the behavior of the granular material under vibration, proposed by Kremer and Fidline [297]. This model (model **C**) can be also referred, to a certain extent, to vibrorheological models. One of the main facts, discovered during numerical modeling, was the following: the frequency of collisions between the particles proved to be much greater than the frequency of vibration. That allowed to perform the statistic averaging of the equations of the transfer of the energy and pulse using the central limit theorem of the theory of probability. The one-dimensional continuum model, built on this principle, leads to a rather complicated nonlinear system of partial differential equations which in some simplest special cases can be solved analytically.

On the basis of the above models it proved possible to describe the chaotic motion of a layer of granular material over the vibrating plane. Such motions, known very well for fluids, were actually observed in the case of granular material as well [181], which is one more confirmation of the possibility of simulating slow actions of the granular material under vibration in form of the motions of viscous fluid (though with the above made reservations and additions).

The described models can be also used when considering the problem of the penetration of vibration into the granular medium which is of great practical importance (see 16.4).

It should be noted that certain difficulties are connected with the generalization of the model **C** for the three-dimensional case and also with the formulating of the boundary conditions, corresponding to that model.

15.2   Some Applications

15.2.1   *The Process of Vibrobunkering of Granular Materials*

In different industries which are processing the granular materials vibrational feeders have a wide application. The tray of such a feeder, placed under the outlet of the bunker with the granular material, can play the role of the bin gate: in the absence of vibrations of the tray it "locks" the bunker, and in case of oscillations it provides the delivery of the material from the bunker (Fig. 15.2,*a*). This situation seems to be quite understandable and natural. It is remarkable, however, that under proper conditions the opposite process is possible: the bunker is filled with the granular material from the vibrating tray, located under it (Fig. 15.2,*b*); such a process is called *vibrobunkering* [83, 84, 114, 503, 545, 564, vol.4]. For the transition from the regime of *vibrodischarge* of the material to the regime of *vibrobunkering* it may be sufficient, for instance, to change the direction of vibration of the tray, as shown in Fig. 15.2. It is not difficult to see that the natural explanation of the described behavior of the granular material under vibration can be given in terms of vibrational mechanics — as a result of the effect of vibrational forces, directed in the adequate way.

Vibrobunkering is directly connected with the effect of storing and forming

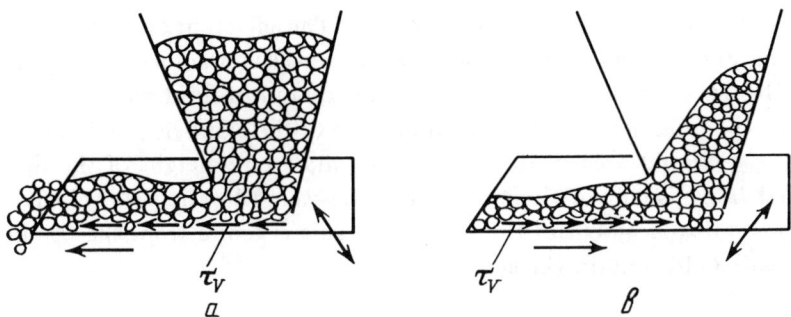

Figure 15.2. Vibrodischarge and vibrobunkering.

a " hill" of the granular material before the wall constraining its displacement along the vibrating surface. Let us find the shape and the height of that hill, making use of the models of 15.1. Consider the simplest case when the vibrating surface is flat, is placed horizontally and performs intensive translational oscillations along the elliptical trajectories (Fig. 15.3).

For the problem under investigation, for both — the model $\mathbf{A}_1$ and the model $\mathbf{B}_1$ the shape of the hill is described by differential equation (15.1.11). Assuming in this equation that $\alpha = 0$, $c = v_\tau hb = 0$ (there being no discharge via the cross-section), we obtain

$$\frac{dh}{ds} = q_1 \tag{15.2.1}$$

so that

$$h = h_0 + q_1 s \tag{15.2.2}$$

where h_0 is the thickness of the layer at $s = 0$, the value q_1 is defined by formula (15.1.9) and the parameters of vibration are assumed to satisfy condition (15.1.7). Thus the "hill" at the wall has a constant slope (Fig. 15.3,a) and the height of the lift of the material is

$$H_{\max} = h_{\max} - h_0 = q_1 l = \frac{\lambda}{2 - \lambda}\frac{1 - R}{1 + R}\frac{A}{B} l \cos \delta. \tag{15.2.3}$$

It is interesting to compare the stated results with those, obtained on the basis of the consideration of a simple model, proposed in [83, 84] (Fig. 15.3,b). This model implies that the granular material is in vessel 1; its bottom is formed by a vibrating plane. At the bottom of the vessel there is a hole through which solid body *2* enters in; part of the material, lying on the vibrating tray, is modeled in the shape of body *2*. The effect upon the body of the vibrating granular mixture which is in the vessel is assumed to be similar to the hydrostatic pressure $T = \rho g h S$ of the column of fluid with the density ρ equal to the bulk density of the medium and of the same height h (S denoting the cross-sectional area of the body). The maximum height of the lift of the material H_{max} is chosen the same as that answered by the zero velocity of the vibrational displacement of the body, that is it is found from the condition of the quasi-equilibrium in the system.

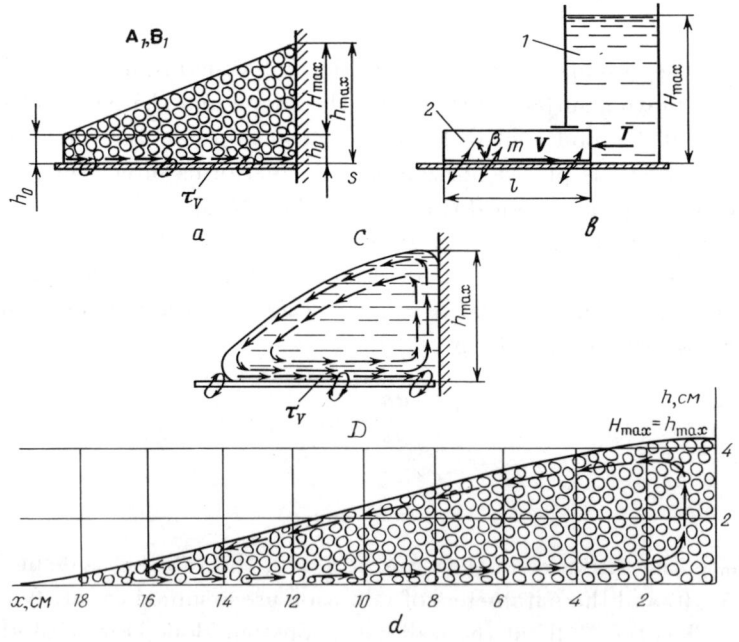

Figure 15.3. Models of vibrobunkering process.

From this condition in [83, 84] a formula was obtained for the regimes with an intensive tossing (the vibration was supposed to be rectilinear, directed at an angle β to the vibrating plane)

$$H_{\max} = \frac{\lambda}{2 - \lambda} \frac{1 - R}{1 + R} l \cot \beta. \tag{15.2.4}$$

As it must be, the same expression is obtained on the basis of the ideas of vibrational mechanics, i.e. from the equality $T = V$ where V is the vibrational force exerted upon the body; for the regimes under consideration this force according to (15.1.1) is defined by the formula $V = \tau_v F$ where F is the area of the base of the body, and the vibrational stress τ_v is calculated according to (15.1.8) and (15.1.9) at

$$\delta = 0, \quad A = A_0 \cos \beta, \quad B = A_0 \sin \beta, \quad F = b\,l, \quad S = h\,b$$

(A_0 being the amplitude of the rectilinear vibration, h being the thickness, and b — the width of the body). It is remarkable that the same formula (15.2.4) is obtained at $\delta = 0$, $A = A_0 \cos \beta$, $B = A_0 \sin \beta$ from formula (15.2.3) corresponding to the models $\mathbf{A}_1$ and $\mathbf{B}_1$.

We will also give here an approximate formula for the maximum value of lifting the material which is obtained on the basis of the adequate considerations for the regimes with the sliding of a body without tossing when $w = A_0 \omega^2 \sin \beta / g < 1$. In this case

$$H_{\max} = f^2 l \tan \beta \tag{15.2.5}$$

where F is the sliding friction coefficient.

A more detailed description of the process of vibrobunkering can apparently be obtained on the basis of considering the model $\mathbf{B}$, which allows to take into account the intra-layer processes. An approximate picture of slow flows of the medium, presented according to this model by viscous fluid under the corresponding boundary conditions, is shown in Fig. 15.3,c. It goes without saying that the corresponding problem can be solved by using modern computing technique.

Fig. 15.3,d shows the results of the experimental investigation of the effect under consideration, performed by Makarov and Fidline. Experiments were conducted with the medium of crude glass balls, 1.8–2.0 mm in diameter, in a horizontal vibrating tray made of acrylic plastic, bounded by a vertical wall in the direction of vibrotransportation. The tray was given rectilinear translational oscillations at an angle $\beta = 60°$ to the surface of the bottom of the tray with the frequencies within the range of $50s^{-1} < \omega < 160s^{-1}$ and the amplitude $A = \sqrt{A^2 + B^2} = 0.15$ cm; one experiment at the frequency $\omega = 108s^{-1}$ was conducted at the amplitude $A_0 = 0.4$ cm. Thus the experiments covered the range of change of the parameter of the overload

$$0.33 < w = B\omega^2/g = A_0 \omega^2 \sin \beta / g < 4.1 \tag{15.2.6}$$

The lower limit of the change of w is answered by the regime of motion without tossing, while the upper limit — with the motion with most intensive tossing. The coefficient of friction of the material against the bottom of the tray f was obtained experimentally, it proved to be equal to 0.36.

As can be seen from the figure, the shape of the "hill", formed near the wall, is very close to a straight line. The remarkable result of the experiment lies in the fact that within the indicated range of the change of the parameters of vibration (15.2.6) the hills proved to be so close to each other that it was pointless to show them in Fig. 15.3,d separately for various frequencies and amplitudes — the picture shows only one "averaged" curve; separate experimental points being in height not more than 1 cm apart from it. When more material was added into the tray, the slope of the hill did not change, only increased its height h_{max} and the length of its base l so that the ratio h/l remained unchanged. In all the experiments a circulating motion of the material was observed; it is shown schematically by the arrows as a projection on the plane of the figure (in reality the spatial picture that took place at the wall of the tray was much more complicated, conditioned by the impact of the sides of the tray)

Comparing the data of the experiment with the theoretical results obtained, we find, that expressions (15.2.4) and (15.2.5) do not depend in the explicit form on the frequency or amplitude of vibration either. According to the second formula at the above given values of the parameters we obtain

$$H_{\max}/l = h_{\max}/l = 0.36^2 \cdot \tan 60^0 = 0.224$$

while experimentally it was found (see Fig. 15.3,d) $H_{\max}/l = 4.2/20 = 0.21$. According to formula (15.2.4), assuming that the coefficient of instant friction at the impact of the material against the tray is $\lambda = 0.6$, and the coefficient of restitution is $R = 0.1$, we find

$$\frac{H_{\max}}{l} = \frac{0.6}{2 - 0.6} \frac{1 - 0.1}{1 + 0.1} \cot 60^0 = 0.203$$

which is also close to the value, obtained experimentally.

15.2.2   Motion of a Layer of Granular Material in a Rectangular Tray Whose Bottom is Vibrating Heterogeneously in the Transverse Direction (to the Theory of Vibrational Screens with an Elastic Resonant Sieve)

In recent years vibrational sieves with an elastic screening surface have been proposed and have gained a certain acceptance [554, 444]. The working head

of such machines (Fig. 15.4,a) is as a rule a rectangular tray *1* whose bottom
is assembled of transversely located deformable rods *2* ("strings"); one of such
rods is shown in the figure. Selecting the stiffness of rods from the condition
of the proximity to the resonance with the working frequency of oscillations of
the frame of the screen, it is possible to provide the transverse oscillations of
the rods with an amplitude, exceeding in their central part the amplitude of
oscillations of the ends of the rods, vibrating together with the frame, at least
twice and may be more. In connection with the theory of such screens and of

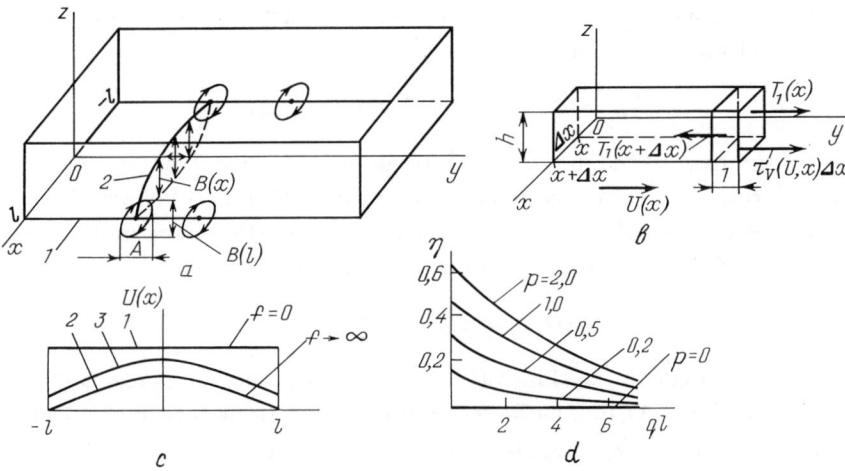

Figure 15.4. Motion of granular material layer in vibrating tray with a flexible bottom.

some other devices, a problem arises concerning the stationary motion of granular material in a tray of a rectangular cross-section, whose bottom performs
vibrations, different in different points of the transverse section of the tray,
while in each such section, i.e. in the longitudinal direction the distribution of
vibration is the same.

   To solve the problem we will make use of the simplified one-dimensional
model **B** [116]. We will assume that the medium is acted upon by a tangent
vibrational stress $\tau_v(U, x)$, exerted from the side of the bottom of the tray and
depending in accordance with what has been said on the velocity of the slow
motion of the medium near the bottom U and on the transverse coordinate

x, but not depending on the longitudinal coordinate y (the coordinate z is counted vertically upward, the origin of coordinates O is on the undeformed surface of the bottom in the longitudinal plane of the symmetry of the tray). We will assume that the thickness of the layer h is a constant value and that it does not depend on x or y and that the velocity of the slow stationary motion of the medium U does not depend on the vertical coordinate z and it doesn't depend on y either. The main aim of the investigation is to find the dependence $U = U(x)$.

We will single out in the moving layer an element in form of a rectangular parallelepiped whose sizes are Δx, 1 and h (Fig. 15.4,b). The lower bound of this element is acted upon by a vibrational force $\tau(x)\Delta x$. The force, acting upon the lateral face of the element with the coordinate x from the adjacent layer will be designated by $T_1(x)$. Then the side with the coordinate $x + \Delta x$ will be acted upon by a force

$$-T_1(x + \Delta x) = -T_1(x) - T'(x + \Delta x)$$

(with an accuracy to the infinitesimals of an order higher than Δx; the derivatives with respect to x are designated by a prime here and below). From the condition of the equilibrium of forces, acting in the stationary motion on the lateral faces of the element, we obtain the equation

$$\tau_v(U, x) = T_1'(x) + \rho g h \sin \alpha, \qquad (15.2.7)$$

where α is the angle of the slope of the tray towards the horizon (not shown in Fig. 15.4,a and b). As for the forces, acting on the front face and on the rear face of the element, they must be mutually balanced.

We will assume that the force $T_1(x)$ within the frames of the model **B** is similar to the viscous resistance i.e. we assume that the force $T_1(x)$ is proportional to the derivative $U'(x)$ with a certain coefficient $hr(x)$, depending on x:

$$T_1(x) = -hr(x)U'(x). \qquad (15.2.8)$$

The function $r(x)$ can then be called the *effective coefficient of resistance*.

Excluding the force T_1 from relations (15.2.7) and (15.2.8), we come to the following differential equation, describing the law of change of the "slow" velocity of motion of the medium across the width of the tray

$$hr(x)U''(x) + hr'(x)U'(x) - \rho g h \sin \alpha + \tau_v(U, x) = 0. \qquad (15.2.9)$$

The function $\tau_v(U, x)$ was already discussed in 15.1. As for the effective coefficient of the resistance $r(x)$, the hypothesis about the constancy of this coefficient in every point of the layer is the simplest:

$$r(x) = r = \text{const.} \qquad (15.2.10)$$

Another possible assumption is that the coefficient $r(x)$ is inversely proportional to the parameter of the overload $w = B(x)\omega^2/g \cos\alpha$ where $B(x)$ is the amplitude of the transverse oscillations of the points of the bottom, which in this case depends on the coordinate x:

$$r(x) = \frac{k}{w} = \frac{k\,g\cos\alpha}{B(x)\omega^2} \qquad (k = \text{const}). \qquad (15.2.11)$$

The function $B(x)$, involved here, i.e. the form of the transverse oscillations of the bottom of the tray, which the stress τ_v may also depend on, can be given, e.g. as the first form of the free oscillations of the bottom or else as a proper expression, approximating this form. To equation (15.2.9) it is necessary to add the corresponding boundary conditions (see below). This equation, generally speaking, does not allow an exact solution in a closed form; its solution can be built in form of series, or obtained by computer.

Let us, however, consider the case when equation (15.2.9) under certain simplifications allows a simple analytical solution; this consideration is mostly of an illustrative nature. Let the tray be horizontal ($\alpha = 0$) and suppose the regime with an intensive tossing takes place (see condition (15.1.7)). Oscillations of all the points of the frame and of the bottom of the tray in the longitudinal with respect to the tray direction will be assumed to be equal, i.e. we will assume that $A(x) = A = \text{const}$; thus the oscillations of the points of the bottom are different at the expense of their transverse (vertical) component $B(x)$. The angle δ in formulas (9.4.33) of chapter 9 defining the law of oscillations of the points is also assumed to be constant.

Under the indicated assumptions according to formulas (15.1.8) and (15.1.9) the expression for $\tau_v(U, x)$ will have the form

$$\tau_v(U,x) = \rho g h q_1 \left(1 - \frac{U}{A\omega\cos\delta} \right), \quad q_1 = \frac{\lambda}{2-\lambda}\frac{1-R}{1+R}\frac{A}{B(x)}\cos\delta. \qquad (15.2.12)$$

Then in view of assumption (15.2.11) equation (15.2.9) will be written as

$$U'' - \frac{B'(x)}{B(x)}U' - q^2(U - A\omega\cos\delta) = 0$$

$$\left(q^2 = \frac{\rho\omega}{k\cos\delta}\frac{\lambda}{2-\lambda}\frac{1-R}{1+R} \right) \qquad (15.2.13)$$

Finally, we will neglect in this equation the term $B'(x)U'/B(x)$, taking into account that for the deformable elements of the bottom, fixed on the frame of the tray, and for the first form of their free oscillations we have $B'(0) = B'(l) = 0$ ($2l$ being the width of the bottom). Then equation (15.2.13) will become a simple ordinary equation with constant coefficients

$$U'' - q^2(U - A\omega\cos\delta) = 0. \tag{15.2.14}$$

Making use of the condition of the symmetry of the profile of the velocity $U(x)$ with respect to the plane $y0z$ and the condition of the equality of the force $T_1(l)$ on the wall of the frame $x = 0$ to the corresponding value of the vibrational force $V(l)$, we will record the boundary conditions as

$$U'(0) = 0, \quad T_1(l) = -V(l). \tag{15.2.15}$$

To obtain an approximate expression of the vibrational force $V(l)$ which in this case will consist of only the vibro-transformed component V_l, it is possible to use equation (10.2.15) of chapter 10. Taking into account that we mean here the vibrational force per unit length of the frame, and also that for this case the weight of the body mg should be changed for the "hydrostatic force"

$$\rho g \int_0^h h_1 dh_1 = \rho g h^2/2$$

and assuming that $\dot{X} = U$, for the equal in both directions values of the sliding friction coefficient $f_+ = f_- = f$ we obtain

$$V(l) = V_1[U(l)] = -\frac{\rho g h^2 f U}{\pi A\omega} \tag{15.2.16}$$

In view of expressions (15.2.8) and (15.2.11) the boundary conditions (15.2.15) will be written in the following final form:

$$U'(0) = 0, \quad -U'(l) = U(l)d \tag{15.2.17}$$

where

$$d = \frac{\rho h f\omega B(l)}{\pi k A} \tag{15.2.18}$$

The solution of differential equation (15.2.14), satisfying the boundary conditions (15.2.17) has the form

$$U(x) = A\omega\cos\delta\left(1 - \frac{d\cosh qx}{d\cosh ql + q\sinh ql}\right) \tag{15.2.19}$$

Let us analyze the expression we have obtained. According to (15.2.18), when there is no friction against the walls of the box ($f = 0$), we have $d = 0$ and

$$U(x) = A\omega \cos \delta \qquad (15.2.20)$$

i.e. the velocity is the same all over the section and it coincides with that of vibrotransportation of an isolated solid body (particle). When the values of f are different from zero, the walls retard the layer, that retardation growing when the layer is approaching the walls. In the limiting, theoretically thinkable case $f \to \infty$ which corresponds to the sticking of the material to the walls of the tray, the braking is especially strong; formula (15.2.19) then acquires the form

$$U(x) = A\omega \cos \delta \left(1 - \frac{\cosh qx}{\cosh ql} \right) \qquad (15.2.21)$$

The profiles of the velocities, corresponding to the limiting cases and to a certain intermediate case, considered above, are shown in Fig. 14.4,c.

Using formula (15.2.19), one can easily obtain an expression for the weight capacity of the vibrational device

$$Q = 2\rho g A\omega \int_0^l U(x)dx = Q_0[1 - \eta(p, ql)] \qquad (15.2.22)$$

where

$$Q_0 = 2\rho ghl A\omega \cos \delta, \, \eta(p, ql) = \frac{p}{p\dfrac{ql}{\tanh ql} + 1},$$

$$p = \frac{ld}{q^2 l^2} = \frac{1}{\pi}\frac{h}{l}f\frac{B(l)}{A}\frac{2 - \lambda}{\lambda}\frac{1 + R}{1 - R}\cos \delta \qquad (15.2.23)$$

The graphs of the dependence $\eta(p, ql)$ are shown in Fig. 15.4,d. The value Q_0 presents the capacity of the device, calculated in the assumption that the velocity of the motion of the material is constant all over the cross-section of the tray and is equal to $A\omega \cos \delta$. Therefore the value $\eta(p, ql)$ will be called the *coefficient of capacity reduction*.

As one can see in the graph and from formula (15.2.23), in the case of a sufficiently intensive transverse vibration of the bottom which is being considered, the reduction of the capacity is conditioned only by the retarding effect of the walls of the tray. Indeed, when there is no friction of the material against the walls ($f = 0$) we have $p = 0$, $\eta = 0$ and $Q = Q_0$. In another limiting case, corresponding to the "sticking" of the material to the walls $((f \to \infty))$ there will be $p \to \infty$, and the reduction of the capacity will be the greatest:

$$\eta = \frac{\tanh ql}{ql}, \quad Q = Q_0 \left(1 - \frac{\tanh ql}{ql} \right). \qquad (15.2.24)$$

In case the relative thickness of the layer h/l is small, and there are real values f, the retarding effect of the walls is inessential, as has to be expected.

It should be noted that when the vibration of the tray and of its bottom is less intensive, when the velocity of transportation of an isolated particle $U_*(x)$ is essentially reduced near the walls $x = \pm l$, the reduction of the capacity may be conditioned by that factor as well. The reckoning of that factor, however, requires, generally speaking, a numerical analysis of differential equation (15.2.9) taking into account the corresponding expression for the vibrational stress $\tau_v(U, x)$.

15.2.3   *Motion of the Loading in Vibrational Mills and Instruments for the Volume Vibrational Treatment of Machinery Parts*

For the theory of the devices indicated in the title it is of interest to solve a problem about slow (circulational) motions of a layer of granular material ("the loading") in a cylindrical or toroidal vibrating drum or in that with an axial line in form of a spiral [34, 261, 323, 324, 417, 564, vol.4]. In a transverse (meridional) cross-section of such a drum under the action of vibration there appears a "slow" circulational motion of the loading, whose approximate character is presented in Fig. 15.5. The technological indices of the process depend on the intensity of that motion. We must mention the qualitative resemblance of the character of the flows in Figs. 15.5 and 15.3,*c*.

A number of publications has been devoted to the creation and investigation of the models of circulational motion of the loading in the devices under consideration (see e.g. [316, 527]. On the basis of a discreet version of a vibrorheological model the problem was investigated by Levengarts and the author [101]. A more exact description of the process can be obtained on the basis of the continual model **B**; but we will not dwell on it here.

15.2.4   *On Supplements to the Theory of Crushers*

In fact item 14.2.2 might be included into this section, because it also deals with modeling slow motions of the granular material (or suspension) in case of vibrational effects.

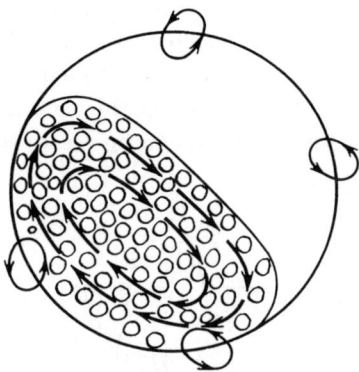

Figure 15.5. Slow circulational motion of granular material layer in a vibrating drum.

15.3   On the Behavior of the Granular Material in the Communicating Vibrating Vessels

Peculiarities of the behavior of the granular material in the communicating vibrating vessels on the one hand are of an essential conceptual interest — as a bright manifestation of vibrorheological regularities; on the other hand, these peculiarities serve as a basis for a number of unconventional technological solutions.

We will dwell here on a purely qualitative description of the corresponding regularities and will refer to more appropriate publications. In the absence of vibration there may exist an infinite number of the continuously distributed equilibrium positions of a certain volume of the material in the vessels — positions, corresponding to different levels (1, 2, 3,...) of the free surface of the medium in the vessels (Fig. 15.6,a). In case of the absolute symmetry of the system — when the vessels are identical and the vibration with respect to the vertical line is symmetrical, for instance when harmonic oscillations are either strictly horizontal or strictly vertical (Fig. 15.6,b), as a rule (but not always, see below), equal levels of the medium are established in the quasi-equilibrium state. In case the vessels are not identical, (Fig. 15.6,d and e) or the vibration is asymmetrical (Fig. 15.6,f), the quasi-equilibrium levels will be different.

The model of the situation, corresponding to Fig. 15.6,d, was studied theoretically and experimentally by Lipovsky [335, 564, vol. 4] who showed that the presence of air plays in this case an important role (in the conditions of vacuum the levels proved to be equal), and who suggested some original devices. A model answering Fig. 15.6,*e* was suggested and studied in [296] in connection with interesting effects discovered in similar systems by Palilov, a gifted engineer, untimely deceased. He established experimentally and in the indicated work proved theoretically that in this case at different frequencies of vibration a higher level of the medium can be established in different vessels.

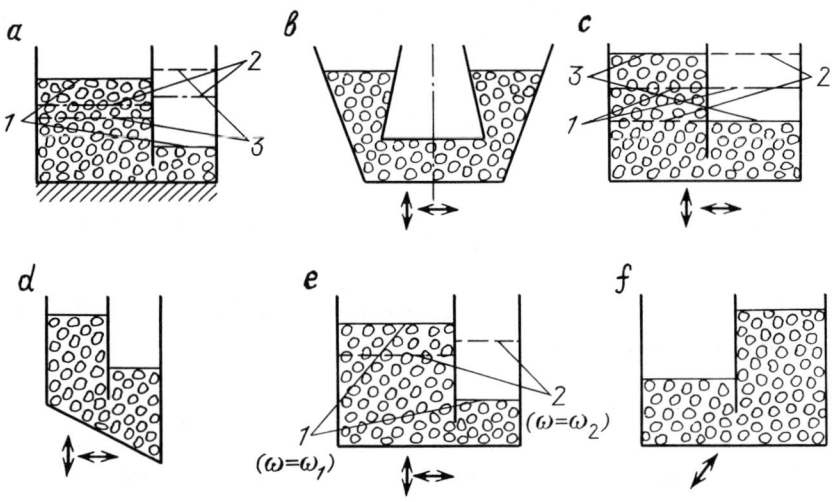

Figure 15.6. Different levels of the free surface are established in the communicating vibrating vessels under the asymmetry of the system or under vibration.

It is still more curious that under certain conditions in an absolutely symmetrical system and under symmetrical vibration two or even three different states may be established in the vessels (Fig. 15.6,c), and in two mutually symmetrical states levels *2* and *3* are different. This is connected with the instability (or the stability only under sufficiently small perturbations) of state *1* with equal levels; which of the two or three states *1–3* will be established in reality — depends on the initial conditions of motion (the possibility of such a situation was mentioned in 9.3. (see Fig. 9.2,*aIV*). As for the case presented in Fig. 15.6,*f*, it corresponds to the effect of vibrobunkering, considered in 15.2.1. It should be noted that the value of the difference of the levels, essential for applications, depends on both — the geometrical parameters of the system, including sizes of the transitional elements, and on the parameters of vibration. It should be also noted that every device, shown in Fig. 15.6,*a, c–f*, can be used for vibrotransportation, provided the granular material is eliminated from the zone close to the highest level of the free surface that is being established.

The above effects once more convince us that vibration leads not only to the fluidization of the granular material, but also to the appearance of vibrational forces. From the position of the observer **V** it is these forces that explain the invalidity in this case of the law of the communicating vessels for fluids. Invalid in this case is also the assertion that the positions of stable equilibrium of the system correspond to the minimum of the potential energy of the gravitational force: at different levels of the medium in the communicating vessels the center of masses is located, as a rule, higher than when the levels are the same. (About the erroneousness of the so called *potential theory* with respect to the systems under consideration see also 10.2 and Fig. 10.4).

Chapter 16

Penetration of Vibration into Certain Media

16.1   Preliminary Remarks

The present chapter is, to some extent, of an auxiliary character: we consider here the "fast" process — the stationary field of vibration in certain media, simulated as continuous, under the action of vibration of solid boundaries. As follows from what was said in chapter 15, the solution of the corresponding problems is quite necessary when studying slow processes in the media, appearing under the action of vibration, i.e. at the vibrorheological approach. Besides, the solution of such problems is of a special interest for the theory of vibrational processes and devices.

Despite the actuality of the question, the problem of the penetration of vibration into different media cannot yet be considered to be solved satisfactorily. Only a number of important special cases have been considered and a few general models have been suggested. Experimental data are being accumulated. This chapter contains the presentation of some of the indicated results.

16.2   Penetration of Vibration into a Viscous Fluid

The problem about the motion of a viscous incompressible fluid, filling the half-space above the plate, performing longitudinal harmonic oscillations, was studied as far back as by Stokes [518]. Due to the linearity of the problem the corresponding solution can be easily generalized for the case of arbitrary periodic oscillations [312]. A similar problem was solved for the cylinder, performing harmonic angular oscillations in an unbounded fluid [328]. Solution of the problem about oscillations of the layer of viscous fluid of a finite thickness, limited by a free surface, was obtained in the fundamental course by Landau and Lifshits [313]. The solutions are also known of certain

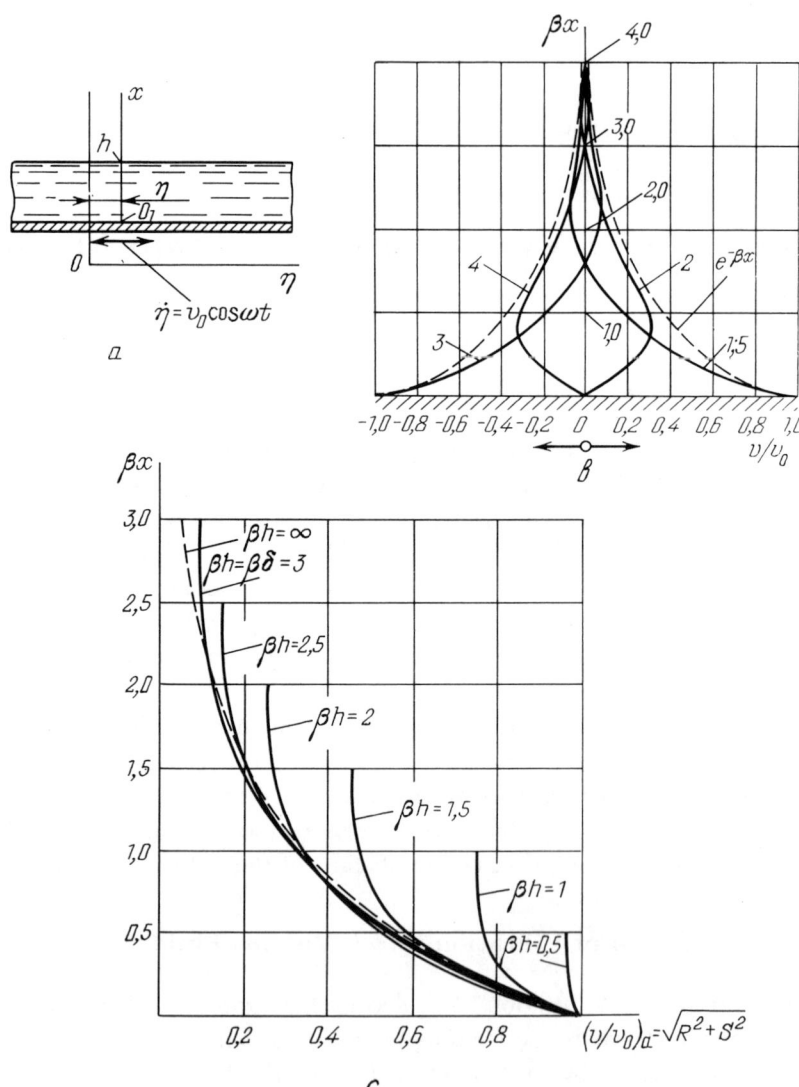

Figure 16.1. Penetration of vibration into the viscous incompressible fluid under the longitudinal oscillations of the plate. a) The scheme of the system; b) distribution of the velocities in the unlimited layer of the fluid ($1,5$— $\omega t = 0$, 2π; 3 — $\omega t = \pi$; 4 — $\omega t = 3\pi/2$; c) distributions of the amplitudes of the velocities in the layers of finite- (solid curves) and infinite (the dotted curve) thickness.

class under consideration for the Newtonian fluids [544, 505]. The presentation given below follows the work by Blekhman, Junior and Kizevalter [131].

Let us first consider the case when the endless plate performs rectilinear translational harmonic oscillations in the same plane according to the law

$$\dot{\eta} = v_0 \cos \omega t \qquad (16.2.1)$$

where $\dot{\eta}$ is the velocity of the oscillations of the plate, v_0 is its amplitude, and ω is its frequency (Fig. 16.1,a). For the velocity of the incompressible viscous fluid over the plate $v = v(x,t)$ from the Navier-Stokes equations the following equation can be obtained

$$\frac{\partial v}{\partial t} = \nu \frac{\partial^2 v}{\partial x^2} \qquad (16.2.2)$$

where ν is the kinematic coefficient of viscosity. Equation (16.2.1) is in form similar to the diffusion equations and heat equations. The boundary conditions of the problem have the form

$$v(0,t) = \dot{\eta} = v_0 \cos \omega t, \qquad \frac{\partial v}{\partial x}\bigg|_{x=h} = 0 \qquad (16.2.3)$$

where h is the thickness of the layer of the fluid above the plate. The first equality presents the condition of sticking of the fluid to the plate, and the second — is the condition on the free surface. The stationary solution of equation (16.2.1), satisfying conditions (16.2.3), has the form

$$v = U(x)\sin \omega t + V(x)\cos \omega t \equiv v_0 \sqrt{R^2(x) + S^2(x)} \, \cos[\omega t + \varphi(x)] \qquad (16.2.4)$$

where

$$U(x) = v_0 S(x), \quad V(x) = v_0 R(x);$$
$$R(x) = K_0(\beta x) + M K_1(\beta x) + 2N K_3(\beta x),$$
$$S(x) = N K_1(\beta x) - 2K_2(\beta x) - 2M K_3(\beta x); \qquad \beta = \sqrt{\omega/2\nu},$$
$$M = \frac{1}{2}\frac{\sin 2\beta h - \sinh 2\beta h}{\sinh^2 \beta h + \cos^2 \beta h}, \quad N = \frac{1}{2}\frac{\sin 2\beta h + \sinh 2\beta h}{\sinh^2 \beta h + \cos^2 \beta h},$$
$$K_0(z) = \cosh z \cos z, \quad K_1(z) = \frac{1}{2}(\cosh z \sin z + \sinh z \cos z),$$
$$K_2(z) = \frac{1}{2}\sinh z \sin z, \quad K_3(z) = \frac{1}{4}(\cosh z \sin z - \sinh z \cos z);$$
$$\sin \varphi(x) = \frac{S(x)}{\sqrt{R^2(x) + S^2(x)}}, \quad \cos \varphi(x) = \frac{R(x)}{\sqrt{R^2(x) + S^2(x)}}. \qquad (16.2.5)$$

For the case when the fluid occupies the unlimited space, i.e. $h \to \infty$, we get

$$R(x) \to \exp(-\beta x)\cos\beta x, \qquad S(x) \to \exp(-\beta x)\sin\beta x,$$
$$v = v_*(x,t) = v_0 \exp(-\beta x)\cos(\omega t - \beta x). \qquad (16.2.6)$$

This formula was obtained by Stokes [494, 518]. The corresponding distributions of the velocities of the fluid at the moments of time separated by a quarter of an oscillation period $\pi/2\omega$ are shown in Fig. 16.1,b (curves $1, \cdots, 5$ respectively); the dotted line shows the amplitude values of the dimensionless velocity. As is marked in the article [131], these graphs are shown in the book [494] inaccurately.

Let us turn to the analysis of the given relations. From formula (16.2.6) it follows that the plate involves the fluid into the oscillatory motion of the same frequency ω with its amplitude quickly (exponentially) decreasing with the growth of the distance from the plate. The thickness δ of the layer of fluid, involved by the plate into oscillations due to the viscosity of the fluid can be characterized by the distance from the plate at which the amplitude of the velocity makes 5% of its value on the plate. This value, determined according to (16.2.6) by the expression

$$\delta \approx 3.00/\beta = 3\sqrt{2\nu/\omega} = 3\delta_0, \qquad \delta_0 = \sqrt{2\nu/\omega} \qquad (16.2.7)$$

is called in the work [131] the *depth of penetration of oscillations*.

When the viscosity coefficient value is $\nu = (1.007 \div 1.519) \cdot 10^{-2}$ cm^2/s, which corresponds to the temperature change of the water from $20°$ to $5°$ C and to the change of oscillation frequency $n = 30\omega/\pi$ from 150 to 300 oscillations per minute (such change ranges of the parameters are characteristic of some ore dressing apparatus), the depth of penetration makes about 1 mm.

Figure 16.1 shows the graphs of the distribution of the amplitudes of the dimensionless velocities of the fluid $(v/v_0)_a = \sqrt{R^2 + S^2}$ in the layers of finite and infinite thickness, plotted according to formulas (16.2.4) and (16.2.6). As one can see in the graphs, when the thickness of the layer is $h < \delta$, the distribution of the velocities in height can differ essentially from the corresponding $h \to \infty$. If $h \geq \delta$ i.e. $\beta h > 3$, then this difference is small; in this case formulas (16.2.4) and (16.2.6) give similar results. Therefore in this case for practical calculations the velocity of the fluid in the layer of thickness δ near the wall can be determined accurately enough by means of Stokes' simple formula (16.2.6), assuming that the fluid beyond that layer is immobile. If $h < \delta$, then it is necessary to use for calculations a more complicated formula (16.2.4). The layers of the fluid for which $h < \dfrac{1}{6}\delta\,(\beta h < 0.5)$ move practically together with the plate (as a solid body).

In the layer of a finite thickness over the harmonically oscillating plate, for which $\beta h \geq c \approx 3.69$ where c is the root of the equation $\sinh^2 c + \cos^2 c = 400$, the depth of penetration, determined in the former way, is somewhat more than the value $\delta = 3/\beta$, but it does not exceed the value c/β, which it takes at $\beta h = c$. In the layers for which $\beta h > c$ the twenty fold decrease of amplitude of the velocity is not achieved.

Due to the linearity of the problem these results can be easily generalized for the cases of arbitrary rectilinear periodic oscillations and of the periodic oscillations in two mutually perpendicular directions, lying in the plane of the plate.

When the motion of the plate is periodic, rectilinear, presented as a sum of harmonics whose frequencies are $k\omega\,(k = 1, 2, \ldots,)$, the depth of penetration for a certain m-th harmonic according to formula (16.2.7) will be $\sqrt{m}$ times less than that for the first. Therefore when being far enough from the plate, the oscillations of the fluid are determined by the first harmonic, i.e. they approach the harmonic oscillations, irrespective of the law of periodic oscillations of the plate (provided, of course, that $h \geq \delta$ and that the amplitudes of the harmonics do not increase with the growth of their number).

If the plate performs arbitrary periodic oscillations in two mutually perpendicular directions, lying in its plane, then the velocities of the fluid in each of these directions are determined irrespective of each other, the qualitative regularities, stated above, remaining valid. In particular, if the periods of the mutually perpendicular oscillations are the same, then the trajectories of the particles of the fluid which are far enough from the plate approach ellipses, the sizes of those ellipses decreasing exponentially. In case the periods are essentially different, the motion of the fluid when it is far enough from the plate approaches rectilinear harmonic oscillations in the direction and with the frequency corresponding to those of oscillations with a great period.

We can say that a layer of fluid possesses the properties of a filter of high frequencies.

It is interesting to compare these results with the solution of the problem about the damping of perturbations, made by the ball, oscillating harmonically along a certain direction in the endless incompressible fluid (Fig. 16.2,a); this problem in the case of Reynolds' small numbers was considered by Granat [217, 218, 221]. "The domain of perturbations", made by an oscillating ball, is defined in her work as a sphere around the ball in which 95% of the total loss of energy is dissipated in the fluid.

Let R_b be the radius of that sphere, and r_0 be the radius of the ball. Then the value $\delta = R_b - r_0$ can be assumed to be the corresponding depth of the penetration of vibration into the fluid. According to [218] a graph, showing

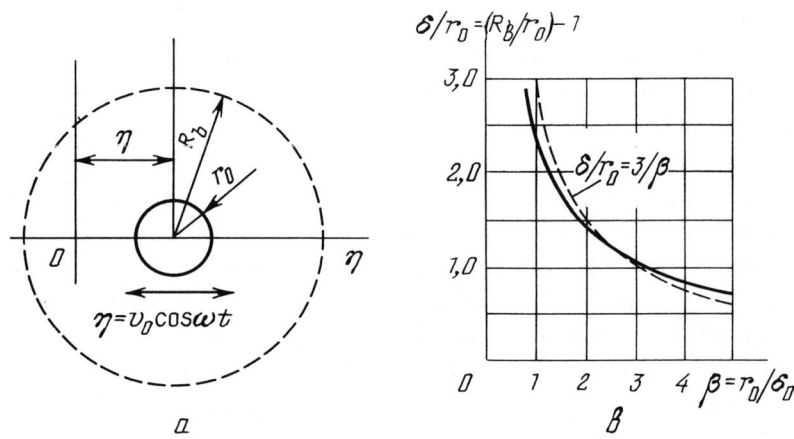

Figure 16.2. Penetration of vibration into the viscous incompressible fluid under translational oscillations of the ball in the unlimited volume of the fluid. a) The scheme of the system; b) the dependence of the depth of the penetration of vibration on the parameter β.

the dependence of the value δ/r_0 on the parameter $\beta = r_0/\delta_0 = r_0\sqrt{\omega/2\nu}$ can be obtained. This graph is shown in Fig. 16.2,b by a solid line ($\delta_0 = \sqrt{2\nu/\omega}$ is the same value with the dimension of length as that in formula (16.2.7)). The same graph within the range of change of β is approximated well enough by the analytic dependence

$$\frac{\delta}{r_0} = \frac{3}{\beta} = \frac{3\delta_0}{r_0} = \frac{3\sqrt{2\nu/\omega}}{r_0} \tag{16.2.8}$$

(the dotted curve in Fig. 16.2,b). After canceling r_0, this dependence exactly coincides with (16.2.7). Thus, there is a practically complete agreement obtained for the case of the oscillations of the plate and of the ball.

As is shown in [218, 221], the result of the investigation made is almost the same if the domain of perturbations, produced by the ball in the fluid, is determined not in accordance with the energy dissipation, but in accordance with leveling the field of pressures or in accordance with the damping of the field of velocities, as it was done above in the case of the plate. In the work [218] it is also marked that the perturbation zone in case of a translational motion of the ball in the viscous fluid is in the context of the problem one order of magnitude greater in size than in case of oscillations. This conclusion agrees qualitatively with formulas (16.2.7) and (16.2.8), according to which the perturbation zone increases as the oscillation frequency ω is lowered.

In conclusion it should be noted that in the assumption of the incompressibility of the fluid, the case of the transverse oscillations of the plate is trivial: the fluid in this direction oscillates together with the plate. In case of compressibility, the process is described by a wave equation, containing the term which takes into account the dissipative factors.

The oscillations of the system comprising a symmetric solid body with an inertial vibro-exciter and the absorbing liquid medium are considered in the article by Kunin and Hon [303].

16.3 On the Penetration of Vibration into Suspensions

It is noted in the same work [131] that in case of suspension the depth of the penetration of vibration can be estimated, assuming that the suspension is equivalent to a certain fluid, more viscous than that, carrying particles of the suspension.

Taking into account that the equivalent viscosity ν_* under certain conditions may exceed the viscosity of the fluid by an order of magnitude (and more), we come to the conclusion that the depth of the penetration of vibration in case of suspension can also be much greater. A number of formulas have been suggested to calculate the effective dynamic viscosity coefficient ν_*. In particular, the Vand formula is well known [268] (also see 17.2). By way of example, in [131] it is noted that when the volume content of solid particles is 30%, which is typical for certain devices of gravitational dressing, the effective dynamic viscosity coefficient, found in that way, proves to be 2.85 times greater than the dynamic viscosity coefficient for water. Accordingly, the kinematic viscosity ν_* for suspension (pulp) is $\sqrt{1.9} = 1.4$ times greater than for water, which, according to formula (16.2.7), leads to the value of the depth of penetration 1.4 times greater than for water.

Possibilities of increasing the depth of the penetration of vibration are investigated by Krasnov [294].

16.4   Penetration of Vibration into Granular Material

16.4.1   The Case of Circular Oscillations of a Horizontal Plate in its Plane

Some very important qualitative regularities of a behavior of the granular material in this simplest case under consideration are brought to light when we consider the solution of the corresponding problem of Zhukovsky about the motion of a plane solid body (material particle) [599]. If the layer of granular material is not very thick, then the indicated solution can be also used when studying its behavior as well, which was done when considering the pseudo-resonance effect in 10.2.4. According to formula (10.2.11) of the indicated chapter, if the acceleration of oscillations of the plate $A\omega^2$ (ω being the frequency, and A — the radius of the circular trajectory of oscillations) does not exceed the values $f\,g$ (f being the sliding friction coefficient, and g being the free fall acceleration), then the body moves together with the oscillating plate. At $A\omega^2 > f\,g$ the acceleration, velocity and the radius of the trajectory of the body in the stationary absolute motion are defined according to the formulas

$$W = g\,f, \quad U = g\,f/\omega, \quad r = g\,f/\omega^2 \qquad (16.4.1)$$

i.e. U and r are quickly diminishing with the growth of ω. Under a sufficiently large oscillation frequency the body remains practically immobile in space. The maximal value of the acceleration of absolute oscillations of the body is $W_{\max} = f\,g$. The indicated regularity is to some extent manifested during the motion of thicker layers of the granular medium: there exists a range of accelerations of oscillations of the plane at which vibration penetrates into the medium in a certain sense most effectively; in particular, it is this circumstance that explains the above mentioned pseudo-resonance effect during the vibrational separation of granular mixtures.

The problem about the motion of the granular medium in the case under study was considered by Gortinsky whose results of the investigations [214, 215] are given below. The main assumption, made by Gortinsky to explain the motion of elementary horizontal layers of the granular material with respect to each other is that the dry friction coefficient f between the layers depends on the normal pressure, i.e. on the weight of the above-lying body, increasing with the increase of the latter. The existence of this dependence follows from some direct and indirect experiments, including those, made by Gortinsky and his colleagues.

Denoting the weight of the above-lying part of the granular body per unit area by G, Gortinsky writes down the relation between the dry friction force

F ("the force of resistance to the relative shift of the layers") and the values G and f in the form

$$F = Gf(G) \tag{16.4.2}$$

where $f(G)$ according to the adopted supposition is a certain increasing function G. Then with the increase of acceleration of the oscillations of the plate $A\omega^2$ the picture of motion looks as follows. At $A\omega^2 = (A\omega^2)_1 < g\,f_0$ where $f_0 = f(0)$ is the value of the friction coefficient on the upper border of the layer, all the granular material moves together with the plate as one solid body. After the acceleration $A\omega^2$ has exceeded the value $g\,f_0$, the upper part of the layer

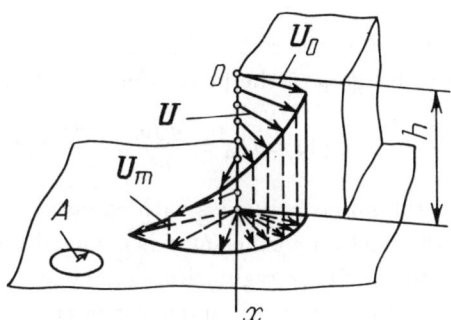

Figure 16.3. The distribution of velocities in the granular material layer on the plate under circular oscillations of the plate in its plane (according to Gortinsky [214, 215]).

begins sliding along the lower part which keeps on moving together with the plate. As the acceleration increases again, the relative motion is spread to the layers, lying below, and at

$$A\omega^2 = (A\omega^2)_2 = \sqrt{F_m\left(\frac{d^2F}{dG^2}\right) + \left(\frac{dF}{dG}\right)^2} > (A\omega^2)_1 = g\,f_0$$

reaches the lower border of the layer (the values, referring to that border are marked by the index "m"). At $A\omega^2 > (A\omega^2)_2$ the whole layer is sliding relative to the plate, in different ways according to the height. In this case the absolute speed of particles is decreasing from the lower to the upper border. The distribution of the absolute speeds according to the height of the layer is

shown in Fig. 16.3. The absolute values of speeds at the points, characterized
by a certain value G is given by the expression

$$U = \frac{g}{\omega}\sqrt{F\frac{d^2 F}{dG^2} + \left(\frac{dF}{dG}\right)^2}. \tag{16.4.3}$$

If the material is homogeneous from top to bottom of the layer, then $0 \le u = G/G_m = x/h \le 1$ is a relative coordinate, counted vertically downward from
the upper border of the layer (x is the absolute coordinate, G_m is the total
weight of the layer per unit area of the vibrating plate, h is the thickness of
the layer; see Fig. 16.3). If further on the dependence of the friction coefficient
f on G, i.e. also on $u = x/h$, is linear, then formula (16.4.3) acquires the from

$$U(u) = \frac{g\,f_0}{\omega}\sqrt{1 + 6u\xi(1 + u\xi)} \tag{16.4.4}$$

where $\xi = (f_m - f_0)/f_0 > 0$ is the relative difference of the friction coefficients
on the lower and upper borders of the layer. On these borders respectively

$$U(0) = U_0 = \frac{g\,f_0}{\omega}, \quad U(1) = U_m = \frac{g\,f_0}{\omega}\sqrt{1 + 6\xi(1 + \xi)}. \tag{16.4.5}$$

The first of these expressions coincides with equation (16.4.1) for the velocity of a solid body if the friction coefficient f is changed for its value f_0 for the
upper border of the layer. Other regularities of the penetration of vibration
for the model of the granular medium under consideration are also similar to
those established for the solid body: the maximum acceleration of vibration of
the medium

$$W_{\max} = U_m\omega = g\,f_0\sqrt{1 + 6\xi(1 + \xi)} = (A\omega^2)_2$$

is achieved on its lower border at the same value of acceleration of the oscillations of the plate $A\omega^2 = (A\omega^2)_2$; with the further growth of $A\omega^2$ the
accelerations of the points of the medium remain unchanged, while the velocity and the radius of the trajectory at the constant amplitude of vibration A
decrease with the growth of the frequency ω in proportion to ω and ω^2 respectively. In other words, with the acceleration of the oscillations of the plate
$A\omega^2 < (A\omega^2)_1 = g\,f_0$ the oscillations penetrate throughout the whole thickness
of the layer, while at $A\omega^2 > (A\omega^2)_1 = g\,f_0$ they penetrate into it only partially,
and with the growth of the value $A\omega^2$ "the penetration of the acceleration" is
stabilized at the level $g\,f_0$ for the upper border of the layer, and "the penetration of the velocity and of the amplitude" decreases with the increase of the
frequency ω in proportion to ω and ω^2 respectively.

It should be noted that for certain applications what is important is actually not the penetration of vibration into a layer of the granular medium, but the provision of a sufficiently intensive relative motion of the medium along the plate. That refers, for instance, to the problems of screening the material through the sieve (see e.g. 10.2). The greatest amplitude of the relative oscillations, close to the amplitude of the absolute oscillations of the plate A, takes place at a sufficiently large value of the acceleration of the oscillations $A\omega^2$ when the medium remains practically motionless in space.

16.4.2   The Case of Rectilinear Longitudinal Oscillations of a Horizontal Plate

The case of rectilinear longitudinal translational oscillations of the plate is characterized by the same regularities of the penetration of vibration, as the case of circular oscillations, considered above. The solution of the problem when a layer of granular material can be simulated by a plane solid body (material particle) was given by Loytsyansky [339] (see also [84, 114]). The investigation of the behavior of the layer on the same assumption as in 16.4.1 was also made by Gortinsky [214, 215]. The solution of the problem is more complicated here than in the case of circular oscillations, especially for the second model, due to the fact that the motion of particles has two stops either of some duration or instant, in every oscillation period. Some results of the investigation made by Loytsyansky are given in 17.3.

16.4.3   The Case of Transverse Oscillations of the Plate

In case of the transverse oscillations of the plate or of the bottom of the vessel, the penetration of vibration is of another character. Experimental data show the exponential character of the decrease of the amplitudes of vibration of the particles of granular material while the distance from the vibrating surface is growing (see e.g. [244, 513, 159]).

A more refined one-dimensional continual model, suggested by Kremer and Fidline (model **C** — see 15.1) [297] was used by the latter for the theoretic solution of the problem. Denoting the pulsating component of the velocity of particles of the medium by v, the author discovered that should we introduce the function

$$u = v_{\sim}^{(1+R)/2R} \qquad (16.4.6)$$

where R as before is the coefficient of restitution of the velocity at the impact, this function will satisfy the equation of the type of heat conduction equation, i.e. it will play the role of temperature. For the pulsational component of the velocity v the following formula will be obtained

$$v_\sim = v_{0\sim} \frac{2(N-x)^{\frac{R}{2}} \left\{ I_s\left(\frac{\sqrt{1-R^2}}{R}N\right)\right\}^{\frac{(1-R)}{(1+R)}} \left\{ I_s\left[\frac{\sqrt{1-R^2}}{R}(N-x)\right]\right\}^{\frac{2R}{(1+R)}}}{N^{\frac{R}{2}}\left[2\frac{\sqrt{1-R}}{\sqrt{1+R}}I_{s-1}\left(\frac{\sqrt{1-R^2}}{R}N\right) - \frac{3-R}{2N}I_s\left(\frac{\sqrt{1-R^2}}{R}N\right)\right]}$$

$$(16.4.7)$$

Here the coordinate x is counted from the vibrating plane vertically upward and presents a "distributed number of the particle", so that $0 < x < N$ where N — the number of particles $s = (1+R)(2-R)/4R$, $I_s(y)$ — is Bessel's modified function of the order of s, and v_0 is a certain typical value of the velocity.

Figure 16.4 shows the graphs of the function $v_\sim(x)/v_{0\sim}$ at different values of the restitution coefficient R; as one can see, the damping of the pulsating

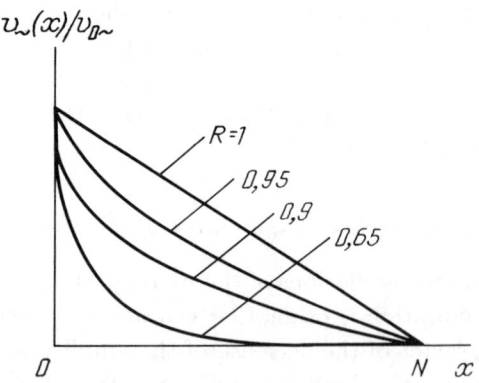

Figure 16.4. The distribution of velocities in the layer of granular material under the transverse oscillations of the plate (according to Fidline [297, 184]).

component of the velocity of the particles of the medium is of an exponential character, which agrees with the above-mentioned experimental results.

The case under consideration is of an essential applied interest in connection with the theory of vibrational mills, as well as of apparatus with a vibro–boiling layer, used in the chemical technology [213, 259, 323, 324, 469, 159, 160]. In case of relatively small particles the "transverse" motion of the layer is affected by the air. The presence of the air leads to a number of remarkable regularities, but we cannot dwell on them here.

16.5   On Penetration of Vibration into Concrete Mixes

The process of penetration of vibration into concrete mixes is quite peculiar. The theoretical and experimental study of this process is of vital importance for such technologies as vibrational compaction and forming concrete and reinforced concrete wares, widely used in the building industry.

The concrete mix is a multicomponent medium consisting of specially selected big and small aggregates, a binding agent and water. When the concrete mix is being prepared, it is inevitable that the air should get into it. Though the amount of air in it is comparatively small (it changes during the processing), the physical and mechanical properties of the mix itself as well as of the hardened concrete depend essentially on its content [475]. As a result, the concrete mix is characterized by rather complicated rheological properties; the most important of which are the resistance of dry friction type, as well as viscous friction and elasticity, the latter being mainly determined by the content of air. Vibrating results in the vibrational transformation of the rheological characteristics and properties of the mix, in particular in its fluidization (see 14.1, Fig. 14.1) which provides effective compaction, necessary for producing concrete with high strength characteristics and a good filling of the formwork when forming.

It is natural that in order to study and ensure the fluidization of the concrete mix it is necessary to have a well-defined idea about the regularities of the penetration of vibration into it. In accordance with what has been said, in theoretical investigations the concrete mix is simulated in form of the continuum, possessing elastic, plastic and viscous properties. It has been established experimentally and proved theoretically that in the process of a vibrational action upon the mix a big role is played by the break-off of the mix from the vibrating element and by the break of its continuity. The review of the present day state of the question and the bibliographical data can be found in the cited above book by Savinov and Lavrinovich [475].

16.6   On the Theory of Vibro-conduction

Palmov [424] has developed an interesting conception of the distribution of vibration in complicated mechanical systems. This theory is quite tempting from the viewpoint of simplicity and clearness of its practical use and it is called by him the *theory of vibro-conduction*. In the joint work by Belyaev and

Palmov [63] this conception was extended to a more general case. The authors cite the articles of the American investigators, in particular those by Davies [165], Eichler [173], Lotz and Crandall [343], Scharton and Lyon [476] where it is shown that the problem of extending random vibration to a wide class of mechanical systems is reduced to the pattern of solving problems on spreading the heat. Further, they suggest an equation of stationary vibro-conduction of the orthotropic macro-heterogeneous object

$$\nabla \cdot \mathbf{K} \cdot \nabla \mathbf{S} - \alpha \mathbf{S} = 0 \qquad (16.6.1)$$

and write down the boundary condition

$$\mathbf{N} \cdot \mathbf{K} \cdot \nabla \mathbf{S} = F \qquad (16.6.2)$$

where ∇ is the Hamilton operator, $\mathbf{K}$ is the vibro-conduction tensor (the analogue of the heat conduction tensor), $\mathbf{S}$ is the analogue of temperature, F is the external flow of vibration inward the body, $\mathbf{N}$ is the vector of the external normal to the surface of the object, α is the parameter, characterizing the intensity of the spatial damping of vibration; dots denote the operation of the scalar multiplication.

Having considered the problem on the distribution of vibration in a certain medium with a complicated structure (an elastic body, every point of which is connected with an infinite set of non-interacting isotropic linear oscillators), the authors come to the conclusion that " the role of temperature in vibrational problems is played by the averaged with respect to frequency mean weighted value of the acceleration of vibration along the axes of orthotropism with weight coefficients, equal to the corresponding velocities of spreading the longitudinal perturbations". In this case "for every frequency component of vibration it is necessary to introduce its own "temperature". Therein lies the difference of the vibrational problem from the thermal". For the flow F an expression has been obtained

$$F = \omega^2 (\mathbf{S}_{NN} + \mathbf{S}_{N1} + \mathbf{S}_{N2})$$

where ω is the frequency, $\mathbf{S}_{NN}$ is the spectral density of stress, normal with respect to the boundary, and $\mathbf{S}_{N1}$ and $\mathbf{S}_{N2}$ are the spectral densities of stresses, tangent to the boundary, acting in two arbitrary orthogonal directions.

Recall that in the problem of section 16.2 of this chapter the role of temperature was played by the velocity of the viscous fluid v, while for the model $\mathbf{C}$ of the granular material (see 16.4) it was played by the function $u = v_{\sim}^{(1+R)/2R}$ where $v_{\sim}$ is the pulsational component of the velocity of particles, and R is the restitution coefficient. It is not contradictory to what was said because it is evident that the classical model of the viscous fluid and the model **C** of the granular material differ essentially from the model of a complicated medium, considered in [63].

It seems that the theory of vibro-conduction still requires (and deserves!) further development and specifications as applied to different media.

Chapter 17

Microvibrorheology: the Behavior of Suspension under Vibration, Effective Viscosity and Effective Density of Suspension

17.1 Preliminary Remarks

As was marked in 12.1, it is natural that the study of the specific features of the behavior of multiphase systems under vibration should be referred to microvibrorheology. This chapter contains a brief presentation of certain results which refer to the behavior of the structural and structureless suspensions i.e. systems which are typical in chemical technologies, when enriching the minerals, and in some other industries. As well as chapter 16, this chapter is of an auxiliary character: first, it considers the "fast" motions and, secondly, the problems under consideration are not necessarily nonlinear.

Microvibrorheological effects play an essential role when solving the problem of controlling rheological properties of nonlinear systems by means of vibration, in particular the problem of creating dynamic materials (see chapter 18).

17.2 Structureless Suspensions — Solid Particles in a Viscous Fluid

In vibrational technology we often have to deal with volumes, filled with suspension, to which harmonic oscillations of a certain frequency ω and amplitude A are imparted (Fig. 17.1,a).

In such cases there arise at least three questions.

1) what are the absolute- and the relative oscillation amplitudes of solid phase particles;

2) what force should be applied to a single volume of suspension to provide the above mentioned oscillations;

3) what is in that case the expenditure of energy.

It is most simple to answer these questions if the particles are balls of the same diameter $d = 2r$, suspended in the viscous incompressible fluid, with their volume concentration c not too great, so that the mutual interaction of particles may be neglected. As was shown in the work by Granat [218], mentioned in 16.2, in that case c must not be more than 5%, which corresponds to the distance between the particles, not less than two or three diameters (it is noted there that in the conditions of a uniform rectilinear motion of the balls in the immobile fluid this distance must be about 30 times greater, which corresponds to considerably lesser concentrations). We will give here the results of the same author [217, 218, 219, 220, 221] that are of interest to us.

The law of oscillations of the bulk of the fluid is assumed to be given in the form $\xi = A \sin \omega t$, the relative oscillations of the particles are obtained in the form $x_r = A_r \sin(\omega t + \varepsilon_r)$ and the absolute — in the form $x_a = A_a \sin(\omega t + \varepsilon_a)$ with

$$\frac{A_r}{A} = |\Delta - 1| \left[\Delta^2 + \Delta \left(1 + \frac{9}{2}\alpha^{-1}\right) + \frac{1}{4} + \frac{81}{16}\alpha^{-4}(1 + 2\alpha + 2\alpha^2 + \frac{4}{9}\alpha^3) \right]^{-\frac{1}{2}},$$

$$\frac{A_a}{A} = \left[1 - \frac{\Delta + 2 + \frac{9}{2}\alpha^{-1}}{\Delta - 1} \left(\frac{A_r}{A}\right)^2 \right]^{\frac{1}{2}}; \quad \varepsilon_r = \arctan \frac{9}{4} \frac{1 + \alpha}{\alpha^2 \left(\Delta + \frac{1}{2} + \frac{9}{4}\alpha^{-1}\right)},$$

$$\varepsilon_a = \arctan \frac{-(\Delta - 1)(1 + \alpha)\alpha^{-2}}{\left[\frac{2}{3}\Delta\left(1 + \frac{3}{2}\alpha^{-1}\right) + \frac{1}{3} + \frac{9}{4}\alpha^{-4}\left(1 + 2\alpha + 2\alpha^2 + \frac{8}{9}\alpha^3\right)\right]}.$$

$$(17.2.1)$$

Here $\Delta = \rho/\rho_0$ is the ratio of the density of the material of the particles to the density of the fluid, and $\alpha = r\sqrt{\omega/2\nu}$ where ν is the coefficient of the kinematic viscosity of the fluid. Graphs of dependences (17.2.1) are given in Fig. 17.1,*b–f*. As follows from the formulas and from the graphs, the greater is the difference of the density of particles from the density of the fluid, the greater is the amplitude of relative oscillations of the particles. The oscillations of the particles, denser than the fluid, lag behind in phase from the oscillations of the fluid — the phase shift ε_r is within the ranges from $-\frac{1}{2}\pi$ to $-\pi$. With the growth of the parameter α these amplitudes increase, and the phase shift ε_r asymptotically approaches $-\pi$. In the absolute motion the particles, denser than the fluid ($\Delta > 1$), will, lagging behind in phase, oscillate with the amplitudes A_a lesser than those of the oscillations of the fluid, and the particles of

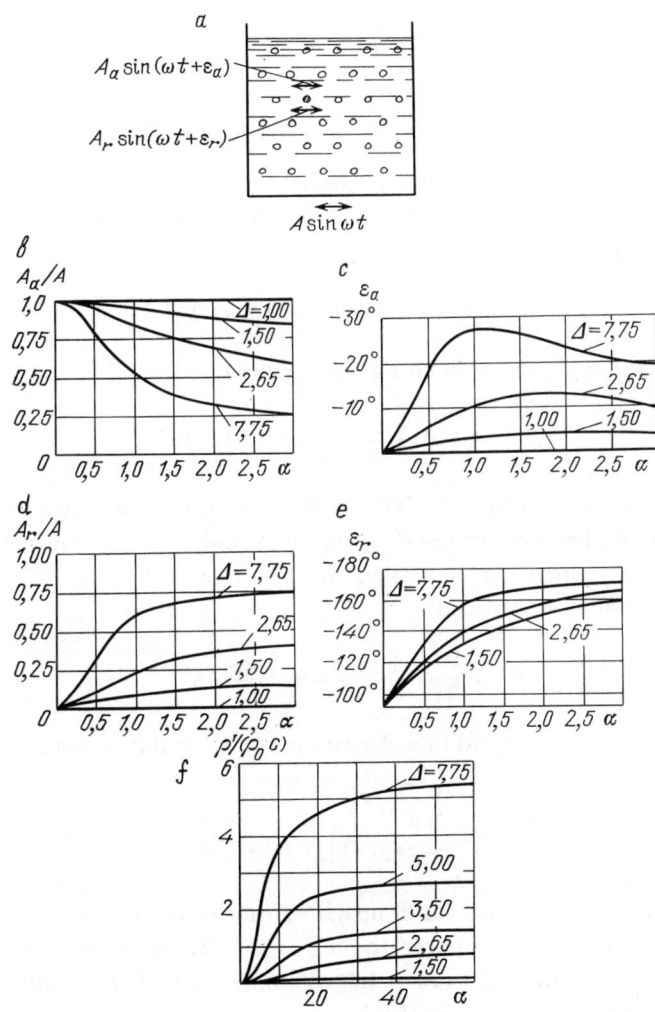

Figure 17.1. If a vessel with a suspension is subject to harmonic oscillations, then the solid particles will perform oscillations with respect to the fluid; the greater the difference between the density of the particles and that of the fluid, the greater the amplitude of oscillations. The effective density of the suspension under oscillations will be always less than the density of the suspension.

lesser density ($\Delta < 1$), being advanced in phase, will oscillate with the absolute amplitudes exceeding the oscillation amplitudes of the fluid.

The force p, necessary to provide harmonic oscillations of a unit volume of the suspension, can be presented in the form

$$p = \rho^* \ddot{x} + k^* \dot{x} \qquad (17.2.2)$$

where

$$\rho* = \rho_s - \rho' \qquad (17.2.3)$$

is the value which may be called the *effective density of suspension under oscillations* [219]; due to the mobility of particles it proves to be less than the density of suspension

$$\rho_s = \rho_0(1 - c) + c\rho = \rho_0[1 + c(\Delta - 1)] \qquad (17.2.4)$$

by a positive value ρ', determined by the formula

$$\rho' = \rho_0 c(\Delta + \frac{1}{2} + \frac{9}{4}\alpha^{-1})(A_r/A)^2. \qquad (17.2.5)$$

Graphs of dependence of the relative effective (seeming!) decrease of density of the suspension under oscillations $\rho'/(\rho_0 c)$ on Δ and α are shown in Fig. 17.1,*f*. For the value k^* which can be called the *effective coefficient of damping* the following expression is obtained

$$k^* = \frac{9}{4}\rho_0 c\omega(1 + \alpha)\alpha^{-2} A_r^2/A^2. \qquad (17.2.6)$$

The power, necessary to spend in order to maintain oscillations of a unit volume of the suspension is

$$N = <p\dot{x}> = \frac{1}{2}k^* A^2 = \frac{9}{8}\rho_0 c\omega(1 + \alpha)\alpha^{-2}A^2(A_r^2/A^2). \qquad (17.2.7)$$

Formulas $(2.1) - (2.7)$ are valid, provided that either the Reynolds number $Re = A_r\omega d/\nu$ is small as compared to one, or the Strukhal number $Sh = d/A_r$ is large enough as compared to one. In [220] one can find the results, referring to a more general case. It should be also noted that within the limits of the validity of the indicated formulas it is not difficult, using the principle of superposition, to obtain the corresponding results for the cases of the arbitrary periodic law of the oscillations of the suspension and also for the suspension, consisting of particles of different sizes and densities. It is also possible to consider the problem about the oscillations in the suspension of solid bodies big as compared to particles of a solid phase and to the distance between them. In this case it is natural that there appears a notion of the *vibroviscosity of suspension* [221].

Finally, it should be noted that in case the viscosity of the fluid can be neglected, i.e. when $\alpha \gg 1$, formulas (17.2.1) and (17.2.5) are much simplified, taking the form

$$\frac{A_r}{A} = |\Delta - 1|/(\Delta + \frac{1}{2}) = 2|\rho - \rho_0|/(2\rho_0 + \rho),$$

$$\frac{A_a}{A} = 3/(2\Delta + 1) = 3\rho/(2\rho_0 + \rho); \quad \varepsilon_r = -\pi, \quad \varepsilon_a = 0;$$

$$\rho'/\rho_0 c = 2(\rho - \rho_0)^2/\rho_0(2\rho + \rho_0). \tag{17.2.8}$$

The described regularities of the behavior of suspensions under vibration are used to intensify the technological processes and to accelerate the chemical reactions [114, 213, 324], and also when elaborating devices for measuring the amount of dust in the atmosphere [272].

17.3 Structurized Suspensions — Particles in the Medium with the Resistance of the Type of Dry Friction

Let us now consider the case when solid particles are not in the viscous fluid, but in the medium with a resistance of the type of dry friction, for instance in a structurized suspension (Fig. 17.2,a). In this case we will assume that the size of the particles is much larger than that of the particles which form the suspension. The differential equation of the relative motion of the particle in such medium which performs oscillations according to the law $\xi = A \sin \omega t$, has the form

$$m_1 \ddot{x} = m_0 (\Delta - 1) A \omega^2 \sin \omega t + F(\dot{x}) \tag{17.3.1}$$

where m_1 is the mass of the particle with the associated mass, m_0 is the mass of the medium in the volume, equal to that of the particle, $\Delta = \rho/\rho_0$ is the ratio between the mean densities of the particle and the medium,

$$F(\dot{x}) = \begin{cases} -F & \text{at} \quad \dot{x} > 0, \\ F & \text{at} \quad \dot{x} < 0, \end{cases}$$

$$-F_1 < m_0 (\Delta - 1) A \omega^2 \sin \omega t < F_1 \quad \text{at} \quad \dot{x} = 0. \tag{17.3.2}$$

Here F_1 and F are the constants, answering the forces of the resistance to the motion of the particle in the medium from the state of a relative rest and in the motion to the medium ($F_1 \geq F$) respectively.

Equation (17.3.1) with an accuracy to the coefficients coincides with the equation of relative motion of a plane solid body (material particle) along a rough horizontal surface, performing longitudinal oscillations according to the law $\xi = A \sin \omega t$ (Fig. 17.2,b):

$$m\ddot{x} = mA\omega^2 \sin \omega t + F(\dot{x}) \qquad (17.3.3)$$

where m is the mass of the particle and

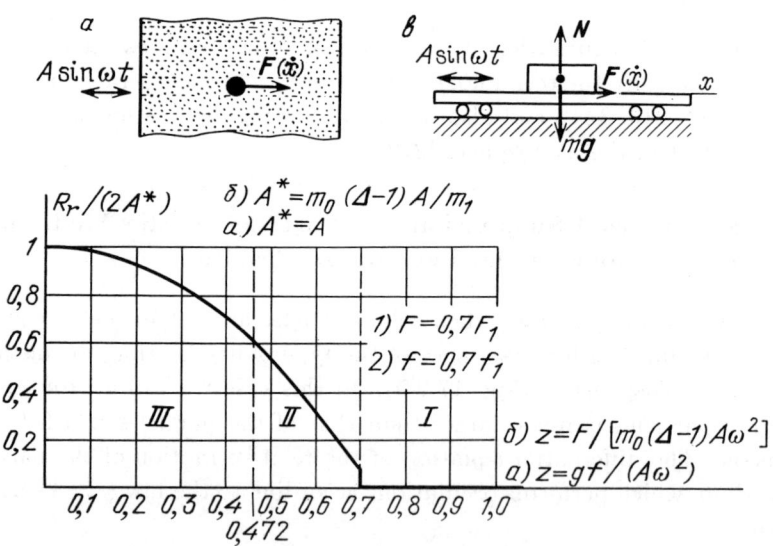

Figure 17.2. A solid particle lying on a rough horizontal surface or placed into a medium with a resistance of dry friction type under sufficiently intensive oscillations of the plane or the medium, perform relative oscillations with the semiswing R_r.

$$F(\dot{x}) = \begin{cases} -mgf & \text{at} \quad \dot{x} \; > \; 0, \\ mgf & \text{at} \quad \dot{x} \; < \; 0, \end{cases}$$

$$-mgf_1 < mA\omega^2 \sin \omega t < mgf_1 \quad \text{at} \quad \dot{x} = 0. \qquad (17.3.4)$$

f_1 and f denote the friction coefficients of rest and sliding respectively. The motions, defined by equation (17.3.3) were studied in detail by Loytsyansky

[339] and later in the book of the author and Dzhanelidze [84], therefore we can make use of the results available.

Figure 17.2 shows the graph of dependence of a semi-swing of relative oscillations of the particle in the medium R_r on the parameters of the system. The value $R_r/2A^*$ where $A^* = m_0(\Delta - 1)A/m_1$ in the case under consideration depends on two dimensionless parameters

$$z = \frac{F}{m_0(\Delta - 1)A\omega^2}, \quad z_1 = \frac{F_1}{m_0(\Delta - 1)A\omega^2}. \tag{17.3.5}$$

When plotting the graph, we assume that $F = 0.7F_1$. At $z > 0.7$ in the conditions of Fig. 17.2 the particle moves together with the medium (region *I*), with $0.472 < z < 0.7$ it slides alternatively forward and backward, stopping at the change of the direction of sliding for the finite intervals of time (region *II*), and at $z < 0.472$ (region *III*) — instantly changing the direction of sliding. The same graph, if z_1 and z imply the values $z_1 = g\, f_1/A\omega^2$ and $z = g\, f/A\omega^2 = 0.7z_1$, characterizes the swing of relative oscillations of the particle along a rough surface.

Chapter 18

The Problem of the Control of Vibrorheological Properties of Mechanical Systems. The Idea of Creating Dynamic Materials

18.1 On the Problem of Forming the Properties of Nonlinear Mechanical Systems by Means of Vibration

Almost everything that has been said refers to the analysis of the behavior of nonlinear mechanical systems under the action of vibration. Meanwhile problems of synthesis are of a considerable interest. By those problems we mean purposeful formation of properties of the mechanical systems by means of vibration. In particular, the problem of controlling vibro-rheological properties of different media and different materials is of great importance. The methods of solution of the indicated problems are beginning to be developed.

In the general case the problem of forming the properties of the nonlinear mechanical system by means of vibration (the problem of synthesis) can be formulated in the following way: let the initial system be described by the equation (see 3.2.2)

$$m\ddot{\mathbf{x}} = \mathbf{F}(\dot{\mathbf{x}}, \mathbf{x}, t). \tag{18.1.1}$$

What kind of function must $\mathbf{\Phi}$ be in the equation

$$m\ddot{\mathbf{x}} = \mathbf{F}(\dot{\mathbf{x}}, \mathbf{x}, t) + \mathbf{\Phi}(\dot{\mathbf{x}}, \mathbf{x}, t, \omega t) \tag{18.1.2}$$

for the system to possess the needed properties?

In the context of vibrational mechanics the problem is put in the following way: what kind of function must $\mathbf{\Phi}$ be for the system, the slow motions in which are described by equation

$$m\ddot{\mathbf{X}} = \mathbf{F}(\dot{\mathbf{X}}, \mathbf{X}, t) + \mathbf{V}(\dot{\mathbf{X}}, \mathbf{X}, t) \qquad (18.1.3)$$

for the system to possess the needed properties? It is natural that in this case the problem is reduced to the adequate formation of the vibrational force $\mathbf{V}$.

The concrete content of the problem of synthesis depends to a great extent on the class of the systems under consideration and on the statement of the technical problem; in particular, in some cases the function $\mathbf{\Phi}$ is given with an accuracy to certain parameters and the problem is reduced to the choice of these parameters.

It is natural that not all the given properties of the system can be provided by means of vibration, i.e. the problem of synthesis does not always happen to have a solution. It should be also noted that this problem can be readily formulated as a problem of the theory of control. In that way this problem was formulated in [129].

In another form the problem of vibrational control is considered in [62] where they solved some interesting problems. See also the proceedings of the IUTAM-symposium mentioned in [124]. Below the problem of forming properties of mechanical systems under vibration is considered as applied to the problem of creating materials possessing the predicted vibro-rheological properties [129, 130].

18.2   On Dynamics Materials

18.2.1   The Idea and Definition of Dynamic Materials

We discuss here a new concept incorporating the inertial, elastic and other material properties that affect the dynamic behavior of the mechanical systems such as deformable solids, liquids, and others. Here we mean a concept of creating the so called *dynamic materials* [129, 130].

Dynamic materials are defined as the media with material parameters alternating in space and time. An important class of dynamic materials is represented by *dynamic composites.* These formations differ from the ordinary composites due to the explicit involvement of time as an additional fast variable participating in the microstructure that becomes spatio-temporal in this case.

The properties of dynamic materials can, naturally, be quite different from the properties of the original constituents. Changing the material parameters of the original constituents the microgeometry of the assemblage, and the character of change of hose parameters in time, we can control the dynamic properties of the materials, obtaining effects, which are absolutely impossible when using ordinary materials (see below).

What has been said refers not only to the mechanical materials, characterized by inertial, elastic, dissipative and other parameters, but also to the electro-technical materials, whose main characteristics are self-induction, capacitance, etc. Aspects of the problem which are of principal importance are also connected with the relativistic effects [350, 348, 349]. However, here we will confine ourselves to the consideration of only the classical mechanical materials.

18.2.2   Two kinds of Dynamic Materials

One can imagine two ways of obtaining dynamic materials, and, accordingly, two types of such materials.

The materials of the first type are obtained by either instaneous or gradual change of the material parameters of different parts of the system (of masses, fluids, self-induction, capacitance, etc.) in the absence of relative shift of those parts.

This way is called *activation* [350, 348], and the corresponding materials are called *dynamic materials of the first kind or the activated dynamic materials.*

The second way presupposes that certain parts of the system are endowed with relative motions, prearranged or excited in some way. That way will be arbitrarily called *kinetization*, and the corresponding materials will be called *dynamic materials of the second kind or the kinetic dynamic materials.*

The dynamic material of the second type can be imagined as two or several media penetrating into each other, filling up a certain domain of space, with every medium performing a certain motion (in particular fast oscillations) with respect to other media. It is quite natural that the material parameters and properties of such a material will differ essentially from the parameters and properties of the original media and that there are great potentialities to control these properties.

18.2.3   About Some Possibilities of the Implementation of Dynamic Materials

A natural question arises about the possibility of the implementation of the above described materials. As for the materials of the first kind, the corresponding ways are well known as applied to electrotechnical materials. Therefore we will dwell here on some ways of implementation of the dynamic materials of the second kind.

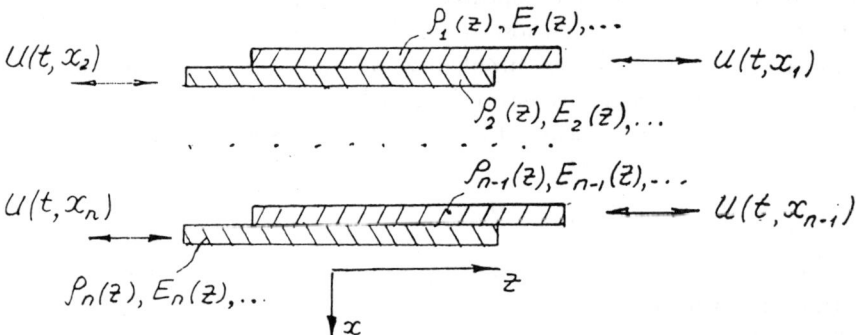

Figure 18.1. One of the ways of implementation of the dynamic material of the second kind.

Figure 18.1 shows a scheme of a device which is an aggregate of plates adjoining each other, each plate being endowed with oscillations according to the periodic law $u(x_s, t) = u(x_s t + T_s)$ with a period T_s, depending on the coordinate x_s (the "number of the plate"). The density, modulus of elasticity, or other material parameters as well as the thickness of every plate can be distributed in a certain way along its length (the coordinate z). Fig. 18.1 presents a one-dimensional case when the material properties are changing "quickly" along the coordinate x; it is not difficult, however, to imagine a more complicated two-dimensional version of this scheme. A simpler version for the implementation can be obtained if all the odd- numbered- and all the even-numbered plates are moving in the same way, i. e.

$$u(t, x_1) = u(t, x_3) = \ldots = u(t, x_{2n-1})$$
$$u(t, x_2) = u(t, x_4) = \ldots = u(t, x_{2n}).$$

In this case every odd-numbered plate can be attracted to one oscillating body while every even-numbered plate will be attached to another.

Figure 18.2 represents a scheme of the device in which the round discs with the odd numbers are fixed on one shaft, and the discs with the even numbers are fixed on the other. The shafts may rotate with certain angular velocities ω_1 and ω_2 or perform rotary oscillations with the frequencies ω_1 and ω_2 and

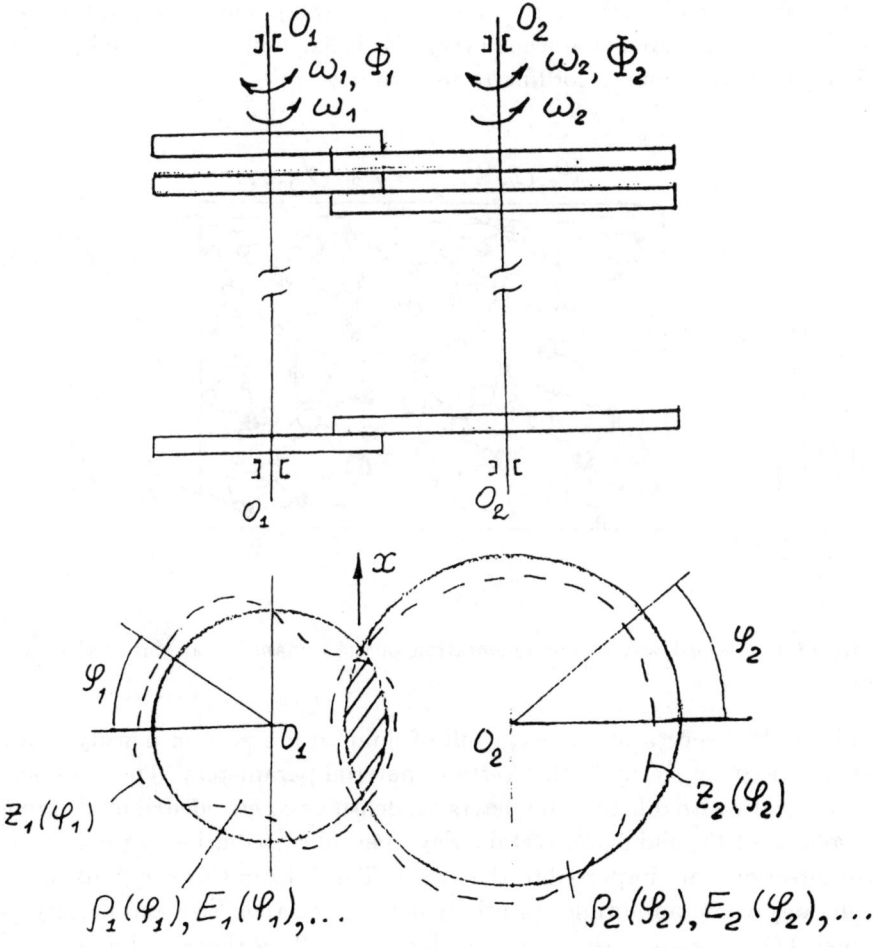

Figure 18.2. Another way of implementation of the dynamic material of the second kind.

with the angular amplitudes Φ_1 and Φ_2. The parameters of the material of the plates can depend on the angular coordinates $\varphi_1 r$ and φ_2 and certainly on the number of the plate. The one-dimensional medium with the changing material parameters presents in this case a rod with an axis x and with a lens-shaped cross-section. It is not difficult to make the area of the cross-section of the rod also variable both in space and time by making the discs have their shapes different from circular (the dotted lines in Fig. 18.2). One can also imagine a much more complicated case when every disc is fixed on its shaft and performs oscillations or rotations according to its own law.

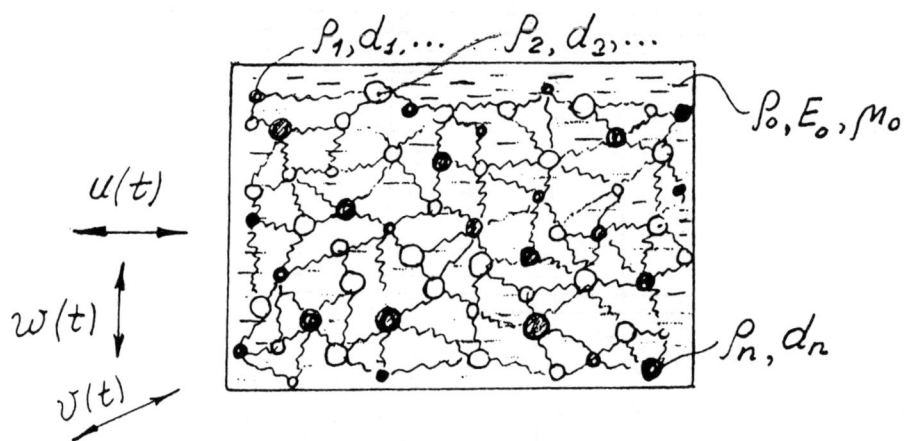

Figure 18.3. A third way of implementation of the dynamic material of the second kind.

Figure 18.3 represents a vessel full of fluid with a certain density ρ, stiffness E_0, viscosity μ_0, and other certain material parameters. There are small spherical balls with different diameters d_s, densities ρ_s etc. distributed through the volume of the fluid in a certain way. Periodic oscillations in one, two or three directions, are imparted to the vessel. The balls in this case will perform oscillations whose amplitude essentially depends on their size and density (see chapter 17). Choosing theese parameters, as well as the coordinates of the balls in the vessel in a certain way, we will obtain a medium whose effective properties will change along one, two or three coordinates. The balls can be deformable, for instance they can be rubber capsules, filled with air, then one can use resonance effects. To preserve the properties of the medium when the oscillations temporarily cease, the balls may be connected by elastic elements, provided a certain distribution of the balls in the static position and, possibly, the resonance effects during the oscillations of the vessel. The homogeneous suspension is the simplest version of the described medium. In this case, one of the mutually penetrating media will be fluid, the other will be the multitude of particles forming the suspension. Instead of balls one can use particles of a more complicated shape. The corresponding problems can be referred to the problems of vibro-rheology. One can find the effective properties of the above-described media using the methods presented in this book and in book [39].

18.2.4   *On Some Potentialities Provided by Using the Dynamic Materials*

Some potentialities, offered by the use of dynamic materials, can be illustrated by a simplest example.

Let us consuder a rod whose effective density ρ, the modulus of elasticity E, and the cross-section area F can be preassigned as functions of the coordinate x, counted along the axis of the rod and of the time t. The motion of such a rod is described by the equation

$$(\varphi u_t)_t = (\psi u_x)_x \qquad (18.2.1)$$

where $\varphi = \varphi(x,t) = \rho(x,t)F(x,t)$, $\psi = \psi(x,t) = E(x,t)F(x,t)$ and the lower indexes t and x designate the corresponding partial derivatives.

The task is to provide the wanted motion of the rod $u(x,t)$ by a proper choice of the function ρ, E, and F (and certainly the initial conditions as well), which is a peculiar inverse problem of mechanics. The solution of equation (18.2.1) depends on two functions φ and ψ and these functions are involved in the equation quite symmetrically. If the function ψ is given, it is quite easy to find from this equation the function φ

$$\varphi = \frac{1}{u_t} \int (\psi u_x)_x \, dt. \qquad (18.2.2)$$

If the function φ is given, then the function ψ is defined by the same equality, substituting φ for ψ, and t for x, and x for t.

Two special cases are of interest when the expression (18.2.2) is still more simplified.

1) The function $u(x,t)$ is given as a product

$$u(x,t) = u_1(t)u_2(x)$$

or as a sum of such products. Then it is natural to find the functions φ and ψ in the same form

$$\varphi(x,t) = \varphi_1(t)\varphi_2(x), \quad \psi(x,t) = \psi_1(t), \psi_2(x).$$

In this case the variables in the equation (18.2.1) are separated and we find

$$\varphi_1 = \frac{\lambda}{(u_1)_t} \int \psi_1 u_1 dt, \quad \varphi_2 = \frac{1}{\lambda}\frac{[\psi_2(u_2)_x]_x}{u_2}$$

$$\psi_1 = \frac{1}{\lambda}\frac{[\varphi_1(u_1)_t]_t}{u_1}, \quad \psi_2 = \frac{\lambda}{(u_2)_x} \int \psi_2 u_2 dx$$

where λ is a constant. Any two functions can be found by these formulas if the other two are given.

2) The function $u(x,t)$ is given as

$$u(x,t) = u(x - vt),$$

i.e. in the form of a wave, running with a velocity v.

In this case it is natural to find the functions φ and ψ also in the form of the running waves ("of the waves of properties")

$$\varphi(x,t) = \varphi(x - vt), \quad \psi(x,t) = \psi(x - vt).$$

And from equation (18.2.1) we obtain an ordinary differential equation

$$v^2(\varphi u')' = (\psi u')'$$

with the independent variable $z = x - vt$ (the prime denoting differentiation with respect to that variable). From this equation it is easy to define one of the functions φ or ψ if the other one is known. As has been shown in [349, 350, 348], by means of arranging "a wave of properties" a certain part of the body can under certain conditions be fully isolated from the long-wave perturbations.

18.2.5   *On Dynamic Surfaces*

It should be noted that there are known two-dimensional analogues of the proposed dynamic materials (they can be called *dynamic surfaces*). The remarkable properties they manifest serve as another confirmation of the considerable technical potentialities of the implementation of dynamic materials. Mark, by the way, that in the descriptions of the above-mentioned surfaces they were never meant as dynamic materials. We will give a brief description of the three mentioned devices.

The first of them represents a plane, formed by two systems of the alternating parallel threads. The threads of the first system (say, the odd-numbered threads) move in one direction, while the threads of the other system (the even-numbered threads) move with the same velocity but in the opposite direction. It is easy to see that an extended body, lying on such a surface, will be affected by forces of the type of viscous friction, while the friction between a single thread and the body is dry (the Coulomb friction) [426, 546].

Another example is the surface, formed by two similar parallel horizontal shafts, rotating with the same angular velocity in opposite directions. An extended body placed on such shafts will be acted upon by a linear "elastic" restoring force. In other words, the body will behave as a conservative linear oscillator, despite the fact that dry friction forces are acting between the body and the surface of the shafts [426, 546].

The third example is given by a plane, formed by two groups of alternating parallel rods. The rods of the first group (say, odd-numbered) are linked with one solid body, and the rods of the second group — with another. Every body performs the preset translational oscillations along certain trajectories [94]. A sufficiently extended body, placed on such a surface, will, under certain conditions, be subjected to impact and periodic force effects with a non-zero average component, which is unattainable or hardly attainable when using a continuous vibrating surface [129, 94]. This makes it possible to obtain a considearble technical effect in the devices for the transportation and screening of the granular materials.

It goes without saying that in the case of space the implementation of the idea of dynamic materials is much more complicated and requires special technological solutions. The ideas of three such solutions were proposed above. Though in general such solutions must be subjects of patents. One of the aims of this chapter is to initiate the appearance of such solutions.

Chapter 19

Supplements

19.1 Vibrorheological Effects in Macroscopic Homogeneous Media (Turbulent Viscosity, Vibrocreep, Vibrorelaxation, Vibroplasticity, Fatigue of Materials)

Uniting in one section five different effects, listed in the title of this section, especially the first and the last — the fundamental ones — may seem unusual and even dubious. All of them, however, refer undoubtedly to vibrorheology in terms of the definition given in 12.2, in all the cases vibration causes a considerable change in the rheologiclal constants of the bodies.

We spoke in detail in 13.3 about a considerable increase of the effective (seemingly-effective, perceived by the observer **V**) viscosity of the fluid as about a vibrorheological effect which takes place when passing from the laminar regime of the flow of the fluid to the turbulent one. The distinction from the other three effects which will be discussed below, lies in the fact that the pulsation of the elements of the fluid is excited not from outside, but appears autonomously in the fluid itself. As was marked, interesting publications have recently appeared which testify to a considerable effect of the external vibration upon the turbulent flow and upon the resistance of the motion of the bodies in the turbulent flow [569, 570].

By *creep* one means the increase, in the course of time, of the deformation, with stress being at a constant level. The *vibrocreep effect* consists in a considerable acceleration of the process of creeping when additional stresses of alternating signs of a relatively small amplitude are imposed on it (Fig. 19.1, a): a certain level of deformation ε_* is achieved much sooner than at the constant stress $\sigma = \sigma_0 + \sigma_a$.

By *relaxation* one means the decrease of stress in the course of time, with the given deformation remaining unchanged. Accordingly, the *vibrorelaxation effect* consists in a considerable acceleration of the process of relaxation with an imposing of additional deformation with alternating signs of a relatively small amplitude (Fig. 19.1,*b*): the level of stresses σ_* is achieved much sooner than at the constant deformation ($\varepsilon = \varepsilon_0 + \varepsilon_*$).

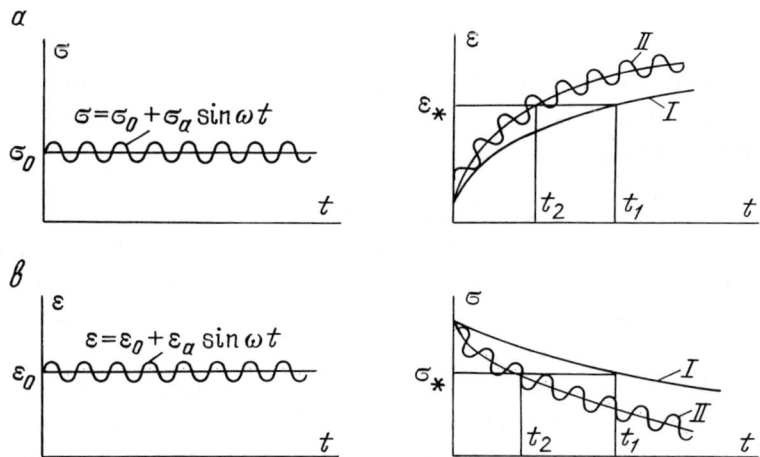

Figure 19.1. Effects of vibrocreeping and vibrorelaxation. The velocity of the processes of creeping and relaxation is much higher when there are additional effects with alternating signs. a) Vibrocreeping (to the left — the program of loading; to the right — the change of deformation in time; *I* - at the constant stress, *II* - at the pulsating stress); b) vibrorelaxation (on the left is the program of loading, on the right — the change of the stress in time: *I* - at the constant deformation, *II* - at the pulsating deformation).

Undoubtedly, the seeming reduction of the plasticity limit of the materials at the imposing of vibration of high frequency, in particular — of the ultrasonic vibration, (the *vibroplasticity effect*) is related to the indicated effects.

The phenomenon of the *fatigue of the material* consists in the fact that the destruction of a detail or a specimen, in case the stresses are pulsating or with alternating signs, takes place at much smaller mean levels of stress than when the stresses do not change in time. Since the enumerated effects are of great applied and fundamental importance, an immense number of experimental and theoretical investigations are devoted to them. We will name just a few of them which refer to relatively new problems, such as vibrocreep, vibrorelaxation, and vibroplasticity [29, 45, 260, 487]. However, the problem of creating models, explaining and describing the mechanism of all those effects is far from being solved. Perhaps the comprehension of their vibrorheological common nature will give the investigators some useful ideas.

Though in many cases it turns out that to explain the peculiar behavior of the bodies under vibration, it is not necessary to create new models which presuppose the change of physical properties of these bodies under the action of vibration, but it is enough to consider very attentively the dynamic behavior of the system within the framework of the existing models. Publications on the theory of vibrosinking of piles, on the effective coefficients of friction under vibration (see 10.3 and 13.1) are the examples of it, as well as the work by Astashev [29] where the effect of vibroplasticity is explained in the context of an ordinary model of an elastically-plastic body.

19.2 Certain General Vibrorheological Regularities

What has been said about the action of vibration on the nonlinear mechanical systems with friction can be summed up in form of the following brief theses.

For systems with dry friction and for vibro-shock systems:

Forces of dry friction type, possibility of repetitive shock interactions	*+ vibration =*	*slow vibrational forces of viscous friction type [of type of vibro-transformed force $V_1(X)$] + additional slow vibrational forces [of type of the driving vibrational force $V(0)$ type] + either random or determined vibration.*

For the nonlinear systems with viscous friction:

Forcers of viscous friction	*+ vibration =*	*slow forces of the transformed viscous friction + additional slow vibrational forces + random or determined vibration.*

The last terms in the right-hand sides of these conventional formulas point to the fact that the resultant oscillations, even under a determined action, are often of a random character, for example, due to the statistic character of the micro-properties of bodies or elements these bodies consist of. Recall that when speaking of the vibrational action, we do not necessarily mean an external effect, but also the effect of either a random or determined vibration, appearing within the autonomous system.

What has been said refers to macrovibrorheology. As for the microvibrorheology, the main thing there is the appearance of more or less intensive relative oscillations of dissimilar components (phases) of a multy-phase system.

Part V

Some Other Problems

Chapter 20

The Motion of the Particle in a Fast Oscillating Nonuniform Field

20.1   The simplest Case: Rectilinear Motion in a Field of a Harmonic Standing Wave

A short review of the main regularities of the behavior of particles in fast oscillating force fields and of the corresponding publications was given in 5.3 in connection with the problem about a pendulum with a vibrating axis of suspension and its generalizations. The most important of those regularities can be established when considering the simplest problem — about a one-dimensional motion of a particle in a fast oscillating stationary force field. The equation of motion of the particle in such a field will be written down as

$$m\ddot{x} = F_0(x) - h\dot{x} + \Psi(x)\sin\omega t. \qquad (20.1.1)$$

Here x is the coordinate of the particle, m is its mass, $F_0(x)$ is the "slow" positional force, h is the viscous resistance coefficient, $|\Psi(x)|$ is the amplitude of the force with the frequency ω, dependent on the coordinate x. Thus, the fast oscillating field is of the character of a standing wave (Fig. 20.1,a).

The points x where $\Psi(x) = 0$, will, as usual, be called *nodes*, and the points x where $\Psi(x)$ has maximums or minimums will be called the *crests* of the wave; at such points we have $\Psi'(x) = 0$ (considering that $\Psi(x)$ as well as $F_0(x)$ is a sufficiently smooth function).

Assume that the frequency $\omega \gg 1$ (see the footnote on page (??), and the force $\Psi(x)$ is large as compared to $F_0(x)$ and $h\dot{x}$, so that equation (20.1.1) can be presented as

$$m\ddot{x} = F(\dot{x}, x) + \omega\Phi_1(x, \omega t) \qquad (20.1.2)$$

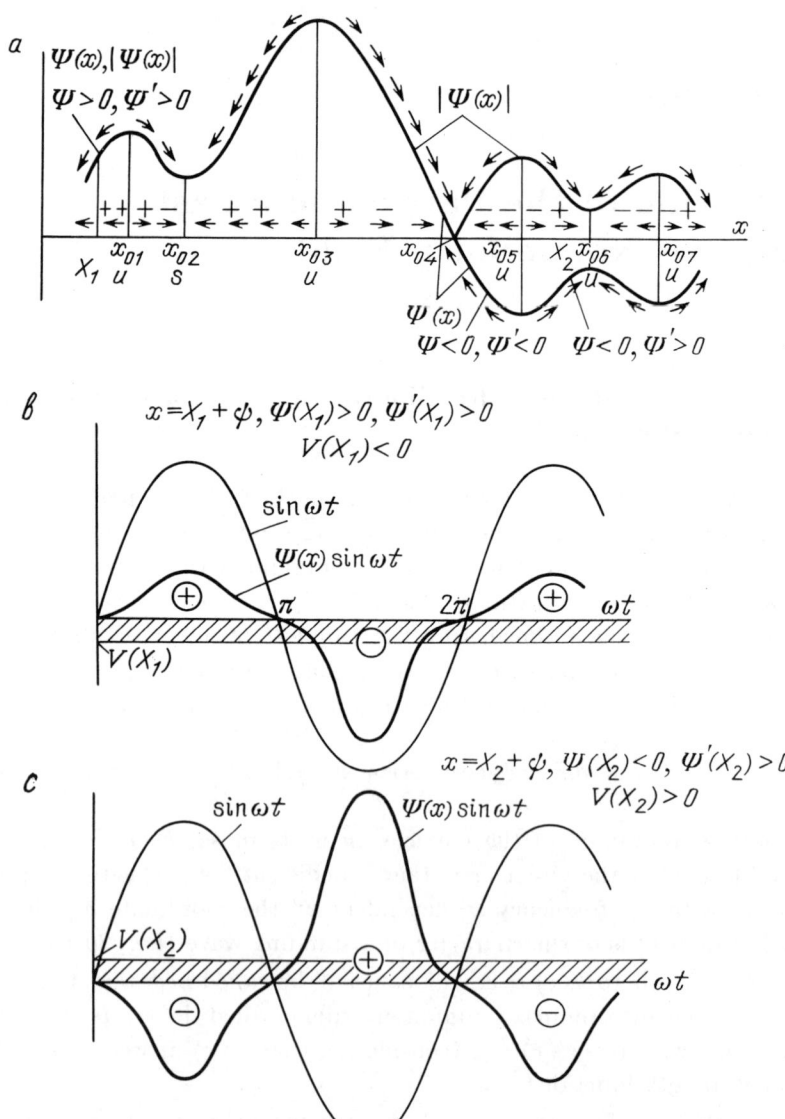

Figure 20.1. The particle in a fast-oscillating nonuniform field "drifts" to the points of the field where the amplitude of the force $|\Psi(x)|$ has a minimum.

where

$$F(\dot{x}, x) = F_0(x) - h\dot{x},$$

$$\omega\Phi_1(x, \omega t) = \Psi(x)\sin\omega t, \quad <\Phi_1>_{x=\mathrm{const}} = 0. \qquad (20.1.3)$$

We are interested in the solutions of equation (20.1.2) of the type

$$x = X(t) + \varepsilon\psi_1(t, \omega t); \quad <\psi_1> = 0, \quad \varepsilon = 1/\omega \ll 1. \qquad (20.1.4)$$

Since relations (20.1.2)–(20.1.4) have the form of equalities (3.2.32)–(3.2.34) of 3.2.7 and assuming that the corresponding conditions of the first and the second theorem of Bogolubov are satisfied, one can use the theorem, given in the indicated subsection. According to this theorem in the first asymptotic approximation the main equation of vibrational mechanics can be presented in the following form:

$$m\ddot{X} = F_0(X) - h\dot{X} + V(X) \qquad (20.1.5)$$

where in view of relations (20.1.3), (20.1.4) and formulas (3.2.35) and (3.2.37) of chapter 3 the vibrational force is

$$V(X) = \frac{1}{\omega} < \Psi'(X)\psi_1^*(X, \omega t)\sin\omega t > \qquad (20.1.6)$$

with ψ_1^* being a periodic solution of the equation of fast motions

$$m\ddot{\psi}_1 = \omega\Psi(X)\sin\omega t, \qquad (20.1.7)$$

found at $X = \mathrm{const}$. In other words

$$\psi_1^* = -\Psi(X)\sin\omega t/m\omega, \qquad (20.1.8)$$

which corresponds to the so called *purely inertial approximation* (see 3.2.8). As a result, after averaging according to formula (20.1.6) we get

$$V(X) = -\frac{1}{2}\frac{1}{m\omega^2}\Psi(X)\Psi'(X) = -[\Psi^2(X)]'\frac{1}{4m\omega^2}. \qquad (20.1.9)$$

If one introduces the "ordinary" potential energy of slow forces Π_F, so that

$$\frac{d\Pi_F}{dX} = -F_0(X), \qquad (20.1.10)$$

the potential energy of the vibrational forces

$$\Pi_V = \frac{1}{4m\omega^2}\Psi^2(X), \quad \frac{d\Pi_V}{dX} = -V(X) = \frac{1}{2m\omega^2}\Psi(X)\Psi'(X), \qquad (20.1.11)$$

and the potential function
$$D = \Pi_F + \Pi_V, \tag{20.1.12}$$
then the main equation (20.1.5) will be presented in the form

$$m\ddot{X} + h\dot{X} = -\frac{dD}{dX}, \tag{20.1.13}$$

typical for the potential on the average dynamic systems when there is dissipation of the energy (see 4.1), while the initial equation (20.1.1) even at $h = 0$ corresponded to an essentially nonconservative system.

Thus, according to above mentioned theorem 6 of 3.2.7, and to the theorems of Thomson-Tait-Chetayev, the points X_0 of the strict minimum of function D are answered by the positions of the asymptotically stable quasi-equilibrium of the initial system, i.e. of the motion of type (20.1.4), for which $X = X_0 = $ const (the notion of quasi-equilibrium is given in 4.3).

Now let us turn to the analysis of the result obtained. First we will consider the case when the force $F_0(x)$ is absent, so that $D = \Pi_V = \frac{1}{4}\Psi^2(X)/m\omega^2$. Then if the field of force is also absent, in that case for the initial, as well as for the transformed system, every point of the axis x is a position of equilibrium of the particle. In case there is a field of force, there appear isolated positions of quasi-equilibrium, corresponding to the nodes and crests of the wave, i.e. points of maximums and minimums of the function $\Psi^2(X)$ — the points where $\Psi'(x) = 0$, $\Psi''(x) \neq 0$ and $\Psi(x) = 0$ (Fig. 20,1,a). In this case points of the minimum of the amplitude of the wave $|\Psi(x)|$ (including the nodes of the function $\Psi(x)$) are answered by the positions of the stable quasi-equilibrium, while points of the maximum are answered by the positions of the unstable quasi-equilibrium of the particle. In other words, in the force field of type of the standing wave particles are "attracted" to the points of the minimum of the amplitude of the field $|\Psi(x)|$ and are repulsed from the points of the maximum $|\Psi(x)|$.

Summing up, we may also say that for the slow motions of the particle the presence of the standing wave leads to the appearance of the "potential pits", corresponding to its stable states. These pits are located at the points of the minimum of the amplitude of the wave $|\Psi(x)|$, including the nodes of the wave where $\Psi(x) = 0$.

The formulated results may also be arrived at by means of rather simple reasoning of a heuristic nature. We will give here this reasoning since it may prove to be useful when analyzing many other problems of this type.

Suppose a particle is near a certain position $x = X_1$ and performs fast oscillations ψ relative to it. These oscillations are conditioned by the action of the force $\Psi(x)\sin\omega t$ and they may be obtained from the equation

$$m\ddot{\psi} = \Psi(X_1)\sin\omega t \qquad (20.1.14)$$

whence

$$\psi = -A\sin\omega t = A\sin(\omega t + \pi) \qquad (20.1.15)$$

where

$$A = A(X_1) = \Psi(X_1)/m\omega^2. \qquad (20.1.16)$$

Thus, ψ presents a harmonic oscillation of the opposite phase with regard to the force $\Psi(X_1)\sin\omega t$. This circumstance is most essential.

Indeed, suppose at the point $x = X_1$, which is under consideration, the inequalities $\Psi(X_1) > 0$ and $\Psi'(X) > 0$ are valid, i.e. the function $\Psi(x)$, being positive, grows (Fig. 20.1,b). In the first half-period $0 < \omega t < \pi$, when the force $\Psi(x)sin\omega t$ is positive, the mean value $< \Psi(x)sin\omega t >_{T/2}$ of this force will be in absolute value less than the mean value of the same force in the second half-period when this force is negative. As a result, the mean value of the force $\Psi(x)sin\omega t$ during the period, — i.e. of the vibrational force — will prove to be negative: $V(X_1) =< \Psi(x)sin\omega t > < 0$.

Analogously, it can be easily established that if the fast motion takes place near the point $x = X_2$, at which $\Psi(X_2) < 0$ and $\Psi'(X_2) > 0$, then (Fig. 20.1,c) $V(X_2) =< \Psi(x)\sin\omega t > > 0$. And in general, if at a certain point $x = X_n$ the signs of $\Psi(X_n)$ and $\Psi'(X_n)$ are the same, then $V(X_n) < 0$, and if the signs are different, then $V(X_n) > 0$. This regularity is illustrated by Fig. 20.1,a where the first sign at a certain interval of the change of the function $\Psi(x)$ corresponds to the sign of $\Psi(x)$, while the second sign corresponds to the sign of $\Psi'(x)$. The arrows over the axis of abscissas and over the graph of the function $\Psi(x)$ show the directions of the action of the vibrational force $V(X)$. But while from Fig. 20.1,a it follows immediately that as a result of the action of the force $\Psi(x)\sin\omega t$ there appears a slow force V which always "attracts" the particle to the points of the minimum of the function $|\Psi(x)|$ and "repulses" it from the points of the maximum of that function.

In other words, due to the action of the force $\Psi(x)\sin\omega t$ the particles will move slowly ("drift") to the points of minimums of the function $|\Psi(x)|$ as to certain stable states of quasi-equilibrium. Among such stable points (in Fig. 20.1,a they are marked by the letter "s", while the unstable points are marked by the letter "u") there will also be points where $\Psi(x) = 0$, i.e. the nodes of the wave $\Psi(x)$. Thus, by means of a simple reasoning which does not claim to be strict we come to the same results as those obtained by means of analytic consideration.

In case of the presence of an "ordinary" force $F_0(X)$, the appearance of the vibrational force $V(X)$ may lead to essential consequences which have already been discussed in 5.3: the former positions of stable equilibrium may disappear and there may appear new positions of stable quasi-equilibrium, the former stable positions of equilibrium may become unstable positions of quasi-equilibrium and vice versa, the former positions of stable equilibrium may get shifted etc. All these regularities may be interpreted from the position of vibrational mechanics as the result of the addition to the "ordinary" potential energy Π_F of the term Π_V, answering the vibrational forces; this was also discussed in 5.3.

20.2   Special Case: a Pendulum with a Vertically Vibrating Axis of Suspension

The problem of the behavior of a pendulum with a vertically vibrating axis of suspension which has already become classical (Fig. 1.2,b) is a special case of the problem considered in 20.1. Indeed, the differential equation of motion of such a pendulum has the form

$$I\ddot{\varphi} + h\dot{\varphi} + ml\left(g + A\omega^2 \sin \omega t\right)\sin \varphi = 0. \tag{20.2.1}$$

It is obtained from equation (5.1.2) of chapter 5 if we assume that $H = 0$, $G = A$, and $\theta = \pi/2$. It differs from the equation of the oscillations of the pendulum with a fixed axis only by the fact that we have added to the acceleration of gravity g the acceleration of vibration (the designations are here the same as in 5.1.2). Comparing equation (20.2.1) with (20.1.1), we come to the conclusion that in this case we can assume that

$$F_0(\varphi) = -m\,g\,l \sin \varphi, \quad \Psi(\varphi) = -mA\omega^2 l \sin \varphi \tag{20.2.2}$$

with the angle of rotation of the pendulum φ playing the role of the coordinate x.

According to what was said in 20.1. the points of the minimum of the function $|\Psi(\varphi)| = mA\omega^2 l|\sin \varphi|$ are the "points of attraction" of the pendulum which are conditioned by the action of vibration. There are only two such essentially different points $\varphi_1 = 0$ and $\varphi_2 = \pi$; the former corresponding to the lower — and the latter — to the upper position of the pendulum. If the force of gravity is absent (the axis of the pendulum is vertical) or small as compared to the force of inertia $mA\omega^2$, then these points are answered by the positions of the stable equilibrium of the pendulum, the latter being located along the direction of vibration. In the other limiting case when $mA\omega^2 \ll m\,g$, the lower position $\varphi_1 = 0$ is stable, while the upper position is unstable. These conclusions, naturally, coincide with those, given in 5.1.5. Also coincide the corresponding expressions for the vibrational torque, the potential energy of the vibrational forces, the equations of slow motions, and the conditions of the stability of the upper position of the pendulum, obtained from the relationships 20.1 in view of equalities (20.2.2).

20.3   Generalization of the Problem

The solution of the problem in a much more general case of the system with an arbitrary number of degrees of freedom n and when the fast-oscillating force field presents a sum of any number k of the standing waves of different frequencies $\nu_k \omega$ has in fact been given in 4.4. The fast generalized force, corresponding to a certain generalized coordinate q_r then has the form

$$\Phi_r(q_1, \ldots, q_n, \mu, \omega t) = \frac{1}{\mu} \sum_k f_{rk}(q_1, \ldots, q_n, \mu) e^{i\nu_k \omega t} \qquad (20.3.1)$$

where $\mu = 1/\omega$ is a small parameter, $\nu_k \neq 0$, $\nu_{-k} = -\nu_k$, $f_{r,-k} = \overline{f}_{r,k}$. As has been shown, if the following relations are satisfied

$$\left. \frac{\partial f_{mk}}{\partial q_r} \right|_{\mu=0} = \left. \frac{\partial f_{rk}}{\partial q_m} \right|_{\mu=0} \qquad (m = 1, \ldots, n; \ r = 1, \ldots, n), \qquad (20.3.2)$$

then there is a potential function

$$D = \Pi|_{\mu=0} + \Pi_V \qquad (20.3.3)$$

whose rough minimums with respect to the generalized coordinates $q_1, \ldots, q_n$ are answered by the asymptotically stable quasi-equilibriums of the system which is described by the differential equations (4.4.1) of chapter 4 (Π being the "ordinary" potential energy of the system, i.e. the potential energy, answering the slow generalized forces).

Thus, in the general case too, the role of the potential energy in the presence of the oscillating field is played by the sum (20.3.3) and therefore everything said in 20.1 and 5.3 about the transformation of equilibrium positions under the action of the field can be repeated here.

The slow motions of the system are described by differential equations (4.4.9) of chapter 4 which like equation (20.1.12) have the form corresponding to the potential on the average system in case of the presence of dissipation.

It should be noted that the results, discussed in this section as well as in 20.1, can be generalized for the case of the "slowly running" waves.

The review of the investigations [30, 66, 112, 114, 190, 191, 192, 193, 194, 170, 239, 240, 282, 291, 332, 333, 334, 344, 351, 358, 499, 148, 149, 150, 151, 152, 153, 496] on the problem considered in this chapter and on the corresponding applications is given in 5.3 where these investigations are considered as the development of the classical investigation of the behavior of the pendulum with a vibrating axis of suspension, and also in 3.2.12 and in 4.4 in connection with the publications on the development of the method of direct separation of motions and on the theory of the potential on the average dynamic systems.

Chapter 21

Resonance (Synchronization) in Orbital Motions of Celestial Bodies

21.1 Preliminary Remarks

The problem of *resonance (of commensurability of periods, of synchronization) in the motion of celestial bodies*, which is one of the main problems of many bodies, has been the subject-matter of a great number of investigations, both classical and modern, e.g. see [14, 55, 56, 57, 58, 59, 102, 103, 206, 223, 598, 393, 392, 441, 147, 141, 421, 264, 197]. In this chapter when studying resonances in the orbital motions of celestial bodies we use the approaches of vibrational mechanics, particularly the theories of potential on the average dynamic systems, as well as certain results of the theory of synchronization of dynamic objects. Of utmost importance in this case is the *integral sign of stability (extremal property) of synchronous motions* (chapter 4), from which it follows, under rather general assumptions that stable resonant motions do exist, i.e. there is a *tendency to synchronization*. That explains, in the general form, the fact that resonances are widespread in the Solar system.

From the integral sign of stability follows also " the rule of selection" of stable phasings in resonant motions; the results, following from that rule, are proved very well by the observed data; in simple cases these results correspond to those obtained analytically.

More specifically, in the case of orbital motions of the system of k-bodies (point masses) under the action of gravitation with respect to a certain body whose mass is much greater, there is a potential function, analogous to the potential energy in problems on the stability of equilibrium (see chapter 4). That function proves to be approximately equal to the value of the potential of the gravitational interaction of those bodies, averaged for the period. Hence it

follows that the heuristic *"principle of least interaction"*, formulated in article
[421](1974), as crudely valid following from the results analytically obtained
before. In addition to the "checking" of the principle given in the above article, this chapter shows that for the motion of Uranus and Neptune, and of
Saturn's satellites Enceladus and Dion and also of the analogous pairs of bodies (resonance of type 1:2), the values of the difference in time which it takes
the bodies to pass across the pericenters in stable motions, found theoretically,
agree very well with the observed data.

The investigation which has been carried out makes it possible to suggest
a classification of resonances in the systems with small eccentricities and small
inclinations of orbits of the revolving bodies according to the "relative force" of
these resonances. *Molchanov's hypothesis about complete resonancity of orbital
motions of big planets of the Solar system* is discussed here on the basis of this
classification and a "weaker" *hypothesis about simple resonancity of the motions
of bodies in the planetary- and satellite systems* is advanced. Consideration is
given to the agreement of this hypothesis with the observed motion of bodies
of the Solar system. Ideas are advanced of a possible way of checking up the
hypotheses about the relative closeness of motions of the celestial bodies to the
resonances and of establishing the direction of evolution of these motions.

A remarkable peculiarity of the system under consideration is that the scale
of fast motions (revolutions of bodies round the central body) is not small, but
rather large, as compared to the scale of slow motions (to the evolution of the
elements of the orbits). As was marked in 3.2.4 and 3.2.9, this circumstance
does not prevent one from using the method of a direct separation of motions,
though it changes the requirements to the relative orders of the right-hand
sides of the equations of slow and fast motions when solving problems by this
method. Though, to prove the validity of all these results, it is not necessary
to analyze the validity of these requirements, since the right-hand sides of the
equations of motion do not depend on slow time, and the motion is being
considered in quite small vicinities of the stable stationary regime (see note 4
in 3.2.4).

In conclusion, attention is paid to the expediency of using the approaches of
vibrational mechanics when solving problems on the motion of celestial bodies
with internal degrees of freedom.

The chapter develops and generalizes the results, presented in publications
[102, 103, 120].

In the article by Hentov [264] it is shown that under certain conditions the
extremal sign of stability is valid for revolutionally-spinning motions of celestial
bodies.

21.2   Main Results of the General Theory of Synchronization and of the Theory of Potential on the Average Dynamic Systems as Applied to the Problem under Consideration

21.2.1   Description of the System. The Initial System and the Frame System.

We will consider a system, consisting of $k + 1$ gravitating bodies, idealized as material points (Fig. 21.1,a). One of the bodies (we will call it "central") has a mass m_0, considerably greater than the masses m_s ($s = 1, \ldots, k$) of all the other bodies. In this case the forces which the bodies m_s are interacting with are considered to be small as compared to the force of their attraction to m_0 (here and further on bodies are designated in the same way as their masses). Accordingly all the bodies are believed to be moving in the first approximation (i.e. in the so called nonperturbed motion) in elliptical (Kepler's) orbits. The motion of each of these bodies in such a fixed orbit is described by T_s — periodic function of time t, as an example of which the argument of the latitude $u_s^0(t+\tau_s)$ may be taken. Due to the autonomy of the equations of the nonperturbed motion of each body, this function has been determined with an accuracy to an arbitrary initial phase τ_s, for which one can take, say, the moment when the body passes the pericenter[a].

Let the frequencies of revolutions of bodies in the nonperturbed motion — the mean motions $n_s = 2\pi/T_s$ — be connected by the relationships

$$n_s = l_s n_0 \tag{21.2.1}$$

where n_0 is a positive number, and l_s are relatively prime numbers, which means that in a nonperturbed motion there is a resonance (synchronization) whose general shortest period of revolution is $T_0 = 2\pi n_0$.

First we will present the result which refers to the simplest analog of the above system. In that case the bodies m_s move not freely but in rigid tubes whose lines of axes coincide with, say, elliptical orbits of the nonperturbed initial system. All the tubes are bound with each other and with the central body by means of a rigid frame, so that they form one solid body (Fig. 21.1,b). Then each of the bodies m_s contributes to the system only one degree of freedom; its position in the tube being fully determined by the angular coordinate u_s. Unlike the initial system which for brevity sake will be called *free system*, this simplified system described above will be called the *frame system*.

---

[a] In this chapter we partly use the symbols, accepted in the investigations devoted to celestial mechanics. For instance the parameters α_s (chapters 4 and 7) are answered here by the parameters τ_s, and the coordinates u_s are answered by the rotational (angular) coordinates ω_s.

Note that a similar idealization, though somewhat different, is in fact used in celestial mechanics when solving problems on the rotational motions of bodies, when the orbit of the center of inertia of the body is considered to be given

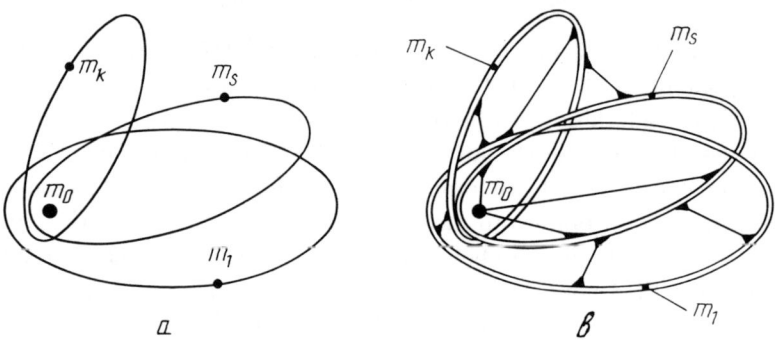

Figure 21.1. Graviating bodies $m_1, \ldots, m_k$ moving in the orbits around the central body m_0. a) The original (free) system; b) the idealized ("frame") system.

(as a rule it is Kepler's orbit) (see, e.g., [56, 59, 61]). This idealization is quite acceptable when one has to solve many questions. It is necessary however, to take into consideration that here we have a case when certain degrees of freedom of the system are ignored, and this may lead (and in fact leads — see below) to the opposite conclusion about the stability of its motions.

Let the coordinates u_s be chosen in such a way that in the *nonperturbed motion*("*generating motion*") they satisfy the differential equations

$$I_s \ddot{u}_s^0 + k_s(\dot{u}_s^0 - l_s n_0) = 0 \quad (s = 1, \ldots, k) \qquad (21.2.2)$$

where l_s and k_s are positive constants, i.e. it means that the nonperturbed motion of the bodies m_s in the coordinates u_s presents a uniform synchronous rotation

$$u_s^0 = l_s n_0(t + \tau_s) \qquad (21.2.3)$$

with certain initial phases τ_s which cannot be determined from the consideration of the nonperturbed motion alone. The nonperturbed motion of the frame corresponds to its small purely forced oscillations in the absence of any resistance forces appearing when the bodies m_s move according to (21.2.3).

21.2.2   *The Potential Function and the Integral Sign of Stability (Extreme Property) of Resonant Motions of the System of Bodies*

The system described refers to *systems with almost uniform rotations*, considered in 4.2.1. It is also somewhat similar to systems with selfsynchronizing vibroexciters, studied in detail in chapter 8. According to the results of 4.2.1, the stable synchronous motions of the system, transformed when there is no interaction between the bodies m_s, into synchronous motions (21.2.3) of the nonperturbed system (21.2.2), may correspond to the points of *rough minimum* (see the footnote on page 78) of the *potential function*

$$D = D(\tau_1 - \tau_k, \ldots, \tau_{k-1} - \tau_k) = -(\Lambda + B) \qquad (21.2.4)$$

where $\Lambda =< [L] >$ is the average value of the Lagrangian of the system for the period $T = 2\pi/n_0$, calculated at the nonperturbed motion (21.2.3), and B is the *potential of the averaged nonconservative forces*. Then the requirements of a rough minimum of the function D with respect to the difference of the phases $\tau_s - \tau_k$ are the main conditions in the sence that they determine the "selection" of phase difference in possible stable synchronous motions.

These conditions are also "roughly necessary", i.e. they are necessary in the sense that the absence of the minimum, which can be discovered by analyzing the terms of the second order in the decomposition of the function D in the vicinity of the stationary point, indicates the instability of the synphaseous motion under consideration. It can be also shown that the presence of a rough minimum of the function D provides for the system the asymptotic stability of the corresponding synchronous motion according to the phase difference $\tau_s - \tau_k$ $(s = 1, \ldots, k - 1)$.

Note that just like it is in problems on self-synchronization of vibroexciters, introducing into the system the small positional forces which provide the distinction of the frequency of free oscillations of the frame from the zero, and also introducing small dissipative forces with a complete dissipation guarantees the asymptotic orbital stability of synchronous motions in all the coordinates, provided there is a minimum of the function D.

Since the Lagrange function L of the system under consideration can be presented as a sum (4.2.18) of chapter 4, the potential function D can be written as follows:

$$D = \Lambda^{(I)} - \Lambda^{(II)} - B. \qquad (21.2.5)$$

Here $\Lambda^{(I)}$ and $\Lambda^{(II)}$ are the averaged Lagrangians of the carrying constraints and carried constraints respectively between the bodies m_s; for this system we have

$$\Lambda^{(I)} = < [T^{(I)}] >, \quad \Lambda^{(II)} = < [U^{(II)}] > \tag{21.2.6}$$

where $T^{(I)}$ is the kinetic energy of the central body m_0, and

$$U^{(II)} = \frac{1}{2}f \sum_{s,j=1}^{k} ' \frac{m_s m_j}{\Delta_{sj}} \tag{21.2.7}$$

is the potential of the gravitational forces of the interaction between the bodies m_s (f — being the gravitational constant, and Δ_{sj} being the distance between the bodies m_s and m_j); the prime at the sign of the sum indicates the omission of the summand, corresponding to $s = j$.

Now, considering the system to be free, we will make use of the results of Nagaev, referring to the problem of synchronization of the weakly connected quasi-conservative objects stated in 4.2.2. Their distinction from the results, obtained for the frame system, lies in the following:

1. For the free system, the presence of a rough minimum of the potential function D gives but the crudely necessary conditions of stability according to the differences of the coordinates $u_s - u_k$. At the same time for the free system too this condition of the minimum of D is the main condition of stability, because like in the case of the frame system, it is determining the selection of combinations of the initial phase differences $\tau_s - \tau_k$ which can be answered by the motions stable with regard to $u_s - u_k$.

The totality of the necessary and sufficient conditions of stability of the free system follows from the results of [397]. It should be noted, however, that these conditions are, generally speaking, not satisfied for it is evident that for some coordinates this system is unstable (see, say, the example in 20.4)

2. The function D for the free system looks as follows:

$$D = D(\tau_1 - \tau_k, \ldots, \tau_{k-1} - \tau_k) = < [U^{(II)}] > - < [T^{(I)}] > +B \tag{21.2.8}$$

which corresponds to the change of sign before the first two summands (as a rule — the main summands) of expression (21.2.5). In other words, if the value B can be neglected, the stable synchronous motions in case of a free system answer the minimum of the function $< [U^{(II)}] > - < [T^{(I)}] >$, while for the frame system the stable motions correspond to the maximum of that function.

This circumstance is a direct consequence of the fact that the bodies m_s in the frame system, just like the unbalanced rotors in the problem on their self-synchronization, are rigidly anisochronous objects, while the free bodies are softly anisochronous objects. It can be easily established by definition and by equation (4.2.41) in 4.2.2. That is why the value σ changes in equations (4.2.40) and (4.2.42) given there. The fact that this situation, connected with ignoring the degrees of freedom of the system, is quite possible was already paid attention to at the beginning of this section.

Thus, the integral sign of stability (extreme property) of synchronous motions can be applied both to the initial system and to its framewise idealization.

Mark that the presence of expressions (21.2.4), (21.2.5) and (21.2.8) for the potential function D makes it possible to deduce the main equations of vibrational mechanics in form of equations (4.1.1) or (4.1.3) of chapter 4.

21.2.3   On the Theoretic Explanation of the Natural Prevalence of Resonance in the Solar System

An important consequence which follows from the validity of the integral sign of stability for the system under consideration lies in the fact that the synchronous resonant motions exist under quite general assumptions [103, 120]. Particularly, the rough minima of the function D are, generally speaking, sure to exist (see, e.g. the example in 20.3) for both the system under consideration and those with the self-synchronized vibroexciter (see 7.3.4), due to the smallness of dissipative forces and to the periodicity of the functions $< [U^{(II)}] >$ and $< [T^{(I)}] >$ with respect to $\tau_s - \tau_k$. Thus the fact that there is a great abundance of resonances in the Solar system can be explained in a general form. Such resonances will be discussed in more detail in 20.5.

21.2.4   On the "Principle of the Least Interaction" by Ovenden, Feagin and Graff and on the Role of Dissipative Forces in the System

If the mass of the central body m_0 is so great compared to the masses $m_1, \ldots, m_k$ and the dissipative forces are so small that in the expression (21.2.8) the last two summands can be neglected, then the potential function D practically coincides with the averaged potential of the mutual gravitation of bodies:

$$D \approx < [U^{(II)}] > . \tag{21.2.9}$$

In connection with this relation it is pertinent to note that Ovenden, Feagin and Graff in [421](1974) set up the so called *principle of least interaction*, according to which the motions of planetary and satellite systems are such that during the greatest period of time in the process of evolution the system moves in such a way that the temporal average of the potential of interaction $< [U^{(II)}] >$ has a minimal value. From the above results and from formula

(21.2.9) it follows that this euristic principle flows out as crudely valid from the integral sign of stability (extreme property) of synchronous motions which had been established before [77, 78, 90, 395](1960–1970).

In connection with the problem under consideration we must point to a very intricate, partly contradictory character of the effect of dissipative forces upon the behaviour of the system under consideration. On the one hand, the tendency to synchronization which takes place in case the averaged dissipative forces are small compared to the potential forces, can be suppressed by the dessipative forces if this condition is not complied with. In other words, the "insufficiently deep" potential pits, answering certain resonances, can disappear as a result of a comparatively strong dissipation (see [103]). Next, as it is known, in some cases the dissipative forces lead to the asymptotic stabilization of motion, in other cases (see the example in 20.4) — to a gradual (very slow) distruction of the temporary stability in the system under consideration.

Finally, the due regard for the dissipative forces in the system under consideration leads, generally speaking, to the change in the period of synchronous (resonant) motions. On the other hand, when the dissipative forces are small, this change, just like the breaking of temporary stability, becomes noticeable only during quite long periods of time, as compared to the typical periods of time during which the properties of motion that interest us are studied.

In accordance with what has been said, the conception of the role of dissipative forces in the process of evolution of the motions of celestial bodies, described, say in [206], seems quite legitimate. Due to the dissipation, the relations between the averaged orbital rotational motions of bodies in the course of the evolution change. The capture into a stable resonant motion and the subsequent continuous "sticking" in it takes place (or, to be more exact, may take place) in cases when these relations approximate the corresponding resonant relations.

21.3   The Case when the Orbits of Bodies Lie in the Same Plane or in Proximate Planes and Have Small Eccentricities. On classification of resonances

This case refers particularly to the motion of bodies in the Solar system. Let the orbits lie first in the same plane. Then the expression for the potential $U^{(II)}$ can be written as follows:

$$U^{(II)} = \frac{1}{2}f \sum_{s,j=1}^{k} {}' \frac{m_s m_j}{\sqrt{r_s^2 + r_j^2 - 2r_s r_j \cos(u_s - u_j)}} \tag{21.3.1}$$

where r_s and u_s are the polar coordinates of the body m_s relative to the pole O, coinciding with the central body (Fig. 21.2). For the Kepler motion we

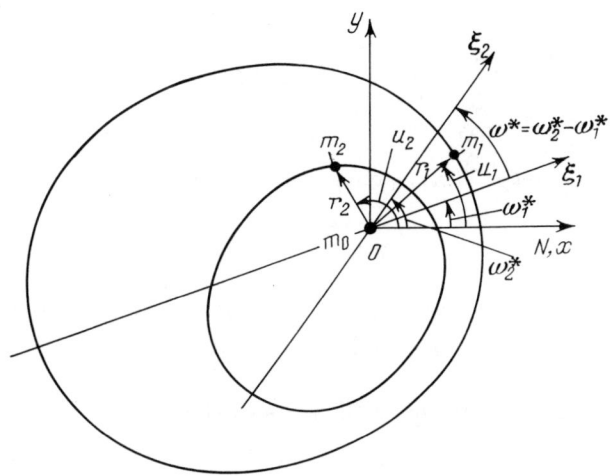

Figure 21.2. The case when the orbits of the bodies lie in the same plane (to simplify the picture only two masses are shown here).

have the well-known decompositions in terms of degrees of the eccentricity of the orbit:

$$r_s = r_s^0 = a_s[1 - e_s \cos n_s(t - \tau_s) + \ldots],$$
$$u_s = u_s^0 = \omega_s^* + n_s(t - \tau_s) + 2e_s \sin n_s(t - \tau_s) + \ldots \qquad (21.3.2)$$

Here u_s is the argument of the latitude of the body m_s; ω_s^* is the angular distance of the pericenter from the node (i.e. from the line ON whose direction in this case is arbitrary); τ_s is the moment of the passage through the pericenter, playing, as it was marked, the role of the initial phase; α_s is the large semiaxis of the orbit, connected with the mean motion by Kepler's third law $n_s = (fm_0)^{1/2}a_s^{-3/2}$.

Substituting expressions (20.3.2) into formula (20.3.1) and averaging it by time, we will get the following expression with an accuracy up to the terms higher than the first order with regard to the eccentricities e_s and with an accuracy to the summand, independent of τ_s.

$$
D \approx < [U^{(II)}] >= \frac{1}{2\sqrt{2}} \frac{f}{a} \sum_{\substack{s,j=1 \\ n_s=n_j}}^{k} {}' \frac{m_s m_j}{\sqrt{1 - \cos(\omega_s^* - \omega_j^* - n_s \tau_s + n_j \tau_j)}}
$$

$$
+ \frac{f}{4} \sum_{\substack{s,j=1 \\ \frac{n_s}{n_j}=\frac{l}{(l\pm 1)}}}^{k} {}' \frac{e_s}{a_s} m_s m_j \{2b_3^{(p_{sj})}(z_{js}) - z_{js}[3b_3^{(p_{js})}(z_{js})
$$

$$
- b_3^{(q_{sj})}(z_{sj})]\} \cos p_{sj}[n_j(\tau_s - \tau_j) - \omega_s^* + \omega_j^*]. \qquad (21.3.3)
$$

Here

$$
z_{js} = a_j/a_s = (n_s/n_j)^{\frac{2}{3}}, \qquad p_{sj} = n_s/(n_s - n_j),
$$

$$
q_{sj} = (2n_s - n_j)/(n_s - n_j) \qquad (21.3.4)
$$

and $b_3^{(r)}(z) = b_3^{(-r)}(z)$ are the coefficients in the decomposition

$$
(1 + z^2 - 2z \cos \lambda)^{-3/2} = \sum_{r \to \infty}^{\infty} b_3^{(r)} \cos r\lambda, \qquad (21.3.5)
$$

called the *Laplas coefficients*. The relation $n_s = n_j$, written down under the first double sum, shows that it involves only those summands which correspond to all the bodies in the system which have equal mean motions n_s (which means they have equal semiaxes a_s), while the relation $n_s/n_j = l/(l \pm 1)$ under the second double sum shows that only those summands are taken into account which correspond to all pairs of bodies in the system whose mean motions n_s and n_j are related as successive integers (l is integer, $l \neq 0$, $l \pm 1 \neq 0$).

From expression (20.3.3) it follows that in the case of a purely circular planary motion the only synchronization possible is that with equal mean motions. When orbits are elliptical with small eccentricities e_s, the consideration of the terms of the first order with regard to e_s in the decomposition of the averaged potential $< [U^{(II)}] >$ makes it possible to take into account the interactions which may result in the synchronization of bodies with mean motions whose relation is that of successive integers (resonances of the type of $l : (l \pm 1)$).

As has been shown by Veretinsky, if in the case of a non-planar system it is possible to find in the decomposition $< [U^{(II)}] >$ terms, linear with respect to $v_{sj} = \sin^2(I_{sj}/2)$, ($I_{sj}$ being the angle between the planes of the nonperturbed orbits of the bodies m_s and m_j), then the interaction will be taken into account which may lead to the resonances up to the type of $l : (l \pm 2)$ inclusively, and, if we also take into account the summands, quadratic with respect to e_s and v_{sj}, then up to the type of $l : (l \pm 4)$ inclusively. It goes without saying that when e_s and v_{sj} are small, the interactions of the latter will be weaker than those caused by the terms linear with respect to e_s and v_{sj}.

Reasoning from what has been said, it seems expedient to suggest a classification of orbital resonances according to the degree of their "relative significance" (with the fixed l). In this classification resonances of the type $l : l$ will be referred to the order zero, resonance of the type $l : (l \pm 1)$ and $l : (l \pm 2)$ will be referred to the first order, those of the type $l : (l \pm 3)$ and $l : (l \pm 4)$ to the second order and so on. Note that unlike our classification, in the classification, suggested by Moser, [393] what is meant by resonance of the n-th order is the resonance of the type $l : (l \pm n)$. Both classifications however prove to be identical, at least up to the resonances of the second order inclusively provided it is not the values $v_{s,j}$, but the angles $I_{s,j}$ that are of the order e_s. It stands to reason that just like any classification based on taking account of the orders of the values, the suggested classification offers but a conventional estimation of the " significance of resonance". In every particular case when analyzing the significance of resonances it is necessary to consider the concrete values of the masses of bodies and of the parameters of their orbits.

21.4   The Case of Two Revolving Bodies: Comparison with the Results of the Direct Analytical Investigation and with the Observed Data

We will consider the simplest system, comprising only two bodies m_1 and m_2, revolving in the same direction.

a) **The case of a circular orbit** (resonance of the order of zero). In this case $n_1 = n_2 = n$, $m_1 = m_2 = m$, $a_1 = a_2 = a$. Then, according to (20.3.3) we have

$$D \approx < [U^{(II)}] >= \frac{1}{2\sqrt{2}} \frac{f}{a} \frac{m^2}{\sqrt{1 - \cos[\omega_1^* - \omega_2^* + n(\tau_2 - \tau_1)]}}. \qquad (21.4.1)$$

The minimum of that expression is answered by the value $\omega_1^* - \omega_2^* + n(\tau_2 - \tau_1) = \pi$. And since in this case we can believe that $\omega_1^* - \omega_2^* = 0$ (any point of the circular orbit can be taken for the pericenter), we can conclude, in accordance with the integral sign, that the necessary conditions of stability will be satisfied for the "antiphase" motion of bodies wherein they are located at the ends of the diameter of the circular orbit (Fig. 21.3).

This special case is of interest because it can be easily considered in a usual way. To this end let us consider the differential equations of the planar motion

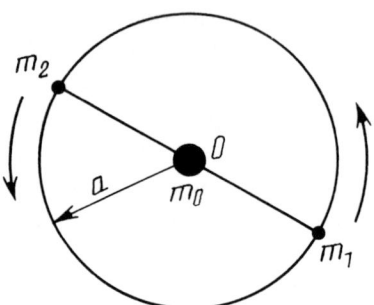

Figure 21.3. The motion of two bodies in the circular orbit (resonance of the zero order), stable with respect to the phase difference.

of the bodies m_1 and m_2 in the polar coordinates $r_1, u_1; r_2, u_2$ (Fig. 21.2; the central body being considered fixed):

$$\ddot{r}_1 - r_1\dot{u}_1^2 = -\frac{fm_0}{r_1^2} + \frac{fm}{\Delta_{12}^3}[r_2\cos(u_1 - u_2) - r_1]$$

$$\ddot{r}_2 - r_2\dot{u}_2^2 = -\frac{fm_0}{r_2^2} + \frac{fm}{\Delta_{12}^3}[r_1\cos(u_1 - u_2) - r_2]$$

$$(r_1^2\dot{u}_1)^{\cdot} = -fm\frac{r_1 r_2}{\Delta_{12}^3}\sin(u_1 - u_2)$$

$$(r_2^2\dot{u}_2)^{\cdot} = -fm\frac{r_1 r_2}{\Delta_{12}^3}\sin(u_2 - u_1)$$

where $\Delta_{12} = \sqrt{r_1^2 + r_2^2 - 2r_1 r_2\cos(u_1 - u_2)}$ is the distance between the bodies m_1 and m_2. As it must be, these equations are answered by the unperturbed motion which interests us.

$$r_1 = r_2 = a \quad u_1 = nt \quad u_2 = nt + \pi,$$

and the relation

$$n^2 = \frac{f}{a^3}(m_0 + \frac{m}{4})$$

is also satisfied. Having made equations in variations for this unperturbed motion, we will come to the following two groups of linear differential equations with respect to the differences x_- and ψ_- and the sums x_+ and ψ_+ according to the variations of the variables r_1 and r_2, u_2 and u_1:

$$\ddot{x}_+ - 3n^2 x_+ - 2na\dot{\psi}_+ = 0, \quad a\ddot{\psi}_+ + 2n\dot{x}_+ = 0; \qquad (21.4.2)$$

$$\ddot{x}_- - \frac{f}{a^3}(3m_0 + \frac{m}{4})x_- - 2na\dot{\psi}_- = 0, \quad a\ddot{\psi}_- - \frac{fm}{4a^2}\psi_- + 2n\dot{x}_- = 0. \quad (21.4.3)$$

Hence it is easy to conclude that the motion under investigation is stable in the first approximation (not asymptotically) with respect to the "phase differences" $u_2 - u_1$ and to the radius-vectors r_1, r_2 whereas it is unstable with respect to the sum $u_1 + u_2$. In other words the necessary (in the above mentioned sense) conditions of stability with respect to phase difference $u_2 - u_1$, i.e. the results, obtained by means of the integral sign of stability and by means of the ordinary investigation, coincide.

It should be noted that in accordance with (21.4.3) the stability of the differences $r_2 - r_1$ and $u_2 - u_1$ is a *temporal stability* because it is provided by gyroscopic terms. As is known (see, say, [372]), such stability can be broken by dissipative forces; it is also known that when the dissipative forces are relatively small, the process of losing stability is quite slow.

b) The case of the elliptical unperturbed orbits with small eccentricities (resonance of the first order of type 1 : 2). Suppose the mean motion of the body m_2 is twice faster than that of the body m_1, in that case we will have

$$n_1 = n, \; n_2 = 2n, \; z_{12} = a_1/a_2 = (n_2/n_1)^{2/3} = 2^{2/3} = 1.59$$
$$z_{21} = 1/z_{12} = 0.629, \quad p_{12} = -1, \quad p_{21} = 2, \quad q_{12} = 0, \quad q_{21} = 3$$
$$b_3^{(1)}(z_{12}) = 1.23, \quad b_3^{(2)}(z_{12}) = 0.913, \quad b_3^{(3)}(z_{12}) = 0.649$$
$$b_3^{(0)}(z_{21}) = 6.03, \quad b_3^{(1)}(z_{21}) = 4.82, \quad b_3^{(2)}(z_{23}) = 3.66 \quad (21.4.4)$$

and expression (20.3.3) will acquire the form

$$D \approx < [U^{(II)}] > = \frac{fm_1m_2}{4a_2}[4.11e_1\cos(2\pi\tau - \omega^*)$$

$$-3.00e_2\cos 2(n\tau - \omega^*)] = \frac{f}{4a_2}M\cos(2n\tau - \delta). \quad (21.4.5)$$

Here

$$\tau = \tau_2 - \tau_1, \quad \omega^* = \omega_2^* - \omega_1^*;$$
$$M\cos\delta = 4.11e_1\cos\omega^* - 3.00e_2\cos 2\omega^*,$$
$$M\sin\delta = 4.11e_1\sin\omega^* - 3.00e_2\sin 2\omega^*. \quad (21.4.6)$$

From (21.4.5) in accordance with the integral criterion it follows that the stable (in the above mentioned sense) synchronous (resonance) solution is answered by the value of the difference of the moments when the bodies pass through the pericenters

$$\tau = \tau_2 - \tau_1 = (\delta + \pi)/2n. \quad (21.4.7)$$

We will consider two still more specific cases. First let us assume that $e_1 = 0$, i.e. the "exterior" body m_1 in the unperturbed motion has a circular orbit (Fig. 21.4,a). Then, just like before, we may assume that $\omega^* = \omega_2^* - \omega_1^* = 0$. Then from the condition of the minimum of expression (21.4.5) we find that

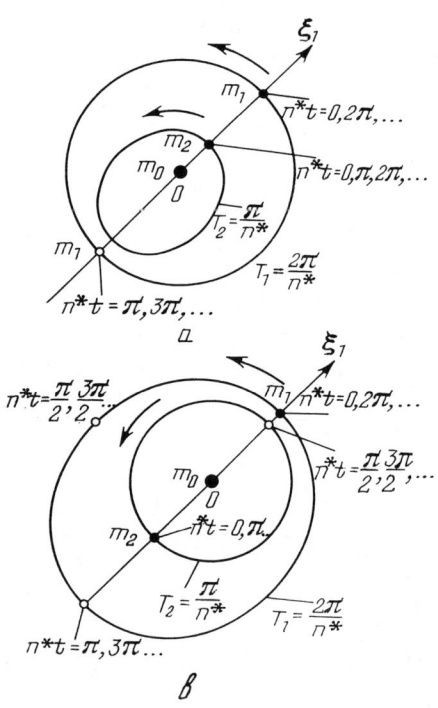

Figure 21.4. The planar motion of two bodies with small eccentricities (resonance of the first order of type 1:2), stable with respect to phase difference.

the stable motion is answered by the value $\tau = \tau_2 - \tau_1 = 0$. In other words, in the stable synchronous motion the juxtapositions and oppositions of bodies take place when the internal body passes its pericenter (Fig. 21.4,a)

Now let us assume that $e_2 = 0$, i. e. the nonperturbed orbit of the "internal" body m_2 is a circumference. Then, believing just like before that $\omega^* = 0$, from the condition of the minimum of expression (21.4.5) we will find that for the synchronous stable motion $\tau = \tau_2 - \tau_1 = \pi/2n$. In this case the juxtaposition of bodies takes place when the external body passes the apocenter, and the opposition — when the internal body passes its pericenter (Fig. 21.4,b). These

regularities agree very well with the observations (see below and also [206]
where the qualitative considerations speak in favour of the stability of the
corresponding motions).

As a more specific example we will consider the data referring to the Neptune and Uranus (Table 21.1). Their mean motions n_2 and n_1 are related
approximately as $2:1$. Since the mutual inclination of the orbits is small, the
motion of the planets is believed to take place in the same plane. Besides, due
to the absence of other planets with the ratio of mean motions of the type of
$l:(l \pm 1)$, the effect of other big planets upon the resonances in the system
the Sun–Neptune–Uranus may be neglected in the first approximation (with
an exception of Pluto, but its mass is more than 7000 times smaller than that
of Uranus or of Neptune). Then we may expect that the results obtained for
the case $n_2/n_1 = 2:1$ can be applied to this system.

Table 21.1

Elements of the orbits of Neptune and Uranus (for January 7, 1979 [552])

Planet	Element					
	Mean motion n_s^*	Eccentricity e_s	Mean anomaly of epoch M_{os}	Longitude of ascending node Ω_s	Longitude of pericenter π_s	Inclination i_s
Neptune $(s = 1)$	0.0060	0.00984	196.89^0	131.53^0	61.63^0	1.772^0
Uranus $(s = 2)$	0.0117	0.0491	52.70^0	74.01^0	170.12^0	0.771^0

Let us verify this assumption. When taking into account the relations
$\omega_s = \pi_s - \Omega_s$ and $M_{os} = n_s(t_0 - \tau_s)$ between the elements of the orbits used
before and those given in table 21.1, we find

$$\omega^* = \omega_2^* - \omega_1^* = \pi_2 - \Omega_2 - (\pi_1 - \Omega_1)$$

$$\approx 170.12^0 - 74.01^0 - (61.63^0 - 1331.536^0) \approx 166^0,$$

$$n\tau = n(\tau_2 - \tau_1) = -\left(\frac{M_{01}n}{n_2} - \frac{M_{01}n}{n_1}\right) = -\left(\frac{M_{02}}{2} - M_{01}\right)$$

$$= -(0.5 \cdot 57.70^0 - 196.89^0) \approx 171^0.$$

While according to the theoretical equations (21.4.6) and (21.4.7), taking into account the data of the table, we obtain

$$M \cos\delta = 4.11 \cdot 0.00984 \cos 166^0 - 3.00 \cdot 0.0491 \cos 332^0 = -0.169;$$

$$M \sin\delta = 4.11 \cdot 0.00984 \sin 166^0 - 3.00 \cdot 0.0491 \sin 332^0 = 0.0789;$$

$$\delta = 155^0, \quad n\tau = 0.5(155^0 + 180^0) = 167.5.$$

Thus, the value of the phase difference when the planets pass through the perihelions, found theoretically, $n\tau \approx 167.5^0$ differs but by 3.5° from the value observed ($n\tau \approx 171^0$). A close theoretic result $n\tau^* = \omega^* = 166^0$ is obtained if the orbit of the Neptune is considered to be approximately circular ($e_1 \approx 0$). (It should be noted that when considering this example we do not assume, as we did before, that $\omega^* = 0$).

Let us consider another example. The mean motions of Saturn's satellites, Dion and Enceladus are related approximately as $2 : 1$ ($n_2 : n_1 = 1.99734$). The orbits of both satellites lie in the same plane, and their eccentricities make $e_1 = 0.0021$ and $e_2 = 0.0045$ respectively. If we assume that the Dion's eccentricity is approximately $e_1 = 0$ (as can be seen from the previous example, the error must then not be too great), we will have the situation adequate to the first example, considered above, when the juxtapositions and oppositions of bodies in the stable motion must be taking place when the internal body is passing the pericenter. As was mentioned, this checks with the observations [206].

Thus, the theoretical results, given above, are in good agreement with the observed data.

21.5   On Resonances in the Solar System. Hypothesis about Simple Resonancity

The results stated, explaining the fact that resonances are quite frequent in the orbitral motions of bodies in the Solar system, make it understandable that the overwhelming majority of "bright" resonances of that type do not have a degree higher than two by the classification proposed in 20.3. So, for instance, for the pairs of the planets Neptune-Uranus and Pluto-Neptune the situation is close to that of the resonance of the first degree of the types 1:2 and 2:3. For the pairs Venus-Mercury, Mars-Venus, Pluto-Uranus, Saturn-Jupiter the situation is close to that of the resonance of the second degree, of the type 2:5, 1:3, 1:3, and 2:5. The same regularity is seen in the zone of asteroids, in the satellite systems and in the rings of Uranus and Saturn (resonance of the type 2:3 in the group of 20 asteroids of Hilda with Jupiter; of the type 2:1 for the satellites of Saturn, Enceladus and Dion, of the type 7:6 and 2:1 in the rings of Saturn etc, see, e.g. [14, 141]).

On the face of it, the presence of Kirkwood's astrohatches in the distribution of asteroids according to the periods of revolution, and the presence of Kassini's slit in the rings of the Saturn, etc. do not comply with the conclusions that have been stated. But that is but a seeming discrepancy:

1. It is not improbable that the presence of astrohatches is conditioned by the "attraction" of bodies to the nearby resonances.

2. It is possible that the pertinent resonant motions in the process of evolution have lost their stability: as was marked in 20.4, here we may deal with a temporary stability, whose comparatively fast destruction in the cases under consideration might have occurred either due to rather strong dissipative forces, or due to relatively small masses of the asteroids,

In the light of the results that have been stated, Molchanov's well known *hypothesis about complete resonancity of the motions of big planets of the Solar system* [392] seems to be theoretically quite acceptable. But is the observed actual motion of planets fully resonant or does it only evolve towards it? For the time being that cannot yet be answered.

Based on the results stated, the author suggests another, "more cautious" *hypothesis about simple resonancity of the motion of a number of bodies in the Solar system* (not only big planets). By simple resonances we mean the resonances of the zero, first and second order, i. e. resonances of the type $l : (l \pm q)$, where $q = 0, 1, 2, 3, 4$. According to what has been said, it is such resonances that are "most probable", and it is well proved by the data observed.

Apparently, it was not without taking into account these results and ideas that Garkavy and Fridman initiated a hypothesis about the resonant nature of the rings of Uranus. According to that hypothesis the positions of those rings are determined by the resonances of the type 1:2, 2:3, and 3:4 from the satellites that had not yet been discovered. Later such satellites were really discovered by "Voyadger-2" [212].

In conclusion we will note that the extreme property of resonant motions makes it possible to carry out a pecular kind of check of the hypotheses of the proximity of the motion of planetary or satellite system to the resonance: it is possible to compute the value of the potential function D for the motion under investigation and compare it with the adequate minimal values of this function. It would be also of interest to define the direction of the evolutionary change of D.

In this chapter we have been discussing only the resonances in orbital motions of celestial bodies, since the integral sign of stability, based on equation (4.2.42) of chapter 4, cannot be used in problems on orbital rotational resonances. It cannot be used because the character of anisochronism while a free body is moving in the orbit differs from that when a body performs a rotational motion around its center of mass (in the first case it is $\sigma = -1$, in the second case it is $\sigma = 1$). While in the extreme sign of stability, suggested by Beletsky, the character of anisochronism does not appear at all (see 4.2.2, and also [55, 57]).

It should be noted that on the basis of some other, non-classical conceptions, the problems of stability, resonance, and extremality, and their application to the motion of celestial bodies have been considered by Chechelnitsky [147].

21.6   On Other Problems (Motion of Celestial Bodies with the Internal Degrees of Freedom)

Problems on the motion of natural and artificial celestial bodies with the internal degrees of freedom are of a considerable interest, both principal and applied. These degrees of freedom can be conditioned by the deformability of elements of the body, by the presence of the cavities with fluid and also by the additional bodies which can move with respect to the main body, etc. In this case it may prove to be necessary to take into account the damping when there is a relative displacement of internal mobile elements.

An important class of problems pertains to the cases where the center of inertia of a body moves in a certain orbit around some central body, or in a nearby trajectory, i.e. when the body under consideration is a satellite of the central body. A cycle of researches by Chernousko and his colleagues Akulenko, Leshchenko and Shamayev and also Nabiullin (see publications [394, 463, 156, 157, 155] with detailed references) have been devoted to these problems. Adequate problems were mentioned by us in 2.2.2 as a problem of the mechanics of systems with hidden motions, and also in 10.5.4 when describing the basic principles of peculiar vibrational coaches — graviflier and magnetoflier.

The aim of this short section is to draw attention to the expediency of using the approaches and interpretations, characteristic of vibrational mechanics, when solving this kind of problems. It seems to us that the advantages of such approaches and interpretations have not as yet been fully made use of. The "fast" character of the motion of the center of inertia of the body in the orbit and of the motions corresponding to the internal degrees of freedom is the necessary pre-requisite for applying the apparatus of vibrational mechanics in such problems.

A peculiarity of the problems under consideration, just like those to which this chapter is mainly devoted, consists in the fact that at least one of the fast motions — the motion of the center of masses in the orbit — is not small by its scale as compared to the slow motions corresponding to the change in the elements of the orbit of the body. Nevertheless, as was pointed in 3.2.4, the approaches of vibrational mechanics prove to be applicable in this case as well. As a result of excluding the fast motions when using this approach, the main equations of vibrational mechanics have been obtained, in which apart from the "usual" slow forces, acting upon the body, there are additional slow forces (vibrational forces), which "integrally" reflect the effect of the presence of the internal degrees of freedom of the body upon the slow motion. In particular, as was marked in 10.5.4, the presence of internal motions can result in the appearance of a force having the character of a thrust or resistance to the motion of the center of masses of the body. It goes without saying that the latter phenomenon proves to be possible due to the presence of a nonuniform field of external force — the gravitational field (see ibid.).

Of special interest is considering the cases of commensurability (resonances) between the frequency of revolution (of the mean motion) of the body and the frequencies of its free oscillations.

Bibliography

[1] Abramovich I. M., I. I. Blekhman, B. P. Lavrov, and D. A. Pliss, The Phenomenon of self-synchronization of rotating bodies, *Otkrytiya, Izobreteniya*, 1 (1988) (in Russian).

[2] Acheson D. I., A pendulum theorem, *Proceedings of the Royal Society of London A* **443** (1993) 239–245.

[3] Acheson D. I. and T. Mullin, Upside-down pendulums, *Nature* **336** (1993) 215–216.

[4] Acheson D. I. and T. Mullin, Ropy magic, *New Scientist* **157** (1998) 32–33.

[5] Afanas'ev M. M., I. I. Blekhman, V. A. Makarov, and A. V. Petchenev, Dynamics of the system with forced synchronization of mechanical exciters with asynchronous drive, *Izv. AN SSSR, Mashinovedenie*, 4 (1983) 3–11 (in Russian).

[6] Afraimovich V. S., N. N. Verichev, and M. I. Rabinovich, Synchronization of oscillations in dissipative systems, *Radiophisics and Quantum Electronics* (Plenum Publ. Corp.) (19??), 795-803 (in Russian).

[7] Agafonov S. A., Effect of stabilizing the equilibrium of Ziegler's pendulum by means of parametric excitation *Izv. Ran. MTT*, 6 (1997) 36-40 (in Russian).

[8] Agranovskaya E. A., The solution of nonlinear differential equations containing periodic functions of depended variable by analog computers, *Izv. AN SSSR, OTN, Mekhanika i Mashinostroenie*, 4 (1963) 175–177 (in Russian).

[9] Agranovskaya E. A., I. I. Blekhman et al., Mechanical and technological peculiarities and certain problems of the theory of vibrational flotation machine of Mekhanobr Institute, *Developing and promoting flowsheets and regimes of non-ferrous metal ore dressing. Proceedings of the Mekhanobr In-te*, issue 144 (Leningrad, 1976), 67–78 (in Russian).

[10] Alabin E. A., V. V. Gortinsky, and A. F. Londarsky, Vibrational displacement in the drive of rotor machines with flexible constraints, *The 2nd All Union Congress on the Theory of Mechanisms and Machines*, Part 1, (Naukova Dumka, Kiev, 1982) 14–15 (in Russian).

[11] Alabuzhev P. M. and S. F. Yatsun, Mathematical modeling of vibrational technological processes, *Abstracts of the 2nd All Union Congress on Nonlinear Oscillations of Mechanical Systems*, Part II, Gorky (1990) 108–109 (in Russian).

[12] Alifov A. A. and K. V. Frolov, *Interaction of nonlinear oscillating systems with energy source* (Nauka, Moscow, 1985), p. 327 (in Russian).

[13] Al'bert I. U. and O. A. Savinov, On rheological models of vibrated concrete mix, *Izv. VNIIGidrotekhniki*, **105** (1974) 138–146 (in Russian).

[14] Al'fven H., and G. Arrenius, *Structure and evolution history of the Solar system* (Naukova Dumka, Kiev, 1981), p. 331 (in Russian).

[15] Ambartsumyan S. A., *Different-modulus elasticity theory* (Nauka. Fizmatlit Publ. Co., Moscow, 1982) (in Russian).

[16] Anahin V. D., D. A. Pliss, and V. N. Monahov, *Vibrational separators* (Nedra, Moscow, 1991), p. 154 (in Russian).

[17] Anahin V. D. and D. A. Pliss, *On the theory of vibrational separators* (Novosibirsk State Univ., Novosibirsk, 1991), p. 124 (in Russian).

[18] Anderson G. L. and I. G. Tadjbakhsh, Stabilization of Ziegler's pendulum by means of the method of vibrational control, *Journal of Mathematical Analysis and Application*, **143** (1989) 198–223.

[19] Andrienko J. A., I. F. Obraztsov, and Yu. G. Yanovsky, Phenomena of resonance type for streams of visco-elastic fluids in pipes, *Doklady RAN* **342** 6 (1995), 765-768.

[20] Andronov A. A. and A. A. Vitt, On mathematical theory of hunting phenomenon, *Journal of Experim. and Theor. Physics*, **7**, Issue 4 (1930) (in Russian).

[21] Andronov A. A., A. A. Vitt, and S. E. Haikin, *The theory of oscillations* (Fizmatgiz, Moscow, 1959), p. 915 (in Russian).

[22] Andronov A. A., *Dynamics of the systems with transformed dry friction*, Doctor's dissertation (Moscow, 1984) (in Russian).

[23] Anisch'enko V. S., T. E. Vadivasova, D. E. Postnov, and M. A. Safonova, Mutual synchronization of chaos. Typical bifurcations and their characteristics, *Proceedings of the 2nd All Union Congress on Nonlinear Oscillations of Mechanical Systems*, Gorky, Part 1 (1990) 17–18 (in Russian).

[24] Arnol'd V.I., *Mathematical methods of classical mechanics* (Nauka, Moscow, 1989), p. 472 (in Russian).

[25] Artobolevskiy I. I., *The theory of mechanisms and machines* (Nauka, Moscow, 1988), p. 639 (in Russian).

[26] Arzamaskov A. M., E. S. Briskin, G. G. Grigoryan, and V.M.Sobolev, On the theory and computation of the walking transport and processing machines of supporting passability, *Abstracts of the 7th All Union Congress on Theor. and Appl. Mechanics* (Moscow, 1991), 20 (in Russian).

[27] Asseynov S. A., *Investigations of the laws of vibrotransportation of a layer of bulk cargo (fish and granular cargo)*, PhD thesis (Moscow, 1973) (in Russian).

[28] Astashov V. K., and M. E. Gerts, On the theory of vibrational transportation, *Izv. AN SSSR, MTT*, 1 (1978) 40–44 (in Russian).

[29] Astashov V. K., On the effect of high frequency vibrations on the process of plastic deformation, *Mashinostroenie*, 2 (1983).

[30] Astashov V. K., V. I. Babitskiy, A. M. Veprik, and V. L. Krupenin, Supression of forced oscillations of strings and rods by a movable washer, *DAN SSSR*, **304**, 1 (1989) 50–54 (in Russian).

[31] Avotinya K. and I. Tomashuns, On synchronization in rolling bearings, *Proceedings of Latvian Agriculture Academy*, Elgava, 109 (1977) 31–36 (in Russian).

[32] Baader W., Das Verhalten eines Schittgutes auf schwingenden Siebrosten, *Grundlagen der Landtechnik*, *13* (1961) (in German).

[33] Babeshko V. A., I am not indifferent to them... *Znanie — sila*, 12 (1990) 63–66 (in Russian).

[34] Babichev A. YA. *Vibrational processing of details of machines* (Mashinostroenie, Moscow, 1974) (in Russian).

[35] Babitskiy V. I. and V. Sh. Burd, Supression of plane oscillations of a platform by unbalanced dampers, *Izv. AN SSSR, MTT*, 6 (1982) 29–33 (in Russian).

[36] Babitsky V. I. and A. M. Veprik, Damping of beam forced vibration by moving washer, *Journal of Sound and Vibration*, **166**, 1 (1993) 77–85.

[37] Babitsky V. I., *Theory of vibro-impact systems and applications* (Springer, Berlin, Heidelberg, New York and others. 1998), p. 291.

[38] Babushkin I. A., M. P. Zavarykin, S. V. Zorin, and G. F. Putilin, Controling the convective stability by vibrational fields, *Proceedings of the 2nd All Union Congress on Nonlinear Oscillations of Mechanical Systems*, Gorky, Part 1 (1990), 22 (in Russian).

[39] Bakhvalov N. S. and G. p. Panasenko, *Averaging processes in periodic media. Mathematic problems of mechanics of composite materials*, (Nauka. Fizmatgiz, Moscow, 1984), p. 352 (in Russian).

[40] Banaszewski T. and A. E. Schollbach, Schwingungsanalyse von Maschinen mit selbstsynchronisierenden Unwuchterregern, *Autbereitungs-Technik* **39**, 8 (1998) 383–393.

[41] Banah L. YA., Investigation of dynamics of regular and quasi-regular systems with rotary symmetry, *Mashinovedenie*, 3 (1984) 9–16 (in Russian).

[42] Banah L. YA., Energy and spectral connections in mechanical oscillatory systems, *Izv. AN SSSR, MTT*, 3 (1988) (in Russian).

[43] Bansevichyus R. YU. and K. M. Ragul'skis, *Vibromotors* (Mokslas, Vilnus, 1981), p. 193 (in Russian).

[44] Bansevichyus R. YU., A. K. Bubulis, R. A. Vodchenkova, and R. E. Kurilo, Vibrational transformers of motion, in *Vibratsionnaya Tekhnika*, 1, ed. K. M. Ragul'skis (Mashinostroenie (Library for Engineers), Leningrad, 1984), p. 64 (in Russian).

[45] Barenblatt G. I., Yu. I. Kozyrev, N. I. Malinin et al., On vibrocreep of polymer materials, *J. of App. Mechanics and Technical Phys.*, 5 (1965) 68–75 (in Russian).

[46] Barkan D. D., *Vibro-method in building* (Gorstroyizdat, Moscow, 1959), p. 315 (in Russian).

[47] Barzukov O. P., Multiple synchronization in the system of weakly connected objects with one degree of freedom, *App. Math. and Mech.*, **36**, 2 (1972) 225–231 (in Russian).

[48] Barzukov O. P., Double synchronization of mechanical vibrators connected with linear oscillatory system, *Izv. AN SSSR, MTT*, 6 (1973) 22–29 (in Russian).

[49] Barzukov O. P. and L. A. Vaisberg, On stability of selfsynchronous rotation of two vibro-exciters in the devices with spatial dynamic scheme, *Vibrational Engineering* (MDNTP imeni F. E. Dzerzhinskogo, Moscow, 1977) 89–94 (in Russian).

[50] Barzukov O. P., N. A. Ivanov, and J. M. Katsman, Precise method of computation of material transportation in cone crusher chamber, *Obogaschenie rud*, 4 (1983) 3–6 (in Russian).

[51] Batalova Z. S., On rotor motion under the effect of external harmonic force, *Engineering Journal, Mechanics of Solid Body*, 2 (1967) 66–73 (in Russian).

[52] Batalova Z. S. and G. V. Belyakova, Diagrams of stability of periodic motion of a pendulum with an oscillating axis, *App. Math. and Mech.*, **52**, 1 (1988) 41–48 (in Russian).

[53] Batalova Z. S. and N. V. Buhalova, On structure of phase space of a parametrically excited rotor, in *Collection of Interuniversity Scientific Proceedings: Dynamics of systems: Qualitative and Numerical Investigations of Dynamic Systems* (Gorky State Univ., Gorky, 1988) 18–33 (in Russian).

[54] Bauman V. A. and I. I. Buhovsky *Vibrational machines and processes in building* (Vysshaya shkola, Moscow, 1977), p. 255 (in Russian).

[55] Beletsky V. V., A. N. Shlyahtin, Extremal properties of resonance motions, *DAN SSSR*, **231**, 4 (1976) 829–832 (in Russian).

[56] Beletsky V. V., *Outlines on motions of cosmic bodies* (Nauka, Moscow, 1977), p. 430 (in Russian).

[57] Beletsky V. V. and G. V. Kasatkin, On extremal properties of resonance motions, *DAN SSSR*, 251, 1 (1980) 58–62 (in Russian).

[58] Beletsky V. V., Extremal properties of resonance motions; in *Stability of motion. Analytical mechanics. Control of motion* (Nauka, Moscow, 1981) 41–55 (in Russian).

[59] Beletsky V. V., Resonance in rotational motions of artificial and natural celestial bodies; in *Dynamics of space apparatus and investigation of cosmic space* (Mashinostroenie, Moscow, 1986) 20 42 (in Russian).

[60] Beletsky V. V. and M. D. Golubitskaya, Stabilization and resonances in a model problem of two-legged walk, Preprint, 14 (The Keldysh Institute of Applied Mathematics, Moscow, 1987), p. 23 (in Russian).

[61] Beletsky V. V. and A. A. Hentov, *Rotary motions of a magnetized satellite* (Nauka, Moscow, 1985), p. 287 (in Russian).

[62] Bellman V. V., J. Bentsman, and S. M. Meerkov, Vibrational control of nonlinear systems: vibrational stabilizability (1), Vibrational controllability and transient behavior (2), *IEEE Transactions on Automatic Control*, **AC-31**, 8 (1986, August) 710–724.

[63] Belyaev A. K. and V. A. Pal'mov, The theory of vibroconduction; in *Problems of Dynamics and Strength*, 36 (Zinatne, Riga, 1980) 138–146 (in Russian).

[64] Belyakova G. V., *Dynamics of systems of pendulum type with parametric excitation*, PhD thesis (physics and mathematics) (Gorky, 1988), p. 194 (in Russian).

[65] Bentsman J, Vibrational control of a class of nonlinear systems by nonlinear multiplicate vibrations, *IEEE Transactions on Automatic Control* **32** (1987) 711- 716.

[66] Bezdenezhnyh N. A., V. A. Briskman, A. A. Cherepanov, and M. T. Sharov, Controling the stability of a liquid surface by means of variable fields, in *Hydromechanics and Transfer Processes in Zero Gravity* (Ural Center of Ac. of Sci. of USSSR, Sverdlovsk, 1983) 37–56 (in Russian).

[67] Blekhman I. I., Investigation of the process of vibroseparation and vibro-transportation, *Inzhenerny sbornik*, **II** (1952) 35–78 (in Russian).

[68] Blekhman I. I., Self-synchronization of certain vibrational machines, *Inzhenerny sbornik*, **16** (1953) 49–72 (in Russian).

[69] Blekhman I. I., Rotation of unbalnaced rotor caused by harmonic oscillations of its axis, *Izv. AN SSSR, OTN*, 8 (1954) 79–94 (in Russian).

[70] Blekhman I. I., Investigation of the process of driving-in piles and sheet piles, *Inzhenerniy sbornik*, **XIX** (1954) 55–64 (in Russian).

[71] Blekhman I. I. and G. Yu. Dzhanelidze, Dynamics of the Boisse-Sarda regulator, *Izv. AN SSSR, OTN*, 10 (1955) 48–59 (in Russian).

[72] Blekhman I. I., Problem of stability of periodic solutions of quasi-linear nonautonomous systems with many degrees of freedom, *DAN SSSR, OTN*, **104**, 6 (1955) 809–812 (in Russian).

[73] Blekhman I. I., Problem of stability of periodic solutions of quasi-linear autonomous systems with many degrees of freedom *DAN SSSR, OTN*, **112**, 2 (1957) 183–186 (in Russian).

[74] Blekhman I. I. and G. Yu. Dzhanelidze, On effective friction coefficients under vibrations, *Izv. AN SSSR, OTN*, 7 (1958) 98–101 (in Russian).

[75] Blekhman I. I., Inertial screen, Russian patent 112448, *Bull. izobret.*, 4, 1958, p. 65 (in Russian).

[76] Blekhman I. I., Dynamics of the drive of vibrational machines with many synchronous mechanical vibro-excires, *Izv. AN SSSR, OTN, Mekhanika i Mashinostroenie*, 1 (1960), 79–89 (in Russian).

[77] Blekhman I. I. and B. P. Lavrov, On an integral sign of stability of motion, *App. Math. and Mech.*, **24**, 5 (1960) 1416–1421 (in English).

[78] Blekhman I. I., Vertification of the integral sign of stability of motion in problems on self-synchronization of vibro-exciters, *App. Math. and Mech.*, **24**, 6 (1960) 1100–1103 (in Russian).

[79] Blekhman I. I. and G. Yu. Dzhanelidze, On stability of vibro-linearized systems, *App. Math. and Mech.*, **25**, 1 (1961) 173–176 (in Russian).

[80] Blekhman I. I., On the critical opening of a vibrational roller mill, *Obogaschenie rud*, 1 (1961) 31–34 (in Russian).

[81] Blekhman I. I., A. D. Rudin, and A. K. Rundkvist, On the conditions of motion with rolling of crushing bodies in vibrational crushing and grinding machines, *Obogaschenie rud*, 3 (1961) 37–41 (in Russian).

[82] Blekhman I. I., V. V. Gortinskiy, and G. V. Ptushkina, Motion of a particle in oscillating medium under resistance of dry friction type (To the theory of vibrational separation of granular mixtures), *Izv. AN SSSR, OTN, Mekhanika i Mashinostroenie*, 4 (1963) 31–41 (in Russian).

[83] Blekhman I. I., Some problems of the theory of vibrotransportation and vibrobunkering of bulk cargo in *Collection on vibroloading machines, vibrobunkering and vibrodischarge of bulk cargo* (TSNIITEI uglya, Moscow, 1963) (in Russian).

[84] Blekhman I. I. and G. Yu. Dzhanelidze, *Vibrational Displacement* (Nauka, Moscow, 1964), p. 410 (in Russian).

[85] Blekhman I. I. and V. Ya. Hainman, On the theory of vibrational separation of granular mixtures, *Izv. AN SSSR, OTN, Mekhanika*, 5 (1965) 22–30 (in Russian).

[86] Blekhman I. I., B. G. Ivanov, N. A. Ivanov, A. K. Rundkvist, and K. A. Rundkvist, A cone inertial crusher, Russian Patent 184125, *Bull. Izobr.*, 14 (1966), p. 140.

[87] Blekhman I. I. and R. F. Nagaev, Optimal stabilization of synchronous motions of mechanical vibro-exciters, *Proceedings of All Union Interuniversities Conferences on Applied Mathematics and Cibernetics*, Gorky (1967), p. 32 (in Russian).

[88] Blekhman I. I. and J. I. Marchenko, The effect of rheological properties of the working medium on self-synchronization of mechanical vibro-exciters, *Izv. AN SSSR, MTT*, 5 (1969) 35–45 (in Russian).

[89] Blekhman I. I. and C. A. Malasyan, On effective coefficients of friction at the interaction of an elastic body with a vibrating surface, *Izv. AN SSSR, MTT*, 4 (1970) 3–10 (in Russian).

[90] Blekhman I. I., *Synchronization of Dynamic Systems*, (Nauka, Moscow, 1971), p. 894 (in Russian).

[91] Blekhman I. I. and K. V. Frolov, Synchronization of parametrically connected vibro-exciters, *Abstracts of the All Union Conference on Oscillations of Mechanical Systems* (Naukova Dumka, Kiev, 1971) (in Russian).

[92] Blekhman I. I, Inertial vibro-exciter, Russian Patent 388947, *Bull. Izobr.*, 29 (1973), p. 63.

[93] Blekhman I. I, The action of vibration on mechanical systems, *Vibrotekhnika*, Issue 3(20) (Mintis, Vilnus, 1973) 369–373 (in Russian).

[94] Blekhman I. I. and G. B. Bukaty, The motion of the body which is in contact in turn with two vibrating planes, *Izv. AN SSSR. MTT*, 2 (1975) (in Russian).

[95] Blekhman I. I, Method of direct separation of motions in problems on the action of vibration on nonlinear dynamic systems, *Izv. AN SSSR, MTT*, 6 (1976) 13–27 (in Russian).

[96] Blekhman I. I. and I. N. Daich, The exciter of oscillations, Russian Patent 539619, *Bull. Izobr.*, 47 (1976), p. 20.

[97] Blekhman I. I. and I. N. Lugovaya, Phenomenon of synchronization and the design problems of group fundations for unbalanced machines, *Proceedings of the All Union Conference on the dynamics of bases, foundations and underground constructions* (FAN UzbSSR, Tashkent, 1977) 241–244 (in Russian).

[98] Blekhman I. I. and N. A. Ivanov, Motion of the material in the chamber of cone crushers as a process of vibrational displacement, *Obogaschenie rud*, 2 (1977) 15–21 (in Russian).

[99] Blekhman I. I., Development of a conception of direct separation of motions in nonlinear mechanics, in *Modern Problems of Theor. and App. Mechanics: proceedings of IV All Union Congress on Theor. and App. Mechanics* (Naukova Dumka, Kiev, 1978) 148–168 (in Russian).

[100] Blekhman I. I. and N. A. Ivanov, On the capacity and shaping of the chamber of cone crusher, *Obogaschenie rud*, 1 (1979) 20–27 (in Russian).

[101] Blekhman I. I. and V. L. Levengartz, Dynamic model of a process of feed motion in working chambers of the machines for vibro-abrasive treatment of machinery parts, in *Problems of Dynamics and Strength*, 36 (1980) 83–93 (in Russian).

[102] Blekhman I. I., Stability of Orbital Systems, in *Stability of Motions. Analytical Mechanics. Motion Control* (Nauka, Moscow, 1981) 55–68 (in Russian).

[103] Blekhman I. I., *Synchronization in Nature and Technology* (Nauka, Moscow, 1981), p. 351 (in Russian);

[104] Blekhman I. I. and O. G. Pirtzhalaishvili, On regularity of synchronization of unbalanced vibro-exciters, *Soobsch'eniya AN Gruz. SSR, Mashinovedenie*, 3 (1982) 569–572 (in Russian).

[105] Blekhman I. I. and L. I. Machabeli, On a model of automaticity and reservation at cardiac rhythm excitation in normal state and in some cases of pathology, *Abstracts of the 3d All Union Conference on Problems of Bio-Mechanics*, 1 (Riga, 1983), p. 30 (in Russian).

[106] Blekhman I. I., Generalization of the Lagrange - Dirichlet theorem on stability of equilibrium positions for certain classes of periodic and rotary motions, *Proceedings of the IX Intern. Conf. on Nonlinear Oscillations*, 3 (Naukova Dumka, Kiev, 1984) 34–36 (in Russian).

[107] Blekhman I. I., N. A. Ivanov, K. S. Yakimova et al. Principles of the profiling of chambers of cone crushers and results of preliminary tests of optimized profiles in field conditions, *Interdepartmental proceedings "Improvement of processes of crushing, grinding, screening and classification of ores and of dressing products"* (Mekhanobr, Leningrad, 1985) 23–33 (in Russian).

[108] Blekhman I. I., Regularities and paradoxes of mechanics of systems with ignored motions and their use in technology, *Abstracts of the VIth All Union Symp. on Theor. and Ap. Mechanics* (Tashkent, 1986), p. 110 (in Russian).

[109] Blekhman I. I. and K. V. Frolov, Nonlinear problems of the theory of inertial excitation of oscillations, *Abstracts of the All Union Conf. on Nonlinear Oscillations of Mechanical Systems*, Part I (Gorky, 1987), p. 134 (in Russian).

[110] Blekhman I. I., Vibrational mechanics and vibrational rheology (applications to technology, new results, unsolved problems), *Abstracts of the All Union Conf. on Nonlinear Oscillations of Mechanical Systems*, Part II (Gorky, 1987) (in Russian).

[111] Blekhman I. I. and O. Z. Malakhova, On potential on the average dynamic systems, *Abstracts of Chetaev's Fifth All Union Conf. "Analytical Mechanics, Stability and Motion Control"* (Kazan', 1987), p. 17 (in Russian).

[112] Blekhman I. I. and O. Z. Malakhova, On quasy-equilibrium positions of Chelomey's pendulum, *DAN SSSR*, **287**, 2 (1988) 290–294 (in Russian).

[113] Blekhman I. I., *Synchronization in Science and Technology*, (ASME Press, New York, 1988) (English translation), p. 255.

[114] Blekhman I. I. *What vibration can do. Vibrational Mechanics and Vibrational Technology* (Nauka, Moscow, 1988), p. 208 (in Russian).

[115] Blekhman I. I., L. A. Vaisberg, and A. I. Korovnikov, Analysis of hydrodynamics of vibrational screen with a sieve oscillating in aqueous medium, *Interdepartmental Proceedings "Investigation of the processes, machines and apparatuses for classifying the material according to size"* (Mekhanobr, Leningrad, 1988) 35–46 (in Russian).

[116] Blekhman I. I., O. A. Vyal'tzeva, L. A. Vaisberg, and A. Ya. Fidline, Capacity of vibrational screens with active working surfaces, *Interdepartmental Proceedings "Investigation of processes, machines and apparatuses for classifying the material according to size"* (Mekhanobr, Leningrad, 1988) 20–35 (in Russian).

[117] Blekhman I. I. and N. P. Yaroshevich, Multiple regimes of vibrational support of rotation of an unbalanced rotor, *Izv. AN SSSR, Mashinovedenie*, 6 (1989) 62–67 (in Russian).

[118] Blekhman I. I., Self-synchronization in nature and technology. Science and Humanity, *Intern. Annual Journal* (Urania Publ. House, Leipzig-Jena-Berlin, 1989) 362–367.

[119] Blekhman I. I., A. D. Myshkis, and Ya. G. Panovko, *Mechanics and applied mathematics: Logic and peculiarities of applications of mathematics* (Nauka, Moscow, 1990, 2nd ed.), p. 356 (in Russian). German translation: Veb Deutscher Verlag der Wissenschaftn, 1989, Berlin, p. 350.

[120] Blekhman I. I. and O. Z. Malakhova, Extreme signs of stability of certain motions, *App. Math. and Mech.*, **54**, 1 (1990) 142–161 (in Russian).

[121] Blekhman I. I. and Y. P. Yaroshevich, Vibro-exciter, Russian patent 1597235, *Otkrytiya, izobreteniya*, 37 (1990), p. 26.

[122] Blekhman I, A. Fedaravicus, and V. Rotovas, *Vlbratio transportavimo harmoningai virpancia horizontalia plokstuma, esant valdornai sausajai trinciai, tyrimas*, 1 (Mokslo darbai, Taikomoji Mechanika) Kaunatechnologijos universitetas Kaunas, "Teclinologija", 1991 (in Lithuanian)

[123] Blekhman I. I., The generalized self- balancing principle, *Proc. of Asia-Pacific Vibrational Confderence, Kitakyshu*, 2 (1993) 509-514.

[124] Blekhman I. I., Self-synchronization of mechanical vibroexciters: control of vibrations, generalized rotors, self- balancing principle, *The active control of vibration, IUTAM Symposium*, (Mech. Eng. Publ. Lim., London, 1994) 169–173.

[125] Blekhman I. I., *Vibrational Mechanics* (Fizmatlit, Moscow, 1994), p. 394 (in Russian).

[126] Blekhman I. I., P. S. Landa, and M. G. Rozenblum, Synchronization and chaotization in interacting dynamical systems,*Appl. Mech. Rev.*, **48**, 11 (Nov. 1995), part 1, 733–752.

[127] Blekhman I. I., A. L. Fradkov, H. Nijmeijer, and A.Y. Pogromsky, On self-synchronization and controlled synchronization, *Systems and Control Letters* **31** (1997) 299–305.

[128] Blekhman I. I, The use of the phenomenon of self-synchronization to cancel vibration, *Mekhanika* **17**, Zeczyt 1 (1998).

[129] Blekhman I. I., Forming the properties of nonlinear mechanical systems by means of vibrations *DCAMM Report*, 616, (Technical University of Denmark, 1999; Proceeding of the IUTAM/IFTAM Symposium on Synthesis of nonlinear Dynamical Systems, Riga, Latvia, 1998. Klüwer Academic Press, 1999).

[130] Blekhman I. I. and K. A. Lurie, Dynamic Materials as the elements of Material Design. *Paper on the EURODINAME'99 Conference "Dynamic Problems in Mechanics and Machatronics"* (Günsburg, Germany, 11-16 July, 1999).

[131] Blekhman L. I. and B. V. Kizevalter, Distribution of flow velocities on the decks of vibrational separators for gravitational dressing, *Obogaschenie rud*, 4 (1981) 21–24 (in Russian).

[132] Bochkovskiy V. M., Stratification as a most important part in the theory and practice of gravitation, *Mining Journal*, 1 (1954) 47–55 (in Russian).

[133] Bogolyubov N. N., *On certain statistic methods in mathematical physics* (AN USSR, Kiev, 1945), p. 139 (in Russian).

[134] Bogolyubov N. N., The theory of perturbations in nonlinear mechanics, *Proceedings of Inst. of Structural Mechanics, AN USSR, Kiev*, 14 (1945), p. 9 (in Russian).

[135] Bogolyubov N. N. and D. N. Zubarev, Method of asymptotic approximation for systems with a rotating phase, *The Ukrainian Mathematical Journal*, **VII**, 1 (1955) 5–17 (in Russian).

[136] Bogolyubov N. N., *Selected works in 3 volumes*, **1** (Naukova Dumka, Kiev, 1969), p. 447 (in Russian).

[137] Bogolyubov N. N. and Yu. A. Mitropolsky, *Asymptotic methods in the theory of nonlinear oscillations* (Nauka, Moscow, 1974), p. 503 (in Russian).

[138] Bogusz W. and Z. Engel, Badania doswiadczalne urzadzcn samosynchronizujacych, *Przeglad mechaniczny*, 8 (1965) 230–232 (in Polish).

[139] Bol'shakov V. M., E. S. Zel'din, R. M. Mints, and N. A. Fufaev, To the dynamics of the system oscillator-rotator, *Izv. Vuzov, radiofizika*, **VIII**, 2 (1965) 359–371 (in Russian).

[140] Bondin V. P., A. I. Vesnitsky, and E. E. Lisenkova, Elementary wave engine, *DAN SSSR*, **318**, 4 (1991) 849–852 (in Russian).

[141] Borderies N., P. Goldreich, and S. Tremaine, Sharp edges of planetary rings, *Nature*, **299**, 5880 (1982) (in English).

[142] Brumberg R. M., On the motion of a solid body along a vibrating tube without breaking away from it, *Izv. AN SSSR, MTT*, 5 (1970) 46–51 (in Russian).

[143] Bubulis A. K., V. P. Yushka and K. M. Ragul'skis, New designs of vibrodevices for the transportation of liquid, *Abstracts of the All Union Conference on Vibrational Engineering* (Kutaisi-Tbilisi, 1981), p. 12 (in Russian).

[144] Bukhalova N. V., *Investigation of the dynamics of a parametrically excited rotor*, PhD thesis (Gorky State Univ., 1989), p. 192 (in Russian).

[145] Bykhovsky I. I., *Basis of the theory of vibrational technology* (Mashinostroenie, Moscow, 1969), p. 363 (in Russian).

[146] Bykhovsky I. I., *Vibrational machines for industry* (MDNTP im. Dzerzhinskogo, Moscow, 1971) (in Russian).

[147] Chechelnitsky A. M., *Extremality, stability, resonancity in aerodynamics and astronautics* (Mashinostroenie, Moscow, 1980), p. 311 (in Russian).

[148] Chelomey V. N., On the possibility of the increase of elastic systems' stability by means of vibrations, *DAN SSSR*, **110**, 3 (1956) 341–344 (in Russian).

[149] Chelomey V. N., Paradoxes in mechanics caused by vibrations, *DAN SSSR*, **270**, 1 (1983) 62–67 (in Russian).

[150] Chelomey S. V., Nonlinear oscillations with parametric excitation, *Izv. AN SSSR. MTT*, 3 (1977) 39–46 (in Russian).

[151] Chelomey S. V., On dynamic stability of elastic systems, *DAN SSSR*, **252**, 2 (1980) 307–310 (in Russian).

[152] Chelomey S. V., Dynamic stability at high-frequency parametric excitation, *DAN SSSR*, **257**, 4 (1981) 853–857 (in Russian).

[153] Chelpanov I. B. The effect of random tossing of a moving ship on gyroscopic devices, *Izv. AN SSSR. MTT*, 4 (1973) 68–70 (in Russian).

[154] Cherednichenko I. I., *Resonance displacement of high-dispersible granular material by a pulsating flow of the carrying gas*, PhD thesis (Moscow, 1989), p. 164 (in Russian).

[155] Chernousko P. L , Evolution of rigid body motions due to dissipative torquer, *Proceeding of IUTAM Symposium (Stuttgart, Germany, 1989) on Nonlinear dynamics in engineering systems* (Springer-Verlag, Berlin-Heidelberg, 1990) (in English).

[156] Chernousko F. L, On the motion of a solid body with elastic and dissipative elements, *App. Math. and Mech.*, **42**, 1 (1978) 34–42 (in Russian).

[157] Chernousko F. L, The motion of a viscous-elastic body relative to the center of masses, *Izv. AN SSSR. MTT*, **15** (1980) (in Russian).

[158] Chervonenko A. G. and D. E. Borokhovich, Peculiarities of vibrotransportation of the bulk cargo with breaking-away from the working surface, *Mashinovedenie*, 4 (1978) 23–29 (in Russian).

[159] Chlenov V. A. and N. V. Mikhailov, *Drying the granular material in a vibro-boiling layer* (Stroiizdat, Moscow, 1967), p. 224 (in Russian).

[160] Chlenov V. A. and N. V. Mikhailov, *Vibro-boiling layer* (Nauka, Moscow, 1972), p. 343 (in Russian).

[161] Chuntz Z. G., *The investigation of dynamics and the improvement of resonance machines for ore processing*, PhD thesis (Riga, 1979), p. 219 (in Russian).

[162] Collins J. J., and I. N. Stewart, Coupled Nonlinear Oscillators and the Symmetries of Animal Gaits, *Nonlinear Science* (1993) 349–392.

[163] Coughey T. K. , Hula-hoop: one example of heteroparametric excitation, *Amer. J. Plys.*, **28**, 2 (1960) 101–109 (in English).

[164] Daich I. M., *Investigation of unbalance and vibrational isolation of settling centrifuges*, PhD thesis (Dnepropetrovsk, 1968) (in Russian).

[165] Davies H., Random vibration of distributed systems strongly coupled as discrete points, *J. Acoust. Soc. Am.*, **54**, 2 (1973) 507–515.

[166] Den-Hartog J. P., *Mechanical oscillations* (Fizmatgiz, Moscow, 1960), p. 580 (Translated from English into Russian).

[167] Detinko F. M., On stability of operation of a self-balancer for dynamic balancing, *Izv. AN SSSR. OTN. Mekhanika i Mashinostroenie*, 4 (1959) 38–45 (in Russian).

[168] Dubrovin B. N. and I. I. Blekhman, On critical opening of inertial cone crushers, *Obogaschenie rud*, 6 (1960) 32–38 (in Russian).

[169] Duckstein H., F. Merten, and L. Sperling, Zum Anlaufproblem beim automatischen Wuchten, *GAMM-Jahrestagung, 6–9 März, Bremen* (1998).

[170] Duhin S. S., Theory of the drift of an aerosol particle in a standing sound wave , *Colloidal Jour.*, **22**, 1 (1960) 128–130 (in Russian).

[171] Dyer F., Reverse classification by crobweed setting in ore-dressing, *Engng. and Mining J.*, **127**, 26 (1929) 1030–1033 (in English).

[172] Dzhakashov A. T. and B. I. Kryukov, Self-synchronization of resonance vibromachines, *Vibrotekhnika*, Issue 60(3) (Mokslas, Vilnus, 1988) 54–59 (in Russian).

[173] Eichler E., Thermal circuit approach to vibration in coupled systems and noise reduction of a rectangular box, *J. Acoust. Soc. Am.*, **37**, 6 (1965) 995–1007.

[174] Entus Ya. B., Motion of a granular material layer under vibrational transportation with tossing, *Izv. AN SSSR. Mashinovedenie*, 3 (1981) 18–22 (in Russian).

[175] Erdelyi A., Über die Kleinen Schwingungen eines Pendels mit oszilieren-den Aufhangepunkt, *ZAMM*, **14** (1934) (in German).

[176] Erdos G. and T. Singh, Stability of a parametrically excited damped inverted pendulum, *Journal of Sound and Vibration* **198** (1996) 643–650.

[177] Erdesz K. and J. Nemeth, Methods of Calculation of Vibrational Transport Rate of Granular Materials, *Powder Technology*, 55 (1988) 161–170.

[178] Ernst H., Automatic precision balancing, *Machine Design, Januar* (1951) 107–114.

[179] Fazullin F. F., On synchronization of mechanical systems with cyclic symmetry, *App. Mechanics*, **12**, 1 (1976), 90-97 (in Russian).

[180] Fedoravichus A. Yu., Method of vibrational transportation and the device for its realization, Russian patent 1022895, *Bull. izobr.*, 22 (1983), p. 45.

[181] Fedorenko I. Ya., Analysis of the behavior of granular material under vibrations on the basis of the theory of Lorentz's attractor, *Izv. Sib. Otd. AN SSSR, series of techniques*, 3 (1990), 112-115 (in Russian).

[182] Fepple A. and L. Fepple, *Force and deformation*, in 2 vols. (ONTI, Moscow-Leningrad, **I**, 1933, p. 420, **II**, 1936, p. 408), (in Russian).

[183] Feuer A. and M. Levine, Vibrational control of a flexible beam, *International Journal of Control* **65** (1996) 803–825.

[184] Fidline A. Ya., Simulation of the behaviour of granular material in a vibrating vessel, *Abstracts of the All Union Conf. on Vibrational Technology* (Tbilisi, 1987), p. 154 (in Russian).

[185] Fidline A. Ya., Forming circular flows of granular material under vibration, *Obogaschenie rud*, 1 (1991) (in Russian).

[186] Fidline A. Ya., On averaging in systems with a variable number of degrees of freedom, *App. Math. and Mech.*, **55**, 4 (1991) 634–638 (in Russian).

[187] Florina N. V., *A stamp on an elastic base under the action of variable loads in case of sliding*, Summary of PhD thesis (Leningrad, 1964), p. 12 (in Russian).

[188] Frolov K. V., *Is vibration a friend or an enemy?* (Nauka, Moscow, 1984), p. 144 (in Russian).

[189] Gaitsgorn V. G. and A. A. Pervozvansky, Separation of motions in Markov's systems, *Interinstitute collection of articles "Dynamics of the systems"*, Issue 6 (Gorky, 1980) 14–45 (in Russian).

[190] Ganiev R. F. and L. E. Ukrainsky, *Dynamics of the particles under the action of vibration* (Naukova Dumka, Kiev, 1975), p. 168 (in Russian).

[191] Ganiev R. F. and V. F. Lapchinsky, *Problems of mechanics in space technology* (Mashinostroenie, Moscow, 1978), p. 119 (in Russian).

[192] Ganiev R. F., N. I. Kobasko, V. V. Kulik et al., *Oscillatory motions in multiphase media and their use in technology* (Tekhnika, Kiev, 1980), p. 142 (in Russian).

[193] Gaponov A. V. and M. A. Miller, On potential pits for the charged particles in a high frequency elecromagnetic field, *Jour. of Experim. and Theor. Phys.*, **34** (1958) 751–752 (in Russian).

[194] Gaponov A. V. and M. A. Miller, On the use of the moving high frequency potential pits for the acceleration of the charged particles, *Jour. of Experim. and Theor. Phys.*, **34**, 3 (1958) 751–752 (in Russian).

[195] Gaponov–Grehov A. V. and M. I. Rabinovich, *Nonlinear physics. Stochasticity and structures; Physics of the XXth Century: development and prospects* (Nauka, Moscow, 1984) 219–280 (in Russian).

[196] Geraschenko E. I. and S. M. Geraschenko, *Method of separation of motions and the optimization of nonlinear systems* (Nauka, Moscow, 1975), p. 295 (in Russian).

[197] Gerasimov I. A. and B. R. Myshailov, Evolution of the orbit in the system Solar-Neptune-Pluto, *Astronomichesky Vestnik* **30**, 2 (1996) 177-182 (in Russian).

[198] Gershman M. D., Vibrational displacement in a medium with nonlinear resistance, *Izv. AN SSSR. Mashinostroenie*, 2 (1981) 8–13 (in Russian).

[199] Gershuni G. Z., *Conventional stability of incompressible liquid* (Nauka, Moscow, 1972) (in Russian).

[200] Gershuni G. Z., E. M. Zhukhovitsky, and A. A. Nepomnyaschy, *Stability of convective flows* (Nauka, Moscow, 1989), p. 318 (in Russian).

[201] Gladun A. D. and Chan V'et Hung., Mathematical model of a friction unit as of nonlinear oscillatory system, *Problems of automatization of design and manufacture in mechanical engineering* (Mosstankin, Moscow, 1983) p. 209 (in Russian).

[202] Goldacre R., The control of rhythm and homeostasis in biology and medicine, *Cibernetica*, 2 (1960) (in English).

[203] Goldin A. V. and N. N. Chirkov, Optimization of the process of vibrational separation of granular materials, *Proizvodstvenno-tehnicheskiy bull.*, 6 (1974) (in Russian).

[204] Goldin A. V. and V. A. Karamzin, *Hydrodynamic basis of processes of fine-layer separation* (Agropromizdat, Moscow, 1985), p. 263 (in Russian).

[205] Goldman P. S. and R. F. Nagaev, Self-synchronization of multidimensional nonconservative objects with one positional coordinate, *App. Mech.*, **23**, 12 (1987) 87–94 (in Russian).

[206] Goldraykh P. V., Explanation of frequent occurrences of commensurable mean motions in the Solar System, *Tides and Resonances in the Solar System* (Mir, Moscow, 1975) 217–247 (in Russian).

[207] Goncharevich I. F., *Dynamics of vibrational transportation* (Nauka, Moscow, 1972), p. 244 (in Russian).

[208] Goncharevich I. F., *Vibrorheology in mining production* (Nauka, Moscow, 1977), p. 143 (in Russian).

[209] Goncharevich I. F., *Vibration as a nonstandard way* (Nauka, Moscow, 1986), p. 209 (in Russian).

[210] Gorbikov S. P. and Yu. I. Neimark, Main regimes under vibrational transportation with tossing, *Izv. AN SSSR. MTT*, 4 (1981) 39–50 (in Russian).

[211] Gorbikov S. P. and Yu. I. Neimark, Results of calculating the mean velocity of vibrotransportation, *Izv. AN SSSR. Mashinovedenie*, 4 (1987) 39–42 (in Russian).

[212] Gorkavy N. N. and A. M. Fridman, On the discovery made by "Voyager-2" of the predicted satellites, determining the resonance nature of the rings of Uranus, *Astr. Let.*, **13**, 3 (1987), 96-99 (in Russian).

[213] Gorodetsky I. Ya., A. A. Vasin, V. M. Olevsky, and P. A. Lupanov, *Vibrational mass-exchange apparatuses*, ed. V. M. Olevsky (Himiya, Moscow, 1980), p. 192 (in Russian).

[214] Gortinsky V. V., Classifying the granular material during its laminar motion along the sieves, *Proceedings of VIM*, **34** (Moscow, 1964) (in Russian).

[215] Gortinsky V. V., A. B. Demsky, and M. A. Boriskin, *Separation processes at the grain processing plants* (Kolos, Moscow, 1973), p. 304 (in Russian).

[216] Gortinsky V. V., E. A. Alabin et al., The drive of reciprocation of the sieve, Russian patent 977060, *Bull. izobr.*, 44 (1982), p. 31.

[217] Granat N. L., The motion of a solid body in a pulsating flow of viscous fluid, *Izv. AN SSSR OTN. Mekhanika i Mashinostroenie*, 1 (1960) 70–78 (in Russian).

[218] Granat N. L., On perturbations made by a body moving in viscous fluid, *Izv. AN SSSR OTN. Mekhanika i Mashinostroenie*, 1 (1961) 86–89 (in Russian).

[219] Granat N. L., Stationary oscillations of the vessels with two-phase mixes, *Izv. AN SSSR OTN. Mekhanika i Mashinostroenie*, 5 (1964) 61–64 (in Russian).

[220] Granat N. L., Tinite oscillations of a ball in viscous fluid, *Izv. Vsesoyuzn. N.-I. In-ta Gidrotekhniki*, **76** (1964) 289–298 (in Russian).

[221] Granat N. L., Energy loss under the oscillations of a ball in two-phase mixes (vibroviscousity and vibrodensity of mixes), *Izv. AN SSSR OTN. Mekhanika*, 1 (1965) 34–41 (in Russian).

[222] Grebenikov E. A., *Method of averaging in applied problems* (Nauka, Moscow, 1986), p. 255 (in Russian).

[223] Grebenikov E. A. and Yu. A. Ryabov, *Resonances and small denominators in celestial mechanics* (Nauka, Moscow, 1978) (in Russian).

[224] Grigorova S. R. and D. M. Tolstoy, On resonance drope of friction force, *DAN SSSR*, **163**, 3 (1966) 562–563 (in Russian).

[225] Grinman I. G. and A. B. Bekbaev, *Control and regulation of processes of crushing and screening the ore* (Nauka, Alma-Ata, Kaz. SSR, 1977), p. 117 (in Russian).

[226] Gudushauri E. G. and G. Ya. Panovko, *Theory of vibrational technological processes at non-Coulombian friction* (Nauka, Moscow, 1988), p. 144 (in Russian).

[227] Gurkov K. S., A. D. Kostylev et al., *Mine vibrational loading machines and feeders* (Nauka, Novosibirsk, 1970), p. 134 (in Russian).

[228] Gurtovnik A. S and Yu. I. Neimark, On synchronization of dynamic systems, *App. Math. and Mech.*, **38**, 5 (1974) 749–758 (in Russian).

[229] Gutman L., *Industrial uses of mechanical vibrations* (London, 1968) (in English).

[230] Guzev V. V., *Reactive stabilization of synchronous rotations of inertial vibroexciters*, Doctor's dissertation (Riga's Polytechnical Ins-te, Riga, 1989) (in Russian).

[231] Haken H., *Advanced Synergetics. Instability hierarchies of self-organizing systems and devices* (Springer-Verlag, Berlin, Heidelberg, 1983), p. 405.

[232] *Handbook on ore dressing. Preparing processes*, 2nd ed (Nedra, Moscow, 1982), p. 361 (in Russian).

[233] Hedaya M. T., An analysis of a new type of automatic balancer, *Journ. Mech. Eng. Sci.***19**, 5 (1977) 221–226.

[234] Hertz H., *Die Prinzipien der Mechanik* (Leipzig, 1894) (in German).

[235] Hapaev M. M., *Averaging in the theory of stability* (Nauka, Moscow, 1986), p. 190 (in Russian).

[236] Hilkevich S. S., *Physics around us* (Nauka, Moscow, 1985) ("Kvant", issue 40), p. 159 (in Russian).

[237] Huygens C., *Three memoirs on mechanics* (AN SSSR, Moscow, 1951), p. 379 (in Russian).

[238] Inoue J., Y. Araki, S. Hayashi, O. Matsushita, S. Miyaura, and Y. Okada, On the self-syncironization of mechanical vibrators, *Trans. Japan. Soc. Mech. Eng.* , Part I, **32**, 234 (1966), Part 2, **33**, 246 (1967), Part 3, **35**, 274 (1969), Part 4, **41**, 350 (1975) (in Japanese).

[239] Iorish Yu. I., One-sided drift and rotation of the pointers of measuring devices under vibration, *Priborostroenie*, 4 (1956) 15–24 (in Russian).

[240] Ishlinsky A. Yu., *Mechanics of hyroscopic systems* (Izd. AN SSSR, Moscow, 1963), p. 482 (in Russian).

[241] Ishlinsky A. Yu., S. V. Malashenko, V. A. Storozhenko, M. E. Temchenko, and P. G. Shishkin, Method of balancing rotatory bodies on string drive, *Izv. AN SSSR. MTT*, 5 (1979) 3–18 (in Russian).

[242] Ishlinsky A. Yu, Motion of a solid body on a string, *Mechanics, ideas, problems, applications* (Moscow, 1985) 380–385 (in Russian).

[243] Ishlinsky A. Yu., V. A. Storozhenko, and M. E. Temchenko, Quasiparadoxal motions of solid bodies, *Abstracts of the VIIth All Union Congr. on Theor. and App. Mechanics* (Moscow, 1991) 174–175 (in Russian).

[244] Islamov M. S., On the motion of granular medium particles under the action of vibration, *Izv. vuzov. Mining Jour.*, 12 (1983) 89–92 (in Russian).

[245] Ispolov Yu. G. and B. A. Smolnikov, Principles of nonholonomic acceleration of movable objects, *Abstracts of the VIIth All Union Cong. on Theor. and App. Mechanics* (Moscow, 1991), 173-174 (in Russian).

[246] Ispolov Yu. G. and V. A. Smolnikov, Skateboard dynamics, *Comput. Methods Appl. Mech. Engrg.*, 131 (1996) 327–333.

[247] Jeffereys H. a. B. S., *Methods of mathematical physics* (Cambridge, 1950) (second edition) (in English).

[248] Jensen J. S., Nonlinear dynamics of the follower-loaded double pendulum with added support-excitation, *Journal of Sound and Vibration*, **215**, 1 (1998) 125–142.

[249] Jensen J. S., Non-trivial effects of fast harmonic excitation, *DCAMM Report*, no. S 83, PhD Thesis (Technical University of Denmark, Lyngby, 1999), p. 57.

[250] Jensen J. S., D. M. Tchernyak, and J. J. Thomsen, Non-trivial effects of fast harmonic excitation: experiments for an axially loaded beam, *DCAMM Report*, no. 601 (1999).

[251] Jones J. C., *Design methods. Seeds of human future* (A Wiley-Interscience Publication, London, 1980), p. 326.

[252] Jordan H., O. Mahrenholtz, G. Röder, and M. Weis, Bedingungen für den selbsttaätigen synchronen Betrieb von Wuchtmassenerregern, Industrie-Anzeiger **89**, 86 (1967) 1889-1892.

[253] Josselin de Jong, *Statics and kinematics in the failable zone of a granular material* (Waltman, Delft, 1959) (in English).

[254] Kadymbekov Z., Earthquakes to order, *Newspaper "Pravda"*, 157(26605) (July 2, 1991) (in Russian).

[255] Kalichuk A. K, Elementary way of studing the dynamic properties of systems, *Journ. of Techn. Phys.*, **4**, 8 (1939) 687–696 (in Russian).

[256] Kanapenas R.-M, *Vibro-supports* (Mosklas, Vilnus, 1984), p. 208 (in Russian).

[257] Kapitsa P. L, Pendulum with vibrating axis of suspension, *Uspekhi fizicheskich nauk*, **44**, 1 (1954) 7–20 (in Russian).

[258] Kapitsa P. L, Dynamic stability of the pendulum when the point of suspension is oscillating, *Jour. of Exper. and Theor. Physics*, **21**, 5 (1951) 588–597 (in Russian).

[259] Karamzin V. D, *Technology and the application of a vibrational layer* (Naukova Dumka, Kiev, 1977) (in Russian).

[260] Karnaukhov V. G., I. K. Senchenko, and B. P. Gimenyuk, *Thermo-mechanical behavior of viscous-elastic bodies under harmonic loading* (Naukova Dumka, Kiev, 1985), p. 288 (in Russian).

[261] Kartynov I. N., *Processing the details of machines by means of free abrasives in vibrating reservoirs* (Vusshaya Shkola, Kiev, 1975) (in Russian).

[262] Katsman J. M, Determination of mechanical parameters of stationary regime of work of the cone inertial crushers, *Obogaschenie rud*, 6 (1984) 34–38 (in Russian).

[263] Keller I. O and E. L. Taruzin, Control of stability of the convective equilibrium of fluid heated from below, *Izv. AN SSSR. Mechanics of fluid and gas*, 4 (1990) 6–11 (in Russian).

[264] Khentov A. A., On one criterion of selection of stable resonance orbital-rotating motions of celestial bodies, *Astronomy Journ.* **73**, 2 (1996) 331-336 (in Russian).

[265] Khodzhaev K. Sh, Synchronization of mechanical vibroexciters connected with the linear oscillatory system, *Izv. AN SSSR. MTT*, 4 (1967) 14–24 (in Russian).

[266] Khodzhaev K. Sh, Integral criterion of stability for systems with quasi-cyclic coordinates and energy relations under the oscillations of current-carrying conductors, *App. Math. and Mech.*, **33**, 1 (1969) 76–93 (in Russian).

[267] Khodzhaev K. Sh, Oscillations of nonlinear electric mechanical systems, in *Handbook "Vibrations in engineering"*, **2** (Mashinostroenie, Moscow, 1979) 331–347 (in Russian).

[268] Kizevalter B. V., *Theoretical basis of gravitational methods of dressing* (Nedra, Moscow, 1979), p. 295 (in Russian).

[269] Kirgetov A. V., On stability of the quasiequilibrium positions of the Chelomey pendulum *Izv. AN SSSR. MTT*, 6 (1986) 57–62 (in Russian).

[270] Kirgetov A. V., *On certain problems of reducibility and stability*, PhD thesis (Moscow State Univ., Moscow, 1989), p. 135 (in Russian).

[271] Klimontovich Yu. L., *Turbulent motion and structure of the chaos: new approach to the statistic theory of the open systems* (Nauka, Moscow, 1990), p. 316 (in Russian).

[272] Kluchininkas A. Yu., The investigation of a vibro-optical method when measuring the parameters of aerosol particles, *Problems of control and protection of the atmosphere from polutions*, 9 (Naukova Dumka, Kiev, 1983) 78–80 (in Russian).

[273] Kobrinsky A. E., *The mechanisms with elastic constraints* (Nauka, Moscow, 1964), p. 390 (in Russian).

[274] Kobrinsky A. E. and A. A. Kobrinsky, *Vibro-shock systems (dynamics and stability)* (Nauka, Moscow, 1973), p. 591 (in Russian).

[275] Kobrinsky A. A. and A. E. Kobrinsky, *Two-dimensional vibro-shock systems* (Nauka, Moscow, 1981), p. 335 (in Russian).

[276] Kogan E. A., Transportation and separation of the material on a rough plane performing longitudinal- transverse elliptical oscillations, *Abstracts of the All Union Conference on Vibrational Engineering* (Kutaisi-Tbilisi, 1981), p. 207 (in Russian).

[277] Kogan E. A., *Regularities of the motion of a particle on a rough plane performing longitudinal- transverse elliptical oscillations and their use in the theory of vibrotransportation and vibroseparation*, PhD thesis (Leningrad, 1984), p. 224 (in Russian).

[278] Kokshaisky N. V., Essay of biological aero- and hydrodynamics, *Flying and swimming of animals* (Nauka, Moscow, 1974), p. 255 (in Russian).

[279] Kolovsky M. Z. and A. A. Pervozvansky, On linearization according to distribution function in problems on the theory of nonlinear oscillations, *Mashinostroenie*, 2 (1962) (in Russian).

[280] Kolovsky M. Z., On the effect of high frequency perturbations on resonance oscillations in nonlinear system, *Proceedings of the Leningrad Polytechnical Institute*, 226 (1963) 7–17 (in Russian).

[281] Kolovsky M. Z., *Nonlinear theory of vibro-protective systems* (Nauka, Moscow, 1966), p. 317 (in Russian).

[282] Kolovsky M. Z., A. D. Sablin, and Z. V. Troitskaya, Oscillations of nonlinear systems with variable and random parameters, *Izv. AN SSSR. MTT*, 4 (1971) 22–29 (in Russian).

[283] Kolovsky M. Z. and Z.A. Sumacheva, On the change of dynamic characteristics of a mechanical system under the effect of vibration, *Electronic engineering*, series 8, Issue 3 (1975) (in Russian).

[284] Kolovsky M. Z., Investigation of the dynamics of the stationary motion of the machine aggregate with an elastic transmitting mechanism, *Izv. AN SSSR. Mashinovedenie*, 2 (1985) 40–47 (in Russian).

[285] Kolovsky M. Z. and A. V. Slousch, *Fundamentals of dynamics of industrial robots* (Nauka, Moscow, 1988), p. 239 (in Russian).

[286] Kolovsky M. Z., Certain nonlinear problems of the dynamics of mechanisms, *Abstracts of the All Union Conference " Nonlinear Oscillations of Mechanical Systems"*, part I (Gorky, 1990) 6–8 (in Russian).

[287] Kondakov V. E., On vibrational fluidization of concrete mix at the state of plane stress, *Izv. Vuzov. Civil Engineering and Architecture*, 4 (1984) 55–60 (in Russian).

[288] Kononenko V. O., *Oscillatory systems with limited excitation* (Nauka, Moscow, 1964), p. 254 (in Russian).

[289] Korotkin A. I., The investigation of hydrodynamic characteristic of vibropump VNL-1, *Krylov's Main Research Institute, Research and Technical Report*, issue 35288 (1993) (in Russian).

[290] Kosolapov A. N., Adaptive property of an oscillatory system with selfsynchronizing vibroexciters, *DAN SSSR*, **309**, 2 (1989) 293–296 (in Russian).

[291] Koshlyakov V. N., A two-rotor pendulum gyrocompass on a vibrating base, *Proceedings* (Mathematical Institute of AN USSR, Kiev, 1988) 47–60 (in Russian).

[292] Kozlov V. V., Averaging in the vicinity of stable periodic motions, *DAN SSSR*, **264**, 3 (1982) 567–570 (in Russian).

[293] Kozlov L. F., *Theoretical bio-hydrodynamics* (Vischa Schkola, Kiev, 1983), p. 238 (in Russian).

[294] Krasnov A. A., Segregation of granular material under share strain of a layer, *Investigation of processes, machines, and apparatuses for classifying material according to size* (Interdepartmental Proceedings, Leningrad (Mekhanobr), 1988) 50–64 (in Russian).

[295] Krasovsky A. A., On a vibrational method of linearization of certain nonlinear systems, *Avtomatika i Telemekhanika*, **9**, 1 (1948) 20–29 (in Russian).

[296] Kremer E. B., V. F. Palilov, and E. B. Shifrina, The behavior of granular material in vibrating communicating vessels, *Abstracts of the All Union Conference on Vibrational Engineering* (Telavi, Institute of Mechanics of Machines of Ac. of Sci. of GSSR, 1984), p. 34 (in Russian).

[297] Kremer E. B. and A. Ya. Fidline, One-dimensional dynamic continual model of granular material *DAN SSSR*, **309**, 4 (1989) 801–804 (in Russian).

[298] Kroll W., Über das Verhalten von Scluttgut in lotrecht schwingenden Gefässen, *Forschung aufdem Gebiete des Ingenieurwesens*, 1 (1954) 2–15 (in German).

[299] Kroll W., Fliesserscheinungen auf Haufwerken in schwingenden Gefässen, *Chemic Ingenieur Technik*, 1 (1955) (in German).

[300] Krylov V. and S. V. Sorokin, Dynamics of elastic beams with controlled distributed stiffness parameters, *Smart Materials and Structures*, 6 (1997) 573–582.

[301] Kryukov B. I., *Forced oscillations of essentially nonlinear systems* (Mashinostroenie, Moscow, 1984), p. 216 (in Russian).

[302] Kumabe D., *Vibrational cutting* (Dzikke Sjuppan,Tokio, 1979), p. 410.

[303] Kunin I. A. and V. F. Khon, On the theory of interaction of a vibroexciter with absorbing liquid medium, *Izv. SO AN SSSR. OTN*, 1 (1961) in Russian).

[304] Kumpikas A. L. and K. M. Ragulskis, On regions of the capture of synchronous rotations of a pendulum with a vibrating axis of suspension, *Vibrotekhnika*, 2(2) (Mintis, Vilnus, 1966) (in Russian).

[305] Kurbatov A. M., S. V. Chelomey, and A. V. Khromushkin, On the problem of the Chelomey pendulum, *Izv. AN SSSR. MTT*, 6 (1986) 63–65 (in Russian).

[306] Kurilo R. E. and V. L. Ragulskene, *Two-dimensional vibrational drives* (Mokslas, Vilnus, 1986), p. 134 (in Russian).

[307] Kuzmina L. K., Methods of the theory of stability in singularly perturbated problems of mechanics, *Abstracts of the All Union Congress on Theoretical and of Applied Mechanics* (Moscow, 1991), p. 215 (in Russian).

[308] Landa P. S., *Self-oscillations in systems with a finite number of degrees of freedom* (Nauka, Moscow, 1980), p. 359 (in Russian).

[309] Landa P. S., *Self-oscillations in distributed systems* (Nauka, Moscow, 1983), p. 320 (in Russian).

[310] Landa P. S., *Nonlinear oscillations and waves in dynamical systems* (Klüwer Academic Publishers, Dordrecht, 1996).

[311] Landa P. S., *Nonlinear oscillations and waves*, (Nauka. Fizmatlit, Moscow, 1997), p. 496 (in Russian).

[312] Landau L. D. and E. M. Lifshits, *Mechanics of the Continuum* (Gostehizdat, Moscow, 1954), p. 795 (in Russian).

[313] Landau L. D. and E. M. Lifshits, *Theoretical Physics*, **1** (Mechanics) (Fizmatgiz, Moscow, 1958 — 1st ed.; Nauka, Moscow, 1965 —2nd ed.), p. 206 (in Russian).

[314] Landau L. D. and E. M. Lifshitz, *Mechanics* (Pergamon, Oxford, 1976).

[315] Lavendel E. E., *Synthesis of optimal vibromachines* (Zinatne, Riga, 1970), p. 252 (in Russian).

[316] Lavendel E. E., A. P. Subach, and G. Yu. Poplavsky, Investigations of motion of the model of load under the volume vibrational treatment, *Problems of dynamics and strength*, Issue 20 (Zinatne, Riga, 1970) 5–19 (in Russian).

[317] Lavrent'ev M. A. and N. N. Lavrent'ev, On one principle of creating a thrust force for motion, *Jour. of App. Mech. and Tech. Phys.*, 4 (1962) 3–9 (in Russian).

[318] Lavrinenko V. V., I. A. Kartashov, and V. S. Vishnevsky, *Piezoelectric engines* (Energiya, Moscow, 1980), p. 110 (in Russian).

[319] Lavrov B. P., Spatial problem on synchronization of mechanical vibroexciters, *Izv. AN SSSR. OTN. Mekhanika i Mashinostroenie*, 5 (1961) 58–68 in Russian).

[320] Lavrov B. P., Vibrational machines with selfsynchronized vibroexciters (Design schemes and specific peculiarities of computation), *Proceedings on the theory and application of the phenomenon of selfsynchronization in machines and devices* (Mintis, Vilnus, 1966) 55–63 (in Russian).

[321] Lavrov B. P., New formulation of the integral criterion of stability of synchronous motions of mechanical vibroexciters and its applications, *Vibrational Technology (Proceedings of Sci. and Thechn. Conf.)* (NIIInfstroidorkommunmash, Moscow, 1966) 306–311 (in Russian).

[322] Leonov G. A., V. B. Smirnova and L. Sperling, A frequency-domain criterion for Global Stability of systems with angular coordinates, *Technische Mechanik* **17**, 3 (1997) 223–229.

[323] Lesin A. D., Moderm comminution equipment. Vibrational mills. Review, *Industry of non-metalloferous materials*, seies 7 (VNIIESM, Moscow, 1989), p. 62 (in Russian).

[324] Lesin A. D., *Vibrational machines in chemical technology* (TSINTIKHIM-NEFTEMASH, Moscow, 1968), p. 80 (in Russian).

[325] Levenshtam V. B., The basis of the method of averaging for the problem of thermal vibrational convection, *DAN RAN* **349**, 5 (1996) 621–623 (in Russian).

[326] Levi M. and W. Weckesser, Stabilization of the inverted linearized pendulum by high frequency vibrations, *SIAM Review* **37** (1997) 219–223.

[327] Levin B. V., Floating up of a heavy ball in vibrating sand, *Journ. Pekladnaya Mekhanika i Tekhnicheskaya Phisika*, 3 (1991) 85–87 (in Russian).

[328] Li K. W. and A. C. Marfatia, Stokes' second problem for the cylinder, *Trans. ASME, 1971*, **93**, 2, D93 (in English; Russian translation: Tr. amer. o-va ingenerov-mekhanikov, **93**, 2 (1971)).

[329] Lighthill M. J., Hydrodynamics of aquatic animal propulsion, *Annual Review of Fluid Mech.*, 1 (1969) 413–446 (in English).

[330] Lighthill M. J., *Mathematical biofluid-dynamics* (Philadelphia (Pa.): Soc. for Industr. and Appl. Math., 1975) (in English).

[331] Lim G.H., On the conveying velocity of a vibratory feeder, *Computers and Structures* **62** (1996) 197–203.

[332] Lin'kov R. V. and Yu. M. Urman, Pondermotor interaction of a rotating conductive ball in an alternating non-uniorm magnetic field, *Jour. of Tech. Phys.*, **44**, 11 (1974) 2255–2264 (in Russian).

[333] Lin'kov R. V. and Yu. M. Urman, Fast rotations of conductive gyroscope in non-uniform alternating magnetic field, *Jour. of Tech. Phys.*, **48**, 6 (1978) 1123–1131 (in Russian).

[334] Lin'kov R. V., On instability of conductive bodies suspended in an alternating magnetic field, *Jour. of Tech. Phys.*, **49**, 5 (1979) 1037–1041 (in Russian).

[335] Lipovsky M. I., On one kind of vibrational displacement of granular material, *Izv. AN SSSR. MTT*, 3 (1969) 3–9 (in Russian).

[336] Lishansky G. Ya., Vibropump, *Bull. izobret.*, 41 (1981) (in Russian)

[337] Lishansky G. Ya., Vibropump, *Bull. izobret.*, 13 (1986) (in Russian).

[338] Logvinovich G. V., Hydrodynamics of the swimming of fish, *Izv. SO AN SSSR*, 8 (1973) 3–8 (in Russian).

[339] Loitsyansky L. G. and A. I. Lurie, *Theoretical Mechanics*, Part III (ONTI, Leningrad–Moscow, 1934), p. 624 (in Russian).

[340] Loitsyansky L. G., *Mechanics of Liquid and Gases*, 6th ed. (Nauka, Moscow, 1987), p. 840 (in Russian. English translation: Begell-Haus Inc., NY, Walligford UK, ed. Robert H. Nann, 1995)

[341] Lokot N. M., Slyusar V. M., Synchronization of self-oscillatory circuit with nonlinear parametric coupling, *Mekhanika Tverdogo Tela* **21**, 5 (1986) 44–48.

[342] Long Y.G., K. Nagaya, and H. Niwa, Vibrational conveyance in spatial-curved tubes, *Journal of Vibration and Acoustic* **116** (1994) 38–46.

[343] Lotz R. and S. Crandall S., Prediction and measurement of the proportionality constant in statistical energy analysis of structures, *J. Acoust. Soc. Am.*, **54**, 2 (1973) (in English).

[344] Lukovsky I. A. and A. N. Timokha, Nonlinear dynamics of the gas-liquid interface with a high-frequency acoustic field in gas, *Preprint*, 88.9–88.10 (Mathematical Institute of AN USSR, Kiev, 1988), p. 39 (in Russian).

[345] Lurie A. I., *Operational calculus in applications to problems of mechanics* (ONTI, Leningrad-Moscow, 1938) (in Russian).

[346] Lurie A. I., *Analytical mechanics* (Nauka, Moscow, 1961), p. 824 (in Russian).

[347] Lurie A. I., Certain problems of synchronization, *Proceedings of the 5th International Conf. on Nonlinear Oscillations*, **3** (Mathematical Institute of AN USSR, Kiev, 1970) 440–455 (in Russian).

[348] Lurie K. A., *Applied Optimal Control Theory of Distributed Systems* (Plenum Press, New York, 1993).

[349] Lurie K. A., Effective properties of smart laminates and screening phenomenon, *Journ. Solids Structures*, **34**, 13 (1997) 1633–1643.

[350] Lurie K. A., The problem of effective parameters of a mixture of two isotropic dielectrics distributed in space-time and conservation law for wave impedance in one-dimensional wave propagation, *Proc. Royal Soc.*, A(454) (London, July, 1998) 1767–1779 .

[351] Lyubimov D. V. and A. A. Cherepanov, Nonlinear stability of liquid-liquid interface in high-frequency fields, *Abstracts of the 7th All Union Cong. on Theor. and App. Mechanics* (Moscow, 1991), p. 234 (in Russian).

[352] Lyul'ka V. A., Hydrodynamical resistance of a periodic structure. Microstructure of the bird's feather, *Abstracts of the 7th All Union Cong. on Theor. and App. Mechanics* (Moscow, 1991), p. 234 (in Russian).

[353] Lyapunov A. M., *General problem on stability of motion* (Gostekhizdat, Moscow–Leningrad, 1950), p. 472 (in Russian).

[354] Machabeli L. I., On the motion of a disc with two pendulums *Izv. AN SSSR. Mekhanika*, 2 (1965) 64–68 (in Russian).

[355] Machiavelli N., *The history of Florence* (First edition — Rome, 1532; the first Russian edition — Nauka, Leningrad, 1973), p. 447.

[356] Machihina L. I., Theoretical pre-quisite to the analysis of the process in stone-selection machines with a vibro-pneumatic principle of operation , *Proceedings of VNIIZ*, issue 60 (Moscow, 1967) (in Russian).

[357] Machihina L. I., *Scientific foundations of rice separation during its industrial processing into grain*, Doctor's dissertation (Moscow, 1989) (in Russian).

[358] Magnus K., *Schwingungen* (Teubner-Verlagsgesellschaft, 1961, Stuttgart), p. 303 (in German).

[359] Mahrenholtz O., R. Strewinski, Zum Synchronisiervorgang angetriebener, rotierender Unwuchterreger mit mechanisher Kopplung, *Zumm* **50** (1970) T237–T239.

[360] Majewski T., Synchronous vibration eliminator for an object having one degree of freedom, *Journal of sound and vibration*, **112**, 3 (1987).

[361] Malakhova O. Z., *Extremal signs of the stability of motion and their use in the creating vibrational devices*, PhD thesis (Leningrad (Mekhanobr), 1990), p. 154 (in Russian).

[362] Malakhova O. Z., On a special case in the theory of self-synchronization of mechanical vibro-exciters, *Izv. AN SSSR. MTT*, 1 (1990) 29–36 (in Russian).

[363] Malkin I. G., *Certain problems of the theory of nonlinear oscillations* (Gostekhizdat, Moscow, 1956), p. 491 (in Russian).

[364] Marinichev B. M. The formation and development of the theory of vibrational processes and devices in the USSR, *Investigations on the history of physics and mechanics* (Nauka, Moscow, 1989) 234–257 (in Russian).

[365] Markert R., H. Pfützner, and R. Gasch, Mindenstantriebsmoment zur Resonanzdurchfahrt von unwuchtigen elastischen Rotoren, *Forsch. Ing.- Wes.* **46**, 2 (1980) 33–68.

[366] Markert R., Resonanzdurchfahrt unwuchtiger biegeelastischer Rotoren, *Fortschrittsbericht VDI-Zeitschr. Reihe 11* **11**, 34 (VDIVerlag, Düsseldorf, 1989).

[367] Martyshkin V. S. A device for researching dynamic characteristics of building materials, *Dynamical properties of building materials* (TSNIIPS, Moscow, 1960) (in Russian).

[368] Mayer E. W., Fundamentals of potential theory of the jigging process, *The VIIth Intern. Mineral. Proc. Congr.* (New York, 1964) (in English).

[369] Mazett R., *Mecanique vibratoire* (Paris, Liege: Libr. politechn. ch. Beranger, 1955), p. 280 (in French).

[370] Menyailov A. I. and A. V. Movchan, On Stabilization of the system pendulum-ring with a vibrating base, *Izv. AN SSSR. MTT*, 6 (1984) 35–40 (in Russian).

[371] Merkin D. R., *Gyroscopic systems* (Nauka, Moscow, 1974, ed. 2), p. 344 (in Russian).

[372] Merkin D. R., *Introduction to the theory of the stability of motion* (Nauka, Moscow, 1976, ed. 3), p. 319 (in Russian).

[373] Merten F., Untersuchungen zum Sommerfeldeffekt mittels Simulation und Experiment, *Preprint, Fakultä für Maschinenbau, Otte- von-Guericke Universität*, 6 (1995).

[374] Merten F. and L. Sperling, Numerische Untersuchungen zur Selbstsynchronisation von Unwuchtrotoren, *Technische Mechanik*, **16**, 3 (1996) 209–220.

[375] Merten F. and L. Sperling, Selbsrsynchronisation von Unwuchtrotoren auf Mehhrkörpersystemen, *GAMM, Wiss. Jahrestagung, 24–27 März, Regensburg* **78**, supplement 2 (1998) 617–618.

[376] Merten F. W. , Untersuchungen zur Synchronisation unwuchtiger Rotoren, *Dissertation, TH Magdeburg, 1999.*

[377] Michalczyk J., Off-axial Synchronous Elimination of Vibrations and Forces in Unbalanced Rotary Machines, *Journal of Theoretical and Applied Mechanics* **31**, 2 (1993) 279–293.

[378] Michalczyk J. and G. Cieplok, Generalized problem of synchronous elimination, *Machine Vibration* **2** (1993) 229–242.

[379] Michalczyk J. and G. Cieplok, Technical system for synchronous elimination of vibrations and forces in unbalanced rotary machines with inaccesible axis, *Proc. of 2nd ENOC, Prague* (1996) 293–296.

[380] Miklaszewski R., On possibility of self-synchronization of rotating eccentric vibro-exciters, *Nonlin. Vibr. Probl.*, 4 (1962) (in English).

[381] Miller M. A., Motion of charged particles in high-frequency electromagnetic fields, *Izv. Vuzov. Radiophysics*, **1**, 3 (1958) 110–123 (in Russian).

[382] Miranda E.C. and J. J. Thomsen, Vibration induced sliding: theory and experiments for a beam with a spring-loaded mass, *Nonlinear Dynamics* **16** (1998) 167–186.

[383] Mirollo R. E. and S. H. Strogatz, Synchronization of pulse-coupled biological oscillators, *SIAM J. Appl. Math.* **50**, 6 (1990) 1645–1662.

[384] Miroshnik I. V. and A. V. Ushakov, Synthesis of the algorithm of synchronous control of system with quasi-uniform objects, *Avtomatika i Telemekhanika*, 11 (1977) 22-29 (in Russian). (1993) 279–293.

[385] Mitropolsky Yu. A., *Problems of asymptotic theory of nonlinear nonstationary oscillations* (Nauka, Moscow, 1964), p. 431 (in Russian).

[386] Mitropolsky Yu. A., *Method of averaging in nonlinear mechanics* (Naukova Dumka, Kiev, 1971), p. 440 (in Russian).

[387] Mitropolsky Yu. A. and O. B. Lykova, *Integral manifolds in nonlinear mechanics* (Nauka, Moscow, 1973), p. 512 (in Russian).

[388] Mitropolsky Yu. A. and A.K. Lopatin, *Theoretical-group approach in asymptotic methods of nonlinear mechanics* (Naukova Dumka, Kiev, 1988), p. 271 (in Russian).

[389] Mitulis A. A., The character of stationary motion of a mathematical pendulum with a vibrating point of suspension, depending on the choice of the initial conditions, *Proceedings on the theory and application of phenomenon of synchronization in machines and devices* (Mintis, Vilnus, 1966) 131–135 (in Russian).

[390] Moiseev N. N., *Asymptotic methods of nonlinear mechanics* (Nauka, Moscow, 1969), p. 379 (in Russian).

[391] Molasyan S. A., The velocity of motion of a two mass system along a rough plane under the effect of periodic pulses, *Izv. AN SSSR. Mashinovedenie*, 1 (1985) 14–17 (in Russian).

[392] Molchanov A. M., On the resonance structure of the Solar system, *Modern problems of celestial mechanics and astrodynamics* (Nauka, Moscow, 1973) 32–40 (in Russian).

[393] Moser J., in *Memories of the American Mathematical Society*, Lectures on Hamiltonian systems, 81 (Courant Institute of Mathematical Science, NY, 1968) 1–60.

[394] Nabiullin M. K., *Stationary motions and stability of elastic satellites* (Nauka. Sib. otd-e, Novosibirsk, 1990), p. 216 (in Russian).

488   *BIBLIOGRAPHY*

[395] Nagaev R. F., General problem on synchronization in an almost conservative system, *App. Math. and Mech.*, **29**, 5 (1965) 953–963 (in Russian).

[396] Nagaev R. F. and I. A. Popova, Selfsynchronization of several mechanical vibroexciters installed on the same working unit of beam type, *Engineering Journal. MTT*, 1 (1967) 29–37 (in Russian).

[397] Nagaev R. F. and K. Sh. Khodzhaev, Synchronous motions in a system of objects with carrying constraints, *App. Math. and Mech.*, **31**, 4 (1967) 631–642 (in Russian).

[398] Nagaev R. F., The case of a generating set of quasi-periodic solutions in the theory of small parameter, *App. Math. and Mech.*, **37**, 6 (1973) 990–998 (in Russian).

[399] Nagaev R. F. and V. Ya. Turkin, Synchronous regime of the operation of shock-vibrational jaw crusher, *Obogaschenie rud*, 2 (1973) 15–16 (in Russian).

[400] Nagaev R. F., *Periodic regimes of vibrational displacement* (Nauka, Moscow, 1978), p. 160 (in Russian).

[401] Nagaev R. F. and E. A. Tamm, Vibrational displacement in a medium with quadratic resistance, *Mashinovedenie*, 4 (1980) 3–8 (in Russian).

[402] Nagaev R. F., *Mechanical processes with repeated damping collisions* (Nauka, Moscow, 1985), p. 200 (in Russian).

[403] Nagaev R. F. and V. V. Guzev, *Self-synchronization of inertial vibroexciters*, ed. K. M. Ragul'skis (Mashinostroenie, Leningrad, 1990), p. 178 (in Russian).

[404] Nagaev R. F., *Quasiconservative synchronizating systems*, (Nauka, St.Petersburg, 1996) (in Russian).

[405] Nayfeh A. H., *Introduction to perturbation techniques*, (John Wiley and Sons, Inc., New York-Chichester, Brisbane, Toronto, 1981).

[406] Nayfeh A. H. and B. Balachandran, *Applied Nonlinear Dynamics: Analytical, Computational, and Experimental Methods* (Wiley, New York, 1995).

[407] Nayfeh A. H. and D. T. Mook, *Nonlinear Oscillations* (Wiley, New York, 1979).

[408] Neimark Yu. I., Theory of vibrational sinking and vibro-extracting, *Ingenerny sbornik*, **XVI** (1953) 13–48 (in Russian).

[409] Neimark Yu. I., N. Ya. Kogan, and V. P. Savel'ev, *Dynamic models of the theory of control* (Nauka, Moscow, 1985), p. 399 (in Russian).

[410] Neimark Yu. I. and P. S. Landa, *Stochastic and Chaotic oscillations* (Nauka, Moscow, 1987), p. 422 (in Russian).

[411] Nepomnyaschy E. A., Mathematical description of the kinetics of granular material separation, *Proceedings of VNIIZ*, Issues 61-62 (Moscow, 1967) (in Russian).

[412] Nigmatullin R. I., *Dynamics of multiphase media*, Part I, p. 464, and Part II, p. 359 (Nauka, Moscow, 1987) (in Russian).

[413] Novozhilov V. V., Two articles on mathematical models in the continuum mechanics, *Preprint*, 215 (IPM AN SSSR, Moscow, 1983), p. 56 (in Russian).

[414] Novozhilov I. V., On the correctness of a number of limiting models in mechanics, *Preprint*, 253 (IPM AN SSSR, Energy In-te, Moscow, 1985) (in Russian).

[415] Novozhilov I. V., The limiting model of a system with elastic elements of great stiffness, *Izv. AN SSSR. MTT*, 4 (1988) 24–28 (in Russian).

[416] Olevsky V. A., Parameters of the regime and capacity of screens, *Obogoschenie rud*, 3 (1967) 31–37 (in Russian).

[417] Opirsky B. Ya. and P. D. Denisov, *Modern vibrational machines. Designs and computation* (Svit, L'vov, 1991), p. 160 (in Russian).

[418] Orlov A. L., Wave models of nonlinear dependences of dry friction forces on the velocity of sliding, *Abstracts of the IInd All Union Conf. "Nonlinear Oscillations of Mechanical Systems"*, Part I (Gorky, 1990), p. 180 (in Russian).

[419] Otterbein S., Stabilisierung des n-pendels und der indische seiltrick, *Archive for Rational Mechanics and Analysis* **78** (1982) 381–393.

[420] Ovchinnikov P. F., *Vibrorheology* (Naukova Dumka, Kiev, 1983), p. 271 (in Russian).

[421] Ovenden M. W., T. Feagin, and O. Graff, On the principle of least inter-action action and the Laplacean satellites of Jupiter and Uranus, *Celestial Mechanics*, **8**, 3 (1974) 455–471 (in English).

[422] Paidoussis M. P. and G. X. Li, Pipes conveying fluid: a model dynamical problem, *Journal of Fluids and Structures* **7** (1993) 137–204.

[423] Pajer G., H. Kuhnt and F.Kurth, *SynchronizFördertechnik, Stetigförderer*, (Verlag Technik, Berlin, 1988).

[424] Pal'mov V. A., Description of a high-frequency vibration of complex dy-namic systems by methods of heat conduction, *Collection of articles de-voted to the 60th anniversary of academician V. N. Chelomey: Selected problems of applied mechanics* (VINITI, Moscow, 1974) 535–542 (in Rus-sian).

[425] Pal'mov V. A., *Vibrations of elasto-plastic bodies* (Springer, Berlin, Hei-delberg, 1998), p. 311.

[426] Panovko Ya. G. and I. I. Gubanova, *Stability and oscillations of elas-tic systems. Modern conceptions, paradoxes and errors* (Nauka, Moscow, 1979), p. 384 (in Russian).

[427] Panovko Ya. G., *Introduction to the theory of mechanical oscillations* (Nauka, Moscow, 1979), p. 270 (in Russian).

[428] Panovko Ya. G., *Mechanics of deformable solid body* (Nauka, Moscow, 1985), p. 287 (in Russian).

[429] Parshin S. V., A. S. Sokolov, and A. G. Tomilin, On the adjustment of elasticity of dolphin flippers by special vascular organs, *DAN SSSR*, 3 (1970) (in Russian).

[430] Patashene L. R., *Elaboration and investigation of vibro-engines with a ring exciter*, PhD thesis (Kaunas, 1979) (in Russian).

[431] Paz M. and J. D. Cole, Self-synchronization of two unbalanced rotors, *J. Vibration and Acoustic* **114**, 1 (1992) 37–41.

[432] Perestyuk N. A., *Oscillatory solutions of differential equations with a pulse effect and their stability*, Doctor's dissertation (Kiev, 1985) (in Rus-sian).

[433] Pervozvansky A. A., *Random processes in nonlinear automatic systems* (Fizmatgiz, Moscow, 1962), p. 351 (in Russian).

[434] Petchenev A. V., On the motion of an oscillating system with limited excitation in the vicinity of resonance, *DAN SSSR*, **290**, 1 (1986) 27–31 (in Russian).

[435] Petchenev A. V., Averaging of systems with a hierarchy of velocities of phase rotation on essentially long time intervals, *DAN SSSR*, **315**, 1 (1990) (in Russian).

[436] Petrova I. M., *Investigations in the field of bionics carried out with the aim of increasing the velocity of ships* (TSNIITEIS, Moscow, 1968) (in Russian).

[437] Petrova I. M., *Hydro-bionics in ship building* (TSNIITEIS, Moscow, 1970) (in Russian).

[438] Pliss D. A. and I. M. Abramovich, Elaboration of the vibrational method of dry classification of ore, *The report of VNII Mekhanobr* (Mekhanobr, Leningrad, 1948), p. 64 (in Russian).

[439] Pliss D. A., On the theory of vibrational separation, *Engineering Journal. MTT*, 4 (1967) (in Russian).

[440] Poduraev V. N., *Machining by cutting with vibrations* (Mashinostroenie, Moscow, 1970), p. 351 (in Russian).

[441] Poincare A., *Les Methods Nouvelles de la Mecanique Celeste, 3 vol.*, **1** (Gauthier-Villars, Paris, 1899; Russian edition: Nauka, Moscow, 1971), p. 771.

[442] Popov E. P. and I. P. Pal'tov, *Approximate methods of investigation of nonlinear automatic systems* (Fizmatgiz, Moscow, 1960), p. 792 (in Russian).

[443] Poturaev V. N., V. P. Franchuk, and A. G. Chervonenko, *Vibrational transporting machines (Fundamentals of the theory and calculation)* (Mashinostroenie, Moscow, 1964), p. 272 (in Russian).

[444] Poturaev V. N., V. P. Ravishin, and A. G. Chervonenko, Method of screening the granular material, Russian patent 761030, *Bull. izobr.*, 33 (1980), p. 29.

[445] Poturaev V. N., A. I. Voloshin, and B. V. Ponomarev, *Vibrational-pneumatic transportation of granular material* (Naukova Dumka, Kiev, 1989), p. 248 (in Russian).

[446] Poturaev V. N., A. G. Chervonenko et al. *Vertical transport at mining enterprises*, ed. V. N. Poturaev, (Nedra, Moscow, 1975), p. 351 (in Russian).

[447] Prigozhin I., *From being to becoming: time and complexity in the physical sciences* (W. H. Freeman and Company, San Francisco, 1980), p. 310

[448] Queck U., Delphin-Luftschiff als Fliegender Kran, *Flieger Revue*, 6 (1974) (in German).

[449] Ragulskis K. M, Mechanisms on a vibrating foundation, *Problems of dynamics and stability* (In-te of Energetics and Electrical Engineering of AN Lietuva SSR, Kaunas, 1963), p. 232 (in Russian).

[450] Ragulskis K. M, I. I . Vitkus, and V.L. Ragulskene, Self-synchronization of mechanical systems, *Self-synchronous and vibro-shock systems* (Mintis, Vilnus, 1965), p. 186 (in Russian).

[451] Ragulskis K. M. and V. V. Naginyavichus, A tube-shaped vibro-valve controlled by oscillations of the tube as an elastic body, *Dep. v Lit. NI-INTI*, 1644 (Vilnus, 1986) (in Russian).

[452] Ragulskis L. K. and K. M Ragulskis, *Oscillatory systems with a dynamically directed vibro-exciter* (Mashinostroenie, Leningrad, 1987), p. 130 (in Russian).

[453] Raskin H. I., The application of methods of physical kinetics to the problems of a vibrational effect on granular media, *DAN SSSR*, **220**, 1 (1975) 54–57 (in Russian).

[454] Rebrik B. M., *Drilling the wells during geologic-engineering investigations* (Nedra, Moscow, 1979), p. 253 (in Russian).

[455] Regirer S. A., About the motion of viscous liquid in the pipe with deforming wall, *Izv. AN SSSR, Fluid and Gas Mechanics* **4** (1968) (in Russian).

[456] Revnivtsev V. I, G. A . Denisov, and L.P. Zarogatsky, Cone inertial crushers KID-300 for disintegration of the granules and cutting the chips of high-speed steel, in *Science and technology achievements. Interindustry*

science and technology collection of publications (Moscow, 1984) 77–81 (in Russian).

[457] Revnivtsev V. I., G. A. Finkelstein, L. P. Zarogatsky, I. I. Blekhman, N. A. Ivanov, Selective liberation of minerals in intertial cone crushers, *Canada Power Technology*, **38** (1984) 195–203 (in English).

[458] Rocard Y., *Dinamique generale des vibrations* (Masson, Paris, 1949), p. 439 (in French).

[459] Romanenko E. V., *Theory of the swimming of fishes and dolphins* (Nauka, Moscow, 1986), p. 152 (in Russian).

[460] Romanovsky Yu. M., N. V. Stepanova, and D. S. Chernavsky *Mathematical simulation in biophysics* (Nauka, Moscow, 1975), p. 343 (in Russian).

[461] Roth W., Eine Theorie über die Schwinimbewegung von Fischen, *Acta Mechanica*, **20**, 4 (1974) 285–301 (in German).

[462] Rozenvasser E. N., *Periodically-nonstationary systems of control* (Nauka, Moscow, 1973), p. 511 (in Russian).

[463] Rubanovsky V. N, Stability of stationary motions of complex mechanical systems, in *The results of science and techniques. General mechanics* (VINITI, Moscow, 1982) (in Russian).

[464] Rudin A. D., Screen for the difficulty screening ores, *Obogaschenie rud*, 3 (Institute "Mekhanobr", Leningrad, 1968) (in Russian).

[465] Rumyantsev V. V, On dynamics of a solid body suspended on a string, *Izv. AN SSSR. MTT*, 4 (1983) 5–16 (in Russian).

[466] Rumyantsev V. V, A. S. Oziraner, *Stability and stabilization of motion relative to some part of variables* (Nauka, Fizmatlit, Moscow, 1987), p. 253 (in Russian).

[467] Rundkvist A. K, Mechanics of inertial crusher of the Mekhanobr Institute, in *Mechanics and computation of the machines of vibrational type* (Izd-vo AN SSSR, Moscow, 1957) (in Russian).

[468] Rundkvist A. K, I. I. Blekhman, and A. D. Rudin, On the theory of a critical slot-opening of inertial crushing and grinding machines, *Obogaschenie rud*, 2 (1961) 37–41 (in Russian).

[469] Ryzhkov A. F., *Hydrodynamics and mass-heat conduction in vibrofluidized disperse systems*, Doctor's dissertation (AN SSSR, Thermal Physic Ins-te, Novosibirsk, 1991) (in Russian).

[470] Ryspekov K. M., *Investigation of operational reliability and the improvement of maintenance service of grinding equipment (Dzhezkazgan GMK dressing plant as an example*, Summary of PhD thesis (Leningrad, 1982) (in Russian).

[471] Samoilenko A. M., *Differential equations with a pulsed effect* (Vischa Shkola, Kiev, 1987), p. 288 (in Russian).

[472] Sanders J. A. and F. Verhulst, *Averaging methods in nonlinear dynamical systems* (Springer-Verlag, New York, 1985).

[473] Savinov O. A. and A. Ya. Luskin, *Vibrational method of driving-in the piles and its application in building* (Gosstroiizdat, Moscow-Leningrad, 1960), p. 251 (in Russian).

[474] Savinov O. A., *Modern designs of foundations for machines and their calculation* (Stroiizdat, Leningrad, 1979), p. 200 (in Russian).

[475] Savinov O. A. and E. V. Lavrinovich, *Vibrational technology of seal-shaping of concrete mix* (Stroiizdat, Leningrad, 1986), p. 278 (in Russian).

[476] Scharton T. and R. Lyon, Power flow and energy sharing in random vibration, *J. Acoust. Soc. Am.*, **43**, 6 (1968) 1332–1343 (in English).

[477] Schigel V. A. and A. S. Grinbaum, Regimes of tossing the particle on a harmonically oscillating plane, *AN SSSR. Mashinovedenie*, 6 (1974) 17–21 (in Russian).

[478] Schmidt B. A., Vibrated pendulum with a mass free to move radially, *J. of applied mech.*, **47**, 2 (1980) 420–430 (in English).

[479] Schmidt W., Delphinluftschiff mit Wellanirieb-Wirkung Hastischer Wellerblatter, *TIZL.*, **10**, 4 (1974) (in German)

[480] Schmidt P. and P. Peltzer, Das Synchronisieren zweier Unwuchtrüttler an Schwingmaschinen *Aufbereitungstechnik* **3**, 1 (1976) 108–114.

[481] Schmitt J. M. and P. V. Bayly, Bifurcations in the mean angle of a horizantally shaken pendulum analysis and experiment, *Nonlinear Dynamics* **15** (1998) 1–14.

[482] Sedov L. I., *Thoughts about scientists and science of the past and the present. Collection of articles* (Nauka, Moscow, 1973), p. 119 (in Russian).

[483] Sedov L. I., *Theoretical models: introduction and a list of publications with annotations by L. I. Sedov and his colleagues on designing models of continuum* (Moscow, 1974; Skople, 1975), p. 219 (in Russian).

[484] Sergeev P. A., I. I. Blekhman, Investigation of the dynamics of centrifuges of types ATS-5 and TSA-5, *Proceedings of VNIIKommunmash*, 2 (Leningrad, 1970) (in Russian).

[485] Sergeev P. A., V. N. Shishatsky, N. A. Shirokov, and Sh. G. Azersky, The hinge arrangement of the support and drive of a drum of a laundry centrifuge, Russian patent 308128, *Bull. izobr.*, 21 (1971), p. 105.

[486] Sergeev P. A., V. V. Ermakov, V. A. Prohorov, V. P. Timofeev, and V. N. Shishatsky, The suspension of a centrifuge for centrifuging the loundry, Russian patent 653319, *Bull. izobr.*, 11 (1979), p. 97.

[487] Severdenko V. P., V. V. Klubovich, and A. V. Stepanenko, *Ultrasound and plasticity* (Nauka and Tekhnika, Minsk, 1976), p. 446 (in Russian).

[488] Shapiro B. and B. T. Zinn, High-frequency nonlinear vibrational control, *IEEE Transactions on Automatic Control* **42** (1997) 83–89.

[489] Sharp R. S., An analysis of a self-balancing system for rigid rotors, *Journ. Mech. Eng. Sci.* **17**, 4 (1975) 186–189.

[490] Shashkov I. P., Vibrational conveyer, The author's certificate for the Russian Invention 146235, *Bull. izobr.*, 7 (1962) 76–77.

[491] Shatalov S. D., *Separation of motions in electromechanical systems*, PhD thesis (Leningrad, 1965), p. 160 (in Russian).

[492] Shekhter O. Ya., On one example of subharmonic oscillations, *Proceedings of the meeting on applying vibrations in building the foundations and in drilling for the purposes of building*, 2 (NTO Stroitel'noy Industrii SSSR, Leningrad, 1959), p. 10 (in Russian).

[493] Shinkin V. N, On searching for the stable resonance regimes by their extremal properties, *Vesti MGU, series Vuchisl. Matematika i Kibernetika*, 2 (1981) 22–29 (in Russian).

[494] Shlikhting G., *The theory of a boundary layer* (Nauka, Moscow, 1974), p. 711 (in Russian).

[495] Shnol' E. E., On synchronization of biochemical oscillations in cells interacting through the surrounding medium, *Preprint* (Science Center of biological investigations of AN SSSR, Puschino, 1985), p. 20 (in Russian).

[496] Shubin I. K, The effect of vibration on a two-mode course pointer, *Applied Mathematics. Interinstitutes collection*, 71 (Leningr. Institute of Engineering and Building, 1979) 105–110 (in Russian).

[497] Signeul R. A. O., Apparat for behandling av et objekt medel riktade vlbrationer, Sverige, patent No. 163270, Kl. 80a: 49, *Patenitid fran den 6 August 1946, pablicerat den 13 Maj 1958* (in Swedish).

[498] Signeul R. A. 0., Vibrating device for a directed vibratory effect by means of rotatable vibratory members, *United States Patent Office*, No. 2, 531, 706, Application May 251948, in Sweden August 6, 1946, Patented Nov. 28 1950 (in English).

[499] Sinel'nikov A. E., Shifts of a pendulum on a vibrating foundation under elliptic vibration, *Izv. AN SSSR. Mekhanika*, 6 (1965) 130–131 (in Russian).

[500] Sliede P. B., Layer-by-layer continuous motion of granular material along the vibro-tray at big friction coefficients, *Problems of dynamics and strength*, issue 23 (Zinatne, Riga, 1972) 69–77 (in Russian).

[501] Sohiele K., On the stabilization of a parametrically driven inverted double pendulum, *Zeitschrift für Angewandte Mathematik und Mechanik* **77** (1997) 143–146.

[502] Sommerfeld A., Beitrage zum dynamischen ausbau der festigkeitslehre, *Zeitschr VDI* , **XXXXVI**, 11 (1902) 391–394 (in German).

[503] Soroko V. V., Works of the Hypronickel Institute in the field of vibro-loading, vibro-bunkering, and vibro-discharge of bulk cargo, in *Collection of papers on vibro-loading machines, vibro-bunkering and vibro-discharge of bulk cargo*, (TSNIITEI uglya, Moscow, 1963) (in Russian).

[504] Sorokodum E. D., On the averaged force acting on an umbrella-like body oscillating in liquid, *Acoustic Jour.*, **27**, 5 (1981) 793–794 (in Russian).

[554] Vaisberg L. A., *Design and calculation of vibrational screens* (Nedra, Moscow, 1986), p. 144 (in Russian).

[555] Valeev K. G. and R. F. Ganiev, Investigations of oscillations of nonlinear systems, *PMM*, **33**, 3 (1969) 401–418 (in Russian).

[556] Valeev K. G., Dynamic stabilization of unstable systems *Izv. AN SSSR, MTT*, 4 (1971) 13–21 (in Russian).

[557] Valeev K. G. and V. V. Dolya, On dynamic stabilization of pendulum oscillations, *Appl. Mechanics*, **10**, 2 (1974) 88–93 (in Russian).

[558] Van der Pol B., Van der Mark M., Le battement du coeur considere comme oscillation de relaxation et un modele electrique de coeur, *L'Onde electrique*, 7 (1928) (in French).

[559] Vasil'ev P. E., *Problems of the theory, calculation and design of parts of vibrational tape-conveying mechanisms*, PhD thesis (Kaunas, 1974) (in Russian).

[560] Vasil'eva L. P., S. N. Gorshkov, A. V. Kozlov, and B. P. Lavrov, The technique of vibrational filling the seamless pipe blank with powder, Russian patent 761208, *Bull. izobret.*, 33 (1980),p. 66.

[561] Veitz V. L., *Dynamics of machine aggregates* (Mashinostroenie, Leningrad, 1969), p. 368 (in Russian).

[562] Veitz V. L. and A. E. Kochura *Dynamics of machine aggregates with internal combustion engines* (Mashinostroenie, Leningrad, 1976), p. 383 (in Russian).

[563] *Vibrational machines in civil engineering and in the production of structural materials*, Handbook, ed. V. A. Bauman, I. I. Bykhovsky, and B. G. Goldshtein (Mashinostroenie, Moscow, 1970), p. 548 (in Russian).

[564] *Vibration in technology*, Handbook in 6 volumes (Mashinostroenie, Moscow, 1978–1981), p. 351 (in Russian).

[565] *Vibrational pumps* (Mashinostroenie, Moscow, 1973) (in Russian).

[566] Vielsack P. and A. Hartung, Orbitaly Stabilität von Bewegungen mit Pausen bei Einwirkung permanenter Störungen, *Universität Karlsruhe, Preprint*, 97/1 (1997).

[567] Vielsack P., Untersuchungen zum Bewegungsverhalten beim Vibrationsrahmen, *Universität Karlsruhe, Insitut für Mechanik, D-76128 Karlsruhe* (1997) 75-78 (in German).

[568] Vinogradov N. N., Statistic method of evaluating the efficiency of stratification the material in jigs, *Dressing and briquetting coal*, 8 (The Main Research In-te of Information and Technique and Economy Investigations of Coal Industry, Moscow, 1963) (in Russian).

[569] Vlasov E. V. and A. S. Ginevsky, Phenomenon of acoustic weakening of turbulence in a subsonic jet, *Discoveries, inventions, industrial models, and trademarks*, 30 (1979) (in Russian).

[570] Vlasov E. V. and A. S. Ginevsky, Coherent structures in turbulent jets and traces, *Results of science and technology. Mechanics of fluid and gas* (VINITI, Moscow, 1986), 20, 3–84 (in Russian).

[571] Volfson S. A., Minkin Yu. G., Energy balance at moving a railway carriage along a rough way, *The increase of service reliability of locomotives in conditions of the Urals and Siberia. Proceedings of the Net Sci. and Tech. Conf.* (In-t Ingenerov Zh.-Dor. transporta, 1973) 139–145 (in Russian).

[572] Volosov V. M. and B. I. Morgunov, *Method of averaging in the theory of nonlinear oscillatory systems* (Moscow State Univ., Moscow, 1971), p. 507 (in Russian).

[573] Voronov S. A., A. M. Gus'kov, and V. A. Svetlitsky, Excitation of lateral self-oscillations in the stalk of the tool under deep drilling, *Strength calculations* (Mashinostroenie, Moscow, 1979), Issue 20, 172–182 (in Russian).

[574] Voronov S. A., Optimization of the process of vibrational drilling, *Proceedings of MVTU im. N. E. Baumana*, Issue 332 (Moscow, 1980) 13–25 (in Russian). ???

[575] Voronov S. A., A. M. Gus'kov, O. S. Naraikin, and G. Ya. Panovko, Vibrational mechanics in technological processes of treatment and assembly, *Abstracts of the VIIth All Union Congress on Theor. and App. Mechanics* (Moscow, 1991), p. 90 (in Russian).

[576] Vorotnikov V. I., *Stability of dynamic systems with respect to some part of variables* (Nauka, Moscow, 1991), p. 284 (in Russian).

[577] Vorovich I. I., Certain problems of the usage of statistic methods in the theory of stability of plates and shells, *Proceedings of the 4th All Union Conference on the Theory of Shells and Plates* (AN ArmSSR, Erevan, 1964) 69–94 (in Russian).

[578] Vorovich I. I., Certain problems of teaching the foundation of classical mechanics in the University course (Institute of Problems of Mechanics of Ac. of Sci., Moscow, 1985), *Preprint*, 252, p. 25 (in Russian).

[579] Vu T. and A. Chuong, Receiving the energy from a wave flow by fishes and birds, *Mechanics*, 23 (1980) (in Russian).

[580] Weibel S., T. J. Kaper, and J. Baillieul, Global dynamics of a rapidly forced cart and pendulum, *Nonlinear Dynamics* **13** (1997), 131–170.

[581] Wolfsteiner P. and F. Pfeifer, Dynamics of a vibratory feeder, *Proceedings of DETC'97, ASME Design Engineering Technical Conferences, September 14-17, Sacramento, California* (1997).

[582] Wu T., Yao-tsu, Hydromechanics of swimming of fishes and cetaceans, *Advances in Appl. Mech.*, **11**, 38 (1971) (in English).

[583] Yakimova K. S., Vibrational displacement of a two-mass oscillatory system, *Izv. AN SSSR. MTT*, 5 (1969) 20–30 (in Russian).

[584] Yakubovich V. A. and V. M. Starzhinsky, *Linear differential equations with periodic coefficients and their applications* (Nauka, Moscow, 1972), p. 718 (in Russian).

[585] Yakubovich V. A. and V. M. Starzhinsky, *Parametric resonance in linear systems* (Nauka, Moscow, 1987), p. 328 (in Russian).

[586] Yudovich V. I., Vibrodynamic of systems with constraints, *DAN*, **354**, 5 (1997), 622–624.

[587] Yudovich V.I. , Dynamics of a material particle on a smooth vibrating surface *Applied Mathematics and Mechanics* **62**, issue 6 (1998) 968–976.

[588] Zaika P. M., *Vibrational grain cleaning machines* (Mashinostroenie, Moscow, 1967), p. 143 (in Russian).

[589] Zaretsky L. B., On self-synchronization of centrifugal vibrational exciters of the vibro-shock mechanism, *Mashinovedenie*, 1 (1967) 40–47 (in Russian).

[590] Zaretsky L. B., Synchronization of centrifugal vibrational exciters in systems with discontinuous characteristics, *Eng. Jour. MTT*, 1 (1968) 18–25 (in Russian).

[591] Zaretsky L. B., Multiple synchronization of centrifugal vibrational exciters, *Izv. AN SSSR. MTT*, 4 (1974) (in Russian).

[592] Zarogatsky L. P., N. A. Ivanov, and G. A. Finkelshtein, Inertial crusher KID-1750 and prospects of the use of crushers of this type, *Obogaschenie rud*, 1 (1982) 18–24 (in Russian).

[593] Zenkovskya S. M. and S. M. Simonenko, The effect of high frequency vibrations upon generation of convection, *Izv. AN SSSR, Fluid and Gas Mechanics* 5 (1966) (in Russian).

[594] Zenkovskya S. M. and S. M. Simonenko, The effect of high frequency vibrations upon filtrating convection, *Appl. Math. and Techn. Phys.*, 5(195) (1992) (in Russian).

[595] Zevin A. A. and L. A. Filonenko, Parametrical oscillations of a pendulum with respect to the upper position of equilibrium, *Izv. AN SSSR. MTT*, 5 (1986) 49–53 (in Russian).

[596] Zevin A. A. and L. A. Filonenko, Stabilization of statically unstable nonlinear systems by means of periodical exciting forces, *Abstracts of the IInd All Union Conf. " Nonlinear Oscillations of Mechanical Systems*, part I (Gorky, 1990), p. 179 (in Russian).

[597] Zhgulev A. S, Trajectory field of a vibrational machine activated by syncronously rotating unbalanced rotors, *Vibrotekhnika*, Issue 4(28) (Mintis, Vilnus, 1979) 69–77 (in Russian).

[598] Zhidkikh S. A., On stability of the boundaries of planetary rings, *Abstracts of the 7th All Union Congress on Theor. and App. Mechanics* (Moscow, 1991) 154–155 (in Russian).

[599] Zhukovsky N. E., Note on a plane screening, *Complete set of works*, 3 (Gostehizdat, Moscow, 1949) 515–522 (in Russian).

[600] Zhuravlev V. F. and A. A. Lapin, Synchronization phenomenon in high-speed gyroscopic supports, *Izv. AN SSSR. MTT*, 4 (1979) 3–10 (in Russian).

[601] Zhuravlev V. F. and D. M. Klimov, *Applied methods in the theory of nonlinear oscillations* (Nauka, Moscow, 1988), p. 325 (in Russian).

Index

[505] Soundalgekar V. M., A. K. Dasgupta, Flow past an oscillating plate in a viscoelastic fluid, *Indian J. Theor. Plys.*, **26**, 2 (1978) (in English).

[506] Sperling L., Selbstsynchronisation unwuchtbehafteter Rotoren in elastischen Ketten, *Internationale Tagung "Verfahren und Gerate der mechanischen Schwingungstechnik"* (Techn. Hochscliile Otto von Guerickc, Magdeburg, 1965), Teil. II (in German).

[507] Sperling L., Beitrag zur allgemeinen Theorie der Selbstsynchronisation umlaufender Unwuchtmassen in Nichtresonanzfall, *Wiss. Zeitschr., Magdeburg: Techn. Hochsdiile Otto von Guericke*, **I**, II (1967) 63–87 (in German).

[508] Sperling L., Selbstsynchronisation statisch und dynamisch unwuchtiger Vibratoren. Teil I:Grundlagen, *Technische Mechanik* **14**, 1 (1994) 61–76.

[509] Sperling L., Selbstsynchronisation statisch und dynamisch unwuchtiger Vibratoren. Teil II: Ausführungen und Beispiele, *Technische Mechanik* **14**, 2 (1994) 85–96.

[510] Sperling L., F. Merten and H.Duckstein, Rotation und Vibration in Beispielen zur Methode der direlten Bewegungsteiung, *Technische Mechanik* **17**, 3 (1997) 231–243.

[511] Sperling L., F. Merten and H. Duckstein, Self-Synchronization and Automatic Balancing in Rotor Dynamics, *ISROMAC-7, 22-26 Februar, Honolulu* **B**, Dynamic II (1998) 785–794.

[512] Sperling L., F. Merten and H. Duckstein, Analytical and Numerical Investigations of Rotation-Vebration-Phenomena, *Vortrag auf der XXVI Summer School "Nonlinear Oscillations in Mechanical Systems"(NOMS'98), 1-10 Juli, Repino(St.Petersburg)*.

[513] Spivakovsky A. O. and I. F. Goncharevich, *Vibrational and wave machines for transportation* (Nauka, Moscow, 1983), p. 288 (in Russian).

[514] Stepanov G. Yu. Why is it impossible to have "Dean's Apparatus"?, *Jour. Priroda*, 7 (1963) 85–91 (in Russian).

[515] Stephenson A. , On a new type of dynamic stability, *Mem. Proc. Manch. Lit. Phil. Soc.* **52** (1908) 1–10.

[516] Stephenson A. , On induced stability, *Phylosophical Magazine* **15** (1908) 233–236.

[517] Stephenson A. , On induced stability, *Phylosophical Magazine* **17** (1909) 765–766.

[518] Stokes G. G., On the effect of the internal friction of fluids on the motion of pendulums, *Trans. Cambr. Phyl.*, **IX** , 8, (1851), P.8.–106; *Math and Phys. Papers*, **III** (1901) (in English).

[519] Storz M., Stabilität der Bewegungen eines Rubschwingers mit Stoß am Beispiel des Vibrationsrammen, *Doktor-Ing.-Dissertation, Univers. Fridericiana zu Karlsruhe* (1994) p.140 (in German).

[520] Strett G. (Lord Rellay) *Theory of sound,* **II** (Gostexizdat, Moscow-Leningrad, 1944), p. 476 (in Russian).

[521] Strizhak T. G., *Methods of investigating dynamic systems of pendulum-type* (Nauka Kaz. SSR, Alma-Ata, 1981), p. 253 (in Russian).

[522] Strizhak T. G., Minimax sign of stability, *Preprint* (Institute of Electrical Dynamics of AN USSR, Kiev, 1981), p. 50 (in Russian).

[523] Strizhak T. G., *Method of averaging in problems of mechanics* (Vischa Shkola, Kiev-Donetsk, 1982), p. 254 (in Russian).

[524] Strogatz S. H. and R. E. Mirollo, Stability of incoherence in a population of coupled oscillator, *Journ. of Statistical Physics* **63**, 3/4 (1991) 613–635.

[525] Strogatz S. H. and I. Stewart, Coupled oscillators and biological synchronization *Scientific American* **269**, 6 (1993) 102–130.

[526] Strygin V. V. and V. A. Sobolev, *Separation of motions by method of integral manifold* (Nauka, Moscow, 1988), p. 256 (in Russian).

[527] Subach A. P., Mathematical models of loading the container by volume vibrational treatment taking into account the additional force field and the motion of feed of layer-by-layer, *Problems of dynamics and strength,* issue 25 (Zinatne, Riga, 1973) 89–96 (in Russian).

[528] Svidersky V. L., *Flight of an insect* (Nauka, Moscow, 1980), p. 137 (in Russian).

[529] Tamm E. A., *Dynamics of self-propelled vibrational apparatus,* PhD thesis (Gorky, 1988), p. 146 (in Russian).

[530] Tanaka H. A., A. J. Lichtenberg and S. Oishi, Self0synchronization of coupled oscillators with hysteretic responses, *Physica D* **100** (1997) 279–300.

[531] Taylor, Sir, Geoffrey I., Analysis of the swimming of long and narrow animals, *Proc. Roy. Soc.* , **A214**, 158 (Lond. 1952) (in English).

[532] Tchernyak D. M. and J. J. Thomsen, Slow effects of fast harmonic excitation for elastic structures, *Nonlinear Dynamics* **17** (1989) 227–246.

[533] Tchernyak D. M. and J. J. Thomsen, Slow effects of fast harmonic excitation for elastic structures, *Nonlinear Dynamics*, **17**, 3 (1998) 227–246.

[534] Tchernyak D. M., The influence of fast excitation on a continuous system, *DCAMM Report*, no. 592 (Technical University of Denmark, Lyngby, 1998).

[535] Tchernyak D. M., Using fast vibration to change the nonlinear properties of mechanical systems *to appear in Proceeding of the IUTAM/IFToMM Symposium on Synthesis of Nonlinear Dynamical Systems, Riga, Latvia, Klüwer* (1999).

[536] Teruo S. and S. Mori, Mutual synchronization of two oscillators, *Summaries of Papers Int. Conf. Microwaves, Circuit Theory and Infarm. Theory*, Part 2 (Tokyo -1964), Part 2 *Inst Electr. Common Eng. Japan* S.a. 111 -112 (in Japanese).

[537] Thearle E. L., A new type of dynamic balancing machine, *Trans. ASME*, **54**, 12 (1932) (in English).

[538] Thearle E. L., Automatic dynamic balancer, *Machine Design*, **22** 9–11 (1950).

[539] Thomsen J. J., Chaotic dynamics of the partially follower-loaded elastic double pendulum, *Journal of Sound and Vibration* **188** (1995) 385–405.

[540] Thomsen J. J., Vibration induced sliding of mass: non-trivial effects of rotatory inertia, *Proceedings of the EUROMECH 2nd European Nonlinear Oscillation Conference, Prague, Czech Republic* **1** (1996) 455.

[541] Thomsen J. J., Vibration suppression by using self-arranging mass: effects of adding restoring force, *Journal of Sound and Vibration*, **197**, 4 (1996) 403–425.

[542] Thomsen J. J., *Vibration and Stability, Order and Chaos* (McGraw-Hill, London, 1997), p. 323.

[543] Thomsen J. J., Vibration-induced displacement using high- frequency resonators and friction layers, *Proceedings of IUTAM/IFTAM Symposium on Synthesis of nonlinear Dynamical Systems*, (Klüwer Academic Press, Riga, Latvia, 1999).

[544] Thrusdell K., *The basic course of the mechanics of continuum* (Mir, Moscow, 1975), p. 592 (in Russian).

[545] Timofeev I. P., Investigation of vibro-bunkering of bulk cargo, *Collection of papers on vibro-loading machines, vibro-bunkering and vibro-discharge of bulk cargo*, (TSNIITEI uglya, Moscow, 1963) (in Russian).

[546] Timoshenko S. P., *The theory of the oscillations in engineering* (ONTI, Moscow, 1934), p. 344 (in Russian).

[547] Tishkov A. Ya., V. M. Grigor'ev, Output of ore by vibro-belt of the feeder "Volna", *Mining Journ.*, 2 (1975) 48–51 (in Russian).

[548] Tolstoy D. M., Eigen-oscillations of the sliding bearing depending on the contact stiffness, and their effect on friction, *DAN SSSR*, **153**, 4 (1963) (in Russian).

[549] Tolstoy D. M. and R. L. Kaplan, On the problem of frictional self-oscillations and speed dependence of the frictional force, *The theory of friction and wear* (Nauka, Moscow, 1965) 44–49 (in Russian).

[550] Tseitlin M. G., V. V. Verstov, and G. G. Asbel, *Vibrational techniques and technology in driving-in piles and a drilling* (Stroiizdat, Leningrad, 1987), p. 261 (in Russian).

[551] Usakovsky V. M., *Inertial pumps* (Mashinostroenie, Moscow, 1973), p. 200 (in Russian).

[552] *USSR Astronomy Annual on 1979*, **58** (Nauka, Moscow, 1976), p. 719 (in Russian).

[553] Uchitel A. D. and V. L. Estraikh, Determining the probability of the particles of arbitrary shape getting into the slotpickers of the screen, *Investigations of processes, machines and apparatuses for classification of materials according to size* (Interdepartmental Proceeding, Leningrad, Mekhanobr, 1988) 64–70 (in Russian).